D1345706

B0145586

RADIATION PROTECTION

RADIATION PROTECTION

A GUIDE FOR SCIENTISTS, REGULATORS, AND PHYSICIANS

fourth edition

Jacob Shapiro

HARVARD UNIVERSITY PRESS | Cambridge, Massachusetts, and London, England   2002

Library of Congress Cataloging-in-Publication Data

Shapiro, Jacob, 1925–
    Radiation protection : a guide for scientists, regulators, and physicians /
Jacob Shapiro.—4th ed.
        p. cm.
        Previously published with subtitle: A guide for scientists and physicians.
        Includes bibliographical references and index.
        ISBN 0-674-00740-9 (alk. paper)
        1. Ionizing radiation—Safety measures.   2. Radioactive substances—Safety measures.
    3. Radiation—Dosage.   I. Title.
    RA569 .S5 2002
    616.07′57′0289—dc21   2002023319

To Shirley with love

Preface

This manual explains the principles and practice of radiation protection for those whose work in research or in the field of medicine requires the use of radiation sources. It provides the information radiation users need to protect themselves and others and to understand and comply with governmental and institutional regulations regarding the use of radionuclides and radiation machines. It is designed for a wide spectrum of users, including physicians, research scientists, engineers, and technicians, and it should also be useful to radiation safety officers and members of radiation safety committees who may not be working directly with the sources. Regulatory officials in governmental agencies responsible for radiation protection will find it useful, both as an introduction to the field and as a reference for the rationale, interpretation, and application of the regulations.

Concern with radiation hazards is not limited to those occupationally exposed to radiation. In fact, by far the largest number of people at risk from radiation exposure are members of the public; this group also receives the largest cumulative population dose. Most of their exposure comes from natural sources in the environment and from the use of radiation to detect and treat disease. Other sources of radiation pollution that cause concern range from those with worldwide consequences, such as the Chernobyl accident and the atmospheric testing of nuclear weapons, to local effects from soil and water contamination by radioactive wastes. Concerns over exposure to radiation are not limited to the high energies associated with nuclear processes, but include the permeation of the environment with low-energy electromagnetic radiations from radio and television antennas and radar installations. A new worry is exposure from close contact with cellular phones, as well as signals sent to and from their communications towers.

The organization of the material covering this vast and complex field of radiation protection was guided by my experience of over fifty years in research, teaching, and management. As Radiation Protection Officer at Harvard University, I conducted training programs and seminars in radiation protection for workers and students in the research laboratories at Harvard University and its affiliated hospitals and was director of the radiation safety office in the Department of Environmental Health and Safety. I also taught or participated in courses for health physicists, industrial hygienists, radiology residents, physicians, regulatory officials, and executives in academic and continuing education programs as a member of the faculty of the Harvard School of Public Health.

The field of radiation protection, as taught to specialists, draws heavily on radiation physics and calculus. A large number of workers who require training in radiation protection, however, have minimal experience in these subjects, and their schedules are usually too full to allow for the luxury of extended reviews of the material. Thus, this manual is designed to obviate the need for reviews of atomic and radiation physics, and the mathematics has been limited to elementary arithmetical and algebraic operations.

Following a historical prologue, Part One introduces the sources of radiation in terms of the energy carried by the radiation, since energy imparted by radiation plays the central role in evaluations of radiation exposure. The coverage in this edition has been expanded to include the entire energy range of radiation exposure, so-called nonionizing as well as ionizing radiation. The introduction of the whole range of energies possessed by radiation particles and electromagnetic waves at the beginning of the text serves to promote a unified view of both ionizing and nonionizing radiation, which are too often considered as two separate disciplines.

Part Two presents the principles of radiation protection against ionizing particles and develops these in the context of the working materials of the radiation user. The central role of the energy imparted by ionizing particles in characterizing radiation exposure is explained and the properties of radiation are illustrated through examples with gamma radiation and beta rays (electrons and positrons) from common radioactive sources in research. Reviews are presented of radiation units, standards, and the significance of various radiation levels, followed by some basic calculations in radiation protection. The heavy ionizing particles—alpha particles, protons, and neutrons—are then introduced. This part concludes with material for users of radiation machines in medical practice and research.

Part Three gives details on the calculation of doses from radiation particles, including dose calculations for some specific radionuclides. Part Four describes detection instruments and their use in making some of the more common measurements on radiation particles. Part Five presents practical information, primarily for users of radionuclides. Part Six is concerned with the public health implications of the use of ionizing radiation in medicine and technology. It contains a detailed treatment of the two subjects that are essential to evaluating the significance of exposure to radiation: the results of studies of the effects of radiation exposure of human populations, and the radiation exposure experience of the population both from manmade sources and from natural sources. It is my hope that the material presented in this part will give individuals who must make benefit-risk decisions a firm foundation for making responsible and ethical judgments regarding the irradiation of other persons, whether these persons are associates working in radiation areas, members of the general public who have no connection with the user, or patients who are undergoing medical diagnosis and treatment.

Part Seven provides background on the nonionizing radiation in the electromagnetic spectrum along with an assessment of exposures from such sources as broadcast antennas, microwave towers, cellular phones, and radar. Nonradiating sources, such as the electromagnetic fields around power lines, are also considered.

Finally, Part Eight presents an overview of the major issues in radiation protection.

The material in Part Two can be used to provide basic information for radiation users. With additional practical training in working with specific radionuclides or other sources, radiation workers would be prepared to handle limited levels of radioactive materials under supervision. Individuals planning to work independently with radiation sources or to administer radiation to human beings must receive additional training, including pertinent material in Parts Three through Six. The book is well endowed with examples to illustrate and expand on the text. For readers who need to prepare for certification or qualification examinations, references to sources that provide training and practice in solving problems are given in the selective bibliography.

My approach to quantifying radiation doses deserves some explanation. The current method in vogue expresses radiation dose as a risk-based quantity, as proposed by the International Commission on Radiological Protection (ICRP) in 1977. By this method a nonuniform distribution of organ doses in the body from a given exposure to radiation is replaced by a single number, which is taken as the uniform whole-body dose with a comparable risk of producing a comparable detriment to health, and the spectrum of risks to health is based primarily on the risk of causing a fatal cancer. The ICRP chose the name "effective dose equivalent" for this substitute dose in 1977, and replaced it by the name "effective dose" in 1990. Although both quantities were based on somewhat different paradigms, they gave quite similar numerical values for the equivalent uniform whole-body dose. Their use has the advantage of simplicity and standardization and is well suited to express dose for regulatory purposes, or whenever a simple expression of the risk of radiation exposure is called for, as in consent forms. However, their value as a measure of individual and population exposure for scientific and epidemiological purposes is, at best, very tenuous. Cancer statistics, on which the formula for their calculation is based, have a high degree of variability; they change not only with better epidemiological studies but with improvements in treatments for the different types of cancer, necessitating changes in the constants in the formula. Given the uncertainty of quantities based on biological effects, I have chosen to present the actual absorbed doses to organs when these data are available. I refrained from presenting effective doses except within a regulatory or nontechnical context, or when they are the only data available for a particular subject.

The risk of lung cancer from exposure to radon gas in the home was a major concern in public health at the time of the third edition. However, this concern has tapered off considerably, possibly because epidemiological studies have not demonstrated a strong relationship between exposure in the home and lung cancer, possibly because the public became accustomed to living with radon, possibly because the homeowner must pay the costs of remediation. Yet radon remains a major source of radiation exposure to the world's population, and its continued epidemiology and dosimetry should shed much light on the risk of exposure to radiation. Accordingly, I have expanded on the detailed discussions of radon presented in previous editions.

The risk of harm from exposure to radiation is not generally considered by physicians when they prescribe radiological examinations, except when the patient is a pregnant woman or a young child. The benefit far outweighs the risk in most studies that have been made. Malpractice litigation and other legal considerations—and possibly economics, as well—are also significant factors in decisions to use ionizing radiation for diagnostic purposes. I have expanded considerably the sections dealing with doses accompanying the use of radiation in medicine to help physicians make these decisions.

I use Standard International (SI) units for dose and activity in this edition, except when reproducing verbatim data in the published literature given in traditional units. However, I have retained the traditional unit for exposure, the roentgen. It is much easier to work with exposures expressed in roentgens than in the SI unit of coulombs per kilogram.

Since publication of the last edition of *Radiation Protection,* the Internet has become an invaluable resource. In this edition, therefore, I have included Internet addresses, where possible, to supplement, enrich, and update the material provided in the book.

This manual originated from a training program in the safe use of radionuclides in research conducted at Harvard University that Dr. E. W. Webster and I developed in 1962. It initially included 10 two-hour sessions of formal lectures and laboratory demonstrations; problem assignments; and a final examination. The program provided training for investigators who wanted authorization to work independently with radionuclides under the broad specific license for the use of radionuclides in research and development granted to Harvard University by the Nuclear Regulatory Commission. The course was later offered also as a self-paced option. The design of training courses has changed considerably in recent years, influenced particularly by the development of computer-based instruction.

I am grateful to the many individuals who offered advice and assistance as *Radiation Protection* evolved through four editions. William Shurcliff carefully read through the first edition. Not only did the manuscript profit immensely from his comments, but his manner of expressing them was a joy to read. I was also fortunate to have the help of James Adelstein, John Baum, Bengt Bjarngard, Robley Evans, Abraham Goldin, Robert Johnson, Kenneth Kase, Samuel Levin, James McLaughlin, Dade Moeller, Frank Osborne, Joseph Ring, Kenneth Skrable, John Villforth, and Michael Whelan in the earlier editions, and of Phil Anthes, William Bell, Frank Castronovo, John Evans, Robert Hallisey, John Little, Richard Nawfel, John Osepchuk, Robert Pound, Joseph Ring, Michael Slifkin, David Spelic, Stanley Stern, and Robert Watkins for this edition. The transformation of my rough sketches into professional illustrations was the result of Bonnie Baseman's expertise in computer graphics and her ability to grasp the intentions of the author, who barely passed drawing class in elementary school. I am thankful to Michael Fisher, science and medicine editor at Harvard University Press, for his interest and support in bringing this edition to fruition. I enjoyed working with my editor, Kate Schmit, as she scrutinized the manuscript, correcting sins of grammar and punctuation and suggesting changes that made the text clearer and more readable, or improved the presentation of data. Also on the *Radiation Protection* team were Sara Davis, who kept the author on track, and Christine Thorsteinsson, who looked after the production of the book. Finally, I am happy to express my appreciation to my daughter, Jean, my wife, Shirley, and my nephew, Mark Shapiro, for editorial comments regarding some of the personal views expressed here, and to my son, Robert, for stimulating me to make a personal statement.

Contents

Historical Prologue

1 IN THE BEGINNING

Although our society has been concerned for some time with the potential radioactive contamination of the environment resulting from the technological exploitation of nuclear energy, the fact is that the universe is and always has been permeated with radiation. At the present time there are about one billion rays traveling through space for every elementary particle of matter. The remarkable set of circumstances arising from the interaction of radiation and matter has a history of some fifteen billion years. Some of the most fascinating theories of physics attempt to explain how we arrived at the present point in time and space, and how to account for the continuing expansion of the universe and for its very low temperature (only 2.7 degrees above absolute zero!). The further backward in time we trace the universe, the smaller and denser and hotter it was, until eventually we come to the moment of birth—time zero—and the initial cosmic fluid.

How did it all begin? Some say the universe burst into existence with a great big bang in which energy, space, time, and matter were created. The progenitors of our entire observable universe, with the mass of a trillion suns, were compressed into the volume of a single proton. In this tiny space a myriad of extremely high energy particles collided with each other at energies characterized by the astronomical temperature of 10^{29} degrees Kelvin (10^{29} K).[1]

---

1. These numbers are so high that we must express them as powers of 10, where the power (the exponent) gives the number of zeroes following the one. For example, $10^4 = 10,000$, $10^{-4} = 1/10,000 = 0.0001$. More will be said about units of energy and their significance later, but for the moment note that the energies associated with radiations from radionuclides are commonly under 2 million electron volts, an insignificant amount compared with the energies released at creation.

This speck of matter began to expand at a fantastic rate, a billion times faster than the speed of light. By 10^{-35} seconds after the Big Bang the rate of expansion of the young universe had slowed down to about the speed of light. The original speck was evolving into a primordial hot soup, a sea of radiation with an admixture of material particles in the form of quarks and electrons and their antiparticles. The antiparticles had the same mass as the particles but carried opposite electrical charges. Particles and antiparticles cannot exist together, however, and their interactions resulted in mutual destruction, both vanishing in a burst of radiation. So long as the temperature remained high enough, they were replaced by the creation of new particles and antiparticles from high-energy radiations, and so continued the cycle of destruction and creation.

As the universe expanded, it cooled rapidly, much as the coolant in a refrigerator gets colder as it expands. After one ten-billionth (10^{-10}) of a second, the temperature of the primordial fluid had dropped so low that the quarks and antiquarks that were annihilating each other were no longer replaced by the creation of new particles. If there had been an exactly equal number of both, their collisions would have been the end of the evolution of the universe as we know it. Fortunately, a slight excess of quarks, about one additional quark for thirty million quark-antiquark pairs, was left over to form all the matter we see in the universe today.

By one-millionth of a second after creation, the temperature and energies had dropped to a point where the remaining quarks were able to combine and produce neutrons and protons and their antiparticles, with a sprinkling of hyperons and mesons. After thirty millionths of a second, essentially all the quarks were gone and the inventory of neutrons, protons, and electrons and their antiparticles was complete. At one ten-thousandth of a second after creation there was still enough heat energy to permit the creation of new particles and antiparticles, and the universe was densely populated primarily with neutrons, protons, electrons, neutrinos, their antiparticles, and photons of electromagnetic radiation. With yet more cooling, there was not enough energy to replenish the supply of protons or neutrons. Electrons and positrons became dominant for a while—for about 10 seconds—until the temperature cooled to the point where they were no longer replenished by the radiation photons. A remnant of electrons survived. At this time, the density of matter was down to ten thousand times the density of water and the temperature was one hundred billion degrees.

By seven days, the universe had cooled to seventeen million degrees, a million degrees hotter than the center of the sun, the density was about a millionth that of water, and the pressure was more than a billion atmospheres.

The enormous amount of radiation energy released by the annihilation of the electrons and positrons began the radiation era. The universe was flooded with electromagnetic radiation for about half a million years, during which the temperature and density continued to drop as the universe expanded and matter existed only as a faint precipitate.

Although the relatively small number of neutrons and protons remaining were capable of combining with each other and forming stable configurations, such as heavy hydrogen nuclei (neutron + proton) and helium nuclei (2 neutrons + 2 protons), these combinations were prevented by the presence of the very energetic photons, which had enough energy to break any bonds that were established. But as the radiation era progressed, and the temperature continued to fall, the mean photon energy dropped below the binding energy of the nucleons, and heavy hydrogen, helium, and lithium built up. (All heavier elements were produced much later, in the interiors of stars, and ejected into space by a variety of processes.)

As the universe expanded, the radiation density dropped more rapidly than the density of matter. When the temperature had dropped below 3000 K, protons and electrons were able to combine, and atoms of matter began to appear. By the time the mean density of the expanding universe had dropped to 10^{-21} g/cm^3, one million years after time zero, matter began to emerge as the dominant constituent of the universe. Figure P.1 illustrates this saga of creation, annihilation, survival, and evolution.

Today, our universe contains about 10^{78} protons and neutrons, a good portion of them aggregated, under the force of gravity, into celestial structures, galaxies, stars, and planets all flying through space at high speeds relative to each other. Extremely energetic cosmic rays, both material particles and electromagnetic radiation, also stream through space, with about one billion photons of electromagnetic radiation and neutrinos for every neutron or proton, and with photon and neutrino densities of the order of 1,000/cm^3. In addition, microwave radiation, left over from the Big Bang, is streaming freely through space at energies corresponding to a temperature of a little less than three degrees above absolute zero.[2]

2 THE DISCOVERY OF INVISIBLE, UNBELIEVABLY ENERGETIC RADIATIONS

We shall dispense with the billions of years of evolution that resulted in the establishment of the solar system, the origin of life, and the emergence of modern man. It took our own species, *Homo sapiens,* at least 25,000 years

2. Great reading on the evolution of the universe includes Guth (1997), Lidsey (2000), Silk (1994), Ronan (1991), and Weinberg (1977).

P.1 Evolution of radiation and matter in the observable universe.

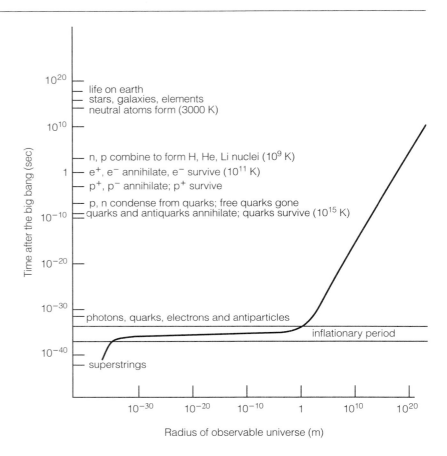

to attain the knowledge and understanding required to discover the existence of highly energetic radiations in the universe. The first clues (in the latter half of the nineteenth century) came from experiments with electrical discharges in vacuum tubes, known as Crookes tubes. The discharges were produced by applying high voltage across the electrodes in the tubes. The observations, for example, of light emitted by gas in the tube and the production of fluorescence in the glass envelope were attributed by Thomson in 1897 to the effects of high-speed negatively charged particles, which he called "electrons." In 1895, Roentgen, experimenting with Crookes tubes, identified penetrating radiations that also produced fluorescence; he named them "x rays." And in 1896, Becquerel discovered that penetrating radiations, later classified as alpha, beta, and gamma rays, were given off by uranium, and thereby opened up a new field of study, the science of radioactive substances and the radiations they emit. Thus, by 1900, scientists had begun to discover and experiment with high-energy radiations of the kind that had dominated the universe during its early history.

At about the same time, the work of Planck in 1900 and Einstein in 1905 showed that many kinds of radiation, including heat radiation, visible light, ultraviolet light, and radiowaves, which had previously appeared to be transmitted as continuous waves of energy, were actually emitted as discrete bundles of energy called photons, and that differences among these types of radiation could be characterized in terms of the different energies of the photons. In time, it was learned that gamma radiation, emitted from the nuclei of atoms, and x rays, produced by the acceleration of electrons (outside the nucleus), were also made up of photons of electromagnetic radiation, but of much higher energies.

The discovery of the new particles and rays led to intense experimentation on their properties and their interactions with matter. The energetic alpha particles (actually helium nuclei) emitted by radioactive materials were directed by Rutherford against thin gold foils. Through analysis of the scattering pattern, he deduced in 1911 that the atom was composed of a tiny central core, or nucleus, containing all the positive charge and almost all the mass of the atom, and a nearly empty surrounding region containing the light, negatively charged electrons, in sufficient number to balance out the inner positive charge. The nucleus of the hydrogen atom, consisting of a single particle with a charge equal in magnitude and opposite in sign to that of the electron, was recognized as a fundamental building block of the nuclei of all complex atoms. It was named the proton (from the Greek *protos,* which means "first"). With the development of the theory of the atomic nucleus composed of protons and other elementary particles, it was possible to visualize how certain types of nuclei could disintegrate and emit particles.

The emitted particles had very high energies, and the source of the energies was a puzzle until the formulation by Einstein in 1905 of the mass-energy equation. This equation expressed in quantitative terms his conclusion that matter could be converted into energy according to the relationship $E = mc^2$, where E was the energy, m the mass, and c the velocity of light. If m was expressed in kilograms and c was expressed in meters per second, the equivalent energy E was given in a unit of energy known as the joule. When it later became possible to determine the masses of individual particles in an instrument known as the mass spectrograph, the relationship between mass and energy was verified experimentally. Whenever a particle was emitted from a nucleus with high energy, it was found that the mass of the nucleus decreased not only by the rest mass of the particles

P.2 Composition of atoms of matter. Atoms are made up of a dense core, consisting of positively charged protons and uncharged neutrons, surrounded by an extended cloud of negatively charged electrons. (The distribution of charge, whether depicted as electron orbits or quantum mechanical wave functions, is much more complex than the circular orbits shown here.) In the lighter elements, the cores of stable atoms contain approximately equal numbers of neutrons and protons. In the neutral atom, the surrounding electrons are equal in number to the protons. The number of protons in the nucleus, called the *atomic number*, symbol Z, uniquely specifies the *element*. The number of protons plus neutrons is called the *mass number*, symbol A. Atoms characterized by their atomic number and their mass number are called *nuclides* and are represented by the notation $_Z^A X$, where X is the element symbol (e.g., $_6^{12}C$). Since Z is known for every element, it may be omitted from the nuclide expression (e.g., ^{12}C). Nuclides with the same number of protons but differing numbers of neutrons are called *isotopes*. All isotopes of a particular element have almost identical chemical properties. Some nuclides are *radioactive* and are referred to as *radionuclides*. Each atom eventually undergoes spontaneous disintegration, with the emission of radiation. The figure shows isotopes of hydrogen (H) and carbon (C).

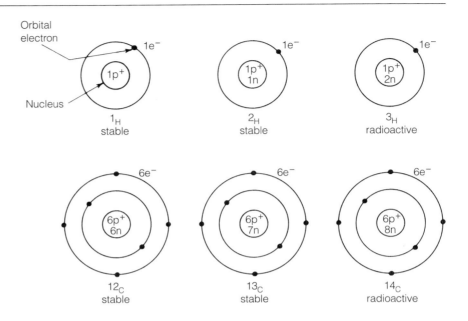

emitted but by an additional mass that was equivalent to the energy carried by the particle, as given by Einstein's equation.

Rutherford bombarded many elements with the energetic particles from various naturally radioactive materials. In 1919 he found that when alpha particles bombarded nitrogen nuclei, energetic protons were released. He had, in fact, produced the first man-made nuclear transformation, by forcing an alpha particle into the nitrogen nucleus, resulting in the emission of one of its fundamental constituents, the proton. The residual atom was deduced to be oxygen. In 1932 Chadwick identified the other basic particle in the nucleus, the neutron. He had ejected it from the nucleus of a beryllium atom by bombarding it with alpha particles. The neutron, unlike the proton, does not have an electrical charge.

The discovery of the neutron gave very strong support to the concept of the atomic nucleus as consisting solely of neutrons and protons, packed very closely together (Fig. P.2). Certain combinations of neutrons and protons are stable: these make up the nuclei of isotopes of elements that retain their identity indefinitely, unless disrupted by nuclear collisions. Other combinations of neutrons and protons do not give stable nuclei. The nuclei eventually undergo a nuclear transformation through spontaneous disintegration processes that result in the alteration of the neutron-proton ratio (Fig. P.3). Some nuclides go through several disintegration processes before finally attaining a stable neutron-proton ratio.

P.3A Decay of the carbon-14 nucleus, $_{6}^{14}$C. The carbon-14 nucleus, with 6 protons and 8 neutrons, is not stable, because the ratio of neutrons to protons is too high. The ratio is changed by the spontaneous transformation of one of the neutrons into a proton and the emission of an energetic beta particle (negative) from the nucleus. The average lifetime of the nucleus is 8,250 years. After its decay, however, the resultant nitrogen-14 nucleus (7 protons, 7 neutrons) is stable.

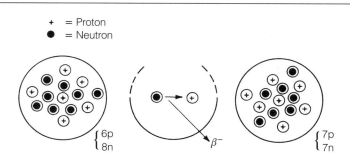

P.3B Decay of the sodium-22 nucleus, $_{11}^{22}$Na. The sodium-22 nucleus, with 11 protons and 11 neutrons, is also not stable. The ratio of protons to neutrons is too high. The ratio is usually changed by the spontaneous transformation of one of the protons into a neutron and the emission of an energetic positron (positive beta particle) from the nucleus. About 10 percent of the time, the ratio is changed by the capture by a proton of an electron from an inner orbit, which transforms it into a neutron. The average lifetime of the unstable sodium-22 nucleus is 3.75 years. The resultant neon-22 nucleus (10 protons, 12 neutrons) is stable.

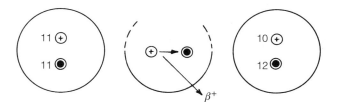

3   THE DEVELOPMENT OF A RADIATION TECHNOLOGY

After the discovery of the neutron, major developments in nuclear research came in rapid succession: the discovery of uranium fission in 1939, and the recognition of the possibility of releasing enormous amounts of energy; achievement of the first self-sustaining fission reaction in a reactor in 1942; explosion of nuclear fission devices in 1945; production of thermonuclear explosions in 1952; commissioning of the first nuclear-powered submarine, the *Nautilus,* in 1954; and the development of high-energy accelerators, with energies over 10^9 electron volts (GeV) in the fifties and exceeding 10^{12} electron volts (1 TeV) at the present time. The result of these developments was the creation of an extensive radiation technology concerned with the production of energetic radiations for use in research, medical treatment, and industry.

4   THE NEED FOR RADIATION PROTECTION

The development of a radiation technology left its occupational casualties—physicists, radiologists, radiation chemists—researchers who investigated the properties and uses of these energetic radiations without appreciating their capacity for destructive effects in living matter. But society soon recognized the harm that energetic radiations could cause when exposure was uncontrolled, and it has worked diligently since to further the understanding of the biological effects of radiation and to establish acceptable limits of exposure.

The development of the nuclear reactor and the production of large amounts of artificial radioactivity created the potential for injury on an unprecedented scale. Governments realized that extraordinary measures were necessary to protect radiation workers and the public from excessive exposure to radiation. The result, as every user of radionuclides knows, was the enactment of extensive legislation, the establishment of regulatory bodies and licensing mechanisms, the setting of standards of radiation exposure, and the requirement of training of radiation workers to conform with accepted practice in working with radiation and radionuclides.

PART ONE

Energy—The Unifying Concept in Radiation Protection

1 RADIATION'S DUAL IDENTITY

Dictionaries often describe radiation as waves or particles that propagate through space. The fact is that any radiation emission can be observed either as a subatomic particle or as a wave, depending on how it is detected. This is a remarkable phenomenon, since the concepts of waves and particles are diametrically opposed. A particle is a discrete object that transmits energy by moving from one point to another. Appropriate measurements give its mass and velocity, or its momentum (which is the product of mass and velocity). A wave is also a means of transmitting energy between two points, but the journey is not made by a discrete identifiable object. Rather, the energy is conveyed through oscillations in time of a physical property at each point along the path. For example, an ocean wave smashing against the coast may have originated hundreds of miles offshore from wind action, but no intact mass of water travels this distance and wind is not needed to propagate the wave. A local circular oscillation of the surface of the water is initiated at some point, and this motion is transmitted in succession to adjacent elements of water until it reaches the shore. Leonardo da Vinci wrote of water waves, "It often happens that the wave flees the place of its creation, which the water does not; like the waves made in a field of grain by the wind, where we see the waves running across the field, while the grain remains in place" (cited in Halliday and Resnick, 1988, p. 392).

If radiation is detected as a wave, measurements give its wavelength and frequency or, in some cases, a group of frequencies. Radiation emitted from antennas is normally treated as waves. High-energy radiation, such as is emitted in the process of radioactive decay or produced by x-ray ma-

chines, is generally observed and described as particles. These dual properties of radiation were characterized by Niels Bohr as being *complementary.*
According to Bohr, the question of whether radiation is really a wave or a
particle is not relevant. Both descriptions are required for a full characterization of the nature of radiation.

From the point of view of radiation protection, a most important characteristic of radiation is the energy carried by the radiation, for the energy
that is imparted to objects struck by the radiation produces physical and
biological effects. Thus, I like to define radiation as *energy traveling through
space, displaying the properties of either particles or waves, depending on how it
is measured and how it interacts with matter.*

Energies at the atomic level are expressed in terms of a unit called the
electron volt. The electron volt (eV) is defined as the energy an electron ac-
quires in going through a voltage difference of 1 volt. For example, the image on a TV screen is produced when electrons traveling at high speed collide with a phosphor coating on the screen of the TV tube, resulting in the
emission of light from the phosphor. The phosphor is at a high positive
voltage with respect to the cathode (from which the electrons are emitted)
and the resulting strong electric field accelerates the electrons toward the
screen. The high voltage is of the order of 25,000 volts, and the energy the
electrons acquire in going through 25,000 volts is thus stated to be 25,000
electron volts.

The range of energy carried by radiations encountered in nature and in
technology is enormous. The photons of electromagnetic radiation from
an FM radio station transmitting at a frequency of 100 million cycles per
second (100 MHz) have an energy of one ten-millionth of an electron volt
(10^{-7} eV). The highest energy ever recorded for a radiation particle was
three hundred million trillion electron volts (3×10^{20} eV), carried by a
cosmic ray from outer space.

1.1   From Energy to Radiation Dose

However energetic a particle may be, that energy is consequential only
to the extent it interacts with an object. The amount of energy imparted to
a target underlies the concept of radiation dose, defined as the energy absorbed per unit mass of the object exposed. Just as the energy carried by radiation particles covers a very wide range, so does the fraction of energy
imparted to a target in any given situation vary widely. Some very energetic
radiation particles have little impact because their interaction with the target is very weak, whereas low-energy particles that have a strong interaction can have a significant impact. Much of the material in the sections
that follow is concerned with the assessment of radiation dose and the ac-

companying effects, but to start, let us review the range of energies that one can encounter in radiation protection.

2 ENERGY RELATIONSHIPS IN THE HYDROGEN ATOM

How does one begin to comprehend this enormous energy range from the viewpoint of radiation protection? The energy relations in the hydrogen atom make a good reference point for comparisons with both higher energies and lower energies.

An isolated hydrogen atom consists of a heavy, positively charged particle at the center, the proton, around which revolves a light, negatively charged particle, the electron, held in orbit by the electromagnetic force. The electron is restricted to specific orbits, as given by the principles of quantum mechanics, and each orbit is associated with a specific energy level. When the electron is in its lowest energy level, that is, at the closest distance it is allowed (by the uncertainty principle) to come to the proton, the atom is said to be in its *ground state*.

It takes energy to move the electron to orbits at greater distances from the proton; when the electron occupies an outer orbit, the atom is said to be in an *excited state*. Atoms revert from an excited state to the ground state with the *emission of radiation* as packets of electromagnetic energy called *photons*, which carry energies equal to the difference in energy levels between the two states.

The greatest energy change occurs when the electron is moved so far from the proton that the electromagnetic force between them is negligible. At this point, the atom is said to be *ionized*. It takes 13.6 electron volts to transfer an electron from the ground state of the hydrogen atom to the ionized state. Radiations with particle energies greater than the energy needed to remove an outermost electron from an atom are called *ionizing radiations*.

For the lower-energy states, the differences in energy levels in hydrogen (and also the energies of the photons emitted) is given by a very simple formula:

$$E = -13.6(1/n_1^2 - 1/n_2^2) \text{ eV} \qquad (1.1)$$

n_1 and n_2 are integers corresponding to an electron's orbit (orbit 1 is nearest the proton, orbit 2 is the next orbit out, orbit 3 is further out still, and so on).

Since it is only the *difference* in energy levels that is relevant, the formula can be rewritten to express energy levels in the hydrogen atom. These are not the actual values of the energy levels, but their differences give the en-

1.1. Energy levels in the hydrogen atom. *(a)* A simplified model of the hydrogen atom, showing the levels as permissible orbits for the single electron. The orbit $n = 1$ is the ground state and the other orbits are excited states. A photon is radiated when the electron drops from a higher to a lower orbit. *(b)* The levels are represented by an energy level diagram. The energy level of the ionized atom is arbitrarily set at 0 eV. The energy level of the ground state is 13.6 eV below the level of the ionized atom and is thus shown as −13.6 eV. All intermediate levels represent excited states.

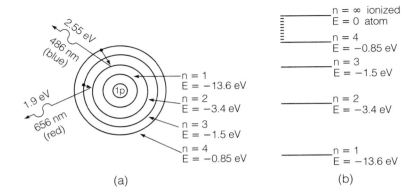

(a)   (b)

ergies of the radiations emitted (when the electron moves from a higher to a lower energy level) or absorbed (when it moves from a lower to a higher energy level).

$$E_n = -13.6/n^2 \text{ eV} \qquad (1.2)$$

When $n = 1$ (the ground state), $E_1 = -13.6$ eV; when $n = 2$, $E_2 = -3.4$ eV; when $n = \infty$ (the ionized atom), $E_\infty = 0$. The amount of energy required to raise an electron from level 1 to level 2 is $-3.4 - (-13.6)$, or 10.2 eV. When an electron drops from $n = 2$ to $n = 1$, a photon of energy 10.2 eV is emitted. These relationships can be shown schematically as orbits of the electron in the Bohr model of the hydrogen atom (Fig. 1.1A) or in the form of an energy level diagram (Fig. 1.1B).

The energy required to ionize the atom (to make the transition from $n = 1$ to $n = \infty$) equals $0 - (-13.6) = 13.6$ eV. This energy is also called the *binding energy* of the electron in orbit 1.

The electromagnetic radiation emitted when an electron drops from $n = 2$ to $n = 1$ may also be expressed in terms of its wave properties. The relationship between the photon energy and the frequency (ν) of the wave is given by the expression

$$E = h\nu \qquad (1.3)$$

where h is Planck's constant, equal to 6.626×10^{-34} joule-second (J-s) or 4.133×10^{-15} electron volt-second (eV-s).

Since for any wave, the velocity (c, for the speed of light) is equal to the product of the frequency (ν) and the wavelength (λ)

$$c = \nu\lambda \qquad (1.4)$$

(and therefore $\nu = c/\lambda$), the relationship between the photon energy and the wavelength of the wave is given by the expression

$$E = h(c/\lambda) \quad \text{or} \quad \lambda = hc/E \tag{1.5}$$

A useful expression gives the wavelength of the radiation in nanometers when the energy of the photons is given in electron volts:

$$\lambda \text{ (nm)} = 1{,}238 \text{ (eV-nm)}/E \text{ (eV)} \tag{1.6}$$

where λ is expressed in nanometers (10^{-9} m) and E is in electron volts (eV). (Note that $c = 3 \times 10^{17}$ nm/s, so hc is 4.133×10^{-15} eV-s times 3×10^{17} nm/s, or 1,238 eV-nm.)

Example 1.1   What is the wavelength corresponding to photons emitted by hydrogen of energy 10.2 eV?

$$\lambda = 1{,}238 \text{ eV-nm}/10.2 \text{ eV} = 121.4 \text{ nm} (121.4 \times 10^{-9} \text{ m})$$

This wavelength, in the ultraviolet region, is one of the spectral lines in the Lyman series, which is produced by transitions of the electron in hydrogen to the ground state ($n = 1$) from the higher orbits. Transitions to $n = 2$ produce the Balmer series, the principal lines of which are blue (2.56 eV, 486.1 nm) and red (1.89 eV, 656.28 nm).

3   ENERGY LEVELS IN ATOMS WITH HIGHER Z

More complex atoms contain a nucleus composed of the positively charged protons and uncharged neutrons, with electrons in orbit equal in number to the protons in the nucleus. These protons, given by the atomic number, Z, produce a central positive charge (and electrical attraction) Z times greater than the charge in hydrogen's nucleus. It requires a greater force (and hence energy) to raise electrons to higher states in atoms with more than one proton. An approximate expression for the energy level, or binding energy, of the innermost electron in atoms of higher atomic number is:

$$E = -13.6(Z - 1)^2 \tag{1.7}$$

Example 1.2   Lead has 82 protons in the nucleus, surrounded by 82 electrons. What is the energy required to eject the innermost electron?

   The approximate value for the binding energy of the innermost electron in lead is $13.6 \times (81)^2 = 89{,}230$ eV. (The handbook value of 87,950 eV is in good agreement with this calculated estimate.)

1.2 Energy levels involved in the production of x rays. *(a)* Depiction of the electrons in an atom (lead is used as an example) as restricted to concentric shells representing the different energy levels in the atom. The two innermost electrons are confined to the K shell, the next eight to the L shell, and so on. X rays are emitted when an electron drops from one shell to a vacancy in an inner shell and the energy of the x ray is equal to the difference in the energy levels of the shells. X-rays emitted in transitions to a vacancy in the K shell are called K x rays; to the L shell, L x rays; and so on. *(b)* Energy level diagram for the innermost levels in lead.

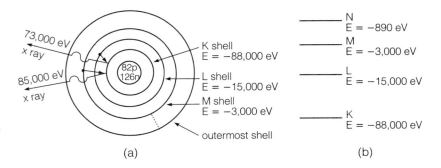

(a)   (b)

When a vacancy that exists at the innermost electron orbits in lead is filled by an electron dropping in from an outer shell or from outside the atom, the energy released may be emitted as an x-ray photon with energy equal to the difference between the two energy levels. X rays produced in this manner are known as *characteristic* radiation (Fig. 1.2).

The binding energy of the outermost electron in lead is 7.38 eV, much lower than that of the innermost electron; in fact, it is in the same energy range as hydrogen's electron. The reason is that the 81 other negative electrons in orbit serve to screen the +82 charge of the nucleus, leaving a net charge in the outermost orbit of approximately +1, as in hydrogen.

4   Energy Levels in Molecules

Molecules contain not only electronic energy levels but also additional levels due to the vibrations of the atoms and rotations of the molecules. Energy transitions between vibrational levels are one-tenth to one-hundredth the transitions between electronic levels. Examples are energy transitions of the order of 0.1 eV, resulting in the emission of near-infrared radiation. Transitions between rotational levels are smaller still: one-hundredth those between vibrational levels. Examples are energy transitions of the order of 0.005 eV, resulting in the emission of radiation in the far infrared.

5   Energies of Motion Associated with Temperature

Atoms and molecules at any temperature above absolute zero are in constant motion, colliding and exchanging energy with each other. Temperature is a measure of the kinetic energy of the atoms and molecules as a result of that motion.

The colliding particles do not all have the same energy; rather, a distribution of energies exists in any collection of molecules or atoms. The energies increase with increased temperature. The distribution $N(E)$ of N molecules with energy E in an ideal gas containing a large number of molecules at absolute temperature T is given by an equation known as the Maxwell-Boltzmann distribution (derived by Clerk Maxwell and Ludwig Boltzmann in 1860)

$$\frac{N(E)}{N}dE = \frac{2}{\pi^{1/2}(kT)^{3/2}}e^{-E/kT}E^{1/2}dE \qquad (1.8)$$

where k is the Boltzmann constant (0.86×10^{-4} eV/K). We need not be concerned with the exact shape of the distribution, except to note that it increases at first with increased energy (given by the $\sqrt{E}$ term in the numerator), reaches a maximum, and then decreases as the negative exponential takes over. The energy at the maximum is the most probable kinetic energy per molecule in an ideal gas at temperature T and is equal to kT. The most probable energy is often used to characterize a distribution of energies at a given temperature, such as the energies of neutrons in a nuclear reactor. The average translational energy, equal to $1.5\ kT$, is also used to characterize the spectrum.

Example 1.3 What is the most probable energy of the molecules in the body at body temperature?

Since body temperature, T, is 310 K, $kT = 0.86 \times 10^{-4}$ eV/K $\times$ 310 K = 0.027 eV.

This is a measure of the kinetic energy of the molecules in the body. This energy is much greater than the energies of photons of electromagnetic energy transmitted in radio and television communications. Physicists often cite this disparity in arguing that the effect of these radiations in producing molecular changes is insignificant in comparison with the energies imparted by the motion of the body's own molecules.

6 BONDING ENERGIES

A fraction of the molecules in a Maxwell-Boltzmann distribution attain energies much higher than the most probable energy, although fewer and fewer molecules do so as the level of attained energy rises. Thus, a fraction of the molecules have energies that can produce significant chemical effects, such as the dissociation of molecules. The likelihood that dissociation will occur depends on the strength of the chemical bonds. For example, the iodine-iodine bond is weak, with a dissociation energy of 1.56 eV per molecule. In contrast, the hydrogen-fluorine bond is very strong with a dissociation energy of 5.9 eV.

7   Energy from Mass—The Ultimate Energy Source

Einstein's mass-energy equation gives a quantitative measure of the tremendous energy obtainable from the conversion of mass into energy and, in particular, from nuclear reactions.

$$E = mc^2 \qquad (1.9)$$

where E = energy in joules, m = the mass in kilograms, and c = speed of light in meters per second.

Example 1.4   What is the energy equivalent of the mass of the electron?

The mass of the electron = 9.1×10^{-31} kg, and $c = 3 \times 10^8$ m/s, so

$$E = (9.1 \times 10^{-31})(3 \times 10^8)^2 = 8.19 \times 10^{-14} \text{ J}$$

To convert joules into mega-electron volts, divide by 1.60×10^{-13}:
8.19×10^{-14} J/1.6×10^{-13} J/MeV = 0.51 MeV.

By a similar calculation, the energy equivalent of the mass of the proton is 931 MeV.

Example 1.5   What is the energy of the beta particle emitted in the decay of phosphorous-32 to sulphur-32?

The difference in mass after a radioactive decay process is what produces the energy released in the decay. The mass of ^{32}P is 31.973910 atomic mass units (u). ^{32}S is stable with a mass equal to 31.972074 u. The difference in mass is 0.001836 u. Since 1 u corresponds to 931.478 MeV of energy, the decay of ^{32}P to ^{32}S represents an energy drop of (0.001836)(931.478) = 1.71 MeV, which appears as the maximum energy of the emitted beta particle. This is a million times higher than the energy associated with chemical reactions.

---

8 SOME INTERESTING ENERGY VALUES

The residual radiation from the Big Bang, the cosmic background radiation that permeates all space, has a temperature of 2.7 K. The most probable energy of the radiation photons is $kT = 0.00023$ eV. The wavelength is 0.538 cm and the frequency is 56 GHz.

The energy of 100 MHz photons from an FM radio station is 4.14×10^{-7} eV.

Photons of infrared radiation have energies ranging from 0.004 to 1.6 eV.

Photons of visible light have energies between 1.6 and 3.3 eV. The corresponding wavelengths are from 760 nm (red) to 380 nm (violet).

Ultraviolet light photons UV-B (radiation in the skin-burn region) range from 3.9 eV to 4.4 eV (320–280 nanometers)

The initial temperature in an H-bomb explosion is 100,000,000 K, so the most probable energy of the photons of thermal radiation, kT, equals 8600 eV, classified as low-energy x rays. These x rays are absorbed in a short distance (within meters) in air, heating the air to very high temperatures and producing the characteristic fireball.

A chest x ray is produced by photons with maximum energies between 75 and 120 keV.

Cobalt-60 emits gamma rays of two energies, 1.18 MeV and 1.33 MeV.

Linear electron accelerators used in radiation therapy typically produce x rays with maximum energies in the range of 18 to 25 MeV.

The Stanford linear accelerator accelerates electrons over a distance of 3 km to achieve energies of 50 GeV for research in high-energy physics.

The Tevatron, located at Fermi National Accelerator Laboratory, is currently the world's most powerful accelerator, producing protons with energies of 1.8 TeV (1.8×10^{12} eV).

The highest energy ever recorded was that of a cosmic ray, at 3×10^8 TeV.

Principles of Protection against Ionizing Particles

1 THE APPROACH

The approach used to address a radiation protection problem has many elements in common with the way other problems in occupational and environmental protection are best handled. In particular, it involves *recognition, evaluation,* and *control* of radiation hazards.

Recognition requires familiarity with all the physical factors that can lead to radiation exposure. This includes knowledge of all the radiation sources and their properties, and the pathways that lead from use of the sources to exposure of the workers and the public. Part Two presents the information needed to recognize the presence and significance of radiation sources.

Evaluation involves comprehensive calculations, radiation surveys, and reference to recognized and authoritative standards to perform a complete exposure assessment of a radiation problem. Part Three covers calculation methods and Part Four covers measurements and surveys.

Control measures against excessive radiation exposure are of two broad types: measures to prevent ingestion of radioactive materials and measures to protect the body from external radiation. The controls against ingestion, including ventilation, filtration, protective clothing, and personal hygiene, are very similar to those employed in other disciplines in occupational protection. Even the protective measures against external exposure have elements in common with control of exposure to other physical agents; for example, limiting exposure time and controlling distance from the source are measures against chemical exposure as well as radiation. Methods of shielding against radiations are unique to radiation control, and these are discussed in Part Two. Part Five covers practical control measures in work with radioactive materials.

2 ENERGY AND INJURY

The production of injury to living matter by ionizing radiation is the result of the transfer of large amounts of energy indiscriminately to individual molecules in the region through which the radiation passes. These large energy transfers cause the ejection of electrons from atoms and initiate a variety of chemical and physical effects, the most critical being those which damage the DNA molecules (Little, 1993). The cell has enzymatic processes for repairing certain types of damage, but if it is unable to repair the damage, the cell may die or be mutated into a malignant cancer cell. Thus the imparting of energy by ionizing radiation to living matter may be characterized as a potentially harmful process, and the greater the energy imparted, the greater is the initial damage produced. Because the transfer of energy plays the key role in the production of injury by ionizing radiation, all measurements and calculations to evaluate the hazard from ionizing particles have as their initial object the determination of the energy imparted by the ionizing particles to the region of concern.

The region affected by the action of a single ionizing particle or ray is small, the damage caused to the person is insignificant, and the risk of induction of any serious delayed effects, such as malignancy, is extremely low. The damage produced by successive particles accumulates, however. Although the effect may be accompanied by some repair for certain types of particles, if enough energy is imparted, the consequences can become serious. To prevent these consequences from developing, limits are set on radiation exposure from ionizing particles. The limits are derived from epidemiologic and laboratory data on the relationship between the energy imparted to the body and injury produced. In essence, they specify the maximum energy allowed to be imparted by ionizing particles to critical regions in the body.

The effects produced by ionizing particles depend not only on the amount of energy imparted to the body but also on the location and extent of the region of the body exposed and the time interval over which the energy is imparted. These and other factors must be taken into account in specifying maximum exposure levels.

While the human body can sense and take measures to protect itself from injury from most physically destructive agents—heat, noise, missiles, and so on—it cannot sense exposure to radiation except at levels that are invariably lethal. Thus we see how important it is to understand how to anticipate radiation problems through calculations and analyses and how to use radiation instruments to monitor the emissions from radiation sources.

3   CHARGED AND UNCHARGED IONIZING PARTICLES

As noted in the previous section, ionizing particles produce damage in matter by ionizing the atoms of which the matter is constituted as they penetrate. The particles that can produce these ionizations are divided into two classes—charged ionizing particles and uncharged ionizing particles. Thus *ionizing radiation* is defined as any radiation consisting of charged or uncharged ionizing particles or a mixture of both. *Charged ionizing particles* are electrically charged particles having sufficient kinetic energy to produce ionization by collision. They include electrons, protons, alpha particles, beta particles, etc. *Uncharged ionizing particles* are uncharged particles that can liberate charged ionizing particles or can initiate a nuclear transformation. They include neutrons, gamma rays, neutral mesons, etc.

There are very basic differences between charged and uncharged ionizing particles in their modes of interaction with matter. The charged ionizing particles produce ionizations at small intervals along their path as a result of impulses imparted to orbital electrons. The impulses are exerted *at a distance* through electrical forces between the charged particles and orbital electrons. Thus, they are *directly ionizing.* Uncharged ionizing particles penetrate through a medium without interacting with the constituents, until, by chance, they make collisions (with electrons, atoms, or nuclei), which result in the liberation of energetic charged particles. The charged particles that they liberate then ionize along their paths, and it is through them that damage in the medium is produced. Thus, the uncharged particles are *indirectly ionizing.* The damage is primarily done by charged particles, even when the incident radiation is uncharged.

The charged particles possess the energy required to produce ionizations by virtue of their mass and motion. (Remember the classical expression for the kinetic energy of a moving body, $1/2mv^2$; kinetic energy equals one-half the product of the mass and the square of the velocity.) As the particles impart energy to the medium through which they penetrate, they lose kinetic energy until they are finally stopped. The more energy they have to start with, the deeper they penetrate before they are stopped. The charged particles emitted from radioactive substances have a limited energy range and are stopped in a relatively short distance, usually less than a few millimeters in the body.

4   ENERGY TRANSFER BY CHARGED PARTICLES

Figure 2.1a shows an energetic free electron passing by a bound electron in an atom. Both the free electron and the bound electron have a single negative charge, *q,* and as a result they repel each other. The repulsive force be-

2.1 Production of impulses by charged particles. *(a)* The incident charged particle (high-speed electron) flies past one of the atoms constituting a molecule. The electric field of the rapidly moving electron exerts impulsive forces on the orbital electrons and imparts energy to them. If an electron is ejected from the atom as a result, the molecule may break up. *(b)* Forces exerted by fast and slow ionizing particles on a bound electron in atom as a function of time. The area under the curve is equal to the size of the impulse. The fast-moving particles (e.g., beta particles) produce much smaller impulses than do slower-moving particles (e.g., protons, alpha particles) that exert equivalent forces at a given distance from an orbital electron.

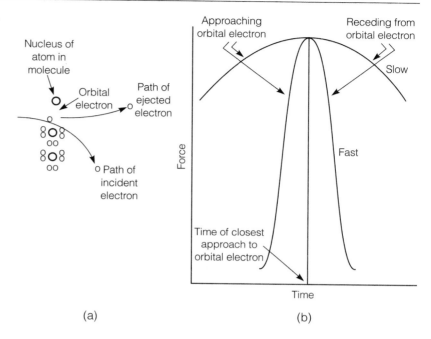

(a)

(b)

tween two particles with charges q_1 and q_2 varies inversely as the square of the distance, r, and is given by Coulomb's law:

$$f = q_1 q_2 / r^2 \qquad (2.1)$$

Assume that you are the bound electron in the atom. When the free, speeding electron is at some distance from you, you hardly feel any repulsive force and it is of no consequence. As the incident electron approaches, however, the force between you increases until, at the closest distance, called the *distance of closest approach,* the force is at a maximum. Then as it continues to fly by, the force quickly decreases.

If you could plot the force you feel as a function of the time, it would look like the curve in Figure 2.1b. Two curves are shown, one for a fast-moving electron, in which case the force rises and falls very quickly, and one for a slower-moving particle, in which case the experience of the force is more spread out in time.

This force you experience, exercised over a period of time, is called an *impulse.* The magnitude of the impulse is the area under the force-time curve, which is given mathematically as

$$\text{Impulse} = \int f \, dt \qquad (2.2)$$

The impulse results in the transfer of energy from the free electron to the bound electron, and a loss of energy by the passing electron.

When the electron is traveling very fast, the area under the curve is small and the impulse is weak. When the electron (or any other charged particle) is traveling relatively slowly, the area under the curve increases, and the impulse is much stronger. The impulse may only be large enough to raise electrons to higher energy levels in the atom—that is, to produce excitation—or it may be large enough to actually knock the electron out of the atom—that is, to produce ionization.

Electrons at energies characteristic of electrons emitted in radioactive decay travel very quickly because of their small mass (many electrons emitted in beta decay travel very close to the speed of light) and produce small impulses. Electrons toward the end of their range in matter, when they are traveling slowly, or heavier charged particles, such as protons and alpha particles (to be discussed later), which travel much more slowly than electrons, produce much larger impulses. The resultant ionizations are also spaced very close together. The closer the spacing of the ionizations, the greater is the extent of the damage for a given energy absorption. Thus the relatively slow moving heavy particles, such as alpha particles and protons, are much more effective in producing biological damage than are fast-moving electrons.

5   THE STOPPING POWER EQUATION

The rate at which charged particles lose energy as they pass through a medium (the energy *absorber*) is given by the *stopping power equation* (Evans, 1955). This equation gives the energy loss per unit distance (e.g., MeV/cm) as a function of the properties of the ionizing particle and the absorber:

$$\frac{dE}{ds} = \frac{4\pi q^4 z^2}{mv^2} NZ \left[\ln \frac{mv^2}{I} - \ln(1 - \frac{v^2}{c^2}) - \frac{v^2}{c^2} \right] \qquad 2.3$$

where

> E = energy of ionizing particle
> s = distance along track of ionizing particle
> z = atomic number of ionizing particle
> q = magnitude of unit electrical charge
> m = rest mass of the electron
> v = speed of ionizing particle
> N = number of absorber atoms per cubic centimeter of medium

Z = atomic number of absorber

NZ = number of absorber electrons per cm^3

c = speed of light

I = mean excitation and ionization potential of absorber atom

The equation shown here was derived for heavy particles, such as alpha particles and protons. The term in brackets is more complex for electrons.

Obviously, the greater the density of intercepting electrons, the greater will be the rate of energy loss per centimeter. Of particular interest is the fact that the rate of energy loss (and therefore energy imparted to the medium) varies as the square of the charge on the ionizing particle and inversely as the square of its velocity. Thus, the charge of $+2$ on the alpha particle has the effect of increasing the stopping power by a factor of 4 (i.e., 2^2) over the stopping power of the electron (with its single negative charge). Note that the stopping power is the same whether there is a repulsive force between the incident particle and the electrons (that is, they have the same sign) or whether there is an attractive force (that is, they have opposite signs).

A property that is closely related to the stopping power is the *linear energy transfer* (LET) of the radiation, the rate at which energy is transferred to the medium along the track of the particle. The LET is equal to the stopping power when all the energy lost by the particle is treated as absorbed locally along the track. This relationship is generally assumed to hold for charged particles emitted by radionuclides. The greater the linear energy transfer, the greater is the damage produced by the particle. The LET of fast electrons is used as the reference LET, and a great deal of radiation research has been done to study the relative effectiveness of radiations with higher LETs, such as alpha particles and protons, in producing damage. As a result, numerical values called *radiation weighting factors (W_R)* or *quality factors (Q)* have been assigned to measures of the biological effectiveness of radiation as a function of the type and energy of the radiation. Beta particles, as the reference radiation, have a radiation weighting factor of 1. Alpha particles cause about twenty times as much damage as beta particles for the same energy absorbed, so their radiation weighting factor is assigned a value of 20.

6 Beta Particles—A Major Class of Charged Ionizing Particles

Carbon-14, tritium (hydrogen-3), sulfur-35, calcium-45, phosphorous-32, strontium-90—these names are familiar to investigators who use radionuclides as tracers in research. All these radionuclides have in common the characteristic that they emit only one type of ionizing particle—a beta particle (β). Beta particles comprise one of the most important classes of charged ionizing particles. They are actually high-speed electrons that are emitted by the nuclei of atoms as a result of energy released in a radioactive decay process involving the transformation of a neutron into a proton (Fig. P.3A). The energy of the beta particle may have any value up to the maximum energy made available by the transformation. The energy difference between this maximum and the actual energy of the beta particle is carried off by another particle, known as the neutrino. The neutrino has virtually no interaction with matter and is thus of no interest from a radiation-protection point of view.

Electrons with speeds and energies comparable to those possessed by beta particles are widely used in technology. The electrons are energized in special machines by applying a high positive voltage between the source of electrons and a collecting terminal (Fig. 2.2). One widespread application of high-speed electrons is in a television tube, as discussed in Part One. Another is in x-ray machines, where electrons are energized with much higher voltages. For example, electrons acquire 70,000–115,000 eV of energy in medical diagnostic machines, and millions of electron volts in some therapy machines. When they strike the heavy metal targets (usually tungsten) in the x-ray tubes, their energy is partly converted into x-ray photons, which are then used to produce x-ray pictures or destroy a cancer. The high-speed electrons emitted by radionuclides have energies within the range covered above, that is, between a few thousand to a little over 2 mil-

2.2 Use of high voltage to produce energetic electrons. The electrons are attracted toward the positive terminal and reach it traveling at high speed. The tube is evacuated in order that electrons will not collide with gas molecules and lose their energy.

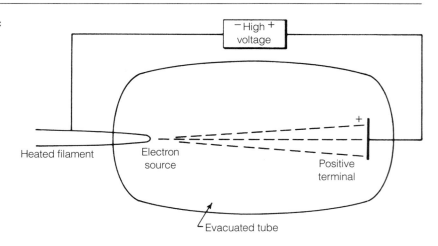

High voltage

Heated filament

Electron source

Positive terminal

Evacuated tube

lion eV, and serve to trace the presence of the nuclei from which they are emitted.

6.1 Properties of Some Common Beta-Emitting Radionuclides

Let us examine some radionuclides that emit only beta particles. Properties of beta emitters that are most commonly used in research at universities are given in Table 2.1. It is not surprising that the radioactive isotopes that trace carbon (carbon-14), sulfur (sulfur-35), calcium (calcium-45), phosphorous (phosphorous-32), and hydrogen (tritium or hydrogen-3) are most popular, as these elements have a basic role in chemical and biological processes. Strontium-90 is included because its decay product, yttrium-90, which is always present with strontium-90, gives off the most energetic beta particle found among the common radioactive nuclides. Hence, this radionuclide is often used as a source of penetrating beta radiation. Let us examine Table 2.1 in detail, discussing the quantities listed and their values.

6.1.1 Half-Lives

Each beta particle given off by a radioactive source results from the transformation or decay of an atom of that source to an atom of another element whose atomic number is greater by one. The rate at which the atoms undergo transformations—and, consequently, the rate of emission of beta particles—is proportional to the number of radioactive atoms present. Thus, as the number of radioactive atoms in the source decreases owing to the radioactive transformations, the rate of emission of beta particles decreases. When half the atoms in a sample have decayed, the rate of emission of beta particles is also cut in half. The time in which half the atoms of a radionuclide are transformed through radioactive decay is known as the half-life of the particular radionuclide.

A radionuclide that is frequently used to demonstrate the nature of radioactive decay is indium-116 (^{116}In). This is produced from indium-115, which is a naturally occurring, nonradioactive metal, by the absorption of neutrons; suitable neutron sources are usually available at research or educational institutions. Figure 2.3 shows the counting rate of a sample of ^{116}In as a function of time, as measured with a Geiger-Mueller counter. The data, when plotted on semilogarithmic graph paper, fall on a straight line. No matter when the measurement (count) is made, the counting rate decreases with a half-life of 54 minutes. Each radionuclide has a unique half-life.

From Table 2.1 we see that the half-lives of commonly used radio-

Table 2.1 Properties of some commonly used beta emitters.

Property	Beta emitter				
	$^3H \rightarrow {}^3He$	$^{14}C \rightarrow {}^{14}N^f$	$^{45}Ca \rightarrow {}^{45}Sc$	$^{32}P \rightarrow {}^{32}S$	$^{90}Sr \rightarrow {}^{90}Y \rightarrow {}^{90}Zr$
Half-life	12.3 yr	5,730 yr	163 d	14.3 d	28.1 yr
Maximum beta energy (MeV)	0.0186	0.156	0.257	1.71	2.27[a]
Average beta energy (MeV)	0.006	0.049	0.077	0.70	1.13[b]
Range in air (ft)	0.02	1	2	20	29
Range in unit-density material (cm)	0.00052	0.029	0.06	0.8	1.1
Half-value layer, unit-density absorber (cm)	—	0.0022	0.0048	0.10	0.14
Dose rate from 100 beta particles/ cm²-sec (mGy/hr)[c]	—	0.56	0.33	0.11	0.11
Fraction transmitted through dead layer of skin (0.007 cm)	—	0.11	0.37	0.95	0.97
Dose rate to basal cells[d] of epidermis					
From 1 Bq/cm² (mGy/day)		0.0091	0.0259	0.0597	0.1103
From 1 μCi/cm² (mrad/hr)	—	1,400	4,000	9,200	17,000[e]

a. From the ^{90}Y decay product. ^{90}Sr emits 0.55 MeV (max) beta. See Part Three, section 5.3.

b. From ^{90}Sr (0.196) + ^{90}Y (0.93).

c. Parallel beam (Jaeger et al., 1968, p. 14).

d. Calculated from Healy, 1971, fig. 1. The dose is from beta particles emitted in all directions equally from contamination on surface of skin. Basal cells are considered to be 0.007 cm below surface.

e. From ^{90}Sr (7,700) + ^{90}Y (8,900). Data for half-lives and maximum and average beta energies taken from MIRD, 1975.

f. The properties of the beta particles emitted by another radionuclide widely used in research, ^{35}S, are almost identical to those emitted by ^{14}C ($^{35}S \rightarrow {}^{35}Cl$, half-life 87 days, maximum beta energy 0.167 MeV.

nuclides cover a wide range. The maximum half-life shown is 5,730 years for carbon-14. The minimum half-life is 14.3 days for phosphorous-32. A source of phosphorous-32 would have half the original number of atoms and would disintegrate at half the initial rate in about two weeks. After several months the disintegration rate or activity would be negligible.[1] On the other hand, the activity of a source of carbon-14 remains essentially unchanged over many years.

Half-lives of radionuclides cover a much wider range than those listed in Table 2.1. Half-lives shorter than 10^{-6} seconds have been measured for some radionuclides. Examples of long-lived radionuclides are uranium-238, 4.5 billion years, and potassium-40, 1.3 billion years. These radionuclides were created very early in the life of the universe, but approxi-

---

1. Methods for evaluating the radioactivity remaining after any decay period are given in section 21.4 of this part and in section 1.3 of Part Three.

2.3 Decrease of counting rate of an indium-116 source after an arbitrary starting time. Measurements were made with a G-M counter. Each count was taken for one minute, starting 30 seconds before the time recorded for the count. When plotted on semilogarithmic paper, the results of the measurements fall close to a straight line drawn as a best fit, with a half-life of 54 minutes. They do not fall exactly on the line because of statistical variations in the number of disintegrations in a fixed counting interval (see Part Four, section 6).

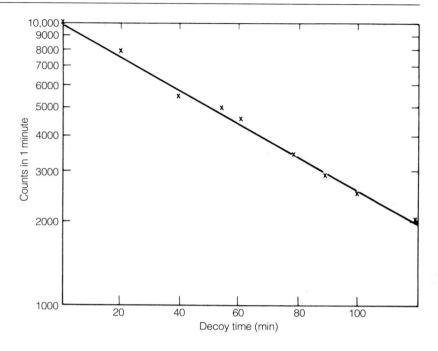

mately 79 percent of the ^{238}U and 99 percent of the ^{40}K have been lost by decay since creation because of their radioactivity.

6.1.2 Maximum and Average Energies of Beta Particles

The energies of the beta particles given off in radioactive decay of the various radionuclides listed in Table 2.1 are expressed in terms of both a maximum energy and an average energy. The maximum energies range from 0.018 million electron volts (MeV) for tritium to 2.24 MeV for strontium-90 (from the yttrium-90 decay product). The tritium betas are very weak, the maximum energy of 0.018 MeV, or 18,000 eV, being less than the energies of electrons hitting the screens of most TV tubes. Average energies are given for the beta sources because, as stated earlier, the nature of beta decay is such that any individual beta particle can have any energy up to the maximum (E_{max}). However, only a very small fraction of the emitted beta particles have energies near the maximum. The frequency with which energies below the maximum are carried by the beta particles is given by energy spectrum curves, which have characteristic shapes. Energy spectra for tritium and carbon-14, low-energy beta emitters, and for phosphorous-32, a high-energy beta emitter, are shown in Figure 2.4. Energy spectra of the type shown in the figure can be used to determine the fraction of the beta particles emitted in any energy range by simply determining the fraction of the area under the curve that lies between the energy points of interest.

Values for the average energies of the beta particles from the radionuclides in Table 2.1 are listed under the values of the maximum energies. In general, the average energy is about one-third the maximum. As we shall see later, the average energy is a basic quantity in determining the energy imparted to tissue by beta emitters and in evaluating the dose from concentrations of these radionuclides in the body.

6.1.3   Range of Beta Particles

If absorbers of increasing thickness are placed between a source of beta particles and a detector, the counting rate falls off steadily until no net count can be detected (Fig. 2.5). The maximum thickness the beta particles will penetrate is called the range. The ranges of beta particles from the radionuclides in Table 2.1 are given for air and for a medium of unit density, characteristic of water or soft tissue. Range is specified for a general medium of unit density because the penetration of the beta particles is primarily determined by the mass of matter that is traversed, and does not depend strongly on other atomic characteristics such as atomic number.[2] For different media, the thickness that will provide the same mass in the path of the particle, and consequently the same amount of attenuation of the beta particles, is inversely proportional to the density of the material. Thus, if we know the range in a medium of density equal to 1, the range in any other medium can be determined by dividing by the density.[3]

Example 2.1   Calculate the minimum thickness of the wall of a glass test tube required to stop all the beta particles from ^{32}P.

From Table 2.1, the range of ^{32}P beta particles in unit-density material is 0.8 cm. The density of the glass is 2.3 g/cm^3. The maximum range in glass is 0.8/2.3 = 0.35 cm. This is the minimum thickness required. (Most of the beta particles would be stopped by a much smaller thickness. See section 6.1.4.)

---

2. Frequently the range is expressed in terms of the density thickness, defined as the mass per unit area presented by a sample of a given thickness. The units are usually in mg/cm^2. For a medium of density ρ g/cm^3, the range in mg/cm^2 is $1{,}000\rho$ times the range in cm.

3. We should expect that the ranges for unit-density material would also apply to water and to soft tissue, since both have densities equal to 1. However, water provides somewhat better attenuation than do other media for equivalent intercepting masses. The reason is that

2.4 Distributions of energies of beta particles (Slack and Way, 1959).

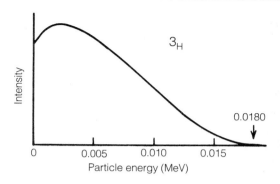

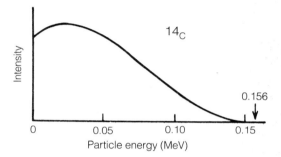

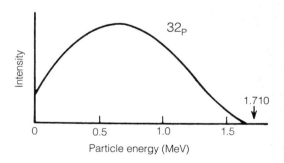

penetrating electrons lose their energy primarily by colliding with electrons in the medium, and it is easy to verify that water has more electrons per gram (because of the hydrogen content) than do other absorbers. For our purposes, however, we can ignore the extra effectiveness of water and use the range and attenuation data in Table 2.1 for all materials, bearing in mind that if a high degree of accuracy is required, special data for the medium in question will have to be used.

2.5 Count rate from ^{32}P beta source as a function of absorber thickness. The diameter of the beta particle beam was limited to 1 mm at the face of the absorber. The thickness is expressed in millimeters of unit-density material and includes an air and detector window thickness of 0.08 mm. The range is approximately 7.4 mm. The count rate appears to level off to a value slightly above background (this count rate is due to the penetrating x radiation produced by acceleration of the energetic ^{32}P beta particles in the vicinity of the nuclei of the absorber atoms).

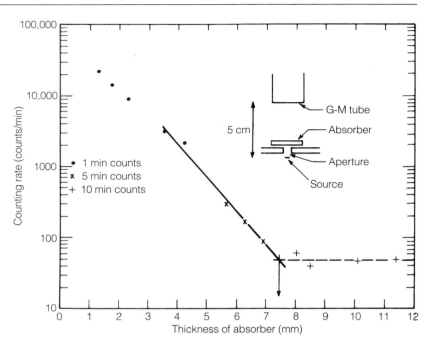

Note that the range of the beta particles is very dependent on E_{max}. Beta particles from ^{32}P (E_{max} = 1.71 MeV) require 0.8 cm at unit density (or 6 m of air) to stop them, while the very weak betas from tritium (E_{max} = 0.018 MeV) are stopped by less than 0.00052 cm at unit density (6 mm of air). A useful "rule of thumb" is that the range in centimeters at the higher energies is approximately equal to the energy in MeV divided by two.

The ranges of beta particles give us an idea of the hazards of the various radionuclides as *external* sources of radiation. For example, the beta particles from tritium are stopped by only 6 mm of air or about 5 μm of water. These particles cannot penetrate the outermost layer of the skin, which contains no living tissue; hence such particles present no hazard when they originate outside the body. They can produce injury only if they originate from tritium inside the body, through either ingestion or inhalation, by diffusion through the skin, or entrance through breaks in the skin. They cannot be detected by Geiger-Mueller (G-M) counters because they cannot penetrate through the window of the counter, even if the window is very thin.

In contrast to the tritium beta particles, the highly energetic particles from ^{32}P will penetrate as much as 8 mm into the body. They are among the most penetrating beta particles used in tracer work, and special care must be taken in working with them. Because of their penetrating power, they are readily detected with G-M counters designed for general radiation monitoring.

6.1.4   Absorption of Beta Particles at Penetrations Less Than the Range

The range for a given source of beta particles is a limiting distance, a thickness of material that no beta particle emitted from a source can penetrate. Actually, most of the beta particles emitted by a source are absorbed in distances considerably less than the range. It turns out that over distances that are a major fraction of the range, the rate of loss of particles is almost constant. Thus we can introduce a concept called the *half-value layer*, which is the distance in which half the particles are absorbed. Half-value layers have been included in the data presented in Table 2.1. Although the maximum range of ^{32}P beta particles is 0.8 cm, about half the beta particles are absorbed in the first 0.1 cm, half of those penetrating the first 0.1 cm are absorbed in the second 0.1 cm, and so on. Also shown in the table are attenuation values for beta particles in the dead layer of the skin, where the thickness of the dead layer is taken as 0.007 cm, a value generally assumed for the thinner portions of the skin. We note that only 11 percent of the ^{14}C beta particles can penetrate this dead layer, while 95 percent of the more energetic ^{32}P particles can pass through.

An approximate formula for the half-value layer (HVL) for beta particles in unit-density material as a function of energy E in MeV is:[4]

$$\text{HVL (cm)} = 0.041E^{1.14} \tag{2.4}$$

6.2   Protection from External Beta Particle Sources—Time, Distance, and Shielding

Three key words are often emphasized when proper working procedures with radiation sources are reviewed. These are *time, distance,* and *shielding.* The *time* refers to the principle that the time spent while being exposed to the radiation from the source should be no longer than necessary. Obviously, the longer the time one is exposed to a source, the greater will be the number of particles incident on the body and the greater the dose. It is necessary to emphasize minimum working times because the worker cannot feel the presence of the radiation or any discomfort from it that would remind him to limit his working time.

Distance refers to the desirability of keeping as much distance as possible between the source and the worker. Distance is very effective in reduc-

4. Evans, 1955, pp. 625–629. The HVL values given by this formula differ from the values in Table 2.1. The data in Table 2.1 are based on dose measurements (see Hine and Brownell, 1956), while equation 2.4 describes results obtained in counting beta particles.

ing the intensity of the radiation particles incident on the body from small sources. The actual relationship follows the inverse square law; that is, for point sources, the intensity varies inversely as the square of the distance from the source. Thus, if we use a distance of 10 cm from a source as a reference point, a distance of 100 cm will be 10 times as far, and the intensity of beta particles incident on the skin will be $1/(10)^2$, or 1/100, the intensity at 10 cm. At 1 cm, the ratio of distance to the distance at 10 cm is 1/10. The ratio of intensities is now $1/(1/10)^2$, or 100 times the intensity at 10 cm. The effect of distance is illustrated in Figure 2.6. The extra distance provided by the use of tweezers or tongs produces a tremendous lowering in exposure rate from the rate occurring when the source is held in the hands. Distance also exerts some protective effect through the interposition of air between the source and worker, but except for the low-energy beta emitters, this effect is small at normal working distances.

Shielding thicknesses greater than the range stop all beta particles, but a complication in the shielding of beta sources arises from the fact that when beta rays strike a target, a fraction of their energy is converted into much more penetrating x radiation called *bremsstrahlung*. The fraction *(f)* is given approximately by the electron energy *(E,* in MeV) times the atomic number of the target *(Z)* times 1/1,000, or $f = E \times Z \times 10^{-3}$. Thus the efficiency of x-ray production increases with increasing atomic number of the target and energy of the beta particles. For this reason, plastic is usually to be preferred over steel and lead to minimize x-ray production in materials for shielding energetic beta emitters. Only a few millime-

2.6 Effect of the inverse square law in reducing intensity of beams of particles. As used here, the intensity is equal to the number of particles crossing unit area per unit time (that is, particles/cm^2−sec) at the indicated location. It is arbitrarily taken to be 100 at a distance of 10 cm. Attenuation in the medium is neglected.

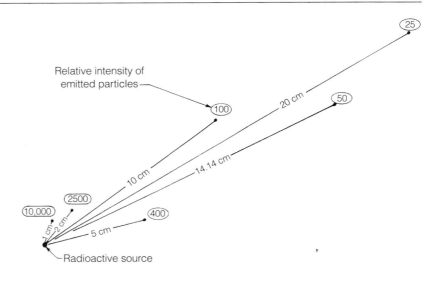

ters are sufficient to stop the beta particles from all commonly used sources.

We have discussed protection from external beta particle sources in rather general terms. An important note to add is that protection measures that are instituted for any specific situation should be commensurate with the hazard. The investigator should not become overconcerned with weak sources that may present trivial radiation hazards. However, it should be reiterated that beta activity is so easy to shield and protect against that complete protection can almost always be obtained with minimal inconvenience, loss of time, or expense.

7 Characteristics of Uncharged Ionizing Particles

Uncharged ionizing particles do not exert electrical forces at a distance and thus do not apply impulses to nearby electrons as charged particles do. Instead, they proceed without any interaction until they undergo a chance encounter with one of the elementary components of the medium through which they pass. The component may be an atom, an electron, or the nucleus of an atom—we shall not be concerned with the nature of the interaction—and as a result of this encounter, energy is transferred from the uncharged ionizing particle to a charged ionizing particle, such as an electron (Fig. 2.7). The electron is liberated from the atom as the result of the energy transfer and proceeds to ionize in the manner characteristic of a charged ionizing particle, as discussed previously. Thus, the net result is that uncharged ionizing particles liberate charged ionizing particles deep within a medium, much deeper than the charged ionizing particles could reach from the outside.[5]

Examples of uncharged ionizing particles are gamma rays, x rays, neutrons, and neutral mesons. Gamma rays are the most important class of uncharged ionizing particles encountered by users of radionuclides.

8 Gamma Rays—A Major Class of Uncharged Ionizing Particles

Gamma rays are electromagnetic radiations emitted by radioactive nuclei as packets of energy, called photons, and often accompany the emission of

---

5. We refer here to charged ionizing particles in the energy range emitted by radionuclides. Particles with very high energies, such as those found in cosmic rays or emitted by high-energy accelerators, will penetrate deeply into matter.

2.7 Imparting of energy to body by uncharged ionizing particles. Unchanged ionizing particle shown is a gamma photon. It penetrates without interaction until it collides with and transfers energy to an electron. The electron is ejected and leaves behind a positively charged atom. The electron continues to ionize directly as it travels through the medium. The gamma photon may transfer all its energy and be eliminated or it may be deflected, in which case it transfers only part of its energy and travels in a different direction at reduced energy. The dots represent potential sites for collision.

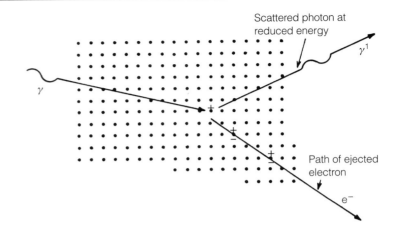

beta particles from the same nuclei. They have energies over the same range as that found for beta particles, that is, from a few thousand electron volts up to several million electron volts. Unlike beta particles, which slow down as they lose energy and finally become attached to an atom, gamma rays of all energies travel with the speed of light. We are quite familiar with other types of electromagnetic radiation, which are classified according to the photon energy. For example, radio wave photons have energies between 10^{-10} and 10^{-3} eV, infrared photons between 10^{-3} and 1.5 eV, visible light between 1.5 and 3 eV, and ultraviolet light between 3 and 400 eV.[6] X-ray photons have energies between approximately 12 eV and the upper limit that can be produced by man-made electronic devices, which has reached over a trillion electron volts.

Gamma rays lose energy through chance encounters that result in the ejection of electrons from atoms.[7] The gamma ray may lose all of its en-

6. A photon will ionize an atom if it has enough energy to remove an electron from the atom. A minimum of 13.6 eV of energy is required to ionize a hydrogen atom. Ionization potentials, expressed in volts, for the outermost electrons of some other elements are: carbon, 11.2; nitrogen, 14.5; oxygen, 13.6. There are no sharp boundaries between the ultraviolet and x-ray regions. The classification of a photon as ultraviolet or x ray in the region of overlap generally depends on the nature of the source or method of detection.

7. Gamma photons with energies greater than 1.02 MeV may also undergo a reaction in the vicinity of an atomic nucleus, known as pair production, in which a portion of the photon energy (1.02 MeV) is used to create two electrons of opposite charges and the remaining energy is divided as kinetic energy between the two particles. However, this interaction is important only at high photon energies and for high atomic number materials. From the standpoint of radiation protection, it is of minor significance for gamma photons from radionuclides (see section 8.3).

ergy in an encounter, or only part. If only part of its energy is removed, the remainder continues to travel through space, with the speed of light, as a lower-energy photon. On the average, the higher the energies of the gamma photons, the higher are the energies of the liberated electrons.

Because gamma rays are uncharged ionizing particles and travel through a medium without interaction until they undergo a chance encounter, every gamma ray has a finite probability of passing all the way through a medium through which it is traveling. The probability that a gamma ray will penetrate through a medium depends on many factors, including the energy of the gamma ray, the composition of the medium, and the thickness of the medium. If the medium is dense and thick enough, the probability of penetration may be practically zero. With a medium of the size and density of the human body, however, gamma rays emitted inside the body by most radionuclides have a good chance of emerging and being detected outside the body. For this reason, suitable gamma-ray emitters are powerful tools for studying body function.

It is important to keep in mind that it is the electrons to which the energy is transferred by the gamma photons that actually produce damage in the medium (by subsequent ionization and excitation of the atoms). Once a photon liberates an electron, the subsequent events depend only on the properties of the electron and not on the gamma photon that liberated it. The ejection from an atom by a photon of an energetic electron, say with an energy of 1 MeV, is only a single ionization. The electron, in slowing down, will produce tens of thousands of ionizations and excitations, and the damage produced will depend on the number and spatial distribution of these ionizations and excitations, rather than on the single ionization produced by the gamma photon.

8.1 Energies and Penetration of Gamma Rays from Some Gamma-Emitting Radionuclides

Data for three gamma-emitting radionuclides are presented in Table 2.2. The radioiodine isotopes listed are probably used more than any other radionuclide in medical diagnosis and therapy, because of their unique value in diagnosis and treatment of conditions of the thyroid gland. We need not discuss the data on half-lives and beta energies, since we have already covered the significance of these quantities in the previous section on beta particles. Note, however, that the gamma rays are generally emitted along with charged ionizing particles and that both types of particles may have to be considered when sources of gamma rays are being evaluated.

Unlike beta particles, which are emitted at all energies up to a maximum characteristic for the radionuclide, gamma rays are emitted at

Table 2.2   Properties of some beta-gamma emitters.

Property	Beta-gamma emitter		
	$^{125}I \rightarrow ^{125}Te$	$^{131}I \rightarrow ^{131}Te$	$^{60}Co \rightarrow ^{60}Ni$
Half-life	60 d	8.1 d	5.27 yr
Maximum beta energy (MeV)	—	0.61	0.31
Average beta energy (MeV)	0.020[a]	0.188	0.094
Gamma energies (MeV)	0.035 (6.7%)[b] 0.027–.032[c] (140%)	0.364(82%) 0.637 (6.5%)[d]	1.17 (100%) 1.33(100%)
Gamma half-value layer Lead (cm) Water (cm)	0.0037 2.3	0.3 5.8	1.1 11
Dose rate from 1 photon per cm²-sec (mGy/hr)	2.0×10^{-6}	6.5×10^{-6}	22×10^{-6}
Specific gamma ray constant, Γ (R/hr per MBq at 1 cm) Γ (R/hr per mCi at 1 cm)	0.0189 0.7	0.0595 2.2	0.3568 13.2

a. Internal conversion and Auger electrons (see Fig. 2.8).
b. Percent of disintegrations resulting in emission of photons.
c. X rays.
d. Also 0.723 (1.7%), 0.284 (5.8%), 0.080 (2.58%), 0.030 (3.7%, x rays). Energies taken from MIRD, 1975.

discrete energies. The nuclides chosen for Table 2.2 provide examples of high-, intermediate-, and low-energy gamma emitters, where the energy classification is roughly indicative of the relative effectiveness of lead shielding in attenuating the gamma rays. The gamma rays from cobalt-60 are of relatively high energy. The listing of a 100-percent value with each photon energy means that each disintegration of a ^{60}Co nucleus, which entails the emission of a beta particle, is also accompanied by the emission of two gamma photons of 1.17 and 1.33 MeV energies.

Radioactive decay data are also presented in the form of radioactive decay schemes. An example of a decay scheme is given in Figure 2.18.

It has been noted that photons, in contrast to beta particles, have a definite probability of passing through any shield without any interactions, and this probability can be calculated or measured accurately. The shielding effectiveness of a material for photons is commonly expressed in terms of the thickness required to reduce the intensity of the incident photons by a factor of two. This thickness is called the half-value layer.[8] The photons from ^{60}Co have relatively high penetration, as indicated by the values for the half-value layers in lead and water, listed in Table 2.2, of 1.1 cm and 11 cm, respectively. It takes 1.1 cm of lead and 11 cm, or 10 times this thickness, of water to provide an attenuation of one-half for the incident photons. Since lead is 11.4 times as dense as water, water appears to be slightly more effective than lead on a mass basis at these energies. However, this effectiveness is only with regard to its attenuation of the incident photons. The attenuation of the incident radiation through a large thickness of water is accompanied by much more intense low-energy secondary radiation than occurs in media of higher atomic number, which must be taken into account in designing protective shields (see section 21.5).

While two gamma energies are usually associated with the decay of iodine-131, the spectrum is quite complex. A 0.364 MeV photon is emitted in 82 percent of the beta disintegrations and a 0.637 MeV photon in 6.5 percent, but small percentages of photons with higher and lower energies are also emitted. The ^{131}I photons have a much lower penetration in lead than the photons from ^{60}Co, since a half-value layer is only 0.3 cm. The thickness of water required is 5.8 cm, or 19 times as great. Thus the lead is much more effective than water in attenuating the incident radiation at these energies.

The gamma photons emitted in ^{125}I decay are of low energy. They do not accompany beta decay, but are emitted in a decay process known as electron capture. In this transformation (Fig. 2.8), an electron in the innermost orbit of the atom is captured by the nucleus, and the energy made available by this reaction is equal to 0.035 MeV. In 6.7 percent of the disintegrations, this energy is emitted as a 0.035 MeV gamma photon. The rest of the time, it causes the release of electrons, known as internal conversion electrons, from the shells surrounding the nucleus.

Electron capture and internal conversion processes are always accompanied by the emission of x rays[9] from the inner shells of the atom. About 1.4 x rays with energies between 0.027 and 0.032 MeV are emitted per disintegration of the ^{125}I nucleus. Note the most unusual correspondence between the energies of the gamma rays (emitted from energy transitions in

8. Note the analogy between half-value layer and half-life (introduced in section 6.1.1). See sections 8.4 and 21.4 for methods of using these concepts in attenuation and decay calculations.

9. Photons originating in the inner orbits of the atom are called x rays, and photons originating in the nucleus are called gamma rays, although they are identical if of the same energy.

2.8 Mechanisms of energy release during electron capture, as illustrated with ^{125}I. *(a)* Nucleus captures electron, usually from inner-most shell, producing 0.035 MeV excess energy in resulting tellurium-125 nucleus. *(b)* In 6.7 percent of the disintegrations, a 0.035 MeV gamma photon leaves the atom. *(c)* In 93 per-cent of the disintegrations, an electron is ejected from one of the inner shells in a pro-cess known as internal conversion. The energy of the internal-conversion electron is equal to the difference between 0.035 MeV and the en-ergy required to remove it from the atom. The energies required for removal from the two in-nermost shells are 0.032 MeV and 0.005 MeV. *(d)* The vacancies left by electron capture and internal conversion are filled by the transfer of electrons from the outer shells. The energy made available by the transfer is emitted as x-ray photons characteristic of ^{125}Te or is used to eject additional orbital electrons, known as Auger electrons.

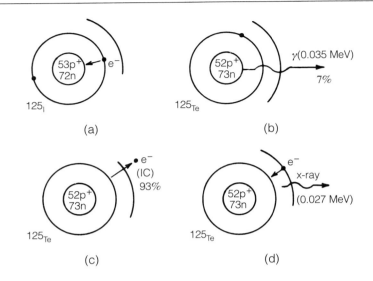

the nucleus) and the x rays (originating from electron transitions in the shells outside the nucleus).

Lead is extremely effective in attenuating the low-energy photons emit-ted by ^{125}I. It takes only 0.0037 cm of lead to reduce them by a factor of 2. Because of their low energies, other materials of higher atomic number, such as steel or brass, are also very effective, and preferable to lead because of its toxicity. Water is much less effective because of its low atomic num-ber and density, requiring 2.3 cm, or 620 times the thickness of lead, for the same degree of attenuation of the incident photons.

8.2   Positron-Emitting Radionuclides and Annihilation Radiation

The electron with a single negative charge has a counterpart in another particle, with the same rest mass but with a single positive charge. This positive electron is called a *positron*. When apart from the electron, the positron is stable, but a free positron quickly comes in contact with an electron—with disastrous consequences for both. They both disappear, their mass being completely transformed into energy as two 0.51 MeV photons, which fly apart in exactly opposite directions.

The electron and positron are antiparticles. Although the positron does not normally exist on earth, it is created in some forms of radioactive de-cay, as well as in reactions involving high-energy radiations. For example, the nucleus of fluorine-18, atomic number 9, contains 9 protons and 9 neutrons. Nuclei with odd numbers of both neutrons and protons are usually unstable, and fluorine-18 is radioactive with a half-life of 110 minutes. Ninety-seven percent of the decays are by positron emission—a proton in the nucleus is transformed into a neutron with the emission of a positive beta particle (positron), maximum energy 0.634 MeV and fluorine-18 is transformed into oxygen-18. The other 3 percent of the time, decay is into oxygen-18 by electron capture, in which an inner electron from the atom combines with a proton in the nucleus to reduce the atomic number by one. This decay process has already been discussed for iodine-125.

Positrons have the same ranges in matter as electrons of the same energies because they undergo similar collision processes. Since the positron must ultimately vanish through annihilation with an electron, positron emitters can be traced not only through the positrons but also through the two 0.51 MeV gamma photons resulting from the annihilation.

Because positron decay in matter is accompanied by the simultaneous emission of two photons moving in opposite directions, the origin of the photons can be determined by appropriate measurements. This has led to the development of a very useful medical imaging technique using positron-emitting radiopharmaceuticals called positron emission tomography, or PET. PET has many applications in the diagnosis of medical conditions, including the visualization and measurement of blood flow. An example of how it is used in both medical treatment and research is the study of blood flow in the brain, as affected by diseases such as epilepsy or by the taking of drugs. PET scans are also used to study blood flow in the heart.

The positron-emitting radionuclides commonly used in medicine and their half lives are fluorine-18 (110 min), carbon-11 (20 min), nitrogen-13 (10 min), and oxygen-15 (2 min). These radionuclides are produced in a cyclotron. Because of their short half-lives, the cyclotron is located at the PET imaging facility. Another positron-emitting radionuclide, rubidium-82 (76 sec), is produced from the decay of strontium-82 (25 days). In clinical practice, it is eluted from a strontium-82/rubidium-82 generator.

8.3   The Three Major Mechanisms Affecting the Penetration of Gamma Radiation

The attenuation of photons traversing a medium depends on the probability that the gamma photons will interact with the atoms, electrons, or nuclei as they pass through. There are three main mechanisms by which the photons can interact: the *photoelectric effect,* the *Compton effect,* and *pair production.* The probabilities of the interactions depend on the energy

2.9 Illustration of Compton scattering, the photoelectric effect, and pair production. The incident photon energies given are those at which each of the interactions is dominant in lead. X denotes the disappearance of incident particles.

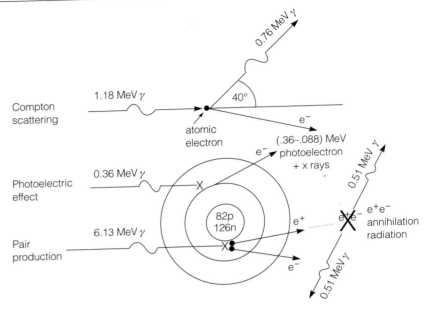

of the photons and on the atomic number of the atoms in the medium. In Figure 2.9, each interaction is illustrated for a gamma ray incident on a lead-208 atom.

At low energies, the *photoelectric effect* dominates. The photon interacts with the atom as a whole and is absorbed by it. Its energy is immediately expended in ejecting an electron with kinetic energy equal to the energy of the absorbed photon less the energy required to remove the electron, its binding energy. As illustrated in Figure 2.9 for a 0.36 MeV photon emitted by iodine-131, 0.088 MeV is required to remove the K electron from the atom, so its kinetic energy is equal to $0.36 - 0.088$ or 0.272 MeV.

Example 2.2 What is the energy of a photoelectron ejected from the L shell of a lead atom as the result of absorption through the photoelectric effect of a 0.36 MeV gamma ray from iodine-131?

In Figure 1.2, the binding energy of an electron in the L shell is given as 15,000 eV. The energy of the photoelectron is $0.36 - 0.015 = 0.345$ MeV.

The vacancy resulting from the ejection of the photoelectron is filled by the transfer of an electron from an outer shell. The energy made available by the transfer is emitted as an x-ray photon (called fluorescence radiation) or is used to eject additional orbital electrons (Auger electrons).

Example 2.3   What is the energy of the fluorescence x rays resulting from the photoelectric effect in lead?

The energy of the x ray depends on the shell from which the photoelectron was expelled and the shell from which the vacancy was filled. The usual case is ejection from the K shell and the filling of the vacancy from the L shell. Since the binding energies of electrons in the K and L shells are 88,000 and 15,000 eV, respectively, the energy of the fluorescence x ray is 88,000 − 15,000 or 73,000 eV. Fluorescence x rays are generally readily absorbed and are not normally evaluated in radiation shielding calculations. The energy made available by the transition could also have resulted in the emission of Auger electrons rather than x-ray photons.

The energy of the incident photon must be greater than the binding energy of the electron that is expelled and the probability of the interaction increases sharply as the photon energy approaches the binding energy of the electron. When its energy drops below the binding energy, it cannot expel that electron, and the probability of the interaction drops and then increases as it approaches the energy required to expel the electron in the next orbit. The effect on the half-value layer is seen in Figure 2.10a, where the half-value layer decreases as the energy decreases until the first discontinuity at the binding energy of the innermost electron (K electron). From inspection of Figure 2.10a at low energies, where the photoelectric effect predominates, it can also be seen that it results in much greater attenuation in materials of high atomic number (lead vs. iron vs. water) as well as at lower energies.

The probability of a photoelectric interaction drops rapidly as the energy of the photon becomes greater than the binding energy of the innermost, most tightly bound electron, and the main interaction then becomes one with the orbital electrons. The theory treats this interaction, called the *Compton effect,* as if the electrons are not bound to the atom. The incident photon collides with the electron as in a billiard ball collision, imparting a fraction of its energy to the electron, and both leave the collision site in different directions. The calculations of the results of the collision are based on the principles of conservation of energy and momentum. Fig. 2.9 illustrates the Compton effect for a 1.18 MeV photon from cobalt-60 incident on an orbital electron of lead-208. The interaction is shown to result in the scattering of a photon of reduced energy (0.76 MeV), with the remain-

2.10 Thickness of a half-value layer of commonly used attenuating materials as a function of energy. The values are based on "good geometry" conditions, that is, no scattered radiation gets to the detector, whether from within the attenuating medium or from surrounding surfaces. Densities of materials (g/cc): H_2O = 1; Fe = 7.85; Pb = 11.3; concrete = 2.36.

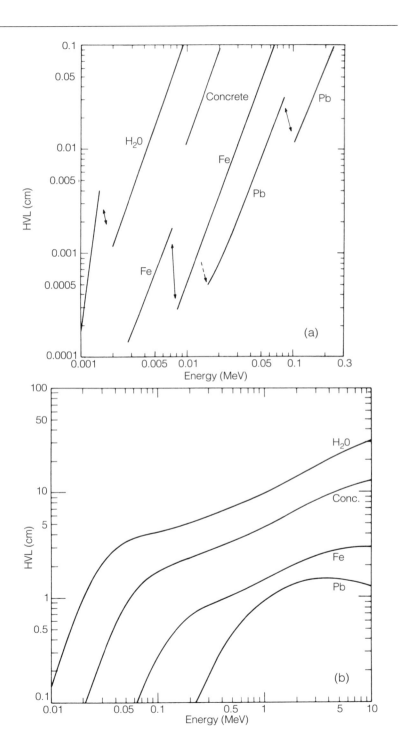

der of the initial energy of the incident photon imparted to the Compton electron.

The energy of the scattered photon *(E_s)* as a function of the energy of the incident photon *(E_i)* and the scattering angle *(θ)* is, in MeV,

$$E_s = \frac{0.5}{\dfrac{0.5}{E_i} + (1 - \cos\theta)} \tag{2.5}$$

Example 2.4   What is the energy of the 1.18 MeV cobalt-60 photon in Figure 2.9 scattered through 180 degrees in the Compton effect (for a back-scattered photon, $\cos\theta = -1$)?

$$E_s = 0.5/[(0.5/1.18) + (1 + 1)] = 0.206 \text{ MeV}$$

The maximum energy of the back-scattered photon approaches 0.25 MeV as the energy of the incident photon increases. It is thus only a small fraction of the incident energy at the higher energies.

As the energy of the photons increases, the probability of the Compton interaction decreases, that is, the half-value layer increases until another interaction begins to operate. In the vicinity of a nucleus and its strong field, the photon disappears and in its place a positive and a negative electron appear. This creation of a pair of electrons is called *pair production.* Since each particle has a mass equivalent in energy to 0.5 MeV, the photon must have an energy of at least 1.02 MeV to produce the pair, and any energy it has above that threshold goes into the kinetic energy of the particles. The energy of the positron is rapidly reduced to thermal energies, at which point it combines with an electron, resulting in the annihilation of both and the transformation of their masses into two 0.51 MeV photons (Fig. 2.9). The probability of pair production increases rapidly as the photon energy rises above 1.02 MeV but it does not act to stop the increase in the penetration (and half-value layer) of photons until an energy of about 3 MeV in lead (Fig. 2.10b) and considerably higher in iron and lower atomic number media. Thus, pair production has little significance in the shielding of gamma rays from radionuclides, although it is important in shielding higher energy radiations, as are produced in high-energy accelerators used in research and in radiation therapy. It is also significant in the attenuation of the 6.13 MeV photons emitted by nitrogen-13 in the coolant of pressurized water reactors. The two 0.51 MeV annihilation photons must also be considered in calculating shielding for high energy photons.

8.4   Attenuation Coefficients of Gamma Photons in Different Materials

To this point, the penetration of gamma rays has been described in terms of the half-value layer, a simple measure of the effectiveness of a shield material. A more fundamental and general approach is through the use of attenuation coefficients, which are derived from basic mechanisms and theory. The attenuation coefficient, designated by the symbol μ, is the probability of an interaction per unit distance. If we consider a very small distance of value x in the medium, then the product of μx gives the probability of interaction in the distance x, *provided μx is much smaller than 1.* (If μx is not much smaller than 1, the probability of interaction is $1 - e^{-\mu x}$, as determined by methods of the calculus.)

Attenuation coefficients can be determined experimentally for a given material by interposing increasing thicknesses of the material between the gamma source and a detector. The setup and data from an attenuation measurement are shown in Fig. 2.11. Special precautions must be taken so photons that are not completely absorbed but only scattered at reduced energy in another direction are not intercepted by the detector.

Attenuation coefficients for different materials and energies of gamma photons are given in Table 2.3. Let us consider the data for a 1 MeV photon, which is representative of fairly energetic photons emitted by radio-

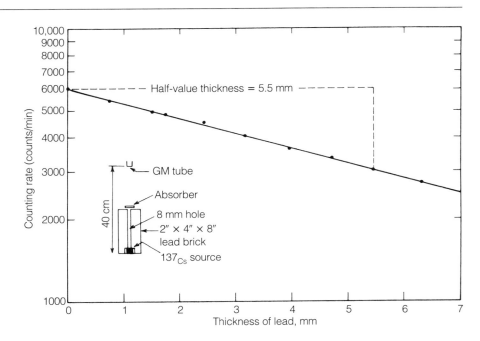

2.11 Attenuation of photons (from cesium-137) in lead. Each count was taken for two minutes. The collimator provides an incident beam of photons that is perpendicular to the absorber, and the large distance from absorber to detector minimizes the detection of photons that are scattered in the absorber. A configuration that minimizes the detection of scattered photons is known as "good geometry."

nuclides. We note that the attenuation coefficient or probability of interaction of this photon per centimeter of travel through lead is 0.776 cm^{-1}. The probability of interaction per centimeter of penetration through water is 0.071. Note that specifying a probability per centimeter of 0.776 is *not* equivalent to saying the probability of interaction in 1 cm is 0.776. This can be seen from the following stepwise calculation in evaluating the penetration of a beam of gamma photons through 1 cm of lead.

Suppose 10,000 photons are incident on the lead, as shown in Figure 2.12. Since the probability of interaction per centimeter is 0.776, the probability per millimeter is one-tenth this, or 0.0776. Let us assume that this is in fact also the probability of interaction in a millimeter, though we shall see shortly that this is not quite correct. We then calculate that the number of photons that interact is the incident number times the probability that an individual photon will interact, or 10,000 × 0.0776 = 776.

Table 2.3   Photon attenuation coefficients.

Energy (MeV)	μ (cm^{-1})			
	Water	Iron	Lead	Concrete
0.01	4.99	1354	1453	62.3
0.1	0.168	2.69	59.4	0.400
1	0.071	0.469	0.776	0.150
10	0.022	0.235	0.549	0.054
100	0.017	0.340	1.056	0.055

Sources: Hubbell, 1969; Storm and Israel, 1970. Coherent scattering is not included (for discussion, see Jaeger et al., 1968, p. 197). See Slaback, Birky, and Shleien, 1997, for a comprehensive table of attenuation coefficients.

Note: The densities of the above materials, in g/cc, are as follows: water, 1; iron, 7.85; lead, 11.3; concrete, 2.35.

2.12   Estimate of transmission of photons through successive layers of attenuating material. The total absorber thickness of 1 cm was divided into 10 fractions, each 1 mm thick. The attenuation coefficient is 0.0776/mm.

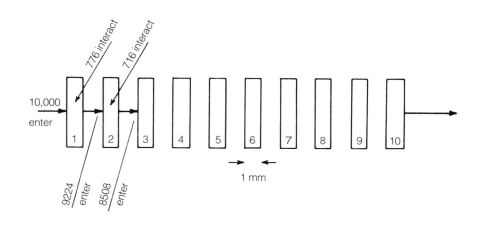

Then 9,224 photons will emerge from the first millimeter without interaction. The number of these that will interact in the second millimeter is evaluated as $0.0776 \times 9,224 = 716$, and 8,508 photons emerge. If we carry this through 10 mm, we calculate that the number emerging is $(0.9224)^{10} \times 10,000$, or 4,459. If we had used the value of 0.776 as the probability of interaction in 1 cm, we would have calculated the number interacting as 0.776 per centimeter times 10,000, or 7,760, leaving 2,240 photons penetrating, a result considerably different from the previous calculation. Even the answer using the stepwise calculation is not completely accurate. The correct answer is obtained by subdividing the absorber into infinitesimally small elements through the use of calculus. (The probability of penetration of a photon through a thickness x is $e^{-\mu x}$ and thus the probability of interaction within the distance x is $1 - e^{-\mu x}$. The probability of interaction in 1 cm determined by this method is $1 - e^{-\mu}$, or $1 - e^{-0.776} = 0.54$. The number of photons that interact is thus 5,400, and the number that emerge is 4,600, which is still almost 10 percent greater than the result of the stepwise calculation. The number emerging could also be obtained directly by multiplying 10,000 by $e^{-0.776} = 10,000 \times 0.46 = 4,600$.)

A frequently tabulated quantity is the attenuation coefficient divided by the density, or the *mass attenuation coefficient*. The contribution to the attenuation coefficient by a given material when it is part of a mixture or compound is then obtained by multiplying the mass attenuation coefficient by the density in the mixture. The contribution to the attenuation coefficient of each constituent is determined separately and all are added to give the total attenuation coefficient for the medium.

Example 2.5   Calculate the attenuation coefficient of silica sand for 0.1 MeV photons. Assume the sand has a porosity of 30 percent and is all quartz (SiO_2, specific gravity of 2.63).

The mass attenuation coefficients at 0.1 MeV for Si and 0 are 0.173 and 0.152 cm²/g, respectively. The density of the sand is $0.70 \times 2.63 = 1.84$ g/cm³. The Si (with atomic mass $A = 28.1$) and O ($A = 16$) constitute 47 and 53 percent by weight of SiO_2 with densities in sand of $0.47 \times 1.84 = 0.86$ and $0.53 \times 1.84 = 0.98$ g/cm³. Thus the attenuation coefficients are: $\mu(Si) = 0.173$ cm²/g $\times 0.86$ g/cm³ $= 0.149$ cm⁻¹; $\mu(O) = 0.152$ cm²/g $\times 0.98$ g/cm³ $= 0.149$ cm⁻¹. The total attenuation coefficient, $\mu_t = 0.298$ cm⁻¹. The corresponding half-value layer = 2.33 cm.[10]

---

10. There is a very simple relationship between the attenuation coefficient (μ) and the half-value layer: HVL $= 0.693/\mu$. Since, by definition, the half-value layer is that value of x for which $e^{\mu x} = 0.5$, $x = (-\ln 0.5)/\mu = 0.693/\mu$.

8.5 Calculation of Attenuation of Gamma Photons by the Half-Value Layer Method

We noted in section 8.1 that the transmission of photons can be described in terms of a half-value layer (HVL),[10] the thickness for which gamma photons have a 50 percent probability of penetration without interaction. Shielding calculations can be made very readily with the use of half-value layer data for the attenuating medium. Thus, if 100 gamma photons are incident on a half-value layer of a material, on the average, 50 gamma photons will penetrate without interaction, that is, they will emerge as if no attenuating medium had been present.

Of the 50 percent that penetrate through the first half-value layer, only half on the average get through a second half-value layer, or only a quarter of the incident photons get through 2 half-value layers without interaction, and so on. The fraction of gamma photons that penetrates a medium as a function of its thickness in half-value layers is shown in Table 2.4.

Using semilog graph paper, it is easy to prepare a curve representing the relationship between fraction penetrating and half-value layers, as shown in Figure 2.13. Because radioactive decay follows the same mathematical relationships as radiation attenuation, both Table 2.4 and Figure 2.13 (or a calculator) can be used for solving decay problems (they have been labeled accordingly). Extensive compilations of values of half-value layers for different materials may be found in radiation-protection handbooks (NCRP, 1976, Report 49; BRH, 1970; Slaback et al., 1997).

Example 2.6   Determine the attenuation of the gamma photons from a cesium-137 source in a container shielded with 3 cm lead.
The HVL of these gamma photons is 0.55 cm. The number of HVL in the shield is $3/0.55 = 5.45$. Thus the attenuation provided by the shield as determined from Table 2.4 or the HVL curve is 0.023.

Table 2.4   Powers of one-half for attenuation and decay calculations, $(1/2)^n$.

n	.0	.1	.2	.3	.4	.5	.6	.7	.8	.9
0	1.000	0.933	0.871	0.812	0.758	0.707	0.660	0.616	0.578	0.536
1	0.500	0.467	0.435	0.406	0.379	0.354	0.330	0.308	0.287	0.268
2	0.250	0.233	0.217	0.203	0.190	0.177	0.165	0.154	0.144	0.134
3	0.125	0.117	0.109	0.102	0.095	0.088	0.083	0.077	0.072	0.067
4	0.063	0.058	0.054	0.051	0.047	0.044	0.041	0.039	0.036	0.034
5	0.031	0.029	0.027	0.025	0.024	0.022	0.021	0.019	0.018	0.017
6	0.016	0.015	0.014	0.013	0.012	0.011	0.010			

Note: n is the number of half-value layers or half-lives. In 2.6 half-lives, the activity or number of atoms will be reduced to 0.165 of the original value. In 2.6 half-value layers, the number of photons will be reduced to 0.165 of the number entering the shield.

2.13 Semilogarithmic plot of penetration of photons against thickness in half-value layers, or fraction remaining in radioactive decay as a function of time in half-lives. The graph is constructed by drawing a straight line between any two plotted points—for example, half-value layer (HVL) = 0, attenuation factor (AF) = 1; HVL = 4, AF = $(1/2)^4$ = 0.0625.

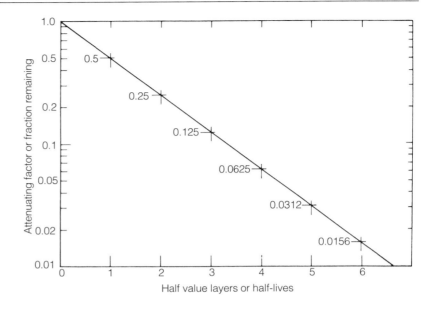

Example 2.7   What is the thickness of lead needed for shielding a container to reduce the gamma photons emerging from the source in the previous problem to 0.012 of the unshielded value?
From the HVL curve, the number of HVL required is 6.4. Multiplying by the thickness of a single HVL, we obtain a required thickness of 6.4 × 0.55 = 3.52 cm.

Figure 2.10 gives the half-value layer as a function of photon energy for several materials. The values given by the curves are for "good geometry" conditions; that is, they give the attenuation if the scattered radiation is not significant at the detector. The augmenting of photons at a point as a result of scattering in a large shield is illustrated in Figure 2.14. Besides removing or deflecting radiation originally headed toward a detector from a source, the shield scatters toward the detector some radiation that was originally traveling in a direction away from the detector. As a result, the radiation falls off more slowly than under "good geometry" conditions and the effective half-value layer is increased. Experimentally determined values for the half-value layer as obtained with large area shields should be used in shielding problems, since they describe actual shield attenuation more accurately than the "good geometry" coefficients (NCRP, 1976,

2.14 Illustration of buildup of secondary radiation. Large area shield scatters toward detector some of the radiation originally traveling away from detector and thus gives less attenuation than is measured in "good geometry" experiment (Fig. 2.11).

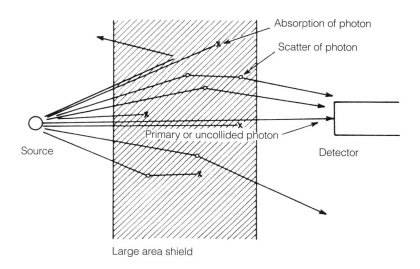

Absorption of photon

Scatter of photon

Primary or uncollided photon

Source

Detector

Large area shield

Report 49; Slaback et al., 1997). There are times, however, when the fall-off may not be describable in terms of a single value for the half-value layer. When relevant experimental data are not available, or when the problem is too complex to be treated by half-value layer concepts, the actual procedure is to make a calculation using the "good geometry" values, and increase the results by a factor that takes account of the production of *secondary radiation* due to scattering. This factor is generally known as the *buildup factor* and its use is described in texts on radiation shielding (Chilton et al., 1984).

The scattered radiation and, accordingly, the buildup factor increases relative to the unscattered or *primary radiation* as the shield thickness increases. It is the dominant radiation in the thicker shields. Some values of the buildup factor are given in Table 2.5.

Note that buildup factors are considerably higher in shields composed of the lighter elements (water, concrete) than the heavier elements (lead). The reason is that the energies of the scattered photons are lower than the energy of the primary radiation, and attenuation and absorption increase as the energy decreases at energies characteristically emitted by radionuclides. The increase in absorption with lower energies is more marked in elements with higher atomic number because of the photoelectric effect; hence the buildup factor in lead decreases much more slowly than the buildup factor in water as the shield thickness increases.

Table 2.5   Values of buildup factors.

E (MeV)	Number of half-value layers in shield[a]			
	1.44	2.88	5.77	10.08
	Buildup factors in water			
0.255	3.09	7.14	23.0	72.9
0.5	2.52	5.14	14.3	38.8
1.0	2.13	3.71	7.68	16.2
	Buildup factors in iron			
0.5	1.98	3.09	5.98	11.7
1.0	1.87	2.89	5.39	10.2
	Buildup factors in lead			
0.5	1.24	1.42	1.69	2.00
1.0	1.37	1.69	2.26	3.02

Source: Jaeger et al., 1968.

a. The number of half-value layers = 1.44 × attenuation coefficient × shield thickness.

8.6   Protection from Gamma Sources—Time, Distance, Shielding

The three key words—*time, distance,* and *shielding*—introduced in connection with protection from beta particles apply equally well to protection against gamma photons. Short working times and maximum working distances are as effective in reducing exposure from gamma photons as from beta particles. As with beta particles, the degree by which the intensity from a small source is reduced at increasing distances is obtained by the inverse-square law. Because gamma photons are so much more penetrating than beta particles, they require more shielding, the amount depending, of course, on the size of the source. If large gamma sources are to be handled, thick shields are required; often special manipulators and lead glass windows are used.

We must keep in mind one basic difference between beta and gamma shielding. Beta particles are charged ionizing particles and have a maximum range. Thus a shield built to stop beta particles from a particular radionuclide will stop the particles from any source consisting of that nuclide, regardless of the source strength.[11] On the other hand, a gamma shield always allows a fraction of the gamma photons to get through, since they are uncharged ionizing particles. The fraction decreases, of course, as the thickness of the shield increases.

11. There may still be a secondary effect from bremsstrahlung.

Suppose the fraction of gamma photons that penetrate a shield is 1 percent. If 1,000 are incident, 10 will penetrate, and if 100 are incident, 1 will penetrate, on the average. These numbers may be insignificant. On the other hand, if 10^{15} photons are incident on the shield, 10^{13} will get through. This could have serious consequences.

With gamma photons, as with all uncharged ionizing particles, a shield that is just thick enough to provide protection for one level of activity will not be thick enough for levels that are significantly higher. The protection offered by a gamma shield must always be evaluated in terms of the source strength, and no shield should be trusted until its adequacy has been verified for the source to be shielded.

9 HEAVY CHARGED IONIZING PARTICLES

The electron is the most common but not the only charged ionizing particle encountered in work with radiation sources.[12] For our purposes, the other particles may be characterized by their charge and their mass when they are at rest. Because they are so small, it is not convenient to express their mass in grams. Instead, we shall express their mass relative to the electron mass (an electron has a mass of 9.1×10^{-28} g) or in terms of the energy equivalent of their mass. We often express the mass in terms of the energy equivalent because it is possible for the rest mass of a particle to be converted into an equivalent amount of energy and vice versa. The equivalence is given mathematically by Einstein's equation, energy = mass × (velocity of light)^2. The rest energy of an electron is 0.51 MeV.

9.1 The Alpha Particle—A Heavy Particle with High Linear Energy Transfer and High Capacity for Producing Damage

The alpha particle is an energetic helium nucleus, consisting of two neutrons and two protons. It is therefore heavier than the electron by a factor of over 7,300 and has double the charge. It is commonly emitted in the radioactive decay of the heaviest nuclides in the periodic table. Examples of naturally occurring alpha emitters are uranium, thorium, radium, and polonium. An artificially produced alpha emitter, plutonium, is likely to be the main component of fuel in the nuclear power plants of the future.

The alpha particles emitted by these nuclides possess kinetic energies ranging between 4 MeV and 9 MeV. The corresponding speeds are be-

---

12. Material in this section on particles not encountered by the reader may be omitted in a first reading.

tween 1.4 and 2.1 × 10^9 cm/sec. They are much less than the speeds of beta particles in the same energy range, which are quite close to the speed of light.

Because of their slower speeds, the alpha particles spend more time than beta particles in the vicinity of the atoms they pass and exert much larger impulses on the orbital electrons. The impulses are increased still more because their charge and the electrical forces they exert are twice as great as those of electrons. As a result the rate at which alpha particles impart energy to the medium along their path is much greater than that of beta particles.

The rate at which charged particles impart energy locally to a medium is known as the linear energy transfer, commonly abbreviated as LET.[13] LET values for charged particles as a function of their energies and speeds are given in Table 2.6. Note how LET increases at lower energies (and speeds).

The range of charged particles is, of course, intimately connected with the LET. The ranges of the high-LET alpha particles are much smaller than the ranges of electrons with comparable energies. An alpha particle with an energy of 5 MeV, typical for alpha particles emitted from radionuclides, travels a distance of 44 μm in tissue, whereas a 1 MeV electron will travel a distance of 3,350 μm. In air, the ranges are 3.5 cm for an alpha particle and 415 cm for a beta particle.

The damage produced in living tissue by absorption of a given amount of energy is generally greater as the distance over which this energy is imparted decreases, that is, as the LET increases. In addition, the ability of the body to repair radiation damage is less as the LET increases. Because of their high LET, alpha particles have long had a reputation for being especially hazardous. They represent the main source of energy from radium and thorium isotopes responsible for the production of bone cancers (in persons making radioluminescent dials) and from radon gas associated with the production of lung cancers (in uranium miners). Because of their short range, however, alpha particles emitted by radionuclides cannot penetrate through the dead outer layer of the skin and thus do not constitute an external hazard. They can cause damage only if the alpha-emitting radionuclides are ingested or inhaled and the alpha particles are consequently emitted immediately adjacent to or inside living matter.

13. In some analyses, only a portion of the LET value is considered, that part which imparts energy less than a stated amount (frequently 100 eV) to electrons. Electrons given higher energies, referred to as delta rays, are not considered part of the track. However, this distinction is not productive in radiation-protection applications. For a detailed discussion of linear energy transfer and its use to specify the quality of radiation, see ICRU, 1970, Report 16.

Table 2.6 Transfer of energy per centimeter in water by energetic charged particles (linear energy transfer).

Particle	Mass[a]	Charge	Energy (MeV)	Speed (cm/sec)	LET (MeV/cm)	Range (microns)
Electron	1	−1	0.01	0.59×10^{10}	23.2	2.5
			0.1	1.64×10^{10}	4.20	140
			1.0	2.82×10^{10}	1.87	4,300
			10.0	3.00×10^{10}	2.00[b]	48,800
			100.0	3.00×10^{10}	2.20[c]	325,000
Proton	1835	+1	1.0	1.4×10^{9}	268	23
			10.0	4.4×10^{9}	47	1,180
			100.0	1.3×10^{10}	7.4	75,700
Alpha	7340	+2	1.0	0.7×10^{9}	1,410[d]	7.2
			5.3[e]	1.6×10^{9}	474	47

Sources: ICRU, 1970, Report 16 (protons and electrons); Morgan and Turner, 1967, p. 373 (alpha particles); Etherington, 1958, pp. 7–34 (ranges for alpha particles, tissue values used).

a. Mass is taken relative to the electron mass.

b. An additional 0.20 MeV/cm is lost by emission of photons.

c. An additional 2.8 MeV/cm is lost by emission of photons.

d. Includes only energies imparted locally. About 24 percent of the total energy lost by the alpha particle is imparted to electrons in amounts greater than 100 eV per collision. These electrons are considered to lose their energy away from the track, and their energy loss has not been included in the local energy loss of the alpha particle.

e. Energy of alpha particle from ^{210}Po.

9.2 The Proton—Another Heavy Charged Particle with High Linear Energy Transfer

We here complete our review of important charged ionizing particles with a discussion of the proton. This particle occurs naturally as the sole constituent of the nucleus of the hydrogen atom and in higher numbers in the nuclei of the other elements. The atomic number of an element *(Z),* which defines its position in the periodic table and is uniquely associated with its identity, is equal to the number of protons in the nucleus.

Protons show evidence of internal structure—they are lumpy. The lumpiness has been associated with still more elementary entities inside the proton, given the name of quarks (Pagels, 1982).

The proton has a mass that is 1,835 times the mass of the electron and has a single positive charge (charge of opposite sign to that of the electron, but of the same magnitude). Protons with energies less than a few MeV travel at velocities that are low in comparison with the velocities of electrons of comparable energies. As a result they impart energy at a high rate as they pass through matter (high LET). Values of linear energy transfer and ranges for protons are given in Table 2.6.

The proton is not emitted in spontaneous radioactive decay of the nuclei of the elements, as are electrons and alpha particles. It must be given a minimum amount of energy, through a nuclear collision with another ionizing particle (for example, a high-energy photon), before it can be expelled.

The human body contains an extremely large number of single protons as the nuclei of hydrogen atoms. When the body is irradiated by neutrons (section 10), most of the incident energy is initially imparted to these nuclei. These, in turn, become the major means (as energetic protons) for transferring the neutron energy to the body tissue.

10 THE NEUTRON—A SECOND IMPORTANT UNCHARGED IONIZING PARTICLE

The neutron is a very common particle, since along with the proton it is a basic constituent of the nucleus.[14] It is almost identical to the proton in mass and size, but it carries no charge. Normally it remains locked in the nucleus along with the proton. The number of neutrons and protons—the mass number—is characteristic for any given nuclide.

10.1 Sources of Neutrons

There are no significant naturally occurring neutron emitters. If a nucleus is unstable because of an excess of neutrons relative to protons, the ratio is changed by the transformation of a neutron into a proton within the nucleus and the emission of a beta particle. Radionuclides that emit neutrons can be produced artificially, but all except one have half-lives that are too short to be useful. The only nuclide source of neutrons that has practical possibilities is californium-252. This is a transuranium isotope with a half-life of 2.65 years that undergoes 1 decay by fission for every 31 decays by alpha-particle emission. The fission of each nucleus is almost always accompanied by the emission of a small number of neutrons, which vary for each individual fission. The average number of neutrons emitted per fission of ^{252}Cf is 3.76.

Aside from the fission of ^{252}Cf, the only way to produce neutron sources is through nuclear reactions, that is, the bombardment of nuclei with projectiles to produce various products. Suitable projectiles for producing

14. Readers who do not encounter neutrons in their work should omit this section in a first reading and continue on to section 11.

2.15 Energy spectrum of neutrons emitted by Am-Be neutron source (Geiger and Hargrove, 1964). *N(E)* gives the relative number of neutrons emitted per MeV of energy range, and the area under the curve between any two energies gives the relative number of neutrons in that energy range.

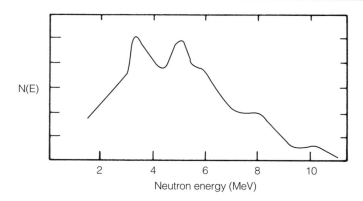

neutrons can be obtained with radiation machines or from some radionuclides. The most common reaction using radionuclides is the bombardment of beryllium with alpha particles. The alpha particles must have an energy greater than 3.7 MeV to initiate the reaction. Suitable alpha-particle sources are polonium-210, radium-226, plutonium-239, and americium-241. The energy spectrum of neutrons produced from an americium-beryllium source is shown in Figure 2.15.

The most powerful sources of neutrons are nuclear-fission reactors. Approximately 2.5 neutrons are emitted per fission of uranium-235, and these can cause further fissions. As a result, a self-sustained fission reaction can be maintained. The fission process releases a tremendous amount of energy, approximately 200 MeV per fission, and it is because of this outstanding energy release that fissionable material has become such a valuable fuel for power plants.

Strong neutron sources are provided by radiation machines known as neutron generators. A reaction which is much used is the bombardment of a tritium target with deuterons (deuterium nuclei) having energies in the 150,000 eV range. The neutrons produced have energies of 14 MeV. These neutron generators are available at a cost within the budgets of many educational and scientific institutions.

The variety of reactions that can produce neutrons increases as the energy of the bombarding particles increases. Neutrons are produced in abundance around high-energy accelerators, as secondary radiations resulting from the interactions of the primary particles with shielding and other material in the environment. The protection of personnel from exposure to these neutrons represents one of the more difficult problems in radiation protection.

10.2   Neutron Collisions

In considering the actions of neutrons as they penetrate a medium, it must be borne in mind that the neutron carries no charge and has a mass only slightly larger than that of the proton. Its mass equals 1.00867 atomic mass units (u),[15] while that of the proton equals 1.00728 u. Because the neutron is not charged, it does not lose its energy in direct, closely spaced ionizations as do charged particles such as protons. It is not electromagnetic (as a gamma photon is) and therefore does not undergo interactions with the electrons in the medium. A neutron travels through a medium without interaction until it collides with an atomic nucleus. The collision (involving two material objects) is governed by the laws of conservation of momentum and energy. The maximum energy transfer that can result occurs when neutrons collide with the nuclei of hydrogen atoms (that is, protons), which are of almost equal mass. The amount of energy transferred depends also on the directions of recoil of the proton and the neutron after the collision. If the proton is ejected directly forward, it will receive all the energy of the neutron, which in turn will come to rest. The maximum energy that can be transferred to an atom heavier than hydrogen decreases as the atomic mass increases. The maximum fractional transfer to a target with atomic mass A is equal to $4A/(A + 1)^2$, which equals 1 for collisions with hydrogen ($A = 1$) and 0.28 for collisions with carbon ($A = 12$). The maximum energy transfer occurs only a small fraction of the time, and on the average the energy transferred is half the maximum.

The energetic recoil atom becomes charged quickly and penetrates through matter as a charged ionizing particle. Because of its large mass, it travels at a relatively low speed and loses energy at a high rate; that is, its ionization is characterized by high linear energy transfer. Energy from recoil atoms is the most important mechanism by which neutrons produce damage in tissue. Hydrogen atoms receive most of the energy from neutrons traveling through tissue and produce most of the damage as recoil protons. This is because of the abundance of hydrogen and the high probability of the interaction of neutrons with the hydrogen nuclei, relative to other nuclei present. Other damage is produced from carbon and oxygen recoils, while a fraction is from gamma rays and protons released by the capture of low-energy neutrons in the nitrogen.

Collisions between neutrons and the light elements found in tissue at neutron energies of a few MeV and lower (generally characteristic of neutron exposure) are elastic; that is, the kinetic energy of the colliding bodies

---

15. This scale is chosen so that the neutral carbon-12 atom has a relative mass exactly equal to 12 (equivalent to 1.6604×10^{-24} g).

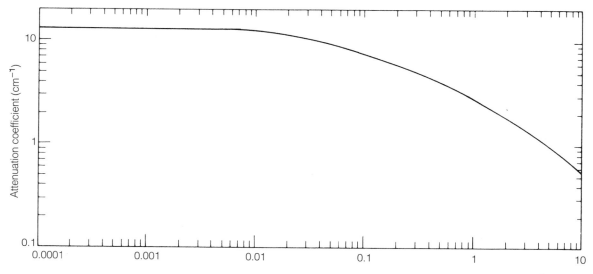

2.16 Attenuation coefficient of hydrogen with a nominal density of 1 g/cm^3 (calculated from data in Hughes and Schwartz, 1958).

is conserved during the collision. In heavier elements, some of the kinetic energy of the neutron may be transferred to the internal energy of the nucleus. In this case, referred to as an inelastic collision, the kinetic energy that can be imparted to the atom will be reduced. The excited nucleus will release the energy of excitation in the form of a gamma photon or other particles. Inelastic collisions have significance in the attenuation of neutrons but do not play an important role in the production of damage in living matter.

10.3   Attenuation of Neutrons

The concepts of the attenuation coefficient and the half-value layer used in the discussion of the attenuation of gamma photons (sections 8.3, 8.4) also apply to the attenuation of neutrons. Since attenuation of neutrons of energies less than a few MeV is most effectively accomplished with hydrogen, the attenuation coefficient of hydrogen is of special interest (see Figure 2.16). The value applies to a medium in which the hydrogen density is 1 g/cm^3. To determine the attenuation from the hydrogen content in any other medium, calculate the hydrogen density in that medium in g/cm^3 and multiply it by the attenuation coefficient in Figure 2.16. Convert the coefficient to the half-value layer by the relationship: half-value layer = 0.693/attenuation coefficient. Calculate the total number of half-value layers for the thickness under consideration and evaluate the attenuation produced, for the number of half-value layers obtained, in the usual manner.

Example 2.8   Calculate the attenuation due to the hydrogen in a water shield, 1.5 m thick, for 8 MeV neutrons.

Water consists of 11 percent hydrogen by weight; that is, the density of hydrogen is 0.11 g/cc. The attenuation coefficient of hydrogen is 0.68 cm^{-1} at unit density (see Figure 2.16), so it is (0.11)(0.68) = 0.0748 for a hydrogen density of 0.11. The HVL for the hydrogen in water is 0.693/0.0748 = 9.26 cm. The number of half-value layers contributed by the hydrogen in the shield is 150 cm/ 9.26 cm = 16.2. The attenuation is therefore $(1/2)^{16.2}$, or 1.3×10^{-5}.

When we are dealing with a complex energy spectrum, such as the one characterizing neutrons produced in fission, the problem can become very complicated, since the attenuation coefficient varies with the energy. It is sometimes possible to use a single value of the attenuation coefficient for the energy distribution considered. A single coefficient which can be used for the whole spectrum is called a removal attenuation coefficient. Values of the removal coefficient for fission neutrons are presented in Table 2.7. They apply to a density of 1 g/cm^3, and the value for a substance in any given shielding material may be obtained by multiplying the value in Table 2.7 by the density of the substance in this material. The number of half-value layers contributed by the substance in the shield may then be determined by dividing the shield thickness by the half-value thickness. The removal coefficient concept applies only if the shield contains an appreciable amount of hydrogen (hydrogen atoms >20 percent) or if the last part of the shield is hydrogenous.[16]

When a shield is composed of several materials, the total attenuation produced by all the materials is obtained by determining the number of half-value layers contributed by each of the materials and adding them. An attenuation factor is calculated for this total number of half-value layers.[17]

16. Price, Horton, and Spinney, 1957; NCRP, 1971b; Shultis and Law, 2000.

17. Shield engineers determine the attenuation from the total number of relaxation lengths, μx, and the value of $e^{-\mu x}$ (see Price, Horton, and Spinney, 1957; Shultis and Law, 2000).

Table 2.7 Removal attenuation coefficients for fission neutrons in attenuating medium with a density of 1 g/cc.

Element	Removal coefficient (cm^{-1})
Iron	0.020
Hydrogen	0.602
Oxygen	0.041
Calcium	0.024
Silicon	0.295

Source: NBS, 1957, Handbook 63.

Example 2.9 Calculate the attenuation provided by a water shield 1 m thick for fission neutrons.

Table 2.7 lists the attenuation coefficient of hydrogen as 0.11 × 0.602, or 0.066 cm^{-1}, and the HVL is 0.693/0.066 = 10.5 cm. Thus the number of HVL in 1 m is 100 cm/10.5 cm = 9.5. The oxygen has a density of 0.89 g/cm^3 in the water. The attenuation coefficient for oxygen in water is 0.041 cm^{-1} × 0.89, or 0.0365 cm^{-1}, and the HVL is 19 cm. The number of HVL contributed by the oxygen in the 1 m water shield is 5.26. The total number of HVL is 9.5 + 5.25 = 14.76, and the attenuation factor is 3.6 × 10^{-5}.

Neutron attenuation problems are concerned with neutron energies down to the average energies of the surrounding atoms, that is, at the energies of neutrons in thermal equilibrium with their surroundings. At normal room temperature, 20°C, these neutrons, known as thermal neutrons, have a most probable energy equal to 0.025 eV. Thus, neutrons from 0.025 eV up are significant in shield penetration and dose evaluation. The consideration of neutron shielding is not terminated with the loss of neutron energy, because all neutrons are eliminated through absorption by the nucleus of an atom after they are slowed down. The nuclear absorption of a neutron results in the release of a substantial amount of energy, generally of the order of 7 MeV. This energy usually appears as very penetrating gamma photons,[18] and so we have the unhappy situation, for anyone trying to provide adequate shielding, of removing the kinetic energy of neutrons by attenuation, only to have more energy emitted in the form of en-

18. This is known as *radiative capture*.

ergetic gamma photons. Neutron attenuation problems are almost always followed by gamma attenuation problems.

Neutron shields always terminate in hydrogenous material, so if lead or some other element with a high atomic number is needed for additional gamma shielding, it should precede the hydrogenous material. In this way the gamma shielding also has maximum effectiveness as neutron shielding and the attenuation is determined with the use of removal coefficients. The design of shields for both gamma photons and neutrons is a complex problem beyond the scope of this text. For details, see shielding texts and handbooks (Jaeger, 1968; Chilton, 1984; and Shultis, 2000).

The concepts of time, distance, and shielding, discussed previously in connection with protection from gamma photons, apply equally to strategies for preventing injury from exposure to neutrons.

11   THE ABSORBED DOSE—A MEASURE OF ENERGY IMPARTED TO A MEDIUM

In previous sections we presented the properties of charged and uncharged ionizing particles, and described the manner in which they imparted energy to a medium. We also emphasized the key role (in damage production) of energy imparted. We now consider how energy imparted is determined and the units in which it is expressed (ICRU, 1980, Report 33; ICRU, 1986, Report 40; NCRP, 1993b, Report 116).

11.1   The Pattern of the Imparted Energy in a Medium

The ionization patterns produced by various particles incident on a medium such as water are shown in Figure 2.17. Beta particles are depicted as continuously producing ionization (from the moment they enter the medium) along a tortuous path marked by occasional large-angle scatterings as well as smaller deflections. The large-angle scatterings occur when the beta particles pass close by an atomic nucleus and encounter strong electrical forces. These large-angle scatterings do not produce ionization, but they cause the beta particle to emit x-ray photons. There are also side tracks of ionization caused by high-speed electrons (known as delta rays) that are ejected from the orbits of atoms. The depths of penetration of individual beta particles vary depending on their energy and the degree of scattering. The maximum penetration is equal to the range of the beta particles in the medium. The particles impart their energy as excitation as well as ionization of the atoms in the medium.

Alpha particles are shown traveling in a straight line and stopping in a

2.17 Ionization patterns produced by various particles in tissue (not drawn to scale).

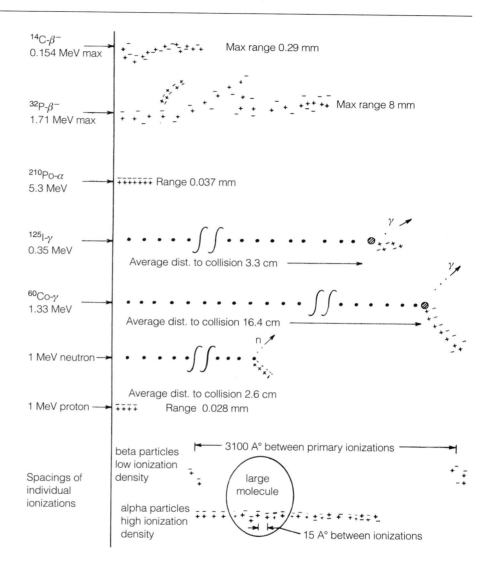

^{14}C-β^-
0.154 MeV max

Max range 0.29 mm

^{32}P-β^-
1.71 MeV max

Max range 8 mm

^{210}Po-α
5.3 MeV

Range 0.037 mm

^{125}I-γ
0.35 MeV

Average dist. to collision 3.3 cm

^{60}Co-γ
1.33 MeV

Average dist. to collision 16.4 cm

1 MeV neutron

Average dist. to collision 2.6 cm

1 MeV proton

Range 0.028 mm

Spacings of individual ionizations

beta particles low ionization density

3100 A° between primary ionizations

large molecule

alpha particles high ionization density

15 A° between ionizations

much shorter distance than beta particles penetrate. The ionizations, always closely spaced, become even more closely spaced near the end of the range. The spacings of the individual ionizations of alpha and beta particles are compared with the dimensions of a large molecule at the bottom of the figure.

Gamma photons are shown as proceeding without interaction until there is a collision in which a high-speed electron is ejected from an atom.

The ejected electrons may have energies up to the maximum energy of the gamma photon minus the energy required to eject the electrons from their shells. The maximum-energy electrons result from complete absorption of the photon, and the lower-energy electrons result from partial loss of the energy of the photon in a scattering. The scattered gamma photon proceeds until it undergoes further interactions.

Photons of two energies are shown for comparison. The gamma photon from cobalt-60 on the average travels a longer distance (16.4 cm) than the lower-energy gamma photon from iodine-125 (3.3 cm) between interactions. The electrons ejected by the ^{60}Co photons are also more energetic, on the average, and travel greater distances before being stopped.

Fast neutrons are shown to proceed without interaction until there is an elastic collision with a nucleus, followed by recoil of the nucleus and scattering of the neutron at reduced energy. The recoil nucleus, because of its large mass, has high linear energy transfer and short range. A track of the most important recoil nucleus in tissue, the proton, is shown diagrammatically under the neutron track.

11.2   Definition of Absorbed Dose

The basic quantity that characterizes the amount of energy imparted to matter is the absorbed dose. The mean absorbed dose in a region is determined by dividing the energy imparted to the matter in that region by the mass of the matter in that region.

From the discussion of patterns of ionization presented in the last section, it is obvious that patterns are not uniform throughout a region, and therefore the absorbed dose is different in different parts of the region. A proper application of this concept requires a knowledge not only of the manner in which the energy is imparted but also of the significance of this manner with regard to the production of injury to tissue. In evaluating absorbed dose, the radiation protection specialist must select the critical region in which he wishes to evaluate the energy imparted. He may average the energy imparted over a fairly large region. If the energy pattern is very nonuniform, he may select that part of the region where the energy imparted is a maximum to obtain a maximum dose. If the region where the dose is a maximum is very small, he may decide that the dose in that region is not as significant as a dose determined by considering a larger region. We shall examine the various possibilities in later examples.

11.3   The Gray—The SI Unit for Absorbed Dose

Absorbed doses are expressed in units of grays or in prefixed forms of the gray, such as the milligray (1/1,000 gray) and the microgray (1/1,000,000 gray).

The gray is equal to 1 joule of energy absorbed per kilogram of matter (1 J/kg). Since we express here the energies of the radiations in terms of MeV, we can also say that the gray (Gy) is equal to 6.24×10^9 MeV/g. The milligray is equal to 6.24×10^6 MeV/g.[19]

The gray was adopted by the International Commission on Radiation Units and Measurements (ICRU) in 1975 as the special name for the standard international (SI) unit of absorbed dose (NCRP, 1985b). It used the SI base units of joules for energy and kilograms for mass in defining the gray. The gray replaced the rad, defined as 100 ergs per gram and now referred to as the traditional unit. The rad is equal to 0.01 J/kg or 1 centigray (cGy).

SI units are used exclusively in all reports of the ICRU, the International Commission on Radiological Protection (ICRP), and the U.S. National Council on Radiation Protection and Measurements (NCRP). Absorbed doses are often expressed in centigrays in the literature to make them numerically equal to the dose values given in rads in the past.

Harmful levels of radiation dose are generally expressed in terms of grays. For example, over a gray must be imparted in a short period over a substantial portion of the body before most individuals will show significant clinical symptoms (Saenger, 1963). Occupational absorbed doses from x and gamma radiation are limited to a maximum of 50 milligrays per year.

12 THE EQUIVALENT DOSE—A COMMON SCALE FOR DOSES TO ORGANS AND TISSUES FROM DIFFERENT RADIATION TYPES AND ENERGIES

Although the injury produced by a given type of ionizing radiation depends on the amount of energy imparted to matter, some types of particles produce greater effects than others for the same amount of energy imparted. For equal absorbed doses, alpha particles produce more injury than protons do, and these in turn produce more injury than beta particles do. The effectiveness of one ionizing particle relative to another may also vary

---

19. We express dose in units of mega–electron volts per gram to simplify the dose calculations for ionizing particles, whose energies are generally given in MeV. The energy acquired by a particle carrying an electrical charge (*q*) and going through a potential difference (*V*) is the product *qV*. If *q* is in coulombs (C) and *V* is in volts (V), then the energy is in joules (J). The handbook value for the electronic charge is 1.602×10^{-19} coulombs. Thus, the energy of an electron that has gone through a potential difference of 1 volt, acquiring an energy of 1 eV, is 1.602×10^{-19} J. Thus $1 \text{ J} = 1/1.602 \times 10^{-19}$ eV $= 6.24 \times 10^{18}$ eV or 6.24×10^{12} MeV. This gives a conversion factor, $1 \text{ mGy} = 0.001 \text{ J/kg} = 10^{-6}$ J/g or 6.24×10^6 MeV/g.

considerably depending on the particular biological material which is irradiated, the length of the time interval in which the dose is delivered, the type of effect considered, the delay prior to appearance of the effect, and the ability of the body of the exposed individual to repair the injury.

An early attempt to account for this difference in the consequences of the same absorbed dose from different radiation particles was the application of a weighting factor labeled the *relative biological effectiveness* (RBE) to the absorbed dose. The RBE of one radiation type in relation to another was defined as the inverse of the ratio of the absorbed doses producing the same degree of a defined effect. In practice, a typical RBE was assigned to a radiation type relative to the effect produced by 250 kV (peak) x radiation as a reference. The choice of the RBE relied on a considerable amount of judgment. The current value of 20 generally used for the RBE of alpha particles was recommended by the NCRP in 1954 and was twice the RBE used in assessment of the dose from occupational exposures to alpha emitters at that time (NCRP, 1954). The ICRP cited the RBE for alpha particles as 20 in 1950, as 10 in reports in 1955 and 1965, and back again as 20 in 1969 (ICRP, 1951, 1965, 1969b; ICRU, 1986). Because of the many factors complicating the designation of a generic RBE, U.S. and international advisory bodies on radiation protection developed alternative approaches to converting absorbed doses from different radiation particles and energies by factors to doses that are roughly equivalent in terms of their effects on humans.

12.1 The Radiation Weighting Factor and the Quality Factor— Measures of the Relative Hazard of Energy Transfer by Different Particles

The physical measure for gauging the relative biological effectiveness of equal absorbed doses from different radiation particles in producing injuries is the rate at which these particles deposit their energy in the body, denoted as the linear energy transfer, or LET.[20] Referring to Table 2.6, a 5.3 MeV alpha particle from polonium-210 in the environment has a range of 47 micrometers in water (essentially the same as in soft tissue); a 1 MeV electron, having an effectiveness that characterizes most of the radiation exposure of human beings from natural radiation, has a range of 4,300 micrometers in water. Tabulated values of the corresponding LET are 474 MeV/cm for the alpha particles and 1.87 MeV/cm for the electrons. The higher the LET of the radiation, the greater is the injury produced for a given absorbed dose.

20. See section 9.1, where the LET was introduced in connection with the penetration of alpha particles. Values of LET as a function of energy for different particles are given in Table 2.6.

Since the physical property of the radiation that drives its biological effectiveness is the LET, adjustments for absorbed doses have been developed as multiplying factors that are a function of the LET and weight the absorbed dose by the relative effectiveness. The weighted doses are then treated as equivalent with regard to their biological effectiveness. Two multiplying factors are used to weight the absorbed dose, the *radiation weighting factor (w_R)* and the *quality factor (Q)*.

The quality factor was introduced first (ICRU, 1986). It is essentially defined for a point and based on the distribution of LET in a small volume of tissue. The assignment of quality factors is based on selected experimental data on the effects of radiations with different LET on various biological and physical targets. The expressions for the quality factor as a function of LET adopted by NCRP and ICRP are identical. For LETs (in keV/μm) < 10, $Q = 1$; for LETs between 10 and 100, $Q = (0.32)(\text{LET}) - 2.2$; for LETs > 100, $Q = 300/(\text{LET})^{1/2}$.

The radiation weighting factor is conceptually different from the quality factor in that it is applied to the absorbed dose averaged over a tissue or organ rather than to the dose at a point; also, it is applied to ranges of energy for the different radiation types rather than derived from a calculation as a function of the LET (ICRP, 1991a, Publication 60). This approach is more consistent with current radiation standards, since they refer in general to average organ and tissue doses rather than to the variations in dose that may occur within the organ. The radiation weighting factors *(w_R)* are given in Table 2.8. They are based on a review of measured values of the relative biological effectiveness (RBE) of the different radiations at low absorbed doses, including such end points as chromosome aberrations in cultured human lymphocytes, oncogenic transformation, specific locus mutations in mice, and tumor induction in mice. Thus they incorporate a considerable amount of judgment and give preference to human data where it exists. Because the concept does not lend itself to a well-defined value, the radiation weighting factors promulgated by the U.S. National Council on Radiation Protection and Measurements (NCRP) and the International Commission on Radiological Protection (ICRP) are identical except for protons with energies greater than 2 MeV (NCRP, 1993b; ICRP, 1991a). Also listed in the table are the values to be used in compliance with the regulations of the U.S. Nuclear Regulatory Commission (NRC). They are the same as the NCRP/ICRP radiation weighting factors for photons and alpha particles, but less by about a factor of 2 for exposures to neutrons. Since w_R equals 1 for all x and gamma radiation and electrons (including beta particles), the numerical values of the absorbed dose will equal the equivalent dose for most users of radiation sources.

When radiation particles and energies are encountered that are not included in the table for w_R, a value for w_R may be obtained by integrating

Table 2.8   Radiation weighting factors (w_R).

	Radiation weighting factors	
	NCRP/ICRP	NRC
X and gamma rays, all energies	1	1
Electrons and muons, all energies	1	1
Neutrons	20[a]	10[b]
Protons > 2 MeV, other than recoil protons	2 (NCRP)	10
	5 (ICRP)	
Alpha particles, fission fragments, heavy nuclei	20	20

Sources: ICRP, 1991a; NCRP, 1993b; NRC, 1991.

a. For general use. Radiation weighting factors adopted by NCRP and ICRP for specific energy ranges of neutrons are: <10 keV, 5; 10–100 keV, 10; 100 keV–2 MeV, 20; 2–20 MeV, 10; >20 MeV, 5.

b. This value is assigned to neutrons of unknown energy. The regulations in 10CFR20 also list a table of radiation weighting factors (which they call quality factors, Q) for monoenergetic neutrons, as a function of energy. They equal 2 for thermal energies, rise to a maximum of 11 for 0.5 and 1 MeV neutrons, and fall to a minimum of 3.5 for neutrons between 200 and 400 MeV.

the quality factor as a function of LET, weighted by the absorbed doses, over the particle LETs. In all exposure situations, the doses are averaged over individual organs or tissues and then multiplied by w_R. The variations of dose within a tissue of uniform sensitivity to cancer induction are not considered to be important.

When the value determined for the absorbed dose is to be used for compliance with the regulations, it must be multiplied by the radiation weighting factor or quality factor. If the multiplier is the radiation weighting factor, the resultant quantity is known as the *equivalent dose* (NCRP, 1993b, Report 116; ICRP, 1991a, Publication 60). If the quality factor is used, technically the product is called the *dose equivalent,* but this distinction in nomenclature can be confusing, particularly to the nonspecialist. The term *equivalent dose* will be used in this text, whether the conversion from absorbed dose is by radiation weighting factor or quality factor. Both equivalent dose and dose equivalent are treated as correlating with the injury produced as a result of the radiation exposure for purposes of radiation protection and compliance with standards.

12.2   The Sievert—The Special Unit of Equivalent Dose

When the absorbed dose in grays is multiplied by the radiation weighting factor or the quality factor, the result is the equivalent dose expressed in sieverts (Sv): sieverts = grays $\times$ w_R.

The traditional unit for the equivalent dose is the rem, i.e., rems = rads

× w_R. Other factors known as modifying factors are also used occasionally in converting from rads to rems, but these are not included when SI units are used.

Standards for radiation protection are given in terms of the sievert unit. Absorbed doses expressed in grays may be compared numerically to limits given in terms of sieverts when dealing with beta particles, x rays, and gamma photons. Absorbed doses for alpha particles and other heavy nuclei are multiplied by 20. Absorbed doses for neutrons are multiplied by 20 (ICRP, NCRP) or 10 (NRC) if the energies are not known. Other multiplying factors for neutrons are applicable, as given in Table 2.8, if the neutron energies are known.

There are situations where available biological data may indicate that the equivalent dose does not express accurately the effectiveness of a particular radiation field for the effect of interest. Under these circumstances, the biological data must take precedence and be used in deciding on maximum permissible exposure levels.[21]

13 TISSUE WEIGHTING FACTORS AND THE EFFECTIVE DOSE—A MEASURE OF RISK AND SEVERITY OF CONSEQUENCES

One of the major tasks in radiation protection is to evaluate the risk of exposure to radiation, whether from environmental sources, medical tests, or occupational exposure. Standards for protection are then developed to control the risk. Thus, it is necessary to translate dose into risk. The equivalent dose, introduced in the previous section, is a quantity designed to adjust for the varying degrees of severity of harm to an organ or tissue produced by different radiation types and energies, but it does not address the risk to the exposed individual as a result of the exposure. This risk usually results from the exposure of a number of organs with different radiation doses and different sensitivities to radiation, particularly with respect to the eventual induction of a cancer. A special quantity, called the *effective dose,* was introduced to give that risk. The effective dose is determined by multiplying doses to specific organs or tissues by tissue weighting factors, which convert them to equivalent whole-body doses that would produce an equivalent detriment to the health of the individual. The equivalent

---

21. In radiobiology, the experimentally determined ratio of an absorbed dose of the radiation in question to the absorbed dose of a reference radiation required to produce an identical biological effect in a particular experimental organism or tissue is called the relative biological effectiveness (RBE). The RBE was originally used as a multiplication factor to convert "rads" to "rem," but this practice was later discontinued.

whole-body doses are then summed and checked for compliance with the protection standards promulgated by regulatory agencies, which generally apply to uniform whole-body radiation and are expressed as limits for effective doses.

The choice of a tissue weighting factor requires a great degree of judgment. One can limit the weighting factor to give the whole-body dose that will produce the same number of fatal cancers as the organ dose, or one can attempt to include a number of additional effects. The International Commission on Radiological Protection (ICRP, 1991a) included four components in the evaluation of the detriment: the probability of fatal cancer, the weighted probability of attributable nonfatal cancer, the weighted probability of severe hereditary effects, and the relative length of life lost. Because of the large uncertainties in the data and the assumptions, the ICRP rounded and grouped the weighting factors into four groups of weights: 0.01, 0.05, 0.12 and 0.20. Table 2.9 gives the values used by the ICRP for the probability of fatal cancer to an organ per sievert, the assigned relative detriment to the health of the individual as a result of the dose to the organ, and the value of the tissue weighting factor. The NCRP accepted the values of the ICRP in the revision of its recommendations published in 1993 (NCRP, 1993b). The values listed in the NRC regulations, which differ somewhat from the ICRP/NCRP values, are also given in the table; these are equal to weighting factors previously recommended by ICRP in 1977 (ICRP, 1977a) and also adopted by the NCRP until the 1993 revision. At that time, the ICRP called the uniform whole-body dose that was considered to be equivalent to the actual exposure in its impact to the health of the individual the *effective dose equivalent (EDE)*. The EDE, conceptually slightly different from the effective dose, carried different tissue weighting factors. The NRC normally uses the recommendations of NCRP, but it takes many years before the recommended values are incorporated into the regulations.

The primary radiosensitive organs, excluding genetic effects, are the bone marrow, colon, lung, and stomach. ICRP gives them a weighting factor of 0.12. Hereditary effects are accounted for by assigning a weighting factor of 0.2 to the gonads. Organs not assigned specific weighting factors are assigned collectively a remainder weighting factor of 0.05. For calculation purposes, the remainder is composed of the adrenals, brain, small intestine, large intestine, kidney, muscle, pancreas, spleen, thymus, and uterus. The ICRP is not clear on how the remainder is partitioned, perhaps reflecting the crudity of this part of the calculation. The weighting factor could be divided equally among the ten organs (0.005 to each), or some other averaging approach could be used that would take into account special circumstances. The NRC does not assign weighting factors to as

Table 2.9 Tissue weighting factors.

Organ	Mortality per 100 population/Sv	Relative contribution to detriment (%)	Tissue or organ weighting factors NCRP/ICRP	NRC
Gonads		0.183	0.20	0.25
Red bone marrow	0.50	0.143	0.12	0.12
Colon	0.85	0.141	0.12	
Lung	0.85	0.111	0.12	0.12
Stomach	1.10	0.139	0.12	
Bladder	0.30	0.040	0.05	
Breast	0.20	0.050	0.05	0.15
Liver	0.15	0.022	0.05	
Esophagus	0.30	0.034	0.05	
Thyroid	0.08	0.021	0.05	0.03
Skin	0.02	0.006	0.01	
Bone surface	0.05	0.009	0.01	0.03
Remainder	0.50	0.081	0.05	0.30
Sum (whole body)	5.00	1.000	1.00	1.00

Sources: NCRP, 1993b, Report 116; ICRP, 1991a, Publication 60; NRC, 1991.

many organs as the NCRP and ICRP do but includes them in the remainder term, which is 0.3 (much greater than the 0.05 for NCRP/ICRP). It assigns a weighting factor of 0.06 for each of the 5 "remainder" organs (excluding the skin and the lens of the eye) that receive the highest doses and neglects the other doses.

The values are based on an average over the whole population, including all age groups and equal numbers of males and females. Accordingly, the probability values for ovarian cancer and breast cancer would be double those listed in the table if the risk is estimated for a woman whose ovaries or breasts received a radiation dose. The coefficients may be higher by a factor of 2 or 3 for young children. For persons of about 60 years, the coefficients may be lower. This approach obviously cannot be used to evaluate the risk to an individual, since the factors are based on population averages.

Risk of prenatal exposure is not reflected in Table 2.9. The major risk of malformations is incurred by radiation exposure during the period of major organogenesis, 4–14 weeks after conception, with organs at greatest risk during their development. The risk of cancer primarily extends from 3 weeks after conception until the end of pregnancy. Values of decreases in IQ have been reported as 30 IQ points per sievert for exposures between 8 and 15 weeks after conception; less marked decreases are reported following exposure in the period from 16 to 25 weeks after conception (ICRP 1991a, Publication 60).

The sequence of operations that go into the evaluation of effective dose may be summarized as follows:

- Identify the source.
- Characterize the radiation particles emitted (numbers, energies, types, etc.).
- Calculate the absorbed doses in all organs irradiated by sources both inside and outside the body.
- Multiply absorbed doses by radiation weighting factors to determine the equivalent doses to the organs in the body.
- Multiply equivalent doses by tissue weighting factors to convert to contribution to effective dose.
- Sum up all the contributions to give the effective dose.

Example 2.10   A female adolescent patient with lateral curvature of the spine (scoliosis) was given a full spine x-ray examination. She was positioned so the x rays penetrated from her back (posterior) to her front (anterior). The radiation exposure at the skin was 145 milliroentgens, closely equivalent to a dose to tissue of 1.45 mGy (see section 2.14). What was the effective dose to the patient from the x ray?

Organ doses in milligrays were determined for this projection (Rosenstein, 1988) and grouped by tissue weighting factor (w):

w = 0.12: lungs (0.54), bone marrow (0.15), trunk (stomach and colon, 0.39)

w = 0.05: thyroid (0.13), breasts (0.12), trunk (bladder and liver, 0.39), esophagus (estimated, 0.39)

w = 0.01: bone surfaces and skin (estimated, 0.15 each)

w = 0.2: ovaries (0.43)

w = 0.05: remainder (0.39)

The effective dose = 0.12(0.54 + 0.15 + 0.39 + 0.39) + 0.05(0.13 + 0.12 + 0.39 + 0.39 + 0.39) + 0.01(0.15 + 0.15) + 0.2(0.43) + 0.05(0.39) = 0.36 mSv.

For purposes of radiation protection, the value for the excess number of cancer deaths to the public associated with an effective dose of 1 sievert is chosen by NCRP to be 5,000 in every 100,000 persons exposed per sievert (NCRP, 1993a).

Example 2.11 What is the nominal risk of fatal cancer in a population of equal numbers of men and women if everyone were subjected to a full spine x ray?

From the previous example, the effective dose is 0.36 mSv or 0.00036 Sv. The increase in fatal cancer would be 0.00036 Sv × 5,000 per 100,000 per sievert, or 1.8 extra cancers per 100,000 population. If only the risk among women were considered, the total probability would be increased to 5,300 per 100,000 per sievert to yield an excess of 1.9 cancers per 100,000 women, although the risk to the ovaries and the breasts would be doubled.

Further examples of the use of the effective dose concept are given in Part Three. This surrogate dose is currently the primary dose quantity in radiation protection. Standards for radiation exposure published by professional standards-setting bodies are given in terms of the effective dose, which serves as the basic quantity for compliance with regulations promulgated by governmental agencies. Effective dose is used in the evaluation and documentation of personnel radiation exposures, in compilations of the significance of doses to whole populations, and in consent forms dealing with exposure to medical radiation. It is useful in any exposure assessment to provide a rough measure of the risk to the individual or to society. Although effective dose provides the regulatory agencies and the public with a practical tool for assessing the risks of exposures to radiation, it should be borne in mind that the risks so derived are for radiation protection and guidance purposes only; the results do not meet the standards of a rigorous scientific methodology.

14 THE ROENTGEN—THE TRADITIONAL UNIT FOR EXPRESSING RADIATION EXPOSURE

One of the most common methods of measuring x and gamma radiation is to measure the electrical charge they produce by ionizing air. The quantity that expresses the ionization (i.e., charge) produced by x rays interacting in a volume element (for example, 1 cc) in air is known as the exposure. The traditional unit in which exposure is expressed is the roentgen (R). It is used for x- and gamma-ray measurements, limited for practical reasons to the energy range from a few keV to a few MeV.

The roentgen was originally defined as the amount of x or gamma radiation that produces 1 electrostatic unit (ESU) of charge per cubic centimeter of dry air at 0°C and 760 mm pressure. The current equivalent definition in SI units is 2.58×10^{-4} coulombs/kg air. An exposure of soft tissue to one roentgen to will produce an absorbed dose of 0.0096 Gy or very nearly 1 cGy in soft tissue. The equivalent dose in centisieverts is also approximately equal to the exposure in roentgens, since x and gamma radiation have a radiation weighting factor of 1. The conversion from exposure to dose is nowhere near as simple if exposure is expressed in terms of coulombs per kilogram.

15 The Significance of External Radiation Levels

Having introduced units for expressing radiation levels, we may now consider the significance of specific values of these levels. It is convenient to separate radiation exposure into two categories: exposure to radiation from sources external to the body and exposure to radiation from radionuclides incorporated inside the body. We shall consider external radiation levels here and deal with the significance of internal emitters in the next section.

Values of radiation levels for various situations of interest are presented in Table 2.10. The table ranges from the very low levels encountered normally as radiation background, through levels defined for purposes of control of radiation exposures, and up to levels that can produce severe injury. The units used are those in which the values were cited in the literature, including roentgens for measurements of x and gamma radiation and mSv for doses in the regulations.

Nominal values for background radiation levels are given in group I. Gamma radiation from radioactive materials in the earth, such as radium and potassium-40, and cosmic radiation coming down through the atmosphere contribute to the dose. The levels are expressed as an annual dose to allow comparison with radiation protection standards.

Note that the background level on the earth's surface increases by a factor of about 2 from 88 mR per year to 164 mR per year in going from sea level to 10,000 feet altitude as a result of the increased intensity of cosmic radiation. The mass of air contained in the atmosphere is equivalent to a shield of water 10.3 m (34 ft) thick and attenuates appreciably the cosmic radiation incident upon it. This shielding effect is lost, of course, when we travel at high altitudes. The increase is dramatic at altitudes at which a supersonic transport would be flown (up to 26,000 mR/yr). Even this level is much less than the level produced during periods of solar flares.

The levels on the ground vary considerably from place to place, de-

Table 2.10 Significance of external radiation levels (excluding neutrons).

Exposure	Significance
I. Radiation background levels	
88 mR/yr, continuous whole body (0.011 mR/hr)	Background radiation, sea level, out of doors, New York City
76 mR/yr, continuous whole body	Radiation inside wooden house at sea level, NYC
164 mR/yr, continuous whole body	Background radiation, altitude of 10,000 ft (ground level)
4,400 mR/yr, continuous whole body	Exposure at cruising altitude of supersonic transport (60,000 ft)
II. Regulatory limits (NRC)	
50 mSv (5,000 mrem) per year effective dose	Occupational exposure limit for whole body
1 mSv (100 mrem) per year effective dose	Whole-body limit for members of public
50 mSv (5000 mrem) per year eff. dose (max. 500 mSv/yr to a single organ)	Single tissue or organ limit if not covered separately
III. Levels requiring posting and control	
<0.02 mSv in an hour	Unrestricted area (no control or sign required)
>0.02 mSv in an hour	Control of area required
>0.05 mSv in 1 hour to major portion of body	Radiation area sign required
>1 mSv in 1 hour to major portion of body	High radiation area sign required
IV. Radiation risks and biological effects	
1 mSv, major portion of bone marrow	Risk of fatal leukemia is about 1 in 90,000 male, 1 in 125,000 female, as extrapolated from epidemiology at high doses
1 mSv, whole body	Risk of eventual appearance of fatal solid cancer is about 1 in 15,000 male, 1 in 14,000 female
100 mSv, whole body	Elevated number of chromosome aberrations in peripheral blood; no detectable injury or symptoms
1 Sv, reproductive system	Dose for doubling spontaneous mutations
1 Sv, single dose, whole body	Mild irradiation sickness
3.5 Sv, single dose, whole body	Approximately 50% will not survive
4–5 Sv, low-energy x ray, local	Temporary loss of hair
5–6 Sv, single dose, locally to skin, 200 keV	Threshold erythema in 7–10 days, followed by gradual repair; dryness and dull tanning

Table 2.10 (continued)

Exposure	Significance
20 Sv locally to skin from 1 MeV x ray at 300 R/day	Threshold erythema
6–9 Sv, locally to eye	Radiation cataract
10–25 Gy, local, at 2–3 Gy/day	Treatment of markedly radiosensitive cancer
15–20 Gy to skin, single dose, 200 keV	Erythema, blistering, residual smooth soft depressed scar
25–60 Gy, local, at 2–3 Gy/day	Treatment of a moderately radiosensitive cancer
V. Diagnostic x-ray exposures	
200 mR (900 mR reported in 1970)	Mean exposure per dental film per examination in 15–29 yr age group
16 mR	Mean exposure per chest x ray (PA, radiographic, 1984)
500 mR	Mean exposure per chest x ray (photofluorographic, 1964)
5,000 mR/min	Output from properly operating fluoroscope without image intensifier (< 2,000 mR/min with image intensifier)

pending on altitude, composition of soil, pressure, solar activity, and so on. The levels in buildings can be appreciably higher or lower than the levels out-of-doors, depending on the materials of construction. One measurement of the radiation level inside a wooden building is given in table 2.10. The level is 12 mR per year less than it would be out-of-doors. There is a small contribution to the environmental radiation dose from neutrons, not shown in Table 2.10. Typical annual doses are 0.0035 mGy at sea level, 0.088 mGy at 10,000 ft, and 0.192 mGy at 60,000 ft. The equivalent dose (in mSv) is about 20 times as high.

The entries in group II list exposure limits promulgated by the U.S. Nuclear Regulatory Commission in the Code of Federal Regulations, Title 10, Part 20 (10CFR20). The limits refer to radiation received occupationally in addition to the radiation from background sources. The maximum level for occupational exposure is 50 mSv per year, whole-body dose or effective dose, if only part of the body is exposed. However, no organ or tissue can receive an annual dose greater than 500 mSv, even if the effective dose turns out to be less than 50 mSv, and the limits for the eyes are lower: 150 mSv per year to the lens. Technically, a radiation worker could receive the whole annual dose at one time without exceeding the standards for exposure, provided it was the only exposure during the calendar year in question. In practice, exposure is controlled to avoid sharp peaks in the dose accumulated in a month. Regardless of limits, doses to individuals should be kept as low as reasonably achievable (ALARA policy).

The maximum level for individuals under 18 years of age and other individuals nonoccupationally exposed is 1 millisievert per year, or 2 percent of the occupational level. The occupational exposure of a woman who has declared her pregnancy must be so limited that the unborn child does not receive a dose greater than 5 mSv during the entire gestation period. Controls must be implemented so a significant fraction of the permitted dose is not imparted over a short period; exposure should be delivered at substantially the average monthly rate that would comply with the 5 mSv limit. To prevent excessive exposure of the fetus before pregnancy is recognized, fertile women should be employed only in situations where the annual dose accumulation is unlikely to exceed 20 or 30 mSv and is acquired at a more or less steady rate.

Group III deals with the posting of areas to warn individuals that significant radiation exposure is possible. Signs are required if the general radiation level exceeds 0.05 mSv in 1 hour. Control of the area is required if the level exceeds 0.02 mSv in an hour. If the level is less than 0.02 mSv in an hour, the area is considered unrestricted. Under these conditions occupants do not need to be warned that radiation is present, but access must still be controlled to limit exposures to less than 1 mSv/yr. If the level is in excess of 1 mSv or more in 1 hour, the area is considered a high radiation area, and regulations specify special control measures.

Group IV describes some of the effects produced in individuals who are exposed to radiation. The effects to the worker of most concern at these levels are the risks of producing leukemia or other cancers. Epidemiological studies, primarily of the survivors of the atomic bombings of Japan, indicate that the lifetime risk of dying of leukemia is of the order of 1 in 100,000 per millisievert; this estimate is based on a direct extrapolation to low doses of the risk at high doses, or about 1 in 200,000 per millisievert to the whole body if a correction factor is used for low doses or dose rates (NAS/NRC, 1990). The chances of any individual's getting leukemia seem to diminish appreciably 10 years after exposure. A major fraction of the bone marrow has to be affected for the person to experience this level of risk. The risk of cancer other than leukemia from whole-body exposure is several times higher. The cancer risk does not peak until more than 20 years after exposure.

Another area of concern is genetic damage. Radiation induces gene mutations, chromosome rearrangements, and losses or gains of whole chromosomes or large segments within them. Virtually all of them are harmful and their frequency of occurrence is related to dose. In assessing the effects of radiation, the concept of doubling dose is helpful. The doubling dose for a population is the dose to the childbearing segment of the population that is believed to double the number of genetic disorders that occur spontaneously in each generation. Both the rate at which new mutations arise in each generation spontaneously and the dose required to double this rate are numbers that are very difficult to derive.

In every million births, about 10,000 children are born with genetic disease as a result of dominant gene mutations. The conditions are clinically severe in 2,500 of these children. About one-fifth of the mutations are new ones, arising in the parents' generation. The rest result from a buildup of mutations in earlier generations that persist for a few generations before they are eliminated.

A conservative estimate of the doubling dose is 1 Sv (NCRP, 1993a, Report 115). Since very few individuals with severe clinical disorders are likely to reproduce and pass these conditions on to a new generation, these conditions all are caused by mutations arising in the previous generation. Thus a dose of 1 Sv, imparted as a dose to the whole parental generation, might result, in every million live births, in the birth of an additional 2,500 children with clinically severe genetic disorders from dominant mutations. Chromosome aberrations also result in very severe consequences, but the effectiveness of x rays or gamma photons in producing chromosome aberrations is believed to be very low for low dose rates.

The problem of the total genetic harm, including the subtle effects of recessive mutations, that might be caused over many generations is extremely complex, and quantitative estimates of consequences are accompanied by a great deal of uncertainty. Accordingly, regardless of the standards that may be set, the only acceptable policy with regard to the exposure of large numbers of individuals is to keep the dose to the gonads to a minimum, well below the dose from natural background radiation. Animal experiments indicate that the effect in females is considerably reduced if there is a delay of several months between exposure and conception. Accordingly, if a substantial dose is incurred by a worker in a short time period, it may be prudent to allow the lapse of several months prior to conception.

The higher exposures shown in group IV produce directly observable effects in the short term. Note the high levels that are used in radiation treatment of cancer. The individual can survive the high exposure because the volume exposed is small. There may be significant risk of cancer induction at a later period from doses imparted at these levels, but if radiation treatments can arrest or cure a cancer that cannot be treated by other techniques, the immediate benefit is more important than any possible later effects.

The entries in group V present levels of exposure associated with diagnostic x rays, the major source of radiation exposure of the population apart from the natural background level. The dose imparted in taking an x ray has decreased tremendously over the years with improvements in film and x-ray equipment. It may be noted, for example, that the lower limit of exposure from a chest x ray using modern techniques is of the order of 10 percent of the annual exposure from external natural background radiation to the same region. On the other hand, even a properly operating fluoroscope can produce exposure rates of the order of 5,000 mR/min. Additional information on medical x-ray exposures is given in section 22 and in Part Six.

16 Exposure from Internal Radiation Sources

To this point we have discussed the imparting of dose to the body from radiation incident upon it from the outside. Our major concern has been with the properties of the radiations and their manner of imparting dose, and we have not considered the characteristics of the source. Now, however, we shall consider radiation exposure resulting from the introduction of radioactive materials into the body through inhalation, ingestion, or medical procedures. We must introduce a new quantity to describe the rate at which the radioactive materials undergo nuclear disintegrations. Knowing the disintegration rate, we can then determine the rate at which the radiation particles are emitted during the course of these disintegrations and evaluate the dose produced when the radiation interacts with and imparts energy to matter.

16.1 The Activity—A Quantity for Describing the Amount of Radioactivity

The basic event that characterizes a radioactive nuclide is the transformation of its nucleus into the nucleus of another element. This transformation is known as decay. The number of nuclear transformations occurring per unit of time is called the activity.

16.2 The Unit of Activity—The Becquerel

The SI unit for expressing the activity of radioactive material is the becquerel.

$$1 \text{ becquerel (Bq)} = 1 \text{ disintegration per second}$$
$$1 \text{ kilobecquerel (kBq)} = 1000/\text{sec}$$
$$1 \text{ megabecquerel (MBq)} = 10^6/\text{sec}$$
$$1 \text{ gigabecquerel (GBq)} = 10^9/\text{sec}$$
$$1 \text{ terabecquerel (TBq)} = 10^{12}/\text{sec}$$

The traditional unit for the activity is the curie (Ci); 1 curie $= 3.7 \times 10^{10}/$ sec $= 3.7 \times 10^{10}$ Bq.

It is important to recognize that the unit of activity refers to the number of disintegrations per unit time and not necessarily to the number of particles given off per unit time by the radionuclide. As an example, consider the decay scheme for bismuth-212, shown in Figure 2.18. This nuclide undergoes alpha decay in 36 percent and beta decay in 64 percent of its disintegrations.[22] A MBq of ^{212}Bi undergoes 10^6 transformations per second but gives off 0.36×10^6 alpha particles per second, 0.64×10^6 beta particles per second, and 0.36×10^6 gamma photons per second. The rate of emission of a particular type of ionizing particle can be equated to the activity only when that particular particle is given off in each disintegration.

The range of activities found in radiation operations varies tremendously, from kilobecquerel quantities in laboratory tracer experiments to millions of terabecquerels inside the fuel elements of nuclear power plants. The distribution of kilobecquerel quantities of most radionuclides is not regulated, since these small amounts are not considered hazardous as sources external to the body. On the other hand, nuclear reactors require very tight governmental regulation because of the potential for serious radiation exposures in the event of reactor accidents.

16.3 The Accumulating Dose from Radioactivity in the Body and the Committed Dose

The dose from radiation emitted by radioactive materials inside the body is evaluated in accordance with the same principles as for radiation from sources outside the body. We must determine the energy imparted per unit mass by radiations emitted by the radioactive materials, and we use the same units that we used for absorbed dose, MeV/g. Maximum levels for activities of radionuclides inside the body are set by requiring that they do not result in doses to organs in excess of those specified for external sources of radiation.[23]

---

22. Very few radionuclides decay, as ^{212}Bi does, by emitting both alpha and beta particles. This example was chosen primarily because of its value in illustrating the meaning of activity as contrasted to particle emission rate.

2.18 Disintegration scheme of bismuth-212. Energy released from alpha decay, $Q_\alpha = 6.09$ MeV. Energy released from beta decay, $Q_\beta = 2.25$ MeV.

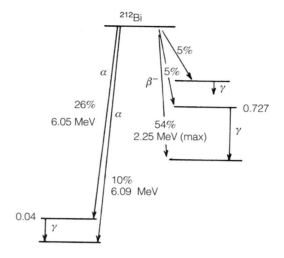

An intake of radioactive material may be a brief incident—an ingestion of a radiopharmaceutical, an accidental inhalation of some airborne radionuclide released in a laboratory synthesis, or ingestion of radioactive material following contamination of the hands and subsequent contact of hand to mouth. The resultant doses, however, are imparted over extended periods, depending on the effective half-life of the material in the body. Half-lives can vary from a few days or less to many decades.

The dose expected to be imparted over the effective lifetime of the radioactivity in the body has a special name in radiation-protection standards. It is called the *committed dose.* Committed absorbed doses are multiplied by radiation weighting factors to give committed equivalent doses, and these in turn are multiplied by tissue weighting factors and summed to give committed effective doses. We shall use the term *dose* for both equivalent and absorbed doses when the type of dose is clear from the context.

The calculation of the committed effective dose requires detailed knowledge of the fate of the radioactivity after it enters the body. Metabolic models are given by the ICRP (1994, 1995b, 1996a). The concepts in the formulation of these models have also been discussed by NCRP (1985a). When exposure is from multiple sources, including external radi-

---

23. Exceptions are the bone-seekers. Because of the extensive epidemiologic data obtained for radium, it has been possible to establish limits for this nuclide based on human exposure. Limits for other bone-seekers are set by consideration of the energy imparted and the LET relative to radiations resulting from radium in the body.

ation, the sum of all the contributing doses, committed and external, must not exceed the basic whole-body limit. However, the annual equivalent dose to a single organ may not exceed 500 mSv, even if higher values would be allowed according to the results of the formula. Examples of calculations of the committed dose are given in Part Three.

17   The Annual Limit on Intake—The Basic Quantity for the Control of Internal Exposures

The approach to the control of internal hazards is to specify limits for the total intake of radionuclides in one year. Designated as the Annual Limit on Intake (ALI), this is the total activity ingested or inhaled by a worker in one year that will impart to the irradiated organs over the total duration of the radiation exposure (or 50 years, whichever is less) an effective dose of 50 mSv. In other words, for determining compliance with the regulations, the committed effective dose received by a worker is assigned to the year the radioactivity is taken into the body[24] and evaluated for a period of 50 years following intake. The limiting dose for members of the public is 1 mSv.

Values of the ALI are published in the standards of the Nuclear Regulatory Commission (NRC, 1991). They are based on recommendations of the ICRP promulgated in 1979. The ICRP reduced its values for the annual dose limits considerably and revised the apportionment of tissue weighting factors in 1990 (ICRP, 1991b). As a result, the values for almost all the revised ICRP ALI were much less (primarily as a result of the lower dose limits) than the NRC values, which remained unchanged. Values of the ALI for selected radionuclides in the NRC regulations (based on the 1979 ICRP recommendations) and in the ICRP 1990 recommendations are given in Part Five, Table 5.3.

Factors entering into the calculation of the ALI include the chemical form, the retention time in the body, and, for inhalation, the aerosol activ-

24. The ICRP previously specified limiting continuous intakes to rates which resulted, at equilibrium, in maximum dose equivalents of 3.75 rem in 13 weeks to any organ except the gonads, blood-forming organs, or lenses of the eyes. These were not allowed to receive more than 1.25 rem (ICRP, 1960). The rate of intake could be varied provided that the total intake in any quarter was no greater than that resulting from continuous exposure at the allowable constant rate. The equilibrium activity in the body resulting in the limiting 13-week dose was referred to as the maximum permissible body burden (MPBB). For radionuclides with very long effective half-lives, which would not reach an equilibrium level in the body during a working lifetime, continuous uptake resulting from inhalation or ingestion was limited so the MPBB would be reached only at the end of a working lifetime of 50 years.

ity median aerodynamic diameter (AMAD, assigned a default value by NRC of 1 μm). The retention time is grouped into three classes, noted as D, W, Y. Class D refers to a retention time of the order of days and applies to a clearance half-time of less than 10 days. Class W pertains to 10 to 100 days (weeks), and Class Y to more than 100 days (years).

18 LIMIT FOR THE CONCENTRATION OF A RADIONUCLIDE IN AIR—A DERIVED LIMIT

Limits for the concentrations of radionuclides in air are calculated by dividing the limit on intake by the volume of air breathed in over the working period. The concentration limits given in the Code of Federal Regulations, 10CFR20, can be converted to permissible intakes by multiplying by the volume of air or water used in the calculations.

Example 2.12   The NRC value for the occupational annual limit on intake by inhalation for ^{125}I is 60 μCi (2.22 $\times$ 10^6 Bq). Calculate the concentration in air that should not be exceeded if the inhalation intake is controlled on a daily basis.

The daily limit on intake is 2.22 $\times$ 10^6 Bq/250 day = 8,880 Bq (based on an annual organ dose ceiling of 500 mSv since the application of the 0.03 weighting factor for the thyroid used by NRC gives a committed organ dose above this limit). The volume of air breathed in by the adult male doing light work during an 8-hr working day is 9,600 liters (ICRP, 1975, Publication 23). The maximum concentration is thus 8,880 Bq/9.6 $\times$ 10^6 cc = 0.000925 Bq/cc or 925 Bq/m^3 (0.025 pCi/cc). The NRC value for the Derived Air Concentration is rounded off to 0.03 pCi/cc.

There is a safety factor in setting allowable concentrations by the method described here, since the controls are applied from the beginning of exposure, when levels in the body are likely to be insignificant, while the limits are derived on the assumption that the body contains the maximum permissible activity at all times. The difference is not important for radionuclides with short effective half-lives, where equilibrium is achieved rapidly. For radionuclides with long effective half-lives, the procedure is very conservative. This is desirable since once these radionuclides are incorporated in the body, there is usually little one can do to reduce the dose rate and there is no margin for unexpected exposures that could be accidental or incurred deliberately in emergencies.

Table 2.11   Concentration guides for radionuclides (in pCi/cc).

Radionuclide	In air, occupational	In air, environmental	In water, environmental
HTO water	20	0.1	1000
$^{14}CO_2$ gas	90	0.3	
$^{14}CO_2$ soluble	4.1	0.003	30
^{32}P soluble	0.4	0.001	9
^{131}I soluble	0.02	0.0002	1
^{125}I soluble	0.03	0.0003	2
^{226}Ra soluble	0.0003	9×10^{-7}	0.06
^{90}Sr soluble	0.002	6×10^{-6}	0.5
^{228}Th insoluble	7×10^{-6}	2×10^{-8}	0.2
^{239}Pu soluble	7×10^{-6}	2×10^{-8}	0.02

Source: NRC, 1991. The limits are those promulgated by the U.S. Nuclear Regulatory Commission. The environmental limits are based on an effective dose of 0.1 rem. The derived concentrations were reduced by another factor of 2 to take into account age differences in sensitivity in the general population. Multiply by 0.037 to convert to Bq/cc.

When radionuclides with very long half-lives are released to the environment it is customary to specify their significance in terms of the dose commitment to the population. This is the total per capita dose to be incurred over the lifetime of the radionuclide in the environment, which may be more than one generation. Doses imparted to individuals throughout their lifetime are evaluated over a 50-year period following intake, in accordance with the procedure for calculating committed dose.

Concentration limits set by the Nuclear Regulatory Commission for radionuclides in air and in water for selected radionuclides as derived from the limits on intake are given in Table 2.11, and a more detailed listing is given in Table 5.4 of Part Five. The environmental levels are 1/300 the occupational levels as a result of the differences in limits (50 mSv vs. 1 mSv), and additional factors of 1/2 to take into account the exposure of children, and 1/3 to adjust for differences in exposure time and inhalation rate between workers and members of the public.

19   Levels of Radioactivity inside the Body—A Useful Benchmark for Internal Exposure

An important benchmark for maintaining perspective on internal exposure is the natural radioactivity of the body. The body contains several radionuclides, most of which have always been contaminants in air, water, and

Table 2.12   Radioactive materials in the body.

Radionuclide	Body content in 70 kg man (Bq)	Annual dose (mGy)
^{40}K	4,433	0.18 (whole body)
^{14}C	3,217	0.011 (whole body)
^{226}Ra	1.48	0.007 (bone lining)
^{210}Po	18.5	0.006 (gonads)
		0.03 (bone)
^{90}Sr (1973)	48.1	0.026 (endosteal bone)
		0.018 (bone marrow)

Sources: UNSCEAR, 1977; NCRP, 1975, Report 45 (^{90}Sr).

food, and others which have appeared only as the result of fallout from the testing of nuclear weapons. Some of the radionuclides present in the body are shown in Table 2.12.

The main source of radioactivity is from the radioactive isotope of potassium, potassium-40. There are 1.12 radioactive atoms of ^{40}K for every 10,000 nonradioactive atoms of potassium. There is about 140 g of potassium in an adult who weighs 70 kg, and 0.0169 g consists of the ^{40}K isotope. This amount of ^{40}K disintegrates at the rate of 4,433 atoms per second. Of every 100 disintegrations, 89 result in the release of beta particles with maximum energy of 1.33 MeV, and 11 result in gamma photons with an energy of 1.46 MeV. All of the beta particles and about 50 percent of the energy of the gamma rays are absorbed in the body, giving annual doses of 0.16 mSv from the beta particles and 0.02 mSv from the gamma rays.

The ^{40}K that decays is not replenished, and presumably all the ^{40}K that exists on earth today was produced at the time that the earth was created. The reason all this activity has not decayed completely over the 10 billion years of the earth's existence is the long half-life of the potassium, 1.26 billion years. Other naturally occurring radionuclides that are still present because of their long half-lives are uranium-238, uranium-235, thorium-232, and a whole series of radioactive decay products.

The second most active radionuclide in the body is carbon-14. The half-life of this nuclide is 5,730 years, so obviously whatever ^{14}C was formed at creation is no longer with us. The ^{14}C is continuously being produced in the atmosphere by the reaction of the cosmic ray neutrons with the nitrogen in the air. In this reaction, the nitrogen absorbs a neutron and releases a proton, yielding ^{14}C. The total content of ^{14}C in the body of an adult undergoes about 3,217 beta disintegrations per second. The dose from absorption of the weak 0.156 MeV beta particle amounts to only 0.01 mSv per year. Thus we see that although there are 0.67 times as many ^{14}C as ^{40}K disintegrations in the body, the dose rate from ^{14}C is only 5 percent of the dose rate from ^{40}K because the energy of the beta particle from ^{14}C is so much less.

The other three radionuclides in the table deposit primarily in bone. ^{226}Ra and ^{210}Po are alpha emitters and the dose in mGy is multiplied by 20 to give the dose equivalent in mSv. ^{210}Po imparts about 0.60 mSv and ^{226}Ra (plus its decay products) about 14 mSv a year to bone cells. The ^{90}Sr activity in bone is the result of fallout during the testing of atomic bombs.

Selective irradiation of the lungs occurs from inhalation of the radioactive noble gases, radon-222 and radon-220 (thoron), and their radioactive decay products, all of which are always present in the atmosphere. These gases are alpha emitters and along with their decay products, they impart most of their energy as alpha radiation. The dose to the public is mainly from ^{222}Rn and its decay products, which impart an average dose to the bronchi of 24 mSv per year (NAS/NRC, 1990), making the lung the organ that receives the highest radiation exposure from natural sources. A detailed discussion on the nature of the dose from radon and its decay products is given in Part Three, section 5.6.

20 Protection from Radioactive Contamination

When we considered protection from radiations emitted by radiation sources, we emphasized *distance* and *shielding* as measures for attenuating the radiation to permissible levels. When we deal with radioactive contamination, the problem is to prevent the contamination from entering the body. This can occur through inhalation, ingestion, absorption through the unbroken skin, or penetration through abrasions, cuts, and punctures. The protective measures needed are similar to those used with any other contaminant that presents an internal hazard and are probably familiar to any person trained to work in the laboratory.

It is general practice to handle significant quantities of radioactive material in a hood to prevent release of these materials to the working environment. The velocity of the air flowing into the hood should be between 100 and 150 linear feet per minute. Gloves should be worn and forceps used to handle sources. Hands should be washed and monitored for contamination after working with these materials. A generally cautious attitude with commonsense precautions should provide adequate protection in most working situations. Of course, as the amount of the active material handled increases, the measures must become more stringent. A detailed discussion of protective measures for individuals working with radioactive materials is given in Part Five.

Questions always arise as to what levels of radioactivity require use of a hood, or a glove box, or gloves. We shall not attempt to specify protection measures for various levels of activity. However, all persons beginning work with radioactive materials should take strict precautions, even at low levels. In this way they obtain valuable training and experience. As proper habits are formed, workers gain an appreciation of the hazards involved. In time they will accumulate enough experience to select the essential precautions and to avoid the application of excessive control measures.

We cannot overemphasize the need always to utilize radiation monitoring equipment for hazard control. Instruments are the only means of evaluating the presence of radioactivity. Because it is not possible to sense the presence of radioactivity, workers cannot be sure that they are in fact working with low amounts when they are expecting to do so. There are instances, for example, where purchase orders were misread and the shipments contained thousands of times the activity ordered, or where shipping containers have unexpected contamination. Workers must be aware of the possibilities of such errors and take such reasonable measures as are necessary to insure that they will be protected.

21 SOME SIMPLE CALCULATIONS IN RADIATION PROTECTION

It is often useful to be able to make a simple estimate of dose. Following are a few elementary calculations utilizing "rules of thumb."

21.1 Dose from Beta Particles

The higher-energy beta particles, in penetrating through water or water-equivalent thicknesses, lose energy at the rate of about 2 MeV per centimeter.[25] This characteristic enables us to make a rough calculation of the dose rate from a beam of beta particles incident upon the body.

25. The energy loss of 2 MeV per centimeter is a *differential* expression, that is, it is the *limit* of the ratio of the energy loss to the thickness ($\Delta E/\Delta x$) as the thickness gets very small. It is *not* the energy loss in 1 centimeter (that is, a 0.5 MeV beta particle loses energy over a small distance at the rate of approximately 2 MeV/cm, but obviously it will not lose 2 MeV in 1 cm; it will lose approximately 0.02 MeV in 0.01 cm). See the end of section 8.3 for additional discussion on differential losses as applied to the attenuation of radiation.

Example 2.13 What is the dose rate at the point where a beam of energetic beta particles is incident upon the body, if, over an area of 1 cm^2, 100 beta particles are crossing per second?

Consider a thin layer of tissue, say 0.01 cm, and assume the tissue is equivalent to water, density = 1 g/cm^3. At an energy loss of 2 MeV/cm, a beta particle deposits very nearly 0.02 MeV in the layer and 100 beta particles/cm^2 deposit 2.0 MeV in a volume 1 cm^2 × 0.01 cm = 0.01 cm^3, with a mass of 0.01 g. The energy deposited per unit mass is 2.0 MeV/0.01 g = 200 MeV/g.

$$\text{Dose rate} = \frac{200 \text{ MeV} / \text{g -sec}}{6.24 \times 10^6 \text{ MeV} / \text{g -mGy}} = 3.2 \times 10^{-5} \text{ mGy/sec}$$

$$= 0.115 \text{ mGy/hr}$$

A more direct and mathematically exact approach to this calculation is: (β particles/cm^2-sec) × (energy loss per cm) ÷ density = energy deposited/unit mass. That is,

$$\frac{(100/cm^2\text{-sec})(2 \text{ MeV/cm})}{1 \text{ g/cm}^3} = 200 \text{ MeV/g-sec} = 0.115 \text{ mGy/hr}$$

As a rough rule of thumb, we may use the relationship 1 beta particle/cm^2-sec = 0.0011 mGy/hr. This relationship holds only for high-energy beta particles. The dose rate increases appreciably at lower energies because the rate of energy loss increases. Values of the dose rate in mGy/hr from 100 beta particles per cm^2-sec incident on the skin are given in Table 2.1 for various radionuclides, and it can be seen that the actual values range from 0.11 for ^{90}Sr to 0.56 for ^{14}C. However, the lower-energy beta particles undergo greater attenuation in the dead layer of the skin, and the net result is to bring the dose rate to the basal cells of the epidermis for the lower energies to within about 30 percent of the dose rate from high-energy particles for a nominal dead-layer thickness of 0.007 cm.

These relationships are useful in estimating the dose if some activity contaminates the surface of the skin.

Example 2.14 What is the dose rate to the cells under the dead layer of the skin if a square centimeter of surface is contaminated with 1 kBq of ^{32}P?

The kilobecquerel emits 1,000 beta particles per second, half of which are emitted in the direction of the skin. To simplify the problem, assume that all the beta particles are incident perpendicular to the surface and deposit energy at a rate characteristic of high-energy beta particles, that is, 2 MeV/cm. Using the results of Example 2.13 and assuming that 95 percent of the beta particles penetrate the dead layer, the dose rate is $(1{,}000 \times 0.5 \times 0.95 \times 0.115)/100 = 0.546$ mGy/hr.

However, the situation is much more complicated. The beta particles enter the skin at all angles and have longer path lengths in the sensitive layer of the skin than at perpendicular incidence. Also, even the higher-energy beta emitters emit a large fraction of low-energy beta particles, which impart more than 2 MeV/cm to tissue. The actual dose rate turns out to be 2.49 mGy/hr, almost five times as high as the simple calculation (Healy, 1971).

In any event, beta contamination on the surface of the skin can produce high local dose rates and must be monitored carefully.

The following example illustrates how an end-window Geiger-Mueller detector[26] can be used to estimate the dose rate from beta particles.

Example 2.15 A thin end-window pancake G-M counter yields a counting rate of 30,000 counts/min when positioned over a source of beta particles. Estimate the dose rate to a hand positioned at the location of the window of the detector, if the window has a diameter of 4.45 cm. Assume every beta particle that passes through the window produces a count.

The cross-sectional area of the window is 15.55 cm^2. If we assume there is no attenuation in the window, then the number of beta particles incident on 1 cm^2 of the counter window per second is $30{,}000/(60 \times 15.55)$, or 32.15. From the results of Example 2.13, an incident beam of 32.15 energetic beta particles per cm^2 per second produces a dose rate of approximately 0.037 mGy/hr.

Thus, a counting rate due to beta particles of 30,000/min when obtained with an end-window pancake G-M tube with a window diameter of 4.45 cm is indicative of a dose rate of 0.0376 mGy/hr. The accumulated dose to the hands of a worker constantly exposed to this dose rate for a working year of 2,000 hr would be 75.2 mGy, only a small fraction of the 500 mGy allowed to the extremities. This example illustrates the high counting rates that can be obtained with sensitive G-M counters from beta particle fields that are within regulatory limits. Even so, high counting

26. See Part Four, section 1.1, for description of the G-M tube.

rates should be reduced when practicable in accordance with the principle that radiation exposures should be kept as low as reasonably achievable.

The dose rate due to a localized beta source falls off sharply with distance as a result of increased separation and increased attenuation by the air. Because of the strong dependence of air attenuation on energy, a simple expression for the dose rate is not possible, but an upper estimate may be made by neglecting air attenuation over a few centimeters. This is a reasonable assumption for the high-energy beta emitters. The expression for the dose rate (in mGy/hr) at a distance from a source of a given activity[27] is (91.5 × activity, in MBq)/(distance, in cm)^2.

> *Example 2.16*   An investigator is evaporating a solution containing 3.7 MBq of ^{32}P in a small beaker. What is the dose rate 10 cm above the bottom of the beaker after the solution is evaporated to dryness?
>
> $$\text{Dose rate} = (91.5)(3.7)/10^2 = 3.39 \text{ mGy/hr}$$

The example illustrates the high dose rates that are possible near a beta source. The dose rate is much less when the activity is in solution, since most of the beta particles are absorbed in the solution.

When beta emitters are taken into the body it is necessary to evaluate the dose rate imparted to organs in which they are incorporated. The average dose rate can be evaluated readily by assuming, because of the short range of the beta particles, that all the beta energy emitted in an organ is absorbed in the organ.[28] Thus the calculation of the average absorbed dose per beta particle is done by determining the average beta energy emitted per gram.

> *Example 2.17*   If iodine is swallowed, about 30 percent ends up in the thyroid gland, which has a mass of 20 g in an adult. What is the initial beta dose rate to the thyroid gland of a person who ingests 1 MBq of ^{131}I? The average energy per disintegration of the beta particles from ^{131}I is 0.188 MeV.

27. This expression is derived from the results of Example 2.13, that is, 1 beta/cm^2-sec equals 0.00115 mGy/hr. There are 79,577 betas/cm^2-sec at 1 cm from a 1 MBq source (see Part Three, section 3.2.2).

28. A detailed treatment of the dose from internal emitters is given in Part Three, section 1.

Of 1 MBq ingested, 0.3 MBq deposits in the thyroid. The rate at which beta energy is imparted to the thyroid is:

$$1\ \text{MBq} \times 0.3 \times \frac{10^6\ \text{dis}}{\text{MBq-sec}} \times \frac{0.188\ \text{MeV}}{\text{dis}} = 56{,}400\ \text{MeV/sec}$$

The dose rate is:

$$\frac{56{,}400\ \text{MeV}}{20\ \text{g-sec}} \times \frac{3600\ \text{sec}}{\text{hr}} \times \frac{1\ \text{mGy-g}}{6.24 \times 10^6\ \text{MeV}} = 1.627\ \text{mGy/hr}$$

The average time the radioiodine is in the thyroid is about 10 days. The total dose imparted to the thyroid is thus of the order of 0.39 Gy following ingestion of 1 MBq. An additional 10 percent is added to the dose from gamma photons emitted by the ^{131}I.

21.2 Exposure Rate and Dose Rate from Gamma Photons

A useful rule of thumb gives the exposure rate (to an accuracy of about 20 percent between 0.07 and 4 MeV) at a distance from a point[29] source of gamma photons. The exposure rate is proportional to the rate of emission of gamma photons from the source and their energy and decreases inversely as the square of the distance. The formula is:

$$\text{Exposure rate, in mR/hr} = \frac{6AEn}{37d^2} \tag{2.6}$$

A = activity of source in MBq (if the activity is in mCi, the factor 37 in the denominator is omitted, and the formula is $6AEn/d^2$)
E = photon energy in MeV
n = average number of photons of energy E per disintegration
d = distance from source to dose point in feet (d may be expressed in centimeters if the coefficient 6 is changed to 5,600)

The factor n is needed if the photon of energy E is not emitted with each disintegration. If photons of several energies are emitted, the contribution of each must be determined separately and the values added.

The numerical value of the exposure rate given by Equation 2.6 when divided by 100 is closely equal to the absorbed dose rate to tissue in mGy/hr and to the equivalent dose rate in mSv/hr.[30]

29. A source may be considered to be a point source if the distance to the source is large compared to the dimensions of the source.
30. Review section 14 for the discussion of the distinctions between these quantities.

Example 2.18   Iodine-131 emits gamma photons of two energies: 0.36 MeV in 87 percent of the disintegrations and 0.64 MeV in 9 percent of the disintegrations. What is the exposure rate at 2 ft from a 370 MBq (10 mCi) source of ^{131}I?

We calculate the contribution to the exposure rate separately for each of the photons, using the following values: $E_1 = 0.36$; $n_1 = 0.87$; $E_2 = 0.64$; $n_2 = 0.09$.

$$\text{Exposure rate from 0.36 MeV photons} = \frac{6 \times 370 \times 0.36 \times 0.87}{37 \times 4}$$

$$= 4.70 \text{ mR/hr}$$

$$\text{Exposure rate from 0.64 MeV photons} = \frac{6 \times 370 \times 0.64 \times 0.09}{37 \times 4}$$

$$= 0.86 \text{ mR/hr}$$

The total exposure rate is $4.70 + 0.86 = 5.6$ mR/hr. This may also be expressed as a dose rate of approximately 0.056 mGy/hr to tissue.

For purposes of calculating exposures from gamma-emitting nuclides in medical applications, the source strength is often characterized by the specific gamma ray constant Γ. This gives the exposure rate in R/hr at 1 cm from a 1 MBq source. With the use of Γ, the exposure rate for a given source activity A (MBq) at any other distance d (cm) may be obtained. The formula is

$$\text{Exposure rate} = \frac{\Gamma A}{d^2}$$

Example 2.19   What is the exposure rate at 61 cm from a 370 MBq source of ^{131}I? Γ for ^{131}I $= 2.18$ R/mCi-hr at 1 cm.

This value for Γ is given in terms of traditional units for activity. We convert from mCi to MBq in Γ by dividing by 37. Thus $\Gamma = (2.18/37)$ R/MBq-hr at 1 cm.

$$\text{Exposure rate} = \frac{2.18 \times 370}{37 \times 61^2} = 0.00586 \text{ R/hr}$$

This is in good agreement with the result obtained by "rule-of-thumb" in Example 2.18. As stated previously, the numerical value divided by 100 may also be expressed as absorbed dose in Gy/hr or equivalent dose in Sv/hr.

Values of the specific gamma ray constant for selected radionuclides are given in Appendix II.

21.3 Reduction of Dose Rate by Both Distance and Shielding

When the exposure or dose rate from a small shielded source is to be evaluated at a point, both distance and attenuation by the medium must be taken into account. We have already looked at examples for treating these effects separately. To evaluate their combined effect, the exposure or dose rate is first calculated on the basis of distance alone, neglecting the shielding effect of the medium. The half-value layers between the source and the dose point are then evaluated, and the exposure or dose rate is reduced by the attenuation factor given for the number of half-value layers encountered.[31]

Example 2.20 Calculate the exposure rate at a point 2 ft from a 370 GBq (10 Ci) ^{131}I source housed in a lead container 2.54 cm thick. The half-value layer of lead is 0.21 cm at 0.36 MeV and 0.54 cm at 0.64 MeV.

Since the activity is 1,000 times the activity in Example 2.18, the exposure rates are also higher by 1,000—that is, 4.70 R/hr from the 0.36 MeV photons and 0.86 R/hr from the 0.64 MeV photons. Because the attenuation in lead is much greater at 0.36 MeV than at 0.64 MeV, the effect of the shield must be determined separately for the two energies.

$$2.54/0.21 = 12.1 \text{ HVL at } 0.36 \text{ MeV}; \quad (\tfrac{1}{2})^{12.1} = 2.28 \times 10^{-4}$$

$$2.54/0.54 = 4.7 \text{ HVL at } 0.64 \text{ MeV}; \quad (\tfrac{1}{2})^{4.7} = 0.038$$

The exposure rate at 2 feet from the shielded source (neglecting buildup)[32] is $(4.7 \times 2.28 \times 10^{-4}) + (0.86 \times 0.038) = 0.0011 + 0.0327 = 0.038$ R/hr.

Thus, while the 0.36 MeV gamma ray contributes most of the exposure rate from the unshielded source, the 0.86 MeV gamma ray contributes most of the exposure rate once the source is shielded. We could also say the dose rate is approximately 0.00038 Gy/hr, or 0.38 mGy/hr.

31. See the end of section 8.5 for comments on cases where the half-value-layer concept is not applicable.
32. The buildup factor is about 2; see Morgan and Turner, 1967.

21.4 Correction for Radioactive Decay

The decay of radioactivity with time follows the same relationship as attenuation through a medium with distance penetrated. This may be inferred from the similarity between the concepts of half-value layer and half-life. One refers to the distance (in an absorbing medium) required to attenuate the radiation by a factor of two; the other refers to the time in which the radioactivity decreases by a factor of two. The fall-off at the same multiples of half-value layers and half-lives is also identical, and for this reason we have shown the data in Table 2.4 and Figure 2.13 as applying to both half-value layers and half-lives.

Example 2.21   A researcher acquired from a national laboratory in June 1996 a surplus ^{60}Co source whose activity was given as 333 GBq on January 1981. The source was shielded with 2 inches of lead. What is the tissue dose rate the user could expect at 12 inches from the source? (Use data in Table 2.2 and Figure 2.13.)

The energies of the two gamma rays emitted from cobalt-60 may be added (1.33 + 1.18 = 2.51 MeV) since they are so close together that an average half-value layer may be used (1.1 cm). The exposure rate at 1 ft, uncorrected for shielding or decay, is given approximately by Equation 2.6 (in R/hr if activity is in gigabecquerels): (6 × 333 × 2.51)/(37 x 1^2) = 135.5 R/hr.

$$\text{Half-value layers} = \frac{2.54 \times 2}{1.1} = 4.62$$

$$\text{Attenuation factor} = (\tfrac{1}{2})^{4.62} = 0.041$$

$$\text{Half-lives} = \frac{15.5}{5.27} = 2.94$$

$$\text{Fraction remaining} = (\tfrac{1}{2})^{2.94} = 0.130$$

The corrected exposure rate (neglecting buildup) = 135.5 × 0.041 × 0.130 = 0.722 R/hr and may also be expressed as 7.22 mGy/hr.

21.5 Shielding of Large or Complex Sources

The calculation of the attenuation of a shield for complex sources follows an approach different from that used in the previous section.[33] The

---

33. This section should be omitted in a first reading. For a thorough treatment, see Jaeger et al., 1968; Shultis and Faw, 2000.

2.19 Exposure rate from 1 photon/cm²-sec. (*Source:* Rockwell, 1956.)

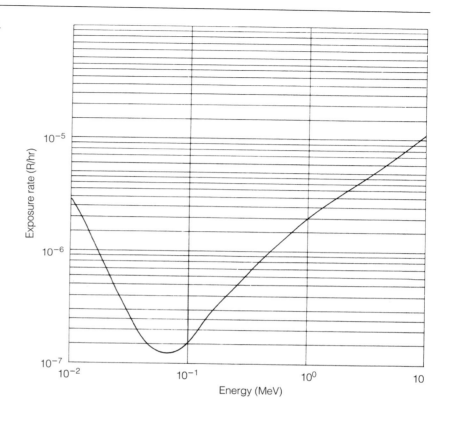

initial quantity determined is the gamma ray flux, the number of photons per square centimeter in a beam passing past a point per second (see Part Three, section 3.2). Fluxes are calculated at the dose point of interest for photons at each of the energies emitted by the source. They are then converted to exposure rates or tissue dose rates by flux–exposure rate (or dose rate) conversion factors, and the dose rates are summed. Flux to exposure rate conversion factors are shown as a function of energy in Figure 2.19.

When the source is of such large dimensions that the distance from the dose point to different points in the source varies considerably, or when there is appreciable attenuation within the source, we can no longer treat it as a point source. We must break it up into subdivisions, chosen so that the distance from the dose point varies little over each subdivision. We evaluate the contribution to the dose rate from each subdivision separately and then add the results. The use of computers makes it possible to define the source elements as small as needed for the desired accuracy. However, useful estimates often can be made with rather large source elements.

Example 2.22   A tank filled with an aqueous solution of ^{60}Co with a total activity of 111 GBq is to be stored behind a lead shield, 4 cm thick, as shown in Figure 2.20. Calculate the exposure rate at points 32 cm and 64 cm from the center of the tank.

All dimensions are drawn to scale in the figure. The tank is subdivided into 6 sources of equal volumes as shown. Therefore, each source has an activity of 18.5 GBq. Distances to the dose point A and distances traveled by the gamma photons in the lead and water are taken off the figure and listed in Table 2.13. The half-value layers in lead and water traversed by the photons from each source are calculated and summed and attenuation factors calculated. The flux without attenuation is evaluated for each source and then multiplied by the attenuation factor. The flux is converted to exposure rate, and the result is multiplied by a buildup factor to account for scattered radiation. The total exposure rate is obtained by summing the contributions from the subdivisions. It is equal to 2,184 mR/hr at 32 cm (and 670 mR/hr at 64 cm by a similar calculation).

Note that the exposure rate at 64 cm from the center of the tank is 0.31 times the exposure rate at 32 cm, whereas it would be only 0.25 times as high if it fell off as the inverse square law with distances referred to the center of the tank. When source dimensions are significantly larger than the distance, the dose rate falls off more slowly than the inverse square because the radioactivity is not all at the same distance. The rate of fall-off is further decreased over the inverse square as the distance from the tank in Figure 2.20 increases because the gamma photons go through a shorter slant distance in the attenuating medium.

When one wishes to calculate the thickness of shielding required to attenuate the dose rate from a complex of sources to a specified value, the calculation in Example 2.17 is repeated for two or more assigned thicknesses that preferably bracket the thickness required and the desired thickness determined by interpolation.

22   X Rays—Radiation Made by Machine

The origin of the x-ray machine can be traced to the discovery by Roentgen in 1895 that electrons striking surfaces within an electron tube at high

2.20 Method of subdividing large source for dose calculations. For simplicity the problem is presented as two-dimensional. The source is subdivided into 6 elements of equal volume, and the activity in each element is considered to be concentrated at the center of the element. The distances are drawn to scale so that the dimensions can be taken directly from the diagram. The use of smaller subdivisions would give greater accuracy.

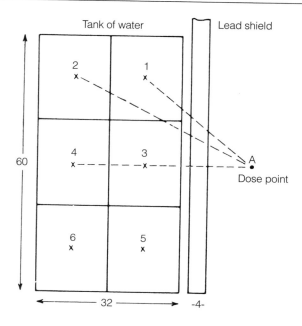

speed generated a penetrating radiation, which he called x radiation. The radiation was detected accidentally when a paper screen washed with barium-platino-cyanide lit up brilliantly in a darkened room in the vicinity of the electron tube, which was covered with a closely fitting mantle of thin black cardboard.

It was soon found that, because of their ability to penetrate matter, x rays could be used to produce pictures of the interior of objects, and, over the years, x-ray machines were developed that could show the interior of objects in great detail. Our concern here is primarily with the use of x rays in examinations of the internal structure of the human body, that is, with photon energies less than 150,000 eV. Because x rays are so valuable in the diagnosis and treatment of disease and injury, they are used routinely in medical and dental practice, and as a result they are responsible for most of the exposure of the public to ionizing radiation, outside of exposure due to the natural radiation background.

In this section, we shall consider the ways in which x rays are used to provide diagnostic information, the dose to the subject that results from such use, and measures for minimizing exposure of both subject and operator. The discussion applies, of course, not only to medical x-ray equipment but also to all other kinds of x-ray machines.

Table 2.13 Organization of calculations of dose rate from large source (Fig. 2.20).

Source point	Distance to A (cm)	Distance in H_2O (cm)	Distance in Pb (cm)	No. of HVL, H_2O	No. of HVL, Pb	No. of HVL, H_2O + Pb	AF
1	32	10	5	0.91	4.55	5.46	0.0227
2	46	27	4.5	2.45	4.09	6.54	0.0108
3	24	8	4	0.73	3.64	4.37	0.0484
4	40	24	4	2.18	3.64	5.82	0.0175
5	32	10	5	0.91	4.55	5.46	0.0227
6	46	27	4.5	2.45	4.09	6.54	0.0108

Source point	Distance squared (cm^2)	Flux at A $(cm^2\text{-sec})^{-1}$	Exposure rate (mR/hr)	Buildup factor	Exposure rate including buildup (mR/hr)
1	1.02×10^3	65,530	154	2.2	339
2	2.12×10^3	15,000	35	2.4	85
3	5.76×10^2	247,400	581	2.0	1,162
4	1.6×10^3	32,204	76	2.3	174
5	1.02×10^3	65,530	154	2.2	339
6	2.12×10^3	15,000	35	2.4	85
				Total:	2,184

Note: Data for HVL are taken from Table 2.2 and exposure rates from Fig. 2.19. An average energy of 1.25 MeV/photon is used. (A more accurate calculation would require separate calculations for the 1. 18 and 1.33 MeV photons from ^{60}Co.) Buildup factors to correct for scattered radiation are taken from table 9 of NCRP, 1964, Report 30, which gives buildup factor for point isotropic source. To use table 9 the half-value layers given above are converted to relaxation lengths (μx) by dividing by 1.44. Since we have a shield of two materials, the number of HVL in the sequence of materials was obtained by adding the values for the separate materials and a buildup factor determined for the summed HVL, assuming the medium was the last material in the sequence, that is, lead. This is a simplified and approximate method. See Jaeger et al., 1968, for a more accurate calculation.

22.1   Production of X Rays

At about the time x rays were discovered by investigators making observations on high-speed electrons in electron tubes, persons working with radioactive material discovered that these materials also emitted penetrating radiations; they called these radiations gamma rays. The penetrating rays emitted by the radioactive material were detected through the unexpected blackening of photographic film developed following storage in a container near the source of the radiation. The names x and gamma ray were assigned to these radiations before their fundamental properties were determined, and it was later found that both types of radiation consisted of photons of electromagnetic radiation and that in fact x rays and gamma photons of the same energy were identical. Thus, the difference in names reflected only the difference in origin of the radiations, rather than any difference in their nature.

We have already seen how gamma photons arise from transitions in the nuclei of the atoms of radioactive substances. On the other hand, x rays are produced in processes that take place outside the nucleus. These processes involve interactions between high-speed electrons and the atom.

There are two different mechanisms through which x rays are produced. These are illustrated in Figure 2.21. The most important mechanism, from the point of view of the use of x rays in radiography, is through a violent acceleration of the electron, resulting in a sharp deflection as it interacts with the electrical field around the nucleus. Such acceleration results in the emission of photons of x radiation. These are uncharged and therefore constitute a penetrating radiation, in contrast to the charged electrons from which they arise (see section 5). The photons are generally referred to as bremsstrahlung ("braking radiation"), because the electrons lose energy and slow down in the process of emitting the radiation. All the kinetic energy of the electron may be converted into an x ray, but this occurs only rarely. Most of the time, only part of the energy is converted. The result is

2.21 Methods of x-ray production by electron bombardment. *(a)* Bremsstrahlung production by acceleration of bombarding electrons. Electrons accelerated (shown here as a change of direction) near the highly charged nucleus of a heavy element may lose all or most of their energy through the emission of photons (called bremsstrahlung, meaning "braking radiation"). *(b)* Characteristic x radiation production by deexcitation of atom energized by bombarding electron. Step 1: Incident energetic electron ejects orbital electron from inner shell of atom, leaving atom in excited state. Step 2: Electron from outer shell drops to vacant shell, resulting in emission of characteristic x-ray photon.

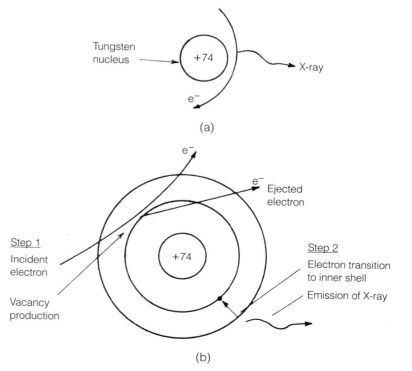

a continuous spread of energies in the x rays produced, up to the maximum energy of the electron. In commercial x-ray units, the x rays are produced by bombarding a target with electrons that receive their energy by acceleration through a high voltage.

The second mechanism of x-ray production is through transitions of electrons in the inner orbits of atoms. These orbital transitions produce photons of discrete energies given by the differences in energy states at the beginning and end of the transitions (Fig. 2.21). Because of their distinctive energies, the photons produced are known as characteristic x rays. Characteristic x rays can be produced only if electron vacancies are created in the inner orbits of atoms to which outer electrons can be transferred. There are several ways in which such vacancies can be produced. One is through the bombardment of an atom with energetic electrons, which may result in the ejection of other electrons from the innermost shells. This can be an important source of x rays in an x-ray tube, although the main source is from acceleration of the electrons in collisions with the target, as discussed in the previous paragraph. Vacancies can also be produced by gamma rays through an absorption process known as the photoelectric effect. There are two forms of radioactive decay that also create vacancies followed by x-ray emission: internal conversion and electron capture. These processes are responsible for the development of radionuclide x-ray sources for use in radiography. The manner in which they give rise to x rays is shown diagrammatically in Figure 2.22.

Although there are many physical processes that result in the production of x rays, the process of most value for medical purposes is electron bombardment of a target through the use of the x-ray tube. The reason for the almost universal use of this method of x-ray generation is that the x-ray tube produces the smallest and most intense x-ray source. The closer an x-ray source can be made to approach the point source of bombarding electrons, the sharper is the image that can be produced in radiography. Also, the higher the intensity of the source, the shorter is the exposure time required.

22.2   Diagnostic Radiology

When an x ray is taken of a person, the part of the body studied is exposed to photons that emanate as rays from the tube target. The fraction of the x-ray photons in an incident ray that penetrates through any portion of the object irradiated depends on the energy of the photons and the type and thickness of material in the path of the ray. If a photographic plate is placed behind the object radiated, and then developed, the darkening is proportional to the amount of radiation intercepted and absorbed by the

2.22 Radionuclide sources of x rays. *(a)* X-ray production by electron capture. Step 1: Electron is captured from inner-most (K) shell by iron-55 nucleus, converting a proton in the nucleus into a neutron and producing a manganese-55 atom with an electron missing in the K shell. Step 2: Innermost orbit of manganese-55 atom is filled by transition of electron from outer shell, accompanied by emission of an x ray. The most favored transition is from the closest shell, which gives rise to an 0.0059 MeV x ray. *(b)* X-ray production by internal conversion. The example used is the isomer of tin-119, a metastable state of the nucleus that is more energetic than the ground state by 0.089 MeV. It reverts to the stable state with a half-life of about 250 days by releasing this energy. Step 1: The 0.089 MeV released when the nucleus reverts to the ground state is imparted to one of the inner electrons and ejects it from the atom as an internal-conversion electron. Step 2: The vacant shell is filled immediately by another electron, a process accompanied by the emission of an x ray. Internal-conversion electrons from the next (L) shell are followed by 0.004 MeV x rays. (A process that competes with internal conversion is the emission of 0.089 MeV gamma photons following deexcitation of the nucleus.)

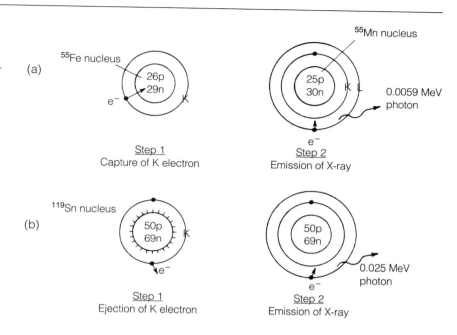

plate. The pattern of darker and lighter areas produces an image related to the internal structure of the object being examined.

It should be noted that the reason x-ray pictures are produced is that a variable fraction of the x-ray energy is absorbed as the x rays penetrate through the body. This absorbed energy produces a dose distribution in the irradiated person and thus has the potential to cause injury coincidental with its use in promoting the health *of* the patient.

22.3   X-Ray Attenuation in the Body

We can understand the nature of the image formed and the dose distribution imparted to an object that is examined with x rays by looking at data describing the penetration of the x rays through matter. Two quantities of particular importance in describing the behavior of x rays are the half-value layer and the mass energy-absorption coefficient. Values of these quantities for muscle and compact bone are presented in Table 2.14.

We defined the half-value layer in connection with gamma photon attenuation (section 8.1), as the distance in which half the photons interact. The number of half-value layers traversed by a beam of photons traveling from the source to the x-ray film determines how many of the photons ac-

Table 2.14   X-ray attenuation data for muscle and compact bone.

Photon energy (MeV)	Half-value layers		Mass energy-absorption coefficient	
	Muscle (cm)	Compact bone (cm)	Muscle (cm²/g)	Compact bone (cm²/g)
0.01	0.13	0.019	4.87	19.2
0.02	0.95	0.14	0.533	2.46
0.03	2.02	0.41	0.154	0.720
0.04	2.78	0.78	0.070	0.304
0.05	3.19	1.15	0.043	0.161
0.06	3.54	1.45	0.033	0.10
0.08	3.84	1.88	0.026	0.054
0.10	4.09	2.14	0.026	0.039
1.00	9.90	5.58	0.031	0.029
10.00	31.3	16.3	0.016	0.016

Source: Attix and Roesch, 1968, vol. I, chap. by R. D. Evans. Half-value layers calculated from data in reference by using density of 1 for muscle and of 1.85 for compact bone. The data are for good geometry (see section 8.4 and Fig. 2.11).

tually reach the film and therefore the degree of darkening that will be produced after development.

The mass energy-absorption coefficient gives the fraction of the photon energy absorbed per unit mass of the medium as a result of interaction with the atoms and electrons. When a beam of photons is incident on a medium, the product of the mass energy-absorption coefficient in units of cm²/g and the photon energy carried in 1 cm² of the cross-section of the beam gives the energy absorbed per gram.[34] This can be converted readily into grays.

Example 2.23   A beam of 10^{10} photons, all with the same energy of 0.05 MeV, is incident on tissue. The beam diameter is 7.62 cm. *(a)* What is the dose to the tissue at the point of incidence? *(b)* What would the dose to bone be from the same beam?

(a) The cross-sectional area of the beam is 45.6 cm². Thus the number of photons passing through 1 cm² (called the fluence) $= 10^{10}/45.6 = 2.19 \times 10^8$ photons/cm². The photon energy passing through 1 cm²

---

34. The mass energy-absorption coefficient should not be confused with the mass attenuation coefficient (section 8.3), which gives the *photons interacting* per unit mass of medium rather than the *energy absorbed.* The mass attenuation coefficient is used in the calculation of the number of photons that reach a point, while the mass energy-absorption coefficient is used to calculate the absorbed dose at a point once the photons get there.

$= 0.05 \times 2.19 \times 10^8 = 1.10 \times 10^7$ MeV/cm^2. From Table 2.14 the mass energy-absorption coefficient in muscle for 0.05 MeV photons is 0.043 cm^2/g. The energy absorbed per gram $= 1.10 \times 10^7$ MeV/cm^2 $\times 0.043$ cm^2/g $= 4.73 \times 10^5$ MeV/g. The dose is

$$\frac{473{,}000 \text{ MeV/g}}{6.24 \times 10^6 \text{ MeV / g-mGy}} = 0.0756 \text{ mGy}$$

(b) The mass energy-absorption coefficient for 0.05 MeV photons is 0.161 cm^2/g in bone, which is $0.161/0.043 = 3.74$ times the coefficient in muscle. Therefore the dose imparted by the radiation is 3.74 times greater in bone than in muscle. The dose to the bone is thus $3.74 \times 0.0756 = 0.283$ mGy.

The values of half-value layer and mass energy-absorption coefficients depend on the energy of the photon and the composition of the medium, in particular on the atomic numbers of the atoms in the medium. Note that the half-value layer data presented in Table 2.14 vary tremendously over the energy range covered. The attenuation presented by the chest of an adult, which is equivalent to a thickness of about 10 cm of muscle, is equal to 10.5 half-value layers for 20 keV photons and 2.8 half-value layers for 60 keV photons. Since most of the energy in a diagnostic x-ray beam is carried by photons with energies below 60 keV, almost all the energy directed against the body is absorbed within the body. The actual dose imparted to any region exposed to a given x-ray intensity depends on the mass energy-absorption coefficient and, as we have seen from Example 2.23, is several times higher in bone than in adjacent tissue at diagnostic x-ray energies.

22.4 Effects of Photon Energy Distribution on Image Formation and Absorbed Dose

Let us consider in detail what happens when a beam of x-ray photons is used to take an x ray. The paths of individual photons are shown in Figure 2.23. The photons that penetrate without interaction will produce a pattern on the film that models the amount of matter through which they had to penetrate to reach the photographic plate. The less the amount of matter of a given type in any particular direction, the higher the intensity of the emergent beam and the greater the darkening produced in the film. *A radiologist adjusts exposure conditions to provide a darkening with a normal pattern that is optimum for showing up small changes in the penetrating radiation produced by abnormalities in the tissue that is traversed.* The

2.23 Image production with x rays. The x rays that reach the film interact with the emulsion and produce a latent image, which is brought out by development. The amount of darkening at any point in the developed image is a measure of the radiation exposure at that point. The darkening, generally referred to as the density, is measured with a densitometer. This instrument measures the transmission of a small beam of light through the negative (see insert) with a photocell. The density is defined as the common logarithm of the ratio of the reading for unexposed film to the reading for exposed film. The figure illustrates the density at various portions of the negative. Note that the density is low in the region where the incident radiation is intercepted by the inclusion and increases where a cavity has been introduced. A density of 1 means that the light transmission is 1/10 the light transmission through an unexposed part of the film.

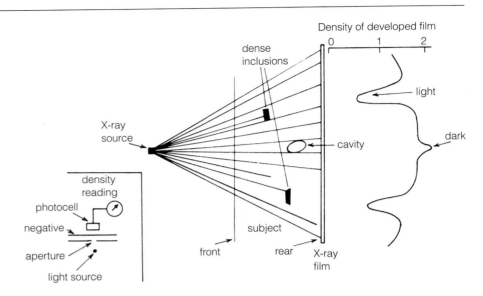

amount of x-ray energy required to produce a response on the film that satisfies the radiologist determines the exposure to that portion of the body of the patient that is adjacent to the film. Data on density versus exposure for x-ray films are given in Figure 2.24. Detailed discussions of the radiologic image and the retrieval of information are given in Hendee and Ritenour, 1992, and Sprawls, 1993.

In considering the exposure[35] and resulting dose to the patient, we must examine the exposure over the entire region of the body traversed by the x-ray beam rather than at the film alone. The dose to soft tissue is highest at the point the photons enter the body and decreases steadily to the point of exit. The dose at the entrance point is higher than the exit dose by a factor

35. Note the use of two terms in this paragraph to describe the level of x radiation: exposure and dose. Exposure refers to the ionization produced by the x rays in a special medium, air. Dose refers to the absorption of x-ray energy in unit mass of the region actually exposed to the x rays. At diagnostic x-ray energies, the absorption of photons is much greater in a material like calcium than in soft tissue, which has a much lower average atomic number. Thus the dose in calcium would be much larger than in tissue, for a quantity of x rays that produced a given exposure in air. When a small mass of soft tissue is contained in a calcium matrix (such as the Haversian cells in bone), the dose to the tissue is essentially the same as that to the surrounding bone, rather than that which would be evaluated if the whole volume were composed of soft tissue. The evaluation of the enhancement of the dose to soft tissue when it is near bone or other material of higher atomic number is complex, and the reader is referred to Attix and Roesch, 1968, for details.

2.24 Relationship between photographic density and radiation exposure. The densities generally acceptable in diagnostic radiology range between 0.25 and 2, corresponding to a range of transmitted light between 56 percent and 1 percent of the incident light. Note the high sensitivity of the medical x-ray film used with an intensifying screen as compared to the dental and personnel monitoring films. The image in the film-screen combination is produced mainly by the light emitted from the intensifying screen rather than from the direct interaction of the ionizing radiation with the photographic emulsion, as in the dental x-ray film.

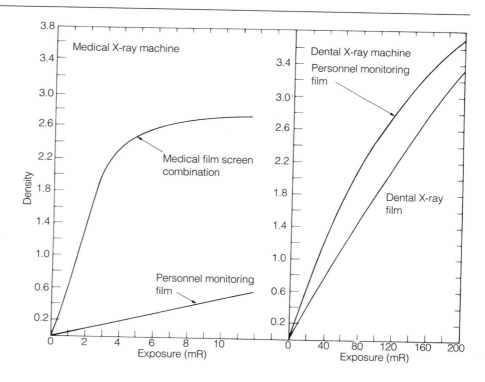

depending on the number of half-value layers through which the radiation penetrates. Thus, in the case of the 10 cm thickness of tissue considered in section 22.3, the attenuation of 20 keV photons (10.5 half-value layers) is $(1/2)^{10.5}$ and the dose to tissue on the entrance side is over two million times the dose on the exit side. If we make the same calculation for 60 keV photons (2.8 half-value layers), we find the entrance dose is seven times as high as the exit dose.[36] Since a minimum number of photons (with a corresponding exit dose to the patient) are required to provide an acceptable radiographic image, the entrance dose required for a given procedure can be reduced by using higher photon energies. The gains increase dramatically

36. The calculations using the half-value layers given in Table 2.14 are based on "good geometry," that is, neglect of the scattered radiation that builds up as the photons penetrate through the medium (see section 8.5 and Fig. 2.14). The effect of the buildup of the scattered radiation is to increase the half-value layer. The extent of the increase depends on the area of the beam and the thickness penetrated. Attenuation calculations that include the effect of the scattered radiation can be quite complicated, and it is often much simpler to make measurements of the effective half-value layer for specific radiation conditions (HPA, 1961).

as the thickness to be traversed increases (that is, for heavier patients). Other considerations must also be taken into account, however, in determining a photon energy.

A very important determinant in selecting a photon energy is the effectiveness with which a given photon energy will reveal an abnormality in the tissue. As an example, consider the effect on the beam of an abnormal mass of tissue that introduces additional mass with a thickness equivalent to 0.5 cm into the path of the photons.

For lower-energy photons, say 0.030 MeV, the attenuation coefficient is 0.34/cm and the fraction of photons removed from the beam in 0.5 cm is approximately 0.5×0.34, or 0.17. At 0.1 MeV, the attenuation coefficient is 0.161/cm and only 0.08 of the photons would be affected by the increased mass. Obviously, the change produced by the additional mass is greater at the lower energies. The results with lower energies are even better when materials with higher atomic number are examined, such as bone. Hence, we conclude that lower-energy photons provide better contrast and more detail in the image.

While the use of higher tube voltages results in lower entrance doses for a given film exposure, the gain is offset somewhat by the greater penetrating power of the higher-energy x rays. This consequence is particularly noteworthy in the case of dental x rays, since the film is placed within the mouth and the tissue behind the film is also exposed to the radiation. Now, one of the measures of the overall damage produced by an x-ray exposure is the total energy imparted by the ionizing radiation in the entire volume irradiated. This is generally referred to as the integral absorbed dose. The irradiation of additional tissue behind the film by more penetrating radiation, increasing the integral absorbed dose, may be more detrimental than the higher skin dose produced at the lower energies, unless the region behind the film is shielded.

Even when the film is located outside the body, as in medical x rays, the lowering of the entrance dose at higher voltages is offset somewhat because of the increased penetration. The total energy imparted to the tissue in the path of the incident beam is proportional to the product of the entrance dose and the half-value layer, when most of the energy is absorbed in the body. If the entrance dose is halved by going to a higher energy, but at the same time the half-value layer is doubled, the total energy imparted to tissue is approximately the same. In certain cases it may be difficult to make an accurate evaluation of the relative hazards of high localized doses versus total energy imparted to the irradiated region, especially when the localized doses are not high enough to produce apparent effects in tissue. However, the major consideration in adjusting the dials of an x-ray machine is to produce an x-ray image to the satisfaction of the radiologist, and imaging equipment should be sensitive enough to make consideration of the fine details of dose distributions primarily of academic interest. In any event, radiologists should be aware of the doses associated with various settings and should check that the entrance doses associated with their procedures do not depart appreciably from the lowest values that produce satisfactory results.

To this point, we have not taken into account the fact that x rays are never produced at a single energy but rather are produced over a whole range of photon energies up to the maximum energy given by the high voltage setting of the machine. For the lowest-energy part of the spectrum, the attenuation of the photons is so great that these photons contribute very little, if anything, to the image. All they do is produce very high doses at the point where they enter the body and for a short distance along their path. Obviously, the numbers of these photons in the beam should be minimized, and we shall show in a later section how filters are used to accomplish this.

22.5   A Description of an X-Ray Machine

The modern medical x-ray machine is engineered to enable the operator to define precisely the region of the subject that is irradiated and to prevent unnecessary exposure of either the operator or the subject. Let us examine the main components of an x-ray machine and their specifications for the production of x rays (Webster, 1995; Hendee and Ritenour, 1992; Sprawls, 1993). A schematic of an x-ray unit is given in Figure 2.25.

Source of electrons. A tungsten filament is heated with an electric current to approximately 2,000°C to emit electrons. The temperature is sufficiently high to produce incandescence. The greater the filament current, the higher is its temperature, and the greater are the number of electrons emitted. The x-ray machine comes equipped with a milliampere (mA) meter that gives the current emitted from the filament. One milliampere is equal to 6.25×10^{15} electrons/sec. Dental x-ray tubes generally operate between 5 and 15 mA, and medical x-ray tubes for radiography between 50 and 1,000 mA.

High voltage supply. This gives energy to the electrons emitted by the source by accelerating them in an electric field. The voltages for diagnostic x-ray machines are obtained by stepping up the voltage from the power company by means of a special transformer and associated electrical circuitry. Depending on the circuitry, the high voltage will vary between 0 and a peak value or be smoothed to a fairly constant value with only a slight ripple. Since the peak voltage is the value given, it is often designated as kV (peak) or kV_p. The higher the voltage applied to the x-ray tube, the

2.25 Components of an x-ray machine.

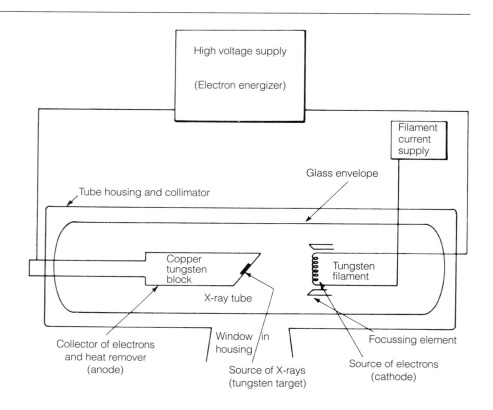

higher the energy achieved by the electrons. The energy of the electrons is specified in terms of the voltage through which they are accelerated. If the voltage is adjusted to 75,000 V, the electrons receive maximum energies specified as 75,000 electron volts (75 keV). X-ray tubes for diagnostic radiology generally have applied high voltages up to about 120,000 volts.

In radiation therapy, it is desirable to use high-energy x rays in treating cancers within the body, energies of millions of electron volts. The machine of choice to produce these energies is the linear accelerator ("linac"). The linear accelerator uses high-frequency (about 3,000 megacycles/sec) electromagnetic waves generated by a magnetron or klystron to accelerate the electrons through an accelerator tube to high energies. Medical linear accelerators are available commercially that produce beams of electrons or x rays up to energies greater than 20 MeV.

Target, including focusing system. The energetic electrons from the source are made to collide with a suitable target, such as tungsten, set in molybdenum or copper to conduct heat away. Source and target are enclosed in a

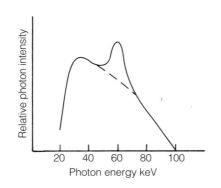

2.26 Spectrum of bremsstrahlung from thick target (constant potential tube, 100,000 V). Peak at 59 keV is due to characteristic tungsten K-radiation (ICRU, 1962).

sealed tube that is held under high vacuum so the electrons accelerated by the applied high voltage will not collide with gas molecules and lose energy or be deflected. The collisions of the electrons and their deflections in the vicinity of the target atoms result in the production of the useful x-ray beam.

The electrons are focused by appropriate shaping of the electrical field at the same time they are accelerated to the target so they will strike the target within as small an area as practicable. The smaller the source of x rays, the sharper the image produced.

Results of the measurement of the spectrum of photons emitted by one x-ray machine are given in Figure 2.26. The high voltage on the machine was 100,000 volts, and the target was tungsten. Note the peak at 59,000 eV caused by the characteristic radiation from tungsten (see section 22.1). The photons contributing to the peak constituted approximately 7.5 percent of the total number of photons emitted by a full wave rectified machine at 100,000 V and 2.5 mm Al total filtration (Epp and Weiss, 1966). At 90,000 V they contributed only 1.4 percent. For machines with a more constant waveform, such as a three-phase, twelve-pulse machine, the contribution of characteristic photons is several times higher.

Only a small fraction of the energy carried by the electrons incident on the target is converted into radiation. The fraction is given approximately by the product: electron energy (MeV) × atomic number of target × 10^{-3}. This formula holds for electron energies up to a few MeV.

As an example, consider the collision of 0.10 MeV electrons with a tungsten target. The atomic number of tungsten is 74. The fraction of energy converted is $0.1 \times 74 \times 10^{-3} = 0.0074$. Less than 1 percent of the electron energy is converted into radiation. The fraction of electron energy not converted and emitted from the target as x-ray photons is absorbed in the target and converted into heat. Targets are designed to accept and dissipate this highly localized heat through the use of features such as rotating anodes and water or oil cooling systems.

The operation of x-ray tubes must be restricted by limitations on tube current, operating time, and frequency of exposures to prevent damage from overheating. The limitations are supplied by the tube manufacturer as maximum heat storage capabilities and special performance and rating charts. There are separate limits for single exposures, a rapid series of exposures or short-term continuous operation of the machine, and long-term operation.

The limits for a single exposure are set to prevent local melting of the target surface. They are given in rating charts which specify the maximum exposure time with given tube current and voltage settings. For example, the rating chart for a Machlett Dynamax "64" x-ray tube with a 0.6 mm (diameter) focal spot, full-wave three-phase rectification and high-speed anode rotation specifies a maximum of 0.1 second exposure at 100 kV_p and 3,300 mA. A maximum exposure time of 3 sec is allowed if the focal spot is 1.2 mm. The limits for a rapid series of exposures or short-term continuous operation are established to prevent excessive temperatures of the anode, particularly at the tube seal or at bearings. The limits for long-term operation, say over a period of an hour, are determined by overall heating of the tube to prevent damage to the glass, insulation, and so on.

The accumulation of heat as a result of operation of the machine is described in terms of "heat units": the number of heat units (hu) is equal to the product of the peak kilovolts, milliamperes, and seconds. Limits in terms of heat units are specified separately for the tube anode and for the tube housing. The specifications for the Machlett tube previously described are 200,000 hu anode heat storage capacity at the maximum anode cooling rate of 54,000 hu/min and 1,500,000 hu housing heat storage capacity.

Example 2.24 An x-ray tube is operated at 70,000 V and 200 mA. A rapid series of 1/2-second exposures is contemplated. What is the maximum number that could be taken in succession without exceeding the anode heat storage capacity of 72,000 hu? How long could the tube be operated continuously at these settings?

Each 1/2 sec exposure adds $70 \times 200 \times 0.5 = 7,000$ hu. A total of 10 exposures could be taken without exceeding 72,000 hu. For continuous operation, 14,000 hu would be added per second, and the operating time would have to be limited to 5 seconds.

Over a longer period, credit can be taken for loss of heat units through cooling, and thermal characteristics charts are supplied with tubes showing the actual accumulation of heat units as a function of time. Housing-cooling charts are also supplied that show the loss of heat units through cooling as a function of time when the machine is not operating.

Medical x-ray machines are designed to give satisfactory cooling for normal operational settings and patient loads. However, one can conceive of situations in which a diagnostic tube would be operated for periods considerably in excess of recommended limits. For example, an inexperienced radiation-protection inspector might convince a medical practitioner of the desirability of measuring exposure rates around the x-ray machine. If neither individual were aware of operating limits, interest in the making of accurate and detailed measurements could lead to long operating times resulting in destruction of the tube—a result that could hardly produce a favorable association with radiation protection in the mind of the owner of the x-ray machine.

Collimator. When human beings are x rayed, the beam must be confined to the region under examination. There is no reason to expose tissue unnecessarily to radiation. Accordingly, modern x-ray machines have special collimators that can be adjusted to limit the beam to the area being studied. A good protective measure is to link the collimator to the film cassette to restrict its maximum area to the area of the cassette. In dental x-ray machines, it is not practical to use adjustable collimators, and standard practice is to collimate the beam merely so its diameter is less than 3 inches at the patient's face. This is done with a long cylinder that collimates the beam by means of a diaphragm located at the end that screws into the machine. The cylinder is open at the end to reduce the scattered radiation. The open cylinder is preferable to the pointer cone, which scatters radiation to parts of the face and body not under examination.

Filter. We have already mentioned that the x rays emitted from the x-ray tube target include many low-energy photons that do not penetrate enough to contribute to the image but can impart appreciable dose to the subject. The dose from these useless photons can be reduced through the use of selective absorbers, or filters, which are relatively transparent to the higher-energy photons, that is, those that actually produce the picture. For example, the addition of 2 or 3 mm aluminum attenuates 30 keV photons by a factor of about 2 while attenuating 60 keV or higher energy photons by less than 20 percent. The actual thickness and material of the filter needed depend on the voltage of the x-ray machine. Radiation-protection standards specify a minimum filter thickness equivalent to 2.5 mm of aluminum for diagnostic x-ray machines operated above 70,000 V peak.

Tube housing. All the components previously described are incorporated within a tube housing that prevents the radiation not emitted in the direction of the subject from irradiating the surrounding area. Radiation-protection standards specify that the leakage from the housing of a diagnostic machine is not to exceed 100 mR in 1 hr at 1 m from the target when the tube is operated at any of its specified ratings.

The amount of radiation actually emitted from an x-ray machine depends on a variety of factors, including operating current and voltage, filtration, absorption in the walls of the tube, and so on. It is important to have data on the exposure rates in the vicinity of the machine in order to know the dose imparted to the patient or to the operator for various procedures. The appropriate data are obtained by measurements around the machine under actual operating conditions, but estimates of the exposure rate from a machine under given operating conditions can often be obtained from tables. The data for any given set of conditions can be extrapolated to

2.27 Exposure rates from diagnostic x-ray equipment for a target-skin distance of 40 inches. Values are for a full-wave rectified single-phase machine. The exposure rates would be about 1.8 times higher if a multiple-phase high-voltage supply (approximately constant potential) were used. The values should also be increased by about 70 percent for fluoroscopic units, since at the low milliamperage the full-wave rectified waveform becomes quite similar to that for a constant potential (McCullough and Cameron, 1971). For comparison, the output of a dental unit at 20 cm is 140 mR per mAs for 70 kV$_p$, 1.5 mm Al and 90 kV$_p$, 2.5 mm Al (NCRP, 1968, 1989a). The rates in the figure apply to a point in air away from any scattering objects. They would be 20 to 40 percent higher at the surface of the body because of backscatter (Johns, 1969).

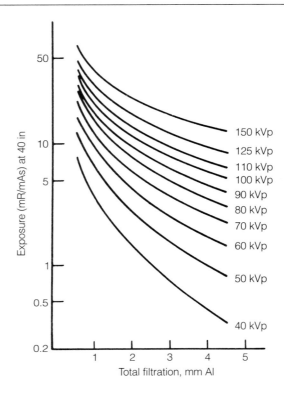

other conditions on the assumption that the exposure rate is directly proportional to the number of electrons hitting the target per second (that is, the tube current), and to the square of the electron energy (measured by the peak voltage). The voltage waveform is also a factor in determining the radiation output for given voltage and current settings. A three-phase or constant-potential machine produces approximately twice the exposure rate of a full-wave machine. The drop-off of the dose in the direct (or useful) beam with distance from the target follows the inverse square law, with an additional loss as a result of attenuation in the air.

Approximate outputs of x-ray machines are given in radiation-protection handbooks. Outputs of radiographic machines are generally expressed in terms of roentgens per 100 milliampere-seconds or milliroentgens per milliampere-second of machine operation at various operating voltages and distances. Outputs of fluoroscopes are given in terms of roentgens per milliampere-minute. Data on outputs of diagnostic radiographic x-ray equipment are given in Figure 2.27.

From the data in Figure 2.27, the exposure incurred in taking an x ray is estimated as follows: Obtain the distance from the x ray target to the skin, the milliampere setting, the voltage, and the duration of the exposure. Determine the milliampere-seconds (mAs) from the product of the current and time. Read the exposure for 1 mAs and the appropriate filtration and kV_p. (If the filtration is not known, use a nominal value of 2.5 mm Al.) Multiply the exposure by the actual mAs used. If the distance is different from the distance given in the figure, also multiply the exposure by the factor, (distance in figure)^2/(actual distance)^2. If the high voltage is outside the range in the figure, also multiply the exposure by the factor, (actual voltage)^2/(voltage in figure)^2. The value for the exposure is given for the condition in which surrounding surfaces do not scatter any significant radiation to the point. However, radiation will scatter from within the body of the patient. The total exposure at the surface, including backscatter, is obtained by multiplying the exposure in the absence of backscattering by the backscatter factor (BSF). The BSF is a function of the size of the field, the radiation energy, and the material of the scattering surface. Some values are given in Table 2.15.

Example 2.25 A modern community health care center reported the following exposure conditions for a chest x ray: target-patient distance, 6 ft; current, 300 mA; peak voltage, 114,000 V; exposure time, 1/120 sec; filtration (aluminum), 3 mm. Estimate the exposure to the patient from a chest x ray.

From Figure 2.27, exposure is 10 mR per mAs at 40 in. Since 6 ft = 72 in and 300 mA × (1/120 sec) = 2.5 mAs,

$$\text{Exposure} = 10 \text{ mR/mAs} \times 2.5 \text{ mAs} \times \frac{(40 \text{ in})^2}{(72 \text{ in})^2} = 7.7 \text{ mR}$$

This value is the exposure in air with the patient absent. The exposure at the surface of the patient would be increased about 40 percent because of backscatter (Johns, 1969, p. 27).

Data on organ doses from medical x rays are given in Part Three, section 9, and Part Six, section 5.3.

22.6   Production of a Photograph of the X-Ray Image

Production of the photographic image on film has come a long way since World War I, when the image was produced on a glass plate. The dose required for a radiograph decreased greatly when an intensifying screen of calcium tungstate was added. From there, double-coated film was

Table 2.15 Backscatter factors.

| | Backscatter factor | |
kVp	8″ × 10″ field	14″ × 17″ field
40	1.16	1.16
60	1.27	1.27
80	1.34	1.35
100	1.38	1.40
130	1.41	1.45
150	1.42	1.46

Source: Trout, Kelley, and Lucas, 1962.

developed along with double screens, and the speed of the process was greatly increased in the middle 1970s when "rare earth" screens, incorporating gadolinium and lanthanum oxysulfides, came into use. It is estimated that patient exposures have probably decreased by a factor of 50 to 100 since x rays were first used in 1896 (Webster, 1995).

With new developments in digital imaging, film is being replaced with nonfilm-based digital detectors, including photostimulable phosphors, charge-coupled device (CCD) sensors coupled to conventional phosphors, and solid-state semiconductor materials such as selenium and silicon (Webster, 1995; Seibert, 1995). The images are displayed on a computer screen, where they can be manipulated to highlight certain features, make quantitative measures, enlarge portions of the image, and even color it to get better views. The most rapid shift from film to digital imaging is occurring in the dental office, where a small, highly sensitive rectangular x-ray receptor is inserted into the mouth instead of the film to produce an instantaneous image that can be viewed both by the dentist and the patient. Dentists who have this equipment are vigorously advertising its virtues, including the reduction in patient dose by an order of magnitude. Film may not long survive in dental radiology with this type of competition.

22.7 Fluoroscopy

Just about a year after x rays were discovered, they were used to produce an image on a phosphor screen that converted the x radiation into light. The screens had low light output and the images had to be viewed in a darkened room with dark-adapted eyes. The images became brighter with improved screens, culminating in the invention of the image intensifier tube. The tube contained a photocathode juxtaposed against the screen, which converted the light from the fluorescent phosphor on the screen into electrons whose number at any point was proportional to the intensity of the light produced at the phosphor. The electrons were accelerated in the intensifier tube to high speeds by an applied high voltage, and focused on to another smaller phosphor, which, because of the high energies of the electrons and the smaller area of the screen, produced an image that was on the order of 500 times the original brightness. Viewing of this image no longer required dark adaptation. The technology continued to evolve with the introduction of improved screens, solid-state detectors, cine camera recorders, closed-circuit TV camera systems, and accessories to take spot films. Recently, computer analysis and manipulation of images introduced a powerful new tool for imaging the body, and there seems to be no end to the contributions that technology is making toward improving the visualization of the interior of the human body in "real time." Unfortunately, certain procedures result in high patient doses that are a cause of concern (Wagner, Eifel, and Geise, 1994).

The simplest parameter to describe the potential for the dose to the patient from a fluoroscopic examination is the exposure time. Fluoroscopy times vary widely for different examinations or procedures, as well as for the performance of the same examination on different patients. For example, one study (in England) reported durations of 3 minutes for a barium swallow exam; 10 to 20 minutes for a coronary angiogram; 36 minutes to insert a single stent into an artery in coronary angioplasty; and 51 minutes to insert 2 stents. To record an image through the cine mode, the operator must activate the fluoroscope with a pedal different from the one used in routine fluoroscopy, without the limitations in dose rate that are automatically imposed on routine fluoroscopy.

In angiography, because of the multiple cine frames exposed and the lengthy fluoroscopy, the exposed area of the patient receives a high dose. Typical entrance exposures are in the range 30 to 40 R/min for a frame speed of 30 per second and twice as high for 60 frames per second. The total cine exposure time in an examination is typically 0.5 to 1 min (Webster, 1995).

Regulatory agencies are concerned about the potential exposure to patients and staff due to increases in the radiation output capability of fluoroscopy systems and to extensive "beam on" times for therapeutic procedures using fluoroscopy for guidance of medical devices (interventional procedures). They report a general lack of awareness of the radiation output capabilities of the machines under different modes of operation and the impact of long "beam on" times and patient positioning on total radiation exposure. Examinations conducted by persons who are not radiologists and have not received special training are of particular concern.

Another area of concern is high-level-control (HLC) fluoroscopy. The high-level setting is an optional control provided on fluoroscopy machines with automatic exposure rate control for producing exposure rates higher than the maximum set by the machine for routine fluoroscopy. It is used, at the discretion of the radiologist, to improve image quality in angiography (reduce quantum mottle) and other procedures. While federal law requires that automatic brightness controls limit the output of the machines to 10 R/min under routine operation at the point where the center of the useful beam enters the patient, there were no limits for high-level fluoroscopy until 1995, although specific means of activation were required to safeguard against inadvertent use. In 1995 regulations were passed to limit the exposure rate for HLC to 20 R/min. Thus, fluoroscopy machines equipped with an optional HLC that were on the market before 1995 were able to exceed the federal limit by substantial amounts. In one study conducted at six academic medical centers in California, the machines had capabilities of maximum radiation exposure with HLC that varied from a low of 21 R/min to a high of 93 R/min, with a mean of 48.7 R/min, almost 5 times the federal limit for routine fluoroscopy (Cagnon et al., 1991). These maximum rates could be produced only under extreme situations, such as occurred with a very large patient, or an extreme projection angle, or both. For an average-sized patient, the exposures increased by 2.3–6.6 times under HLC. The institutions reported that HLC fluoroscopy was used less than 5 percent of the time, and in nearly all cases it was used only during percutaneous transluminal coronary angioplasty procedures. An angioplasty procedure that took 65 minutes would result, at the normal rate of 3.4 R/min, in an exposure at the position of the skin of 221 R, a significant amount. Under HLC, the fluoroscopy could result in doses that produced observable injury to the patient. In one case that was reported, a patient received an exposure of 800 R after going through three cardiac catheterization procedures to remediate a blockage within the heart. After two failed attempts, the third procedure was successful in repairing the heart, but some time later, skin reddening appeared on the upper right area of the back. The wound became infected and required a skin graft (Angelo et al., 1994).

The Food and Drug Administration Center for Devices and Radiological Health (CDRH/FDA) receives reports of radiation-induced skin injuries from fluoroscopy, primarily under mandatory reporting requirements imposed on manufacturers and users of medical devices under the Safe Medical Devices Act of 1990 (FDA, 1990). Serious skin injuries reported include moist desquamation and tissue necrosis, and all resulted from interventional procedures requiring periods of fluoroscopy longer than the duration of typical diagnostic procedures. Of 26 reports, 12 resulted from radiofrequency (RF) cardiac catheter ablation, 4 from coronary angioplasty, and the remainder from a variety of other procedures (Shope,

1995). A Public Health Advisory of the FDA addressed to healthcare administrators, risk managers, radiology department directors, and cardiology department directors was concerned with the need for physicians to be aware of the potential for serious, radiation-induced skin injury caused by long periods of fluoroscopy (FDA, 1994). It included the following recommendations to reduce the potential for radiation-induced skin injuries:

- Establish standard procedures and protocols for each procedure, including consideration of fluoroscopy exposure time.
- Determine the radiation dose rates for specific fluoroscopy systems and for all operating modes.
- Assess each protocol for the potential for radiation injury to the patient.
- Modify protocols, when appropriate, to minimize cumulative absorbed dose to any specific skin area and use equipment which aids in minimizing absorbed dose.

The FDA also recommended recording absorbed dose to the skin for doses above a prescribed level. The FDA suggested a threshold absorbed dose to the skin of 1 Gy. Typical threshold doses for various effects are about 3 Gy for temporary epilation, about 6 Gy for main erythema, and 15 to 20 Gy for moist desquamation, dermal necrosis, and secondary ulceration.

It has been suggested that the operating physician should be able to see a digital readout of the dose being accumulated by the patient during a procedure so that he could take measures, if possible, to prevent unusually high exposures without jeopardizing the success of the procedure.

22.7.1 Training Requirements for Operators of Fluoroscopy Machines

Some states have specific educational requirements for operators of fluoroscopic equipment. Even in states that do not regulate the use of fluoroscopy by physicians, medical institutions may require that any physician who uses fluoroscopy and cine machines undergo training as part of a credentialing program. Topics covered in such training should include how to operate the machine, properly position the patient, minimize the use of radiation, minimize dose to the patient in the operation of the machine, properly use shielding devices and personnel monitoring devices, and be aware of the radiation levels in the treatment room. The training session should also provide the physician with enough understanding of the operation of the machine, the production of the image for diagnosis or guidance of interventional procedures, and the biological effects of radiation to make consideration of the radiation protection of the patient by the operator of the machine an integral part of the examination. Of course, the successful performance of the procedure is the primary concern in the use of fluoroscopy, and it should be noted that serious radiation injury has been produced in only a very small fraction of all fluoroscopy examinations (Wagner and Archer, 2000; Miller and Castronovo, 1985).

22.8   Mammography

X-ray mammography utilizes equipment and techniques especially developed for the detection of breast cancer. The goal is the identification of small abnormalities, including areas of calcification in soft tissue, and the technique requires the use of low-energy photons.

Routine mammographic screening has become a key element in public health programs for the early detection of breast cancer. Following are some of the considerations that enter into the performance of a modern mammography examination (Hendee, 1995):

Breast compression. Breast compression produces a greatly improved image by increasing contrast, minimizing blurring of the image caused by motion, spreading tissue structures, and minimizing the distance between breast tissue and the image receptor. It also can reduce absorbed dose by 25–50 percent.

X-ray machine. Mammography requires a special x-ray machine to show up the changes in the breast that accompany the beginning stages of a tumor. To be affected by the small changes in breast tissue, the incident radiation must be very low energy. Thus, the x-ray tubes are operated at voltages below 28 kV. They have typically a molybdenum ($Z = 42$) target and molybdenum filter (0.03 mm), in contrast to the tungsten ($Z = 74$) target of a conventional x-ray machine. Figure 2.28 (Hendee, 1995) shows the spectra for an x-ray machine with a molybdenum target and 0.03 mm molybdenum filter operated at 25 kV_p, and for an x-ray machine with a tungsten target and aluminum filter operated at 35 kV_p. The curves are adjusted so that each spectrum is produced by the same number of photons. The spectrum for the molybdenum target machine shows two sharp peaks for the K x rays (see Fig. 1.2) of 17.9 and 19.5 keV, and produces excellent contrast for breasts compressed to a thickness of 3–4 cm. Other filters work better for thicker breasts.

Film screen package. Intensifying screens and film designed specifically for mammography produce much better resolution than those used in routine radiography. In contrast to the double-emulsion double-screen used in routine radiography, mammographic systems use only a single-emulsion x-ray film held in close contact with a single high-resolution intensifying screen within a vacuum cassette to give much sharper images.

2.28 X-ray spectra (normalized to equal area) for a molybdenum-target dedicated x-ray unit and a tungsten-target nondedicated x-ray unit for mammography. The dedicated unit employs a molybdenum (Mo) target, 0.03 mm Mo filter, and 25 kV$_p$; the nondedicated unit uses a tungsten target, aluminum filtration, and 35 kV$_p$. (*Source:* Hendee, 1995).

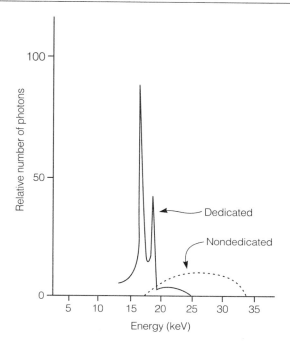

Grids. Mammographic units are normally equipped with moving grids to reduce scattered radiation and significantly reduce image contrast. The grid is composed of very thin (<0.1 mm) lead strips with separators between that are essentially transparent to the radiation. While the use of grids improves image quality, they also cause an increased radiation dose to the patient, often by a factor of 2 or more. Even greater reduction of the scattered radiation can be obtained with two sets of moving slits, one between the breast and the x-ray tube, and the other between the breast and the image receptor. This arrangement results in the scanning of the breast by a number of long and narrow x-ray beams to obtain the image, so that only a very small volume of breast tissue is exposed by any single x-ray beam, but the method is still experimental (Hendee, 1995).

Film processing. Film processing is critical for obtaining optimum results in mammography. Factors requiring care and attention include the use of proper chemicals, replenishment of developer, control of development time, maintenance of constant developer temperature, and a program of quality control and processor maintenance. Certain combinations of developer temperature and increased film-processing time result in an increase in film speed and contrast and a corresponding decrease in radiation dose by as much as 35 percent.

Quality control. Even the use of the best equipment will not produce reliable performance in a mammography program without close attention to quality control. Mammography is technically very demanding and requires close attention to every step in its performance. Accrediting organizations and regulatory agencies are important elements in implementing effective programs. The American College of Radiology has developed a Mammography Accreditation Program (MAP) to certify sites with approved installations, practices, and controls. The MAP program takes a close look at the installation, including the credentials of the staff, the equipment and its use, the quality control procedures, typical mammograms for evaluation by a panel of radiologists, and radiation doses (Hendrick, 1993). The federal government is empowered to set standards for mammography programs through the Mammography Quality Standards Act (MQSA) of 1992. The Center for Devices and Radiological Health of the U.S. Food and Drug Administration (FDA) is responsible for implementing a federal accrediting program and accepts accreditation awarded by the American College of Radiology as well as several states.

Radiation doses in mammography. Patients who are followed carefully over many years in order to detect breast cancer as early as possible are subject to repeated mammographic examinations, which add to the radiation dose accumulated by the breast. The low-energy photons used to produce a mammogram are attenuated rapidly in the breast, and entrance exposures must be much higher than would be needed using conventional x-ray technique. The dose is increased further if film is used alone to get as sharp an image as possible than if high-sensitivity film-screen combinations are used. Thus mammography examinations can result in exposures at the skin of many roentgens, if special measures are not taken to reduce doses without compromising the effectiveness of the examination. Considerable development work has been done on methods to reduce these exposures. Rare earth screen-film combinations reduce the dose to the breast per examination dramatically, to the milligray range. Xeromammography, in which the film is replaced by a positively charged selenium plate similar to that used in a Xerox machine, was also used in the 1980s, but it is no longer a significant imaging mode because the doses are considerably higher than those imparted in mammography. Breast compression, used to improve contrast and sharpness by preventing motion, also can reduce absorbed dose by 25 to 50 percent. While the different systems and techniques used in taking mammograms produce varying levels of dose to the breast of the patient, the nature of the images may differ significantly, and the radiologist's choice of machine settings that give the best image for detecting breast cancer is the major factor in determining the patient dose. While there may be a small risk of inducing cancer from repeated mammograms, studies have concluded that the benefits of screening in detecting and treating early cancers considerably outweigh the radiation risk to the patient.

22.9 Computed Tomography: Taking Cross Sections with X Rays

Each point on a conventional radiographic image represents the penetration of x rays from a point source through the various media encountered on the way to the film, and is thus the composite effect of the different media along the path of the beam. It often requires consummate skill on the part of the radiologist to piece together the actual anatomical detail from the two-dimensional image presented on the film. Frequently two or more views are taken, typically at 90° to each other, to help identify features that may be superimposed and therefore obscured in a single view, but some subtle abnormalities may be impossible to detect. It took the advent of computerized tomography to provide radiologists with an instrument that provided them with essentially a three-dimensional view of the interior of the body.

Computerized tomography had its beginnings in 1957, when Allen Cormack, in the physics department at Tufts University, analyzed the problem of determining the interior of an object from measurements of the transmission of a beam of radiation through it at multiple angles. He derived equations that provided the details of the image and followed his theoretical analysis with confirmatory experiments in 1963. In 1972, largely through the development work of Godfrey Houndsfield at Electro-Music, Inc. (EMI) in England, the first x-ray machine based on computed tomography made its appearance. It involved mating x-ray scanning and digital computer technology. Cormack and Houndsfield shared the Nobel Prize in 1979 for the development of a technology that was as revolutionary for medical imaging as the original discovery of x rays by Roentgen.

The first CT units incorporated a single x-ray beam and a single aligned scintillation detector that rotated around the sample and then moved in translation in small increments to accumulate the data to visualize a slice. Improvements throughout the years produced wide angle beams that covered the whole slice and detectors surrounding larger and larger portions of the object to minimize rotation time and increase resolution. In the latest units, only the x-ray fan beam rotates and the scintillation detectors completely surround the object and are stationary. Scan times per slice are typically 1 second (Webster, 1995).

A diagram of one type of scanner is given in Figure 2.29 (Swindell and Barrett, 1977). The scan is made on a slice through the body 3 to 5 mm thick. Each element of area in that slice is traversed by pencils of x rays en-

2.29 A third-generation CT scanner. The x rays are formed into a fan of pencil beams that encompasses the section of interest. With the source-detector assembly rotating uniformly about the patient, the required set of exposures is obtained by flashing the x-ray tube on at the appropriate angular positions. In some versions only the source moves. (*Source:* Swindell and Barrett, 1977.)

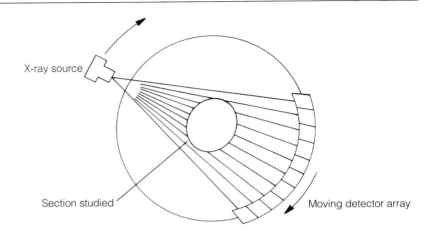

tering from more than 100 different angles. The attenuation presented by the slice to all these rays is measured with a battery of about 300 electronic detectors, each only a few millimeters in diameter. Corresponding data are obtained for all the elements of area in the slice, and the data are incorporated into a series of simultaneous equations from which one can solve for the density of every element of the slice. The results are then plotted by computer to give an accurate rendition of the cross section through which the tomograph is taken.

The challenge to produce dynamic CT scans to follow events in electrocardiography was met with a unit that could perform the scan in 50 milliseconds with a repetition rate of about 20 per second. For this specialized use, such sequential images would be required for about eight 10 mm slices, to cover, for example, the left heart ventricle. In these scans, a focused electron beam is rotated magnetically to hit the x-ray targets and produce a rotating x-ray beam.

Another major innovation in CT machines is the spiral or helical computerized tomography, such as used in pulmonary studies. In spiral volumetric CT, the usual serial scanning of adjacent anatomic slices is replaced by a continuous rotating scan of the tube and detector as the patient is slowly transported through the scanning beam..

Doses are difficult to specify because of the complex motions and the geometry of the beam. The exposure dose for a complete head scan study is between 1 and 2.5 roentgens and compares in value to a single skull film (McDonnel, 1977). Since a complete skull series includes about seven films, the CT dose is less than the dose in film radiography for this examination. Body scans include perhaps four to six slices per patient and deliver about 3 roentgens per scan, although this varies considerably among different machines.

The regulation of computed tomography falls under the diagnostic x-ray standard, published in the Code of Federal Regulations, Title 21, Part 1020.30 (21CFR1O20.30), but the specifications were not designed for the control of a beam that has a spatial extent of the order of millimeters at distances of the order of a meter from the source, and special standards and considerations will have to be developed.

22.10 Technical Approaches for Minimizing the Doses Required to Record an X Ray

Methods for minimizing dose in photographic recording systems are obvious. Increase the sensitivity of the emulsions to either the x radiation or the light, depending on the system used; increase the light output from the fluorescent screen; find faster developers for the film. The fastest film-screen combinations can produce chest x rays with a skin dose to the patient of the order of a tenth of a milligray. However, the quality of the image obtained with the most sensitive techniques is not as good as that obtained with some of the slower methods, and so the choice of the film cannot be made on dose considerations alone.

If the cost or complexity of the recording system is not a factor, it is possible to use electronic recording and data processing systems to minimize dose still further and to increase the amount of information extracted from the beam. Examples of sophisticated electronic readout systems include:

Image intensifier tubes with fast photographic film. Image intensifier tubes are electronic devices that amplify the light signal by transforming the pattern of light photons to a pattern of electrons, accelerating the electrons to high speeds, and producing another light image of greater intensity than the original by directing the more energetic electrons against a fluorescent screen. The resulting image may then be recorded on fast film (16, 35, or 70 mm).

Image intensifier tubes with television recording. The recording of the output of an image intensifier tube on videotape by a television camera provides the most sensitive system for taking x rays. Radiographs of the trunk can be made with less than 1 mR of exposure.

Image enhancement of grossly underexposed radiographs with flying spot scanner. The x ray is deliberately underexposed to reduce the patient dose. The desired information is still in the image but requires processing by special electronic techniques to extract it from the negative. The negative is scanned by a very narrow light beam and the resultant pattern is processed by a computer programmed to provide the desired information. This technique has been employed extensively in the study of aerial photographs for military purposes and, in principle, can be used effectively in radiography.

The methods of image intensification apply to fluoroscopy as well as radiography. The recording of the image on TV tape is especially powerful for fluoroscopy, since it provides a permanent record for continued study with a minimum of exposure. Certainly as resources become available, the "one-time" viewing of fluorescent screens by the physician in fluoroscopy should be replaced by permanent records on TV tape.

The use of pulsed methods provides even more dramatic reductions in dose. The fluoroscope is pulsed briefly and the x-ray image is converted to a television frame and electronically stored on a magnetic disc (expensive and extremely delicate) or a silicon image storage tube (inexpensive, reliable, but image not permanent). It is then instantly replayed as a frozen image over the television monitor with the x-ray beam off. It can also be photographed on a remote TV monitor. The rate of pulsing the fluoroscope can be tailored to the dynamics of the study. A rapidly changing process can be examined with more frequent pulses than one changing slowly. For example, barium enema studies of certain segments of the G-I tract characterized by little motility require exposures only every second or so to provide adequate information concerning the flow of contrast medium. Radiation exposures can be matched to the detail and contrast needed (to the extent that these are determined by photon statistics). If only gross structure is of interest (such as in x-ray images of the G-I tract of a child to search for a swallowed safety pin), much fewer photons (and much lower exposures) are needed than in the search for a possible tumor in the breast.

The possibilities for electronic processing are unlimited. Stored images taken previously can be superimposed on live images. In selective catheterization procedures, electronic radiography has resulted in a much lower radiation dose, reduced procedure time, and reduced amounts of injected radiopaque contrast as compared to conventional fluoroscopy, which can give patient exposures as high as 100 R. There are corresponding substantial reductions in the exposures of radiologists and cardiologists, who receive as much as 5 R/yr occupational exposure in the course of performing these examinations with conventional fluoroscopy.

Extremely low dose systems for medical radiography are being developed based on x-ray techniques used for screening baggage at airports. The patient is scanned with a fine (that is, 1 mm) beam and the transmitted photons are detected with a collimated NaI scintillator. The data are processed and stored by means of a mini-computer, and the contrast and magnification can be adjusted upon displaying the image. The entrance exposure required to x ray humans is about 0.3 mR and the exit exposure is less than 0.03 mR. This is between 100 and 1,000 times smaller than the exposure required for screen-film radiography. The system does not have the resolution of the conventional radiography systems and the exposure time is much longer, so it has limited applicability in routine medical radiography at present.

Doses to the patient in dentistry have been reduced substantially over the years. Technical improvements to reduce dose have included use of faster x-ray films, smaller beam areas, beam collimation, added filtration, increased target-to-skin distances, leaded aprons, thyroid shields, and optimal kilovoltage. Dental radiographic film has a standard film backing of 0.001 mm of lead, which significantly reduces the dose to tissues behind the film. Studies have been made of the effectiveness of increasing the thickness of lead. Tripling this thickness by adding another couple of layers of lead behind the film reduced the dose to the tongue from $4.49\ \mu$Gy to $1.53\ \mu$Gy in a dental x-ray taken at 70 kVp, 10 mA, and 0.6 sec with rectangular collimation (Bourgeois et al., 1992). The additional protection actually provided to the patient would depend on how much radiation is not intercepted by the film with its additional shielding as compared with the radiation penetrating the region shielded by the film backing. The most exciting advance in dental radiography is the use of digital scanning systems, which are gaining in popularity. Dentists who purchase them can justifiably extol their advantages in view of the very low doses they impart, the instant images they provide, and the elimination of the use of dental film and film processing.

It is obvious that the amount of radiation required to produce a suitable image cannot be reduced indefinitely. The limit would be zero exposure, and one cannot take an x-ray picture without an x-ray beam. For meaningful results, a minimum number of x-ray photons must penetrate through the patient. The actual number depends on the size of the area that is examined and the minimum contrast that is acceptable. Studies of image quality, image clarity, and visual perception are an important area of the physics of medical imaging (Hendee and Ritenour, 1992; Sprawls, 1993).

22.11   Impact of the Digital Computer in Radiation Medicine

The digital computer has provided diagnostic radiology with an incomparable resource for imaging the human body. The information provided by an analogue image is replaced by a multitude of electrical signals that are quickly and easily stored, manipulated, analyzed, and retrieved. Images can be clarified and selectively portrayed with a procedure called digital subtraction angiography, in which the unwanted signals can actually be subtracted out of the image. Use of multiple energies and their selective absorption in tissue can also cancel out unwanted interferences in radiological images. The film/screen receptor system, which is the backbone in current diagnostic radiology, continues to lose ground to other technologies. Will it go the way of the slide rule? Its main advantage right now is its cost. Should that advantage disappear, the x-ray film may well join the slide rule.

On the other hand, statistics show a dramatic increase in the number of high-dose examinations in the United States that result from x-ray examinations conducted so as to exploit to the full the wealth of information that can be provided by the computer. Are all these high-tech exams warranted by the requirements for the diagnosis? Will their increased use have a positive impact on health that exceeds the possible detriment of the increased dose? That question still needs an answer.

23   Dose Measurements in Diagnostic Radiology

Many studies have been made of exposures and doses to patients in the course of medical examinations and procedures (UNSCEAR, 2000). In the United States, the Food and Drug Administration (FDA) has assessed radiation exposure levels in diagnostic x rays since 1973. The data are obtained under the Nationwide Evaluation of X-ray Trends (NEXT) program, a joint effort of the Center for Devices and Radiological Health (CDRH) of the FDA and the Conference of Radiation Control Program Directors (CRCPD), an association of state regulatory officials. The NEXT program concentrates on a single radiological examination each year, obtaining up-to-date information on clinical practice, patient workload, and patient dose. The states provide the personnel to do on-site surveys and perform the measurements on the x-ray equipment.

Surveys performed by medical physicists in radiology departments provide another source of exposure data. The doses are often measured in response to inquiries from the Radiation Safety Officer, patients, or the Radiation Safety Committee and may be published in professional journals or on the Internet.

A common measurement of radiation associated with the taking of an x ray is the exposure at a point in air at which the x rays would enter the patient, the skin-entrance plane, but with the patient absent. This is called the Entrance Skin Exposure (ESE) Free in Air and is intended to represent as closely as possible the output directly from the machine without the addition of any radiation scattered from the patient or nearby surfaces. The exposure is recorded along with the relevant operating data for the machine, the kilovoltage, milliamperes, exposure time, filtration, and so on. In a laboratory environment, the exposure is generally measured with an air ionization chamber. Measurements in the field are most practical using thermoluminescent detectors that are closely equivalent to air ionization chambers in their response to radiation. The detector may be calibrated to give the exposure in milliroentgens, or the air kerma, the dose imparted to the air, expressed in milligrays. The two values differ by a constant: the dose to air in mGy = exposure in mR $\times$ 0.00873. Some measurements of the exposure are reported in terms of the actual charge produced per unit mass in the air of the detector, that is, in coulombs per kilogram.

X rays may be taken "automatically"—not by the manual setting of controls, but by the use of automatic exposure controls that set the time to produce an acceptable image irrespective of changes in the attenuation of the x-ray beam in the patient. To obtain a nominal value of exposure under these conditions, a "phantom" (a physical model of the body or body part to be x-rayed) is placed in the x-ray beam to simulate a patient and the actual exposure time is noted. This time is then used to calculate the free-in-air exposure for the simulated x ray of the patient from the free-in-air exposure vs. time data for the machine.

Measurements are also made by placing the detector on the phantom, thus recording the actual dose to the skin. This method is used in the NEXT program in its surveys of doses imparted in mammography. A standard breast phantom is used, representing a breast compressed to 4.2 cm and consisting of 50 percent adipose and 50 percent glandular tissue. The skin dose is then used to calculate the mean glandular dose to the breast, which is the quantity usually reported for the patient dose. An actual mammogram of a 4 cm "average breast" using the screen-film technique might produce an exposure at the skin of 1.27 R. From this figure would be derived an actual dose to adipose tissue of 4.5 mGy (0.45 rad), an average glandular dose of 0.61 mGy (0.061 rad), and a glandular dose at midplane of 0.46 mGy (0.046 rad: NCRP, 1986, Report 85).

It is more difficult to specify typical doses imparted in fluoroscopy than in the taking of a simple x ray because the nature of the examination varies depending on the information required by the radiologist or the surgical procedure guided by the fluoroscopy. Fluoroscopy machines are normally operated using automatic brightness controls, in which the output of the machine changes automatically as the examination progresses to maintain an image of consistent quality. Other factors—such as continuous changes in beam direction, field size, and positioning area on patient viewed; taking of spot x-ray films; use of cine (recording of the fluoroscopy presentation on film); and the actual viewing time—also operate to cause considerable variability in patient doses for a given procedure.

A basic quantity used to characterize radiation exposures in fluoroscopy is the exposure rate at 1 cm above the table top, with a phantom representing the patient in place. It is calculated from measurements made in air with an ionization chamber, placed about 3–4 cm from the phantom to minimize the contribution of backscatter (21CFR1020.32). In addition to the exposure rate obtained with the presence of a standard phantom, the maximum exposure rate delivered by the machine is determined by adding a sheet of lead or copper in the beam to absorb the radiation and force the automatic brightness control to set the machine controls to produce maximum radiation output in striving to obtain an image. The machine high voltage and current and other relevant operating data are also recorded. The results are used to assess doses imparted to the patient in various fluoroscopy examinations.

The radiation imparted in computed tomography also presents a very considerable challenge in dose assessment because of the many variables associated with the performance of the examination. Two quantities usually determined are the *computed tomography dose index* (CTDI) and the *multiple scan average dose* (MSAD). The CTDI is defined as the average dose along the central slice for a series of 14 contiguous scans and its determination is required of manufacturers of CT systems under federal regulations. The MSAD is the average dose along the central slice for a series of scans in an actual examination. The MSAD does not differ greatly from the CTDI, even if it includes less than 14 slices, if the outermost slices in an examination do not contribute significantly to the center dose. The measurements are made with the use of a phantom that simulates the patient while the designated number of slices is performed. Two positions used are a point near the surface of the phantom and one in the center. The average dose over the central slice can be determined with an array of thermoluminescent dosimeter (TLD) chips placed along the slice that record the total exposure as the beam revolves around the patient and as successive incremental scans are taken, until the total region of interest is scanned. The average central slice dose for a multiple-scan series can also be derived from the complete dose profile produced by a single scan. This profile is duplicated at each of the displaced slices that contribute to a complete exam. The contributions from each of the displaced profiles to the central slice are then added to give a composite curve at the central slice identical to the curve produced from the contributions from the complete multiple-scan exam (AAPM, 2001). The results may be reported as exposure in milliroentgens or as air kerma in milligrays.

There is an easy way to determine the MSAD with reasonable accuracy from just a single scan. This is by measuring the exposure produced in the single scan with a pencil ionization chamber, normally with a sensitive length *(L)* of 10 cm (Knox and Gagne, 1996; Shope, Gayne, and Johnson,

1981). This technique is based on the relationship that the area under the multiple scan dose profile over the width *(T)* of the central scan is essentially equal to the total area under a single-scan dose profile, provided that the first and last scans of the series do not contribute any significant dose over the width *T* of the central scan of the series and the distance between scans is equal to the slice thickness. The pencil ionization chamber is placed in the phantom aligned in the direction of the successive slices of a multiple scan and with the center of the chamber positioned at the central scan. The single scan is performed, exposing the chamber to both the direct radiation delivered to that portion of the chamber in the central slice beam and the scattered radiation covering the remainder of the chamber. The charge produced in the whole chamber by the single scan is equal to the charge that would have been produced in the portion of the ion chamber covered by the central scan in a multiple scan examination along the length of the chamber. Since the pencil chamber is calibrated in a uniform radiation field along the whole chamber, the reading equals the actual exposure only if the entire chamber is exposed uniformly. Because the reading is produced by an incident beam whose width is equal to the slice thickness *(T)* that covers only a portion of the chamber, the exposure reading produced by the single scan must be multiplied by the number of scans required to irradiate the whole chamber, *L/T,* to give the actual exposure to the central slice from a multiple scan examination. The exposure reading (in roentgens) may then be converted to the multiple scan average dose (MSAD) by the equation

$$\text{MSAD} = f \times C \times E \times L/T$$

where *f* is the chamber exposure to tissue dose conversion factor (mR to mGy), *C* is the calibration factor for the pencil chamber in a uniform beam (chamber reading to mR), *E* is the chamber reading, *L* is the active length of the chamber, and *T* is the width of the slice.

Exposures and doses obtained from the NEXT program are given in Table 2.16. Additional exposure data may be obtained on the Internet (for example, at *www.mayohealth.org*). Doses to organs in the body calculated from the exposure measurements are given in Part Three.

24 EXPOSURE GUIDES AND REFERENCE LEVELS IN DIAGNOSTIC RADIOLOGY

In addition to giving information on the dose imparted to patients from diagnostic x rays, measurements of exposure are used to monitor trends in doses to the population and to establish reference values for medical exposures made in accordance with a specified protocol. Exposures that are

Table 2.16 Exposures in radiological examinations in the United States (mean values) from NEXT Program data.

Body area examined	Exposure "free in air" at skin entrance (mR)[a]	Comments
Chest, 1970	47	
Chest, radiographic, 1984	16	Hospitals
Chest, radiographic, 1994[b]	16.1	All facilities
Chest, radiographic, 2001	14.2	All facilities (preliminary, D. Spelic, pers. comm., 2002)
Chest, pediatric, 1998	5.9	Preliminary (D. Spelic, pers. comm., 2000)
Abdomen, 1970	910	
Abdomen, 1983	500	
Abdomen, 1995	321	All facility types; 100% grid usage by radiologists; 99% by hospitals; 92% by private practitioners
L/S spine, 1987	424	Hospitals; grid use over 98%; film speed of 400
L/S spine, 1995	367	All facility types
Extremities, 1970	100	
Dental, 1970	900	
Dental, 1983	300	
Dental, 1993	218	Intraoral bitewing
Dental, 1999	209	Intraoral bitewing (preliminary, D. Spelic, pers. comm, 2000)
Mammography, 1988	MGD=1.30 mGy	Mean glandular dose (MGD), screen-film, 4.2 cm phantom, up from 0.93 mGy (4.7 cm phantom) in 1985
Mammography, 1992	MGD=1.50 mGy	Screen-film, 4.2 cm phantom
Mammography, 1997	MGD=1.60 mGy	Screen-film; all facility types; data obtained from inspections conducted
2001	MGD=1.76 mGy	under the Mammography Quality Standards Act; data for 2001 preliminary (D. Spelic, pers. comm, 2002)
Xeromammography, 1988	MGD=4.0 mGy	
Fluoroscopy, upper GI, 1991	4,900 mR/min	
Fluoroscopy, upper GI, 1996[c]	5.200 mR/min	Exposure rate 1 cm above table top (using fluoro phantom)
CT Head, 1990	MSAD=46 mGy	Multiple scan average dose (MSAD)obtained from measurements with
CT Head, 2000[d]	MSAD=50 mGy	pencil ionization chamber in central interior hole of head phantom

Source: Data from 1984 or later taken or calculated from Suleiman, Stern, and Spelic, 1999; 1983 data from Johnson and Goetz, 1986; 1970 data from BRH 1973.

a. Where data are given as average skin entrance kerma, they are converted to mR by the relationship, mR=114.5 × mGy.

b. R. V. Kaczmarek, B. J. Conway, R. O. Slayton, and O. H. Suleiman. Results of a nationwide survey of chest radiography. *Radiology 215*:891-6(2000).

c. From draft of "Summary of 1996 Fluoroscopy Survey," R. Kaczmarek, Center for Devices and Radiological Health, September 2000.

d. Preliminary, S. H. Stern, R. V. Kaczmarek, D. C. Spelic, and O. H. Spelic, poster presentation at Annual Meeting (2001) of Radiological Society of North America. Included calculations of effective doses of 2 mSv for head CT; 14 mSv for abdomen-pelvis CT; 7 mSv for chest CT and 7 mSv for abdomen CT.

above the reference values should be investigated to determine if they can be reduced by changes in the operation of the equipment. However, reference values are not meant to be used to indicate whether a procedure is acceptable or unacceptable. The radiologist has the final say and may find that the higher values are necessary in a particular procedure to provide the desired diagnostic information.

Reference values have been published in Europe by the European Com-

Table 2.17   American Association of Physicists in Medicine reference values for
diagnostic x-ray examinations.

PA chest	25 mR
AP cervical spine	125 mR
AP abdomen	450 mR
AP lumbar spine	500 mR
Dental bitewings, 70 kVp, E speed	230 mR
Dental bitewings, 70 kVp, D speed	350 mR
Cephalometry	25 mR
CT head	6,000 mR
CT body	4,000 mR
Fluoroscopic rate	6,500 mR/min

Source: AAPM, 2001.

Notes: The reference values selected by the American Association of Physicists in Medicine (AAPM) were at the 75–80 percentile of the NEXT survey results, so 20–25 percent of the facilities are likely to exceed the reference values for a particular procedure. These facilities need to investigate the reasons for the higher exposures.

The exposures for all diagnostic x rays are entrance skin exposures (ESE) free in air (without backscatter), except for CT. The CT value is the computed tomography dose index (CTDI), measured in the center of a specified phantom for the head CT and in the outermost hole of the phantom for the body CT.

munity, the National Radiological Protection Board of Great Britain, the Royal College of Radiologists, and the International Atomic Energy Agency. In the United States, the Conference of Radiation Control Program Directors has issued reference values as Patient Exposure Guides. Reference values have also been incorporated by the American College of Radiology into its accreditation programs and some states have included maximum exposure levels in their radiation-control programs. The American Association of Physicists in Medicine (AAPM) has established reference values based on the NEXT data. Some values are given in Table 2.17.

25   PROTECTION OF THE PATIENT IN X-RAY DIAGNOSIS

The key rule in the administration of radiation for diagnostic purposes is to obtain the information required with minimum risk of harm from exposure to the radiation. There is no consensus on the magnitude of the risk at doses encountered in most diagnostic examinations that use x rays or radiopharmaceuticals. The literature provides evaluations that range from quantitative assessments of risk as a function of dose to articles that write the exposures off as of no consequence. These matters are discussed in other parts of this book. The fact is that professional and governmental agencies concerned with radiation protection have, over the years, expressed concern over the dose to the public from diagnostic radiation and the need to keep it at a minimum, consistent with providing the physician with the information required to treat the patient. In January 1978 the president of the United States issued a memorandum entitled "Radiation Protection Guidance to Federal Agencies for Diagnostic X Rays" (Carter, 1978). Along with recommendations for reducing exposure of the population, it contained a set of entrance skin exposure guides (ESEG) as indicators of maximum exposures to be imparted where practicable in certain routine nonspecialty examinations. Twenty-one years later, the theme of the annual meeting of the National Council on Radiation Protection and Measurements was "Radiation Protection in Medicine" and the participants were still faced with the problem of increased radiation doses to the public from medical procedures. Major sources included high-dose fluoroscopic procedures, expanded applications of computerized tomography, and the potential for unnecessarily high doses from digital imaging systems in radiography (Kearsley, 1999; NCRP, 1999). The proper operation of these systems requires a high level of training of the professional and technical staff.

Internationally, the Council of European Communities and the World Health Organization give much attention to protection of the patient exposed to diagnostic radiation. In 1984, the Commission of the European Community (CEC) issued a directive with regard to incorporating provisions for the radiation protection of the patient into the legislation of the member states. The directive had two major elements: improving the quality of the examinations and treatment and reducing the number of exposures (Courades, 1992). The first called for gathering dosimetry data, training, surveillance of installations, and limiting of direct fluoroscopy without the use image intensification. The second stated that exposures have to be medically justified and they have to be kept as low as reasonably achievable. Suggestions for reducing the number of exposures included the use of alternative techniques, such as ultrasound, whenever possible, and the exchange of information between doctors. In addition, steps should be taken to avoid the repetition of radiological examinations, such as by making available existing radiological and nuclear medical records.

The World Health Organization promotes the concept of "rational use" of diagnostic radiology. The elements of a rational use of diagnostic radiology include abandonment of radiological examinations performed for administrative purposes or as "routine" medical practice; selection of the patients to be submitted to radiological investigations according to well-defined clinical criteria when possible, or at least according to clinical signs or symptoms when clinical criteria are not formulated; and choosing the sequence of the diagnostic imaging technologies used in each clinical case as appropriately as possible (Racoveanu and Volodin, 1992).

25.1 Principles

The principles of protection of the patient in x-ray diagnosis as presented in reports of national and international radiation-protection organizations have undergone very little change over the years (NCRP, 1981b, 1982; Kramer and Schnuer, 1992; ICRP, 1993c, 1996b). The first element in a program to protect the patient is to maintain the x-ray machine in proper operating condition through a continuing quality control program involving measurements of output, beam quality, collimator performance, timer operation, and so on (BRH, 1974; Hendee and Ritenour, 1992). The remaining elements are concerned with minimizing patient dose without adversely affecting the objectives of the examination and avoiding the taking of unnecessary x rays. The following recommendations are basic to good practice in radiology from the viewpoint of radiation protection.

1. Minimize field size with accurate collimators. Do not expose parts of the body that are not being examined. (An additional safety measure, but not a substitute for adequate beam collimation, is to use leaded cloth or other shielding material to protect the gonads and possibly other regions not being examined.) Use film larger than the required x-ray field to verify that collimation is being done properly.

2. Use maximum target-patient distance. This decreases the difference between dose at the entrance side relative to the exit dose because of differences in distances to the target. (As an example, it can be easily verified from the inverse square law that when the x-ray film is 1 ft from the point where the beam enters the body, the entrance dose for a 3 ft separation between the target and film is over 6 times the entrance dose for a 6 ft separation.)

3. Make sure that proper filtration is used.

4. Use a setting for the high voltage that will give minimum absorbed dose consistent with a satisfactory picture.

5. Use the fastest film-screen combinations and shortest exposures that give satisfactory results.

6. Pay careful attention to processing procedures to allow minimum exposures (that is, by using full-strength developer, proper temperatures, and so on).

7. Use fluoroscopy only when radiography cannot give the required information. Use image intensification in fluoroscopy.

8. Use all the planning necessary to prevent faulty pictures and the need

for retakes. Retakes are a major source of excessive x-ray exposure. Of all films that need to be repeated, 50 percent are due to under- or overexposures (NCRP, 1989).

9. Do not prescribe the x ray unless it is necessary.

25.2   Policy of the International Commission on Radiological Protection

The International Commission on Radiological Protection (ICRP) addresses recommendations to both the referring physician and the radiologist (ICRP, 1993c).

Recommendations for the referring physician include:

> The referring physician should refrain from making routine requests not based on clinical indications. To achieve the necessary overall clinical judgment the referring physician may need to consult with the radiologist . . .
>
> Before prescribing an x-ray examination, the referring physician should be satisfied that the necessary information is not available, either from radiological examinations already done or from any other medical tests or investigations.

Recommendations for the radiologist include:

> To achieve the necessary overall clinical judgment the radiologist may need to consult with the referring physician . . .
>
> If two or more medical imaging procedures are readily available and give the desired diagnostic information, then the procedure that presents the least overall risk to the patient should be chosen.
>
> The sequence in which x-ray examinations are performed should be determined for each patient. Preferably, the results of each x-ray examination in a proposed sequence should be assessed before the next one is performed, as further x-ray examinations may be unnecessary. On the other hand, the availability and convenience of the patient, as well as the urgency for the clinical information, have to be considered.

A decision by a physician to forgo prescribing an x-ray examination takes a certain amount of courage in our society, considering the doctor's vulnerability to a malpractice suit in the event a diagnosis is missed. It is much safer to continue past practices of routinely prescribing x-ray examinations even if the physician honestly and justifiably believes a clinical diagnosis can be made without their use. Yet the weighing of benefit vs. risk to the patient in prescribing x rays would most certainly result in the elimination of some x-ray examinations and the curtailment of unnecessarily extensive ones (ICRP, 1970, Publication 16). Fortunately, public awareness and concern over the effects of radiation on the fetus has resulted in a sharp curtailment of examinations of pregnant women. Children and even adults, if somewhat less sensitive, are also entitled to the same conservative approach.

The excessive use of x rays in dentistry is also a concern of the ICRP. It has published the following policy statement with respect to dental examinations (ICRP, 1993c).

> Dental radiography requires particular consideration because it is carried out so widely by non-radiologists and because many dental x-ray examinations consist of a series of x-ray fields which are partially superimposed. In addition, many of the patients are children or young adults. Although x-ray examinations are an important component of dental care, dental radiographs should be taken only after a thorough clinical examination and consideration of the dental history, preferably including study of any previous dental radiographs. Dental radiographs should not be performed routinely at every visit, but should be based on definite indications.

25.3 Studies in the United Kingdom

The Royal College of Radiologists and the National Radiological Protection Board in the United Kingdom conducted studies to determine where the dose to the population from medical x rays could be reduced without jeopardizing their potential clinical benefit. The studies looked at the extent of clinically unjustified examinations, repeated examinations that could have been avoided, and examinations that could have been conducted with a reduced exposure to the patient. The College of Radiology surveys indicated that at least 20 percent of x-ray examinations carried out in Britain were clinically unjustified in the sense that the probability of obtaining useful diagnostic information that could have an impact on the management of the patient was extremely low. Some examinations had to be repeated because the original films were not sent on by general practitioners. Retakes of radiographs because the image quality appeared unsatisfactory occurred in from 3 to 15 percent of the radiographs at the various hospitals surveyed. Surface entrance doses per film for nominally the same type of radiography ranged over factors of between 5 and 20 when averaged over a number of patients. It was concluded that tighter control over the range of doses delivered for each examination could reduce the population dose by almost half (Wall and Hart, 1992).

Are measures proposed or implemented for control of exposure to medical x rays excessive and unwarranted? Are the hazards of so-called low-level radiation minimal, as indicated by some studies? Perhaps, but there is a lot to be learned, and in any event, a conservative approach without compromising health care is hardly an unreasonable approach. Furthermore, there is ample basis for avoidance of unnecessary exposure, if not for reasons of adverse health effects, then for economic reasons. X rays are not cheap, and the costs of any programs to reduce their delivery will be more than compensated by the savings incurred.

25.4 Radiography of the Spine in Scoliosis Patients—A Prime Candidate for Dose Reduction

Scoliosis is an abnormal curvature of the spine that requires for its management multiple full-spine x rays throughout childhood and adolescence. Over the years, significant radiation doses can be imparted to the bone marrow and several organs in the body. Radiation dose to the breast in female patients is a main concern, not only because of its magnitude, but because it is imparted at a time of life when the risk of producing cancer in the breast is greatest.

As a result, institutions have adopted measures for reducing the dose to a very low level. Following are some of the measures used and the results obtained:

- The posterior-anterior projection produces a much lower dose to the breast than the anterior-posterior projection and thus is standard in all examinations. In conjunction with high kilovoltage, a film–focus distance of at least 3 m, a fast screen-film combination (up to a speed of 1,200 at one institution), lead shielding (particularly of the breast in lateral projections), and a collimator with rotating compensating filter, the dose to the breast is reduced by a factor in the range of 88–97 percent. Further dose reduction might be achieved by removal of the antiscatter grid and use of the air-gap technique (Andersen et al., 1982).
- Digital radiography resulted in a reduction of at least one-half with improved image quality and good contrast over a wide range of x-ray exposure. The repetition of examinations because of unsatisfactory films was also reduced (Kling et al., 1990).
- A digital radiographic system using laser-stimulated luminescence resulted in exposure reductions of 92–95 percent (Kogutt et al., 1989).
- Scoliosis examinations performed with air-gap technique using stimulable phosphor imaging plates provided satisfactory images with mean entrance doses in the central beam of 0.05–0.12 mGy

and skin doses on the breasts in the range of 0–0.03 mGy (Jonsson et al., 1995).

25.5 Screening for Specific Diseases

Mass screening of the population for various diseases or conditions has always been a controversial policy, in view of the large population dose. The ICRP (1993c) states that the justification for mass screening for particular diseases should be based on a balance between the advantages implied for the individuals examined, as well as advantages for the population as a whole, and the disadvantages, including the radiation risk of the screening. Since the benefits of screening are not always the same for all groups making up the population, screening is often justified only if limited to specified groups of individuals.

The Commission notes that with the techniques currently available in mammography, the number of breast cancers that can be detected and successfully treated if women obtain annual mammograms, beginning at about 50 years of age, has been shown to be significantly higher than the likely number of radiation-induced breast cancers.

As alternate imaging methods to conventional radiology are developed, dependence on the x ray may decrease. Endoscopy, ultrasonography, and magnetic resonance tomography all can provide diagnostic information without the use of ionizing radiation, but their selection over CT is limited to specific applications. Computed tomography remains the most frequently performed x-ray examination in hospitals in the United States and the number of CT exams per year continues to increase, although the rate of rise is probably slowed by the availability of suitable alternatives.

26 Radiation Levels in the Working Areas around X-Ray Machines

The installation and operation of an x-ray machine must be such as to provide adequate protection for both operating personnel and any individuals in the vicinity of the machine (NCRP, 1989b, Report 102; NCRP, 1989c, Report 105). This is accomplished by providing shielding where necessary to attenuate the radiation traveling in unwanted directions or by restricting the occupancy of specified regions. After a radiation machine has been installed and shielded, radiation surveys are made to ensure that the radiation levels are below the limits specified by radiation advisory groups or governmental agencies. If there is a reasonable possibility that exposures can be received by personnel in the course of using the machine that are in excess of 25 percent of the occupational dose limits (that is, in excess of 12.5 mSv/yr for whole-body radiation), personnel monitoring devices must be worn (see Part Five, section 5, for a discussion of personnel monitoring).

The radiation produced at any point by an operating x-ray machine may consist of one or more of the following components:

Useful beam. This consists of photons coming directly from the target and through collimating devices that direct the beam to the region of the patient to be irradiated.

Scattered radiation. Since the useful beam in a properly designed machine is limited to the region of the patient being examined, the scattered radiation originates primarily in the body of the patient. This radiation can also undergo subsequent scattering from the walls and other objects in the examination room.

Leakage radiation. This is radiation that penetrates the x-ray tube housing and collimator and is not part of the useful beam. It is, of course, attenuated very markedly by the tube housing and collimating devices.

The degree of protection required by an operator of an x-ray machine varies considerably, depending on the radiation procedure being performed. Only low radiation levels are found in the vicinity of a dental x-ray machine because of the small size of the beam and the small amount of scatter produced. It is possible for the operator of a dental x-ray machine to stand in the same room with the patient and take pictures without exceeding the allowable exposure, as long as he or she is positioned properly. Because of the low energy of dental x rays, however, they are very easily shielded, and any operator who has any significant work load should be protected by a thin lead screen.

Medical diagnostic x-ray machines create larger environmental levels than dental x-ray machines relative to the dose at the film because the larger beams used produce more scattered radiation. The x-ray equipment must be installed in rooms with adequate shielding, and the radiographer must stand behind protective barriers during radiographic procedures.

Needless to say, the radiographer must not expose any part of his or her body to the direct beam. Most of the severe consequences of exposure to radiation developed in operators who deliberately allowed parts of their own bodies to be irradiated by the useful beam. Many dentists developed atrophy and cancer in exposed fingers as a result of holding the film in the patient's mouth while taking x rays, a folly that is no longer practiced.

When a patient must be held in position by an individual during radiography, the individual holding the patient must be protected with appropriate shielding devices, such as protective gloves and apron, and should be

2.30 Radiation levels (mR/hr) measured during fluoroscopy in hospital x-ray examination room. X-ray tube is under examination table and beam is directed upward through patient to viewing screen. Operating conditions, 90 kVp, 3 mA; all dose measurements made with condenser ion chambers (R. U. Johnson, pers. comm., 1972).

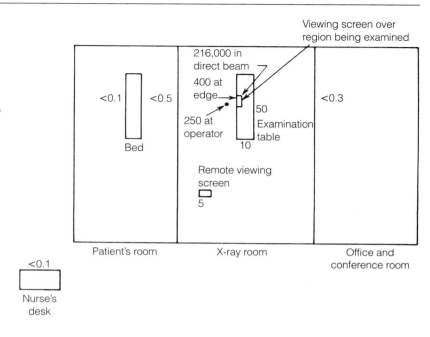

positioned so that no part of the body encounters the useful beam (preferably as far from the edge of the useful beam as possible). The task of restraining a child should be assigned to the patient's parent.

Fluoroscopic examinations pose a problem in radiation protection because the examining physician often stands close to the patient and the useful beam. The results of a survey made during operation of a conventional fluoroscope are presented in Figure 2.30. By far the highest exposure rate (3.6 R/min) was in the useful beam. Thus the most important precaution an operator of a fluoroscope unit can take is not to expose any unshielded part of the body to the useful beam. With the use of image intensifiers, it is possible to locate the viewing screen at a distance from the patient (as long as proximity to the individual is not required). However, some diagnostic procedures require that the physician or other medical personnel perform operations on the patient during the exposure. The potential for receiving high occupational exposures from repeated examinations of this type are great, and special control measures are required. Detailed monitoring of physicians' hands or other critical regions should be employed to prevent overexposure. The monitoring results should be appraised continually in an effort to develop procedures or shielding techniques for lowering the dose as much as feasible.

The maximum environmental levels occur in the vicinity of machines used for therapy. The doses imparted to the patient are high and the leakage and scattered components are correspondingly high. The operator must always remain in a shielded booth or outside the shielded treatment room while operating a therapeutic x-ray machine.

26.1   Shielding the X-Ray Beam

X-ray shielding is based on principles the same as those for gamma shielding discussed in section 8.5. To specify the correct thickness of shielding, it is necessary to know the half-value layer for the x rays of interest. Generally, one determines this value for the highest energy at which the machine will be operated. Although half-value layers may be calculated theoretically, the calculation is complicated because of the complex nature of the x-ray energy spectrum. The HVL changes as the beam penetrates through the medium, because the absorption is different for different energies. Also, the half-value layer depends on the width of the beam and other factors contributing to scatter of radiation in the attenuating medium. Accordingly, it is desirable to select values for half-value layers and other attenuation coefficients that were determined for conditions similar to those encountered in a specific design problem. Data useful in x-ray shield design are given in Table 2.18 for several materials. Also included in the table is an equation that gives the transmission *(B)* of x radiation through several construction materials. Values of three coefficients (α, β, and γ) in the table fit the equation to experimental attenuation curves. Because the half-value layer increases with penetration, the values in the table are given for deep attenuation and broad beam conditions. Even these values are approximate, because the energy spectrum of the x rays differs for different waveforms generated by the high-voltage supply—that is, single phase vs. three phase vs. constant potential. In any event, shields are not designed to limit the dose to a precise number but to reduce the dose below a design number, the lower the better, and a great deal of conservatism is generally built into the design process.

X-ray shielding is designed to limit the maximum exposure at specified locations and over a given period of time, usually a week. The limits differ for controlled and uncontrolled areas. The evaluation of the exposure produced by a machine is based in practice on specification of the output and three other factors: the workload, *W,* defined as the degree of use of the machine and usually expressed in milliampere-minutes per week; the use factor, *U,* defined as the fraction of the workload during which the radiation under consideration is pointed in the direction of interest; and the occupancy factor, *T,* defined as the factor by which the workload should be

Table 2.18 Shielding data for diagnostic x rays.

Shielding material	Half-value layers at high attenuation (mm)			
	50 kVp	70 kVp	100 kVp	125 kVp
Gypsum wallboard	14.45	26.15	39.29	42.65
Steel	0.34	0.78	1.42	2.15
Plate glass	5.75	9.41	15.83	18.05
Concrete			12.90	15.66
Concrete[a]	4.30	8.40	16.00	20.00
Lead[a]	0.06	0.17	0.27	0.28

Parameters α, β, and γ at peak kilovoltage for empirical equation for transmission,
$$B = \left[(1 + \beta/\alpha) \times e^{\alpha \gamma x} - \beta/\alpha\right]^{(-1/\gamma)}, \text{ through thickness } x \text{ cm}$$

Shielding material	α (kV$_p$)				β (kV$_p$)				γ (kV$_p$)			
	50	70	100	125	50	70	100	125	50	70	100	125
Wallboard[b]	0.473	0.245	0.136	0.140	1.094	0.729	0.227	0.164	0.773	0.828	0.625	0.811
Steel[b]	20.66	8.812	4.880	3.226	74.80	51.52	30.53	23.08	0.487	0.608	1.059	1.171
Glass[b]	1.208	0.733	0.427	0.378	2.679	1.611	0.952	0.764	0.652	0.911	1.009	1.293
Concrete[b]			0.537	0.443			2.119	1.475			1.110	1.400

Source: Rossi et al., 1991.
a. Data in NCRP, 1976, Report 49, for single-phase x-ray machine.
b. Average densities, in g/cm³: gypsum wallboard, 0.70; steel, 7.60; plate glass, 2.20; concrete, 2.11.

multiplied to correct for the degree of occupancy of the area in question while the machine is emitting radiation. The product of the output specified for the distance from the target to the point of interest and the three factors *W, U,* and *T* is divided by the design weekly dose to give the attenuation factor required, and the shield thickness is then calculated in the usual way from the number of half-value layers required to produce the desired attenuation. Typical values of *W, U,* and *T* are given in NCRP, 1976, Report 49, for use as guides in planning shielding when complete data are not available.

Example 2.26 Determine the thickness of lead required on the floor of a radiographic installation directly above a waiting room. The maximum high voltage on the x-ray machine is 150 kV. The distance from the source to the floor is 7 ft.
   The shielding thickness is determined through the following sequence of steps:
   (a) Determine the exposure or exposure rate at the dose point in the ab-

sence of any shielding. It would be nice to have this information from measurements made directly on the machine, but when this is not available, we have to turn to handbooks or other sources. One source of data is Figure 2.27. The output of a full-wave rectified, single-phase machine, chosen for a total filtration of 3 mm aluminum, is read off the graph as 18.6 milliroentgens per milliampere-second at 40 inches for a peak kilovoltage of 150 kV. The exposure rate at 7 feet is then (40 in/84 in)^2 × 18.6 mR/mA-sec = 4.22 mR/mA-sec.

(b) Determine the exposure in a week of continuous operation of the tube. Report 49 of the NCRP (1976) suggests a default value for the weekly workload *(W)* of 200 mA-min, or 12,000 mA-sec, for a busy general radiography installation. It also recommends that the machine be considered as pointing to the floor 100 percent of the time, that is, the use factor *(U)* equals 1. The occupancy factor *(T)* for a waiting room is cited as 1/16. The exposure is thus 4.22 mR/mA-sec × 12,000 mA-sec × 1 × 1/16 = 3,165 mR.

(c) Determine the added shielding required. We would like to reduce the weekly dose to 1.92 mR, giving 100 mR per 52 weeks. The attenuation factor is 1.92/3,165 = 0.0006066.

We need to determine the thickness of shielding material we will use that will have an attenuation factor of 0.00061. We can look up tables of half-value layers, but a much easier approach is to read the thickness off an applicable curve of attenuation vs. thickness. We realize, again, that the differences in spectra among different types of x-ray machines will tend to produce some inaccuracy in our calculation, but it should not be critical.

Sources of attenuation curves include NCRP Report 49 (NCRP, 1976) and the *Handbook of Health Physics and Radiological Health* (Slaback, Birky, and Shleien, 1997). We turn to Figure 6.13 in the latter, which gives a thickness of 2.9 mm for an attentuation factor of 0.00061 for 140 kV$_p$, the maximum energy cited in the table. This should be satisfactory for 150 kV$_p$, as machine operation is normally well below these kilovoltages. NCRP's Report 49 gives a half-value layer of 0.3 mm lead for 150 kV$_p$ x rays. It takes 10.68 half-value layers to produce an attenuation of 0.00061 (0.5$^{10.68}$ = 0.61) or 10.68 × 0.3 = 3.2 mm lead. The report has a table listing the thickness of lead for conditions given in the table. A lead thickness of 1.5 mm is given here for 2.1 m distance where *WUT* = 12.5, the same values given in the problem. However, inspection of the conditions of the calculation shows that the design radiation exposure is 10 mR per week, rather than the 1.92 mR in our calculation. It would take another 2.32 half-value layers, or 0.7 mm, to bring the NCRP limit down to the one in

> our example. This shows the need to know the assumptions that led to the results presented in the table. Other differences in assumptions made for the calculation, including the machine output and differences in spectra, would also be responsible for differing results. Accordingly, some judgment, in addition to quantitative analysis and calculation, is involved in selecting a design value for the thickness.

The tube potential used in radiographic exams is typically considerably less than the maximum potential generated by the x-ray machine, and this difference might be taken into account in the shielding design (provided regulatory shielding requirements were met). A weekly workload of 400–500 mA–min is probably more realistic currently than the 200 mA–min used in the problem. Values of tube output for design calculations might have to be modified from tabulated values if waveforms differed (voltage generators that are three-phase or "high frequency" have an output 20–30 percent higher than the output of single-phase generators). The greatest reduction in the shielding needed occurs if it can be shown that the scattered radiation rather than the primary radiation is the radiation that controls the design, since the primary radiation typically is greatly attenuated by the patient and imaging equipment before it strikes a barrier.

Since an exact calculation of scattered radiation is very complex, empirical factors are often used. One approximate relationship for determining the dose scattered from a surface is: scattered dose at one meter from scatterer = 1/1,000 × incident dose at scatterer for a beam with a cross-section equal to 20 cm × 20 cm. Since the scattered radiation is proportional to the cross-sectional area of the beam, a correction for beam areas other than 400 cm^2 may be made by multiplying the estimate of scatter by the ratio (actual area of beam)/400. The attenuation coefficients used in determining the thickness of shielding for scattered x rays in diagnostic installations are usually taken to have the same values as those given for the useful beam. The scattered radiation is less penetrating for x rays generated at voltages greater than 500 kV. Design data are given in NCRP, 1976, Report 49, and Slaback, Birky, and Shleien, 1997).

No credit is taken in Example 2.26 for attenuation by materials of construction such as gypsum wallboard and concrete. The thickness of these materials should be converted to a number of half-value layers and subtracted from the number of half-value layers required for the installed shielding.

If a receptionist's desk will be placed in the waiting room, shielding will have to be based on personnel exposure for a full work week rather than the shorter waiting time assigned to patients.

Accurate and comprehensive shielding calculations are particularly important if the exposure levels are close to regulatory limits, that is, if there is some question as to whether special shielding has to be installed. The incorporation of shielding can add significantly to the cost of a radiographic facility. If shielding is needed, the most economical approach may be to choose a standard "off-the-shelf" thickness that provides the needed protection. A typical shielding design for radiographic installations is 1/16 inch (or 1.6 mm) of lead in all walls and doors up to a height of 7 feet from the floor.

27   D**OSE** R**EDUCTION IN** N**UCLEAR** M**EDICINE**

The most effective method for reducing dose in nuclear medicine diagnostic procedures is to use radionuclides with short half-lives. A spectacular success in this area was the replacement of ^{131}I with its half-life of 8.1 days by ^{123}I with a half-life of 13 hours, or ^{99m}Tc with a half-life of 6 hours. ^{131}I was the only radionuclide used for thyroid function studies prior to 1960. In 1989, only 8 percent of thyroid scans used ^{131}I, the remainder being divided equally between the two short-lived radionuclides. The thyroid dose, 650 mSv from 2 MBq of ^{131}I, was reduced to 39 mSv with the use of 10 MBq of ^{123}I and 2.6 mSv with the use of 75 MBq of ^{99m}Tc (NCRP, 1991b, Commentary 7).

28   E**XPOSURE OF THE** E**MBRYO,** F**ETUS, OR** N**URSING** C**HILD**

When pregnancy is discovered following an x-ray examination, the question may arise as to whether the pregnancy should be terminated. There are no definite rules. The opinion of the National Council on Radiation Protection and Measurements is that the decision should properly be left to the patient, with arguments and recommendations supplied by the physician. The most sensitive period in the development of the fetus is considered to be between the third week and the tenth week post conception. The risk affecting the development of the embryo is considered to be less than 1 percent at a fetal dose of 50 mGy (NCRP, 1994, Commentary No. 9). This dose is well above the level usually received by the fetus in diagnostic procedures. When a concern arises following an x-ray examination of a pregnant woman, an estimate of absorbed dose and the associated risk to the fetus should be made by a qualified expert. Equipped with all the relevant facts, the patient should be in a position to make her own decision regarding continuation of pregnancy.

Table 2.19 Doses (mGy) to ovaries/uterus from various diagnostic procedures.

Procedure	Dose (mGy)
Diagnostic x rays	
Lumbar spine	7.2
Abdomen	2.2
Upper G.I. tract	1.7
Barium enema	9.0
Intravenous pyelogram	5.9
Nuclear medicine	
Brain scan, ^{99m}Tc DTPA, 740 MBq	5.8
Bone scan, ^{99m}Tc phosphate, 740 MBq	4.5
Thyroid scan, ^{99m}TcO$_4$, 185 MBq	1.1
Renal scan, ^{99m}Tc DTPA, 740 MBq	5.8
Abscess/tumor scan, ^{67}Ga citrate, 111 MBq	8.8

Source: NCRP, 1994, Commentary 9. When doses to ovaries and uterus differ, the higher value is given.

Radiography of areas remote from the fetus, such as the chest, skull, or extremities, is safe at any time during pregnancy, provided that the x-ray equipment is properly shielded and the x-ray beam is collimated to the area under study.

The administration of ^{131}I to a pregnant mother will cause a high dose to the fetus, since the iodine readily crosses the placenta and the fetal thyroid accumulates the radioiodine at even a higher rate than the adult thyroid. It is estimated that the fetal thyroid dose will be 7–200 mGy per 37 kBq (1 μCi) administered to the mother over 10–22 weeks of gestational age (Mettler et al., 1985). Much lower doses are imparted by ^{123}I and ^{99m}Tc, which should be used instead if the test is essential. The NCRP recommends that ^{131}I not be used for radiation therapy, such as in cases of hyperthyroidism or thyroid cancer, while the mother is pregnant. If nursing mothers cannot postpone ^{131}I radiotherapy, the nursing should be discontinued (NCRP, 1994).

Table 2.19 gives doses greater than 1 mGy to the ovaries or the uterus in various procedures involving x radiation or radiopharmaceuticals.

29 PROTECTION OF THE PATIENT IN RADIATION THERAPY

The incidence of cancer in industrialized countries is about 3,500 cases per million population per year, half of which are treatable by radiation (UNSCEAR, 1993). The use of radiation to treat disease involves radiation-protection considerations quite different from those accompanying diagnostic radiation, where the radiation dose received by a patient is incidental to obtaining the diagnostic information. Thus the approach to protecting the patient in a radiological examination is to minimize dose consistent with obtaining a satisfactory image. Radiation dose in radiation therapy, on the other hand, is the means through which cancer cells are destroyed, and the objective there is to deliver the high doses required to treat the disease while limiting the damage to healthy tissue as much as possible.

29.1   Treatment with External Radiation Beams

The increasing computational power available with modern high-speed computers is producing dramatic changes in the administration of x radiation to treat disease. The early methods involved laborious and time-consuming calculations by hand of radiation dose distributions for proposed beam configurations. Now radiation physicists determine optimal beam conditions for a desired dose distribution in the patient's body with high-speed computers. Three-dimensional treatment planning is giving radiation oncologists the ability to perform 3D conformational therapy with either multiple "stationary" non-coplanar beams or "dynamic" methods. The computers have made possible the development of stereotactic radiosurgery, in which small lesions in the brain are treated with the use of multiple non-coplanar beams or arcs. The method requires extreme precision, attained by using a head frame securely fastened to the scalp prior to localizing the lesion through CT scans. Fractionation of the administered dose has replaced single-dose regimes, and fractionation regimes are being developed with increasing sophistication to improve treatment outcomes. They range from conventional once-per-day treatments at constant dose-per-fraction to as many as three fractions per day, with sometimes varying dose per fraction (Orton, 1995). One intriguing possibility is determining the patient's sensitivity to radiation by genetic sequencing technology and adjusting the target doses accordingly. The goal of all these developments is to increase the efficacy of the treatment while mitigating radiation-induced complications after treatment.

Irradiation with protons and other heavy charged particles can deliver high tumor doses by exposing the tumors to the particles at the end of their ranges, where the ionization density becomes very high, with a much lower dose to healthy tissue than is possible with x rays. Dramatic successes have been achieved, particularly in the treatment of brain and eye tumors, but also in treatment of tumors in other organs, such as the prostate. Only a few installations have the synchrocyclotrons or other machines capable of producing the radiation beams and energies required and the expertise to deliver the radiation precisely to the targets.

Certain noncancerous but debilitating diseases respond to radiation therapy. The decision to use radiation is a difficult one because of the risk of causing cancer. Patients must weigh the efficacy and risks of radiation and alternative treatments available against their ability and willingness to tolerate the disease. The frequency of use of radiation therapy for "benign" conditions varies in different countries, reported to be about 1 percent in a survey of a major radiotherapy center in the United Kingdom, 2 percent in Japan, and 4 percent in the United States.

Quality assurance programs are critical in the operations of a radiation therapy department. The potential for errors in the performance of the computers and in the administration of the beam is real and significant, and the consequences to the patient are serious, sometimes fatal. In any event, because the high doses used in the treatment of cancer can cause severe and painful injury to normal tissue, the administration of radiation therapy requires a high level of expertise.

29.2 Brachytherapy

Brachytherapy refers to radiation therapy administered by placing sealed radioactive sources inside or on the surface of the patient. The sources may be placed on the surface of the body, within body cavities (intracavitary), or within tissues (interstitial). Brachytherapy was first used in 1901 by a dermatologist who used radium as a surface treatment for a cutaneous lupus lesion; it was used for the treatment of uterine cancer in 1908. However, it was not a popular method of treatment until the 1950s because of concerns about exposures to personnel performing the procedure and reservations about the safety of the radium sources. The development of techniques in which special tubes or holders for receiving the source were positioned in the patient followed by insertion of the source allowed precise positioning of the holder (and source) with minimum exposure of personnel. The use of such "afterloading" systems, the production of alternative sources to radium (such as cesium-137, iridium-192, iodine-125, and palladium-103), and the introduction of computerized treatment planning resulted in the acceptance of brachytherapy as an important modality in the treatment of cancer (ICRP, 1985; Sankey, 1993).

Interstitial brachytherapy entails the insertion of needles, seeds, or wires containing the radioactive material directly into tumors in geometrical arrangements conducive to delivering the desired dose. If short-lived radionuclides are used, the implants may be allowed to decay in the body. Removable implants usually use cesium-137 for intracavitary treatment and iridium-192 for interstitial treatment. Permanent implants use seeds containing iodine-125, palladium-103, or gold-198 because of their short half-lives.

Cesium-137 sources consist of labeled microspheres, approximately 50 μm in diameter, which are doubly encapsulated in stainless steel needles (3–8 mCi) or tubes (13–100 mCi). The beta particles and low-energy characteristic x rays are absorbed in the source wall. Iridium-192 is usually obtained as seeds placed at 0.5 or 1.0 cm intervals in a thin nylon ribbon. The seeds are 3 mm long and 0.5 mm in diameter. Iodine-125 may be obtained as 4.5 mm long seeds, with a wall thickness of 0.05 mm titanium. For seeds in strengths up to 0.5 mCi, the iodine-125 is absorbed on ion-exchange resin, and it is absorbed on silver rod in strengths between 5 and 40 mCi. Palladium-103 seeds are an alternate to iodine-125. The palladium is uniformly distributed on two palladium-plated graphite pellets separated by a lead marker and encapsulated in a titanium tube, 4.5 mm long and 0.8 mm in diameter, with a 0.05 mm wall thickness (Sankey, 1993).

29.3   Therapeutic Use of Radiopharmaceuticals

The main therapeutic use of radiopharmaceuticals is in the treatment of various thyroid conditions with iodine-131. Also used, but much less frequently, are phosphorous-32 in the treatment of polycythemia vera and yttrium-90 for hepatic tumors and arthritic conditions. The use of monoclonal antibodies labeled with yttrium-90 or iodine-125 is seen occasionally in radioimmunotherapy, but the technique is still basically in the development stage. There is some question as to the dose imparted by bremsstrahlung in the case of yttrium-90. Radionuclide therapy in pregnant women, particularly in those with an unsuspected early pregnancy, may result in much higher fetal doses than currently cited.

30   Misadministrations in the Medical Use of Radiation and Radioactive Material

Misadministrations, as defined by the Nuclear Regulatory Commission, include using a radiopharmaceutical other than the one intended; treating the wrong patient; giving the patient a dose of a radiopharmaceutical that differs from the prescribed dose by more than 50 percent in a diagnostic procedure and by more than 10 percent in therapy; or, in the case of a treatment by an external radiation source, imparting an absorbed dose that differs from the prescribed total treatment dose by more than 10 percent.

Misadministrations of licensed radioactivity must be reported to the regulatory agency. The estimated error rate per patient for teletherapy misadministrations is about 1.5 in 10,000, and for diagnostic misadministations about 1 in 10,000 (NCRP, 1991b). The causes of diagnostic misadministrations in one study were: administration of the wrong radiopharmaceutical (77 percent); examination of the wrong patient (18 percent); greater than 50 percent error in the administered dose (4 percent); and wrong route of administration (1 percent).

The NCRP (1991b, Commentary 7) believes that reporting requirements for misadministrations involving radioactive materials in nuclear medicine should be based on a reasonable appraisal of effects likely to result if a misadministration occurs. It notes that most diagnostic tests involve doses of the order of a few milligrays (for whole-body exposures) and a few tens of milligrays (for organ doses) and that these doses entail only a very small element of risk in the event of a misadministration.

31 OCCUPATIONAL EXPOSURES INCURRED IN THE MEDICAL USE OF RADIATION

The medical uses of radiation, particularly in the administration of radiation therapy but also, to a lesser degree, in diagnostic radiology, exposes practitioners, nurses, and technicians to significant radiation fields. Most of the potential exposure is from external radiation, but there is also a potential for intake of radioactivity from aerosols or contaminated surfaces. The most important and effective method for the control of exposure is the use of personnel monitoring to identify significant instances of exposure and to signal the need to impose or improve control methods and working habits. The time-honored methods of control—time, distance, and shielding for protecting against external radiation and ventilation, air cleaning equipment, and protective clothing for protecting against internal exposure—are then implemented as needed.

Occupational exposures are least in dental radiology. The fraction of monitored dental workers receiving an annual dose in excess of 15 mSv is very small, significantly less than one in a thousand workers (UNSCEAR, 1993). Little is generally required in the way of shielding, and careful attention to the orientation of the direct beam plus simple precautionary measures on the part of the workers should provide ample protection. Workers should be trained in the principles of radiation protection as applied to dental radiology (NCRP, 1970).

Table 2.20   Doses imparted to a nuclear medicine technologist from routine procedures.

Procedure	Agent	Activity (MBq)	Dose from single procedure (μGy)			
			Prep.[a]	Admin.[b]	Imaging	Total
Bone	^{99m}TcDiphos	555	0.2	0.1	5.4	5.7
Brain	^{99m}TcO$_4$	740	0.4	0.2	2.2	2.8
Cerebral blood flow	^{99m}TcO$_4$	740	0.4	0.2	0.3	0.9
Infarct	^{99m}TcPyro	555	0.1	0.1	0.2	0.4
Liver	^{99m}TcSC	148	0.1	0.2	0.3	0.6
Thyroid	^{99m}TcO$_4$	74	0.2	0.1	0.4	0.7

Source: NCRP, 1989c.

Note: Divide μGy by 10 to convert to mrad. Divide MBq by 37 to get mCi.

a. Dosage preparation and assay.

b. Administration of radiopharmaceutical to patient.

Most occupational doses in diagnostic nuclear medicine are incurred in the preparation of radiopharmaceuticals. Administration of pharmaceuticals by injection can result in relatively high doses to the hands of the workers. Following injection, the patient is also a source of radiation exposure and nurses must be provided with adequate training to limit exposure. Some results given in published studies of personnel doses received in the use of radiation in medicine are given in the next section and provide a measure of potential exposures in a variety of situations. However, where the potential for exposure is significant, personnel monitoring devices should be used to characterize and document the nature of the exposure. Training programs that include simulations of the medical procedures in all respects except for the presence of the radiation can be very effective in reducing occupational exposures.

31.1   Studies of Occupational Exposures in the Conduct of Specific Procedures

Diagnostic procedures in nuclear medicine. Table 2.20 lists the doses received by a nuclear medicine technologist in the performance of various studies.

Characteristic annual exposures to technologists received at one large nuclear medicine department, as recorded on personnel whole-body monitors, were 1 mSv in injection/imaging and between 3.2 and 4.0 mSv in the nuclear pharmacy. Finger rings gave exposures of 9.5 mSv in injection/imaging and between 18.8 and 226 mSv in the pharmacy (Vetter, 1993).

Cardiac catheterization. Many procedures for the diagnosis and treatment of heart disease require that a catheter be inserted into an artery. The catheter is used to inject contrast agents needed to visualize the heart and circulatory system as well as to insert balloons, stents, or other devices to treat the condition. The catheter is usually introduced through a femoral artery and the physician manipulates it through the arterial system, using a fluoroscopic image as a guide. The x-ray examination is conducted after contrast material is injected through the catheter, and it may include additional fluoroscopy, digital radiography, the use of numerous x-ray films with a rapid film changer, and the taking of motion pictures (cineradiography). Needless to say, the patient is exposed to a high level of radiation, but personnel in the examination room who participate in many procedures also incur high occupational doses from the radiation scattered off the bodies of patients.

In a detailed study of radiation exposures incurred by a surgeon during a comprehensive cardiac examination, time-lapse photography and computer modeling were used to allocate exposures incurred during the various phases of the procedure (Reuter, 1978). In addition, TLD chips were placed on the back of the hands, over the thyroid, over the bridge of the nose, on the forehead, and at locations used in personnel monitoring (collar outside apron, waist, outside and behind lead apron). Some results, averaged over three cardiac catheterization procedures, are presented in Table 2.21.

The cine portion of the procedures accounted for 47 percent of the total exposure. The TLD readings indicated that the dose to the eyes was higher than the dose given by the collar badge, although the variation in the data was high. The eye dose is limiting in this study, although on the

Table 2.21 Radiation exposures during various phases of cardiac catheterization.

Procedure	Time (min)	Eye exposure (mR)	Thyroid exposure (mR)
Right heart catheterization	15.3	4.9	4.2
Left ventriculography	9.4	3.3	2.8
Right coronary angiography	10.5	9.1	7.7
Left coronary angiography	7.0	6.5	6.2
Total procedure	50.4	31.1	27.6
TLD		19.9	16.0
		10.0 (collar)	

Source: Reuter, 1978.

Notes: Calculation by computer modeling is average of 3 examinations. The TLD data is average of 13 procedures and falls within a standard deviation of computer calculations.

basis of a current annual limit of 150 mSv to the eye, the surgeon could perform about fifteen procedures per week without exceeding the eye limit. In view of the doses received, however, prudence would advise that measures be taken to reduce them as much as is practical.

Doses to adult patients receiving a cardiac catheterization averaged over a number of studies were 2.5 mSv to the thyroid, 11 mSv to the chest, and 0.12 mSv to the gonads. Doses to children were 0.26 mSv to the eye, 4.3 mSv to the thyroid, 75 mSv to the chest, 1.5 mSv to the abdomen, and 0.1 mSv to the gonads (Miller and Castronovo, 1985).

A five-year study of occupational doses received during cardiac catheterization procedures reported group averages for annual doses to collar dosimeters of 25 mSv to physicians, 40 mSv to physicians-in-training, 8–16 mSv to nurses, and 2 mSv to technologists (Renaud, 1992). The doses to physicians-in-training given by the collar dosimeters, as well as by whole-body dosimeters, approached regulatory limits. These data were helpful in encouraging cardiologists to use boom-mounted shielding, thyroid shields, lead-impregnated glasses, and a second dosimeter on the collar (Howe, 1993).

Doses imparted to the main operating physician in individual pediatric cardiac catheterization procedures, averaged over 18 procedures, were 0.088 mSv to the lens of the eye, 0.18 mSv to the thyroid, and 0.008 mSv effective dose. Doses to participating assistant physicians were about one-fourth as much. Doses to technicians were comparable to those to the assistant physicians, but in some procedures, the dose to the technician's hand approached 1.5 mSv (Li et al., 1995).

Brachytherapy. Brachytherapy is a major source of occupational exposures. Radiation oncologists and medical radiation physicists handle up to a thousand megabecquerels at short distances in preparing gamma sources and implanting them in patients. One study examined exposure rates to the radiation oncology staff administering the treatment, exposure rates at frequently occupied points in the patient room, and exposure to the nursing staff (Smith et al., 1998). The measurements provide insight into the radiation exposures imparted in the course of treatment of a patient. Following are some of the results:

1. *Intracavitary treatment, using ^{137}Cs tubes*
    a. Dose rates measured at 10 cm from a ^{137}Cs tube typically handled and prepared for loading into a Fletcher-Suit intracavitary applicator in the patient. Measurements made unshielded (for dose rate to hands) and behind a conventional L-shaped materials-handling lead shield (for body dose

rate). Source activity of single tube, 1,340 MBq (36 mCi); dose rate to hands at 10 cm, 0.2 mGy/min; handling time, 5 min; dose to hands, 1 mGy. Dose rate to body, shielded, 0.01 mGy/hr.

b. Dose rates while loading prepared sources into Fletcher-Suit applicator in patient. Source activity handled at one time, 3,080 MBq (83 mCi); dose rate to hands at 10 cm, 0.46 mGy/min; time to load sources, 2 min. Hand dose, 0.92 mGy.

c. Dose accumulated in removing and storing ^{137}Cs tubes about same as in loading process, or about 2 mGy to hands.

d. Total dose in treatment, 4 mGy. Division of labor significantly lowers dose to individual worker.

2. *Interstitial treatment, using ^{192}Ir*

a. Dose rates measured at 10 cm from a single ^{192}Ir ribbon to be loaded into a Syed-Neblett interstitial applicator in the patient. Source activity, 146 MBq; dose rate to hands at 10 cm, 0.28 mGy/min. Loading into interstitial applicator requires placing each ribbon into the appropriate needle, cutting the ribbon, and capping each needle to prevent movement of the ribbon. Loading time is approximately 30 min for all the ribbons, so hand dose can be 8.4 mGy

b. At the conclusion of the interstitial treatment with ^{192}Ir, the removal of the applicator with the sources should take about 15 min, resulting in an additional hand dose of 4.2 mGy, and a total dose from the procedure of 12.6 mGy.

Patients treated with cesium-137 and iridium-192 implants are typically hospitalized from 48 to 72 hours, during which time nurses and other care-givers can receive significant exposures. The patient room should be situated at the end of a corridor, and preferably at a corner with two outside walls, eliminating possible exposures in two adjacent rooms and simplifying control problems in the event of an accident.

Average dose rates next to and 1 meter from the bed of the patient receiving intracavitary irradiation were 0.77 and 0.26 mGy/hr, respectively. A nurse's dosimeter read 0.59 mGy. Corresponding dose rates for a patient receiving interstitial irradiation were 1.05 and 0.30 mGy/hr, respectively, but the nurse's exposure was less, 0.44 mGy. This reflected different degrees of patient care required for the two treatments. It should be noted, however, that there is considerable variability in this type of study and the results should be used only as a general indicator of the doses associated with the procedures. Doses to care-givers in the patient's room could be reduced considerably by the use of mobile shields.

32   Comments for Users of X-Ray Diffraction Machines

X-ray diffraction machines are designed for routine analytical work and, in theory, do not present any radiation hazards to the user if simple precautions are observed. However, they make use of beams of extremely high intensity, and although the direct beams emanating from the machines have diameters generally not exceeding a centimeter, they can produce severe and permanent local injury from only momentary irradiation of the body (Weigensberg et al., 1980). The direct beams also generate diffuse patterns of scattered radiation in the environment; although this radiation does not present a hazard of serious accidental exposures, it can produce harmful somatic or genetic effects to individuals exposed over an extended period of time. The magnitude and extent of radiation fields found typically around x-ray analytical instruments are illustrated in Figure 2.31.

Note that two principal modes of operation are shown, diffraction and fluorescence. In diffraction, the primary beam from the target of the x-ray tube emerges from the machine through a collimator and strikes the sample, which diffracts the beam in a characteristic manner. The diffraction pattern is measured with a photographic film or a radiation counter. In fluorescence, the primary radiation beam strikes the sample inside a shielded enclosure, and only scattered radiation and secondary radiation excited in the sample as a result of the irradiation emerge from the machine for analysis. Consequently, external levels are much lower in the fluorescence mode than in the x-ray diffraction mode.

The greatest risk of acute accidental exposures occurs in manipulations of a sample to be irradiated by the direct beam in diffraction studies. Exposure rates of the order of 10,000 R/sec can exist at the tube housing port. Erythema would be produced after an exposure of only 0.03 sec, and in 0.1 sec severe and permanent injury could occur. The fingers, of course, are the parts of the body most likely to receive these high exposures. Environmental levels near the machine from scattered radiation can also be quite high, perhaps a few millisieverts per hour. Maximum occupational exposures for a three-month period would be attained within a few hours at these exposure rates.

Modern x-ray diffraction machines incorporate shielding and safety features to prevent both acute local accidental exposures and chronic long-term irradiation of the body. Operators should be fully cognizant of the

2.31 Radiation fields around x-ray analytical equipment. In x-ray diffraction studies, the intense primary beam passes through an open area before striking the sample. In x-ray emission (fluorescence) studies, the primary beam and sample are completely enclosed and shielded; only scattered and secondary fluorescence radiation emerge to open areas. Note that typical exposure rates around the fluorescence setup are orders of magnitude lower than around the x-ray diffraction setup. The figures for exposure rates are based on published data and the author's experience. (For additional survey results, see McLaughlin and Blatz, 1955.) Points identified on diagram are: (a) at tube port with shutter open; (b) at specimen chamber when tube not seated properly or sample holder removed and shutter interlock not completely effective; and (c) in vicinity of shielding.

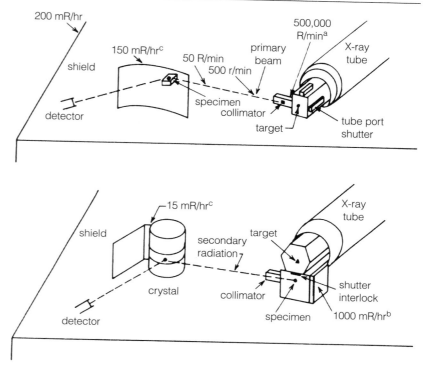

protective devices incorporated into their machines and the possibilities for failure or malfunction. Operators working with older machines must be especially careful of the possibilities of receiving excessive amounts of radiation.

It is very important that x-ray diffraction machines be surveyed for excessive radiation levels on a regular schedule, at intervals not exceeding a year. They should also be surveyed every time a modification in measurement techniques affecting the radiation pattern is introduced. All measurements, except those of the extremely localized and intense direct beam, can be performed satisfactorily with an ionization-chamber type of survey meter equipped with a thin window for the low energies encountered in x-ray analytical work. Photographic film works well for delineating the direct beam and for searching for other small intense beams that may penetrate through holes in the shielding. The exposure due to the direct beam is usually measured with a small condenser-type ionization chamber (R meter) or a lithium-fluoride thermoluminescent detector. A Geiger counter is a very effective instrument for searching for excessive scattered radiation that must then be evaluated accurately with dose or exposure measuring devices.

The need for personnel monitoring is debatable when well-designed instrumentation is used routinely, but personnel monitoring is necessary when machines are used in experimental configurations. The effectiveness of film badges and other personnel monitoring devices for monitoring the primary beam is limited because of their small size. When film badges are used, they should be worn on the wrist to monitor the region most vulnerable to exposure. When high exposures to the fingers are possible, it is very desirable to supplement the standard personnel dosimeters with finger monitors. The detectors may be incorporated into a ring or merely taped to the fingers.

When excessive radiation levels are found around x-ray analytical equipment, they can be easily reduced because of the very low energies of the x-ray photons. Any convenient structural material can be used. Frequently a thin sheet of steel, copper, or brass will suffice. Essentially complete attenuation of the highest intensities normally encountered can be accomplished with a thickness as small as 3 mm brass, or the equivalent.

Following is a list of safety recommendations applicable to persons working with analytical x-ray equipment (Lindell, 1968).

1. Warning signs, labels, and lights should be used at working stations.
    a. Labels bearing the words "Caution Radiation—This equipment produces radiation when energized" should be attached near any switch which energizes a tube. They should contain the radiation symbol and be colored magenta and yellow.
    b. Highly visible signs bearing the words "Caution—High Intensity X-Ray Beam" or other appropriate warning should be placed in the area immediately adjacent to each tube head.
    c. Warning lights that light up whenever the tube is delivering x rays should be placed at tube on-off switches and at sample holders. The installation of the lights should be "fail-safe," or two lights should be installed in parallel to provide a warning even if one of the bulbs burns out.
2. Shutters should be used that cannot remain open unless a collimator is in position. The only equipment failures that have been reported as resulting in radiation injury have involved defective shutters over the tube head ports. Accordingly, even when shutters are provided, they must be inspected and monitored regularly.

3. Shutter interlocks should be used to cut off the beam when samples are changed. Modern fluorescence instrumentation comes equipped with shutters that prevent the beam from entering the sample chamber when samples are removed. However, there are reports of poorly designed shutters that allow significant amounts of radiation to leak through. Accordingly, the presence of these shutters cannot be taken as prima facie evidence of effective radiation protection, and they should be monitored routinely.

4. No repair, cleaning work on shutters and shielding material, or other nonroutine work that could result in exposing anyone to the primary beam should be allowed unless it has been positively ascertained that the tube is completely deenergized.

5. Active educational and indoctrination programs in radiation protection should be conducted for users of the x-ray equipment. The most frequent cause of radiation accidents leading to severe tissue injury is human error, involving either carelessness or ignorance on the part of the operator in performing adjustments or repairs when the tube was energized. Well-planned formal training and indoctrination sessions will help reduce the number of accidents. Lack of training programs is evidence of neglect of responsibilities by the user and the owner of the x-ray equipment.

6. Equipment should be secured so it cannot be used or approached by unauthorized personnel. The most obvious method of preventing unauthorized use is to locate the equipment in a locked room that cannot be entered except by authorized users. If equipment must be located in unrestricted areas, appropriate barricades should be installed and a key required to turn on the equipment.

33 PARTICLE ACCELERATORS—THE UNIVERSAL RADIATION SOURCE

The x-ray machines described in the last section actually represent a special class of particle accelerator. In x-ray machines, electrons are accelerated to energies which produce x rays suitable for radiology. The energies delivered by the machines range from a few thousand electron volts to several million electron volts, depending on the application. Other particles, such as protons and deuterons, can also be accelerated in special machines and have various technological and medical applications.

In physics research, particles at the highest energies attainable are the basic experimental tool in studies on the nature of matter, energy, and nuclear forces. Very large and complex machines are needed to produce the energies desired, and present designs are looking toward several thousand billion electron volts. The field of science concerned with the development and use of these machines is known as high-energy physics, and the growth of this field has generated specialists in radiation protection concerned with protection of personnel from the new and energetic particles produced.

Because particle accelerators increase the velocities of particles through the application of electrical forces, only charged particles can be accelerated. These particles may then be used directly for various applications, or they may be used to collide with atoms or nuclei to produce other particles, including uncharged particles. We have already cited the x-ray machine as the most common application of the latter procedure, where a small fraction of the energy of directly ionizing electrons is converted to the energy of indirectly ionizing photons. This is a process that takes place outside the nucleus of the atom. In other procedures, charged particles are given enough energy so they can penetrate into the nucleus. This results in the release of a considerable amount of nuclear binding energy—usually of the order of 7 or 8 MeV—that is added to the kinetic energy of the incident particles and is available to initiate further reactions.

33.1   History of Particle Accelerators

The potential for learning about the structure of the nucleus by bombarding it with energetic particles gave much impetus to the development of machines for accelerating suitable projectiles. The first machines used simple electrical circuits to produce high voltages, the most successful of which was the voltage-doubling type of circuit developed by Cockroft and Walton. The voltage source was made up of electrical transformers, condensers, and rectifier tubes in an arrangement that resulted in the production of up to 700,000 volts. The first nuclear reaction induced by artificially accelerated particles was obtained with this machine. Protons were accelerated through 150,000 volts and used to bombard a lithium target. The lithium target was split into two alpha particles, each of which carried over 8 MeV, as a result of the nuclear energy released in the process.

The maximum voltages attainable with the Cockroft-Walton circuit were limited, and the next step came with the development of the Van de Graaff electrostatic generator. This utilized a moving belt to deliver electrical charge to a sphere whose voltage increased with the accumulation of charge up to a maximum limited by the ability of the insulators to support the voltage without breaking down. From initial generators of 1.5 million volts, the technology developed to the point where the machines could generate potentials up to 10 million volts. The Van de Graaff generator is the favored accelerator in its energy range because of the accuracy and precision of particle energies obtainable by this means.

The next step forward in the production of higher energies came with the development of the cyclotron. An ingenious method was invented for imparting very high energies to particles without the use of corresponding high voltages. The particles were made to go around in a circle, and with each half-revolution they passed through a high voltage that gave them an increment of energy. The circular orbit resulted from the imposition of a magnetic field perpendicular to the plane of motion of the particles. By use of bigger magnets, the particles could be accelerated to higher energies. Typical cyclotrons have diameters of 60 inches and accelerate protons to about 10 MeV. A diagram of a cyclotron is given in Figure 2.32.

The basic cyclotron design could not be used when the velocities of the particles approached the speed of light, and most of the energy imparted to the particles served to increase the mass rather than the speed. The synchrocyclotron, by gradually reducing the frequency of the accelerating field, overcame the problems of acceleration at higher energies. The 184-inch synchrocyclotron at Berkeley produced proton energies up to 350 MeV.

The attainment of higher energies was prevented by the cost of the con-

2.32 Simplified diagram of the cyclotron. Paths of the charged particles (ions) introduced at the center are bent into near-circles by the vertical magnetic field. A rapidly alternating horizontal electric field applied between hollow electrodes (dees) accelerates the particles each time they complete a half circle and cross the gap between the electrodes. This causes the particles to travel in ever-widening orbits, until they are extracted by deflecting electrodes and aimed at an externally placed target. The dees are enclosed in a vacuum chamber (Kernan, 1963).

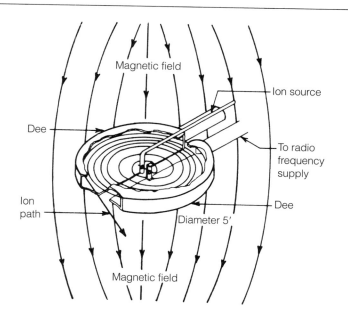

struction and operation of the massive magnets required to keep the particles traveling in the circular orbit. The magnet for the cyclotron at Berkeley weighed 4,000 tons. A new principle was needed if further progress toward higher energies was to be made. The solution to this problem was the development of the synchrotron. The single magnet and radiofrequency field were replaced by a large ring consisting of a sequence of magnets for bending the beam and radiofrequency fields for accelerating it. This arrangement allowed the establishment of the very large orbits required to contain the energetic particles with deflecting magnets of practical sizes. A portion of the ring of the proton synchrotron at the Fermi National Accelerator Laboratory in Batavia, Illinois, is shown in Figure 2.33. This machine is called the Tevatron because it accelerates protons to 1,000 billion electron volts (1 TeV). The proton orbit has a diameter of 2,000 meters. The Tevatron also can produce colliding beams of high-energy nucleons. The use of two beams of particles hurtling at each other at the speed of light instead of single beams colliding with a stationary target increases very substantially the energy available for the production of reactions, allowing exploration of a new range of phenomena in particle physics.

Electrons traveling in a circular orbit emit electromagnetic radiation and lose energy that must be made up by supplying additional power. Radiation losses impose a practical limit on the maximum energy that can be attained. These losses are eliminated by accelerating the electrons along a straight line, but the distance of travel required to attain the desired energies at first discouraged the serious consideration of a linear accelerator. This and other obstacles were finally surmounted, however, and in 1966 the Stanford Linear Accelerator came into operation. This two-mile-long machine accelerates electrons in a straight line to energies of 50 billion electron volts (50 GeV). The electrons are accelerated by traveling electric fields that accompany them and continuously impart energy to them.

It appears that as long as scientists demonstrate their ability to build more powerful and energetic particle accelerators and can use them to push forward the frontiers of knowledge, society will support them in their endeavors. Just when a terminal point based on maximum attainable energy will be reached is an open question, but as long as the answer is pursued, the frontiers of high-energy research will constitute one of the frontiers of research in radiation protection.

33.2   Interactions of High-Energy Particles

The collisions made by ionizing particles with energies less than a few MeV are fairly uncomplicated, at least from the viewpoint of energy trans-

2.33 A portion of the ring of magnets of the 1 TeV proton accelerator, called the Tevatron, at the Fermi National Accelerator Laboratory (Fermilab) in Batavia, Illinois. It has a radius of 1,000 meters and a circumference of 6.3 kilometers (3.9 miles). To bend proton beams around the ring, the Tevatron has 774 superconducting dipole magnets, kept at a temperature of 3.5 to 4.2 Kelvin by liquid helium. More than 2000 quadrupole magnets keep the beams focused. Protons are accelerated in radiofrequency-powered cavities to energies of 980 GeV, at which they travel only 300 miles per hour slower than the speed of light. Experimenters use the Tevatron for studies of secondary particles produced when the high-energy protons strike a target or when beams of protons and antiprotons (also accelerated in the Tevatron, but circulating in the opposite direction) are made to collide, enabling a collision energy of almost 2 TeV. The top quark, one of the twelve fundamental building blocks (6 quarks, 6 leptons) of the universe, was discovered at Fermilab in 1995. It has about the same mass as a single gold atom (175 GeV/c^2). (Courtesy Fermi National Accelerator Laboratory.)

fer to the medium in which they travel. Gamma photons give rise to electrons of relatively short range, which dissipate their energy close to the point of origin. Neutrons collide with and impart energy to the nuclei of atoms, energizing them into projectiles with very short ranges. In those cases where neutrons enter into the nucleus, they liberate additional nuclear energy, which is generally released from the nucleus in the form of gamma photons of moderate energy. Neutron reactions can also result in the emission of charged particles that lose their energy locally.

Both neutrons and gamma photons may either disappear completely after individual interactions or continue on at reduced energies. In any event, we have seen how the reduction of the intensity of neutrons or gamma photons of any given energy in a medium can be specified rather simply in terms of the half-value layer concept.

As the energies of the interacting particles increase, more complex interactions occur. Gamma photons at high energies become readily materialized in the vicinity of the nuclei of atoms into high-energy electrons and positrons (a process known as pair production). The energetic charged particles so produced lose their energy in turn by radiating high-energy photons, as well as by ionization. The result is the creation of what is known as an electromagnetic cascade, a continual shuttling back and forth of energy between material and photon forms that results in deep penetration of the radiation energy in a medium. The photons in the cascade also undergo reactions with nuclei that result in the ejection of considerable numbers of neutrons.

High-energy material particles undergo complex nuclear interactions that result in the transfer of a tremendous amount of energy within the nucleus and the emission of a variety of particles, primarily high-energy protons (with long ranges) and neutrons, but sometimes heavier fragments. The high-energy protons and neutrons are ejected as a result of a series of collisions of individual particles within the nucleus in a process known as an intranuclear cascade. This cascade process begins to be important above 50 MeV. When the energies of incident particles exceed 400 MeV, production of mesons begins to compete with other processes.

After the cascade process, some energy remains distributed in the nucleus, and this produces what may be described as a nuclear boiling process, with the subsequent "evaporation" of low-energy neutrons and protons.

The nuclear interactions undergone by high-energy protons occur at separate points. In between the nuclear collisions, the protons continue to lose energy continuously by ionization in the manner characteristic of directly ionizing particles.

33.3   Shielding High-Energy Particles

Because of the many complex interactions that occur when high-energy particles penetrate matter, including the production of very penetrating secondary radiations, a shielding design is a very difficult problem. The simple half-value layer concept, which was effective for the lower energies discussed previously, is generally not applicable at these high energies. Many of the data for the design of accelerator shields are obtained from measurements in actual shield configurations of the fall-off of dose with distance through a shield. From such experimental data, curves of the attenuation of high-energy radiations through different materials are prepared. An example of attenuation measurements is given in Figure 2.34.

One of the major difficulties in providing accelerator shielding is the extended nature of the sources of high-energy radiation. The accelerating particles cover very large distances in these machines. For example, in one

2.34 Dose measurements on the axis of an electromagnetic cascade produced by a beam of 5×10^{11} 6300 MeV photons. Measurements were made with an airfilled ionization chamber (Bathow et al., 1967).

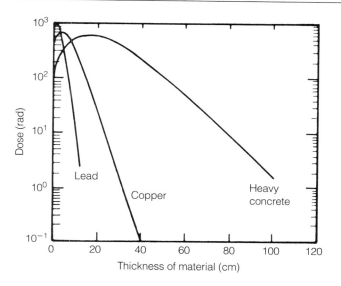

6 GeV electron synchrotron design, each particle travels a distance of 1,400 miles as it makes 10,000 turns around the orbit within an evacuated pipe. The large circulating currents produce very high radiation sources whenever the electrons collide with the walls of the pipe or the surrounding magnets and other equipment. It is not always practical to provide sufficient shielding to protect against continuous exposure from all possible sources of radiation from a misdirected beam, and control of radiations from these machines requires that the accelerator designer and operator confine the beam to its orbit.

In practice, the main sources of exposure to personnel working near accelerators occur not during operation of the machine but from activation by the beam of the materials of construction. The radiation levels from production of radioactive nuclides can become quite high and pose a real problem when work has to be done on the activated components. The technical problem of specifying the degree of shielding required is not difficult, since the sources are radioactive nuclides. The major problem is to install shielding adjacent to the radioactive components in a manner that will provide the required radiation protection and still allow close and intricate operations.

33.4 Particle Accelerators in Radiation Therapy

Particle accelerators, developed originally by physicists to produce the high voltages and particle energies needed for the performance of research in nuclear physics and known popularly as "atom smashers," found wide favor among radiation oncologists for their effectiveness in smashing cancer cells. They produced much higher voltages and particle energies than could be obtained with the high-voltage transformers used in x-ray machines. The first accelerator for radiotherapy was developed and sited in Boston. It was a 1.25 MeV Van de Graaff generator and was used to treat the first patient in 1937. The first betatron, built in 1940, was followed by a 24 MeV machine for radiotherapy at the University of Illinois in 1948. The linear accelerator, the workhorse in modern radiation therapy departments, evolved from the development of the magnetron and the klystron for use in radar systems in World War II. The first patient was treated with an 8 MeV "linac" in London, England, in 1953. Current linacs have dual x-ray energy as well as electron capability, for example, 6 megavolts (MV) and 24 MV. They have multileaf collimation and real-time portal imaging along with computerized treatment planning and control of machine operations (Williams and Thwaites, 1993). This allows for dynamic conformational therapy, which involves the use of multiple concurrent treatment machine and couch motions during the rotational treatment of the patient and state-of-the-art production of dose distributions which conform to the shape of the target volume and limit the irradiation of normal tissues.

The shielding of the x rays produced by a medical linear accelerator can follow the same script used for the more conventional, lower-energy machines, starting with such inputs as the workload *(W)*, the use factor *(U)*, and the occupancy factor *(T)*. The degree of attenuation needed is determined, and a shielding configuration is designed that produces the required protection. As a threshold energy of about 10 MeV is exceeded, however, production of photoneutrons commences. These are a significant radiation source at the higher energies whose characterization presents a formidable challenge. Fortunately, the designer of treatment rooms has considerable assistance at hand in handbooks of the NCRP (NCRP, 1984b) and other professional bodies, in published reports giving details of constructed rooms and the results of radiation surveys, and in analytical techniques developed from shielding experiments and theoretical analyses.

The use of a maze for the entrance to a therapy room may lessen considerably the weight of shielding required on the door or even obviate the need for shielding. To investigate the need for shielding, it is necessary to calculate the neutron and gamma doses at the entrance of the maze. This is a difficult exercise, but an estimate can be based on empirical data or "rules-of-thumb." The target is enclosed in a massive shield, much of which consists of a heavy metal, such as tungsten or lead, which acts as a collimator to shape the useful beam directed at the patient and to limit leakage radiation in other directions. The interactions of the bremsstrahlung with the target—and, to a lesser degree, with the structural materials in the head—produce a source of neutrons emitted fairly uniformly in all directions. The shielding provided in the head is very effective in absorbing the bremsstrahlung but does not absorb neutrons, only reducing them in energy through inelastic scattering. Thus the head constitutes roughly a point isotropic source of neutrons with an average energy somewhat less than 1 MeV. The dose produced by the neutrons from the source is proportional to the therapy dose to the patient, and the relationship can be determined experimentally for any given situation. For example, a measurement of the neutron dose at an acceleration voltage of 18 MV gave an estimate of 4 millisieverts per therapy gray at a distance of one meter from the target. Other experimental data reported that the neutron dose decreased exponentially down the maze with a half-value distance of 1.51 meters. With these data, it is possible to provide an estimate for the neutron dose at the entrance of a simple maze for a specified work load. Examples of good and poor maze designs are shown in Figure 2.35 (NCRP, 1984b, Report 79).

Example 2.27 Estimate the neutron dose produced at the entrance to the maze in the radiation treatment room shown in Figure 2.35a for an accumulated dose imparted to patients of 100 Gy of 18 MeV x radiation. The distance from the neutron source to the inside entrance to the maze is 4.2 m and the length of the maze is 5.3 m.

The dose at the inner entrance of the maze produced by a 100 Gy therapy dose is $(100 \text{ Gy})(4 \text{ mSv/Gy})(1/(4.2\text{m})^2) = 22.7$ mSv. The number of half-value layers down the maze is $5.3/1.51 = 3.5$. The attenuation through the maze is $(0.5)^{3.5} = 0.088$ and the dose at the entrance is $22.7/\text{mSv} \times 0.088 = 2.0$ mSv. Depending on the actual workload and occupancy conditions, this simple maze will probably not be adequate, and either it will need to be redesigned or a shielded door installed (NCRP, 1984b).

The neutron flux at any point in the treatment room not only comes from the head of the machine but also includes fast neutrons scattered from the walls and a lesser contribution from thermal neutrons, both of which are fairly constant throughout the room. This distribution modifies the distance fall-off of the neutron dose from the accelerator head so it is somewhat less than the inverse square law would predict.

Radiation shields for accelerators are primarily made from concrete, but lead shielding may be added to decrease the thickness of shielding required

2.35 Examples of maze configurations showing (a) a poor maze design and (b) a good maze design. Source: NCRP 1984.

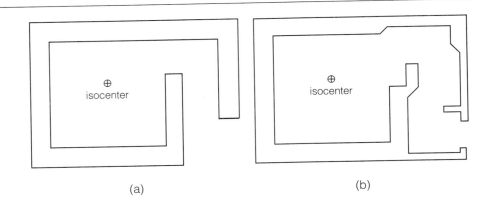

(a)   (b)

for photons. However, the lead can contribute an additional neutron dose outside the shield due to the production of photoneutrons when the primary x-ray beam is aimed at it. The minimum photoneutron dose was achieved when the metal part of the shield was positioned inside the treatment room in front of the concrete and also when steel was used in place of lead (McGinley, 1992; Agosteo et al., 1995).

Measurements of neutron doses to operators and patients are an important component of radiation-protection surveys around medical accelerators. A study of doses incurred in the operation of a 25 MV linac found an annual dose of 6.9 μSv (0.7 mrem) to an operator present for all the treatments in 1995 (Veinot et al., 1998). A person positioned just outside the treatment room door during each treatment would have received 310 μSv (31 mrem). The doses were calculated from measurements of neutron spectra by the multisphere moderator technique. The data were originally recorded in terms of beam output in Monitor Units (MU), where 1 MU corresponded to approximately 10 mGy at the depth of maximum dose in water.

34   Regulation of Radiation Sources and Uses

Regulatory programs for protection of the worker and the public exist at both the federal and state level. These programs start with legislation that establishes a regulatory framework for the establishment of programs in radiation protection. The Atomic Energy Act of 1954 and later amendments established the Nuclear Regulatory Commission (NRC). Other federal agencies assigned major responsibilities by the Congress in regulating the use of radiation include the Environmental Protection Agency (EPA), the Food and Drug Administration (FDA), the Department of Energy (DOE), the Department of Transportation (DOT), and the Occupational Health and Safety Administration (OSHA, in areas of radiation control of workers not covered by the other agencies). Detailed information on the rules and regulations of these agencies can be obtained at their sites on the Internet. The individual states have their own legislation and regulatory departments that cover all aspects of radiation use. More than half have requested and been authorized by the NRC to take responsibility for those areas not concerned with the production of nuclear power or nuclear weapons.

The Conference of Radiation Control Program Directors (CRCPD) was established to interchange information on state programs, help coordinate state programs, and suggest regulations for the control of radiation for all sources of radiation, but adoption of these regulations is voluntary.

While protection of the worker is well covered in most instances under existing federal legislation, the protection of the patient exposed to medical radiation is left primarily to the states, and the scope of state programs for regulation of the medical uses of ionizing radiation varies widely. Some of the regulations focus on medical radiation device efficacy and others focus more directly on user activities and resultant patient protection.

Extension of regulatory control over medical radiation gets an impetus every time a misadministration or accident is made public. Following a series of articles in the *Cleveland Plain Dealer* addressing patient deaths, injuries, and overexposures, an NRC task force addressed the question, "Does the current allocation of authority and responsibility among Federal and State regulatory bodies meet the nationwide goal of ensuring adequate protection of the radiological health and safety of the public, including patients and health care workers, in the medical uses of ionizing radiation?" (Vollmer, 1994). The task group received input from many governmental, professional, and industrial sources, as well public interest and health care advocates. It noted inconsistencies in reporting requirements that resulted in insufficient data to assess the effectiveness of the regulatory framework to protect the public health and safety. However, the task force did not find that the current framework was inadequate for public protection, or that federal and state programs, combined with professional medical practices and voluntary professional standards, did not serve the pubic health and safety. On the other hand, the task force did not have the data to attest to the adequacy of the regulatory framework for medical radiation. It concluded that it was necessary to acquire performance data to quantitatively evaluate options for making cost-effective changes in the regulation of medical radiation.

34.1 Regulatory Measures for Medical Radiation Programs

Two main areas of concern characterize the enforcement activities of regulatory agencies in the policing of medical radiation programs. The first type is concerned with the implementation of quality assurance or quality management programs (QMP) to assure the proper use of diagnostic and therapeutic equipment. The second type covers enforcement actions following the misadministration of radiation to a patient.

The Quality Management Programs relative to radiation therapy developed by the Nuclear Regulatory Commission emphasize *written* directives, verification of the patient's identity, confirmation that final treatment protocols are in accordance with the written directives, determination that the administration is in accordance with the written directive, and identification, evaluation, and appropriate action when there is an unintended deviation from the written directive.

The NRC enforcement policy attaches higher penalties to violations that indicate programmatic deficiencies that are preventable rather than isolated mistakes. This policy is intended to focus the user's attention on self-evaluation of the program and implementation of corrective actions when deficiencies are identified. Examples of programmatic failures include failure to register all users requiring permits to work with radioactive materials, failure to train and train adequately all employees to follow established procedures, and repeated occurrence of closely related violations.

The penalties imposed for a violation of regulations are related to the severity level assigned by the Commission. An example of the maximum severity level, Severity Level I, is the failure to follow the procedures of the QMP that results in a death or serious injury (such as substantial organ impairment) to a patient. A substantial programmatic failure in the implementation of the QMP that results in a misadministration is classified as Severity Level II. A substantial failure to implement the QMP that does not result in a misadministration; or failure to report a misadministration or programmatic weakness in the implementation of the QMP that results in a misadministration is classified as Severity Level III. Failures that are isolated and do not demonstrate a programmatic weakness and have limited consequences if a misadministration is involved; failure to conduct a required program review; or failure to keep records required are considered the lowest severity level, Severity Level IV. Programmatic failures include failure to implement one or more of the QMP objectives, failure to check dose calculations prior to administering the teletherapy treatments, and repetitive failure to prepare written directives. Isolated failures include a single occurrence of an error, mistake, or omission in following an established QMP procedure that resulted in insubstantial or transient medical consequences (Santiago, 1994).

PART THREE

Radiation Dose Calculations

In order to evaluate the hazard of a radiation exposure, it is necessary to determine the energy imparted to critical tissue in the person exposed. The quantity of interest is the absorbed dose, and it is determined in practice by evaluating the energy imparted to a definite mass of tissue. If the dose is imparted by high-LET radiations, it must, in addition, be converted to the equivalent dose by multiplying by the appropriate quality factor, so that a comparison may be made with permissible limits.

In our treatment of dose calculations, we shall evaluate absorbed dose in units of MeV of energy imparted to a gram of tissue. We shall express the absorbed dose in units of grays or milligrays:[1]

$$1 \text{ gray} = 6.24 \times 10^9 \text{ MeV per gram}$$
$$1 \text{ mGy} = 6.24 \times 10^6 \text{ MeV per gram}$$

One factor that complicates dose calculations is that the dose generally is not uniform over the body. If the radioactive material is inside the body, it may be taken up selectively in different organs and tissues. If the source of radioactivity is at a distance from the region of interest, it will have a different effect on different parts of the region, depending on their distance from the source and on attenuation of the radiation by matter between the source and the dose point.

When the dose pattern is nonuniform, we may have to exercise some judgment in making the dose evaluation. Shall we determine the maximum dose that any point gets? Suppose only a very small region gets a high dose, for example, the region near a highly localized source of beta parti-

---

1. The concepts of absorbed dose and the gray unit are introduced in Part Two, section 11.

cles. Under these circumstances, is it more realistic to obtain an average dose over a critical region?[2]

Generally speaking, we shall be concerned with situations where it is accepted practice to average the dose over a fairly large region, say over an organ. We would not usually average over an area smaller than 1 cm^2 or a volume smaller than 1 cm^3. We shall, however, consider examples where it is preferable, on the basis of available data, to average the dose over smaller regions. In particular, we shall discuss doses from tritiated precursors of nucleic acids that tend to concentrate in the chromosomes of cells. Here the dose is calculated by averaging the energy imparted over the volume of the nucleus rather than the whole cell. We shall, however, consider another situation involving extremely localized exposures—the dose from highly radioactive particles in the lungs—where there is little experimental or theoretical basis for averaging and interpreting the exposure.

1 DOSE FROM BETA-EMITTING RADIONUCLIDES INSIDE THE BODY

The fact that beta particles have very short ranges in tissue, of the order of millimeters or less, simplifies considerably the calculation of the dose from beta particles emitted by radionuclides in the body. We can assume that the beta particles impart their energy essentially at the point where they are emitted. This means that the rate at which beta particle energy is imparted per gram is equal to the rate at which the energy is emitted per gram in the medium containing the radioactive material.[3]

To calculate the dose, we first determine the initial dose rate, that is, the rate at which the energy is imparted to unit mass of tissue where the radionuclide has been taken up.[4] Then we calculate the dose over a specific time interval, that is, day, month, quarter-year, or the entire period the material is in the body. The calculation of the dose is complicated somewhat because in most cases the initial dose rate decreases over a period of time as

2. See NCRP, 1993b, Report 116. Most controlled radiation exposures are in the range where the effect is considered proportional to the dose, and under these conditions, it is justifiable to consider the mean dose over all cells of uniform sensitivity as equivalent to the actual dose distribution.

3. The rigorous definition of absorbed dose is the limit of the quotient of the energy imparted divided by the mass of the region under consideration as the mass approaches zero; that is, it refers to a value at a point. This is the sense in which it is generally used with reference to exposure from external sources of radiation, as in exposure to x rays. However, in considering internal emitters, we will use the term to refer to average doses over an extended region. It would be more precise, but also unnecessarily wordy, to use the term *average organ dose* or *average whole-body dose*.

4. Anatomical and physiological data for calculating doses may be found in ICRP, 1975, Publication 23.

a result of the turnover of the radionuclide in the tissue and its physical decay. We shall first see how to calculate the initial dose rate and then how this is used to evaluate the absorbed dose over an extended period.

1.1 Calculating the Initial Dose Rate

The following steps are followed in calculating the initial dose rate from radioactivity in a tissue or organ:

1. Select the region where the dose rate is to be determined.
2. Determine the activity in the region, expressed in terms of becquerels (Bq). Divide by the mass of matter in the region to give the concentration in Bq/g.
3. Determine the energy emitted per second per gram of matter. For beta emission, this also essentially equals the energy imparted per second per gram to the medium.
4. Convert to dose rate in appropriate units, for example, mGy/hr.

Example 3.1 Calculate the initial dose rate to the body from the ingestion of 100 MBq of tritiated water.

Mass of region irradiated (body water): 43,000 g
Concentration: 100 MBq/43,000 g = 0.0023 MBq/g
Energy emitted/sec-g: 0.0023 MBq/g × 10^6 dis/sec-MBq × 0.006
 MeV/dis = 13.8 MeV/sec-g
Dose rate:

$$\frac{13.8 \, \text{MeV/sec-g}}{6.24 \times 10^6 \, \text{MeV/g-mGy}} = 2.21 \times 10^{-6} \, \text{mGy/sec} = 0.0080 \, \text{mGy/hr}$$

1.2 Dose Calculations for a Decaying Radionuclide

If the initial dose rate remains constant, the total dose equals the product of the initial dose rate and the time interval. An initial dose rate of 8 μGy/hr will give 80 μGy in 10 hr and 800 μGy in 100 hr. For many situations, however, the dose rate does not remain constant over the period of interest. If a radionuclide is giving off radiation, it is losing atoms through radioactive decay. If it is losing atoms, there are fewer remaining to decay. The fewer atoms remaining, the lower the rate at which they are decaying. When half the atoms have disappeared, the remaining atoms will decay at half the initial rate.

How do we calculate the total dose when the activity of a nuclide is decreasing over the period in which we are interested? We must determine the total number of disintegrations during that time period and then calculate the total dose from the total number of disintegrations in the same way we determined the initial dose rate from the initial disintegration rate. In the case of simple radioactive decay—that is, when the activity is decaying with a constant half-life—the calculation of the total disintegrations from the initial disintegration rate can be derived very simply from the analysis of basic radioactive decay relationships.

1.3   Some Relationships Governing Radioactive Decay

A radioactive substance is continuously losing atoms through decay and loses the same fraction of its atoms in any particular time period. Thus, if a particular radionuclide loses 10 percent of its atoms in a day, we will find it will lose 10 percent of its remaining atoms in the next day, and so on.

We call the fraction of the atoms that undergo decay per unit time the decay constant and give it the symbol λ.

$$\lambda = \text{decay constant}$$

For example, iodine-131 undergoes decay at the rate of 8.6 atoms per hundred atoms per day, or $8.6/24 = 0.36$ atoms per hundred atoms/hr, so $\lambda = 0.086/\text{day} = 0.0036/\text{hr}$.

Note that 0.086/day is an instantaneous rate. If we start with 100,000 iodine atoms, they will begin to decay at the rate of $0.086 \times 100,000$ or 8,600/day. This does not mean that we will have lost 8,600 atoms by the end of the day. The reason is, of course, that the actual rate of loss is only 8,600/day when all the 100,000 atoms are present. As the number of atoms decreases, the absolute rate of loss decreases, even though the fractional or relative rate of loss remains constant. At the end of a day, the actual number of atoms that have decayed is equal to 8,300; 91,700 remain, and they decay at the rate of $0.086 \times 91,700 = 7,900/\text{day}$.

We can express the fact that the number of atoms that decay per unit time is proportional to the number of atoms present in equation form:

$$\text{Disintegrations/unit time} = \text{decay constant} \times \text{number of atoms}$$
$$A = \lambda \times N$$

(Using calculus notation with disintegrations/unit time given by dN/dt, $dN/dt = -\lambda N$. The negative sign reflects the fact that the number of atoms decreases with time.)

If the number of atoms is doubled, the activity doubles. Halving the number of atoms halves the activity.

Since the disintegrations result in the emission of radiation that we can detect, we can determine the disintegration rate at any time simply by counting and by using suitable factors for converting counting rates to activity. If we know the activity, we know the number of atoms in the sample.

$$\text{Number of atoms in sample} = \frac{\text{activity}}{\text{decay constant}}$$

$$N = \frac{A}{\lambda}$$

Example 3.2   Calculate the number of ^{131}I atoms in a sample, if their disintegration rate is 300/min and $\lambda = 0.000060$/min.

$$N = 300/0.000060 = 300/6 \times 10^{-5} = 5 \times 10^{6} \text{ atoms}$$

Eventually, of course, all the radioactive atoms will disintegrate. Thus N gives the total number of disintegrations that will be undergone by the sample during its life.

Replacing N by the total disintegrations, we can write:

$$\text{Total disintegrations} = \frac{\text{activity}}{\text{decay constant}}$$

It can be shown (see note 5) that the reciprocal of the decay constant, $1/\lambda$, equals the average life of the atoms in a radioactive sample. Thus,

$$\text{Total disintegrations} = \text{initial activity} \times \text{average life}$$

$$N = AT^{a}$$

$$T^{a} = 1/\lambda$$

For ^{131}I, $\lambda = 0.00006$/min.,

$$T^{a} = \frac{1}{0.00006/\text{min}} = 1.67 \times 10^{4} \text{ min} = 11.6 \text{ days}$$

An ^{131}I solution with an initial disintegration rate of 300/min will undergo $300 \times 1.67 \times 10^{4}$ or 5×10^{6} disintegrations before all the iodine is lost by decay.

We have already introduced the concept of half-life, T^{b}. It can be shown that the half-life is 0.693 times the average life.[5] It follows that:

$$T^{b} = 0.693\,T^{a}$$

$$T^{a} = 1.44\,T^{b}$$

$$N = 1.44 A T^{b}$$

---

5. If N_0 = number of atoms at a time $t = 0$, integration of $dN/dt = -\lambda N$ gives the number of atoms N at any time t as $N = N_0\, e^{-\lambda t}$. In one half-life (T^b), $N/N_0 = 0.5 = e^{-\lambda T^b}$. $T^h = -\ln 0.5/\lambda = 0.693/\lambda = 0.693\,T^a$.

For samples with long half-lives, it would take a long time for essentially all the disintegrations to occur. Suppose we wanted to know the number of disintegrations occurring over a short period of time, for example, in order to evaluate the dose imparted over that time.

We calculate the number of disintegrations occurring over a period of time by determining the number of atoms remaining at the end of the period and subtracting from the initial number of atoms. The number remaining equals the fraction remaining at the end of the time period times the initial number. The fraction of atoms remaining is evaluated by converting the time period to a number of half-lives, n, and calculating $(\frac{1}{2})^n$.

There are other ways of presenting the fraction remaining versus time relationship, for example: $f = e^{-\lambda t}$ or $f = e^{-0.693 t / T^b}$.

Note that the decay of the activity follows the same relationship as the decay of the number of atoms, since both are directly proportional to each other.

Example 3.3   What is the fraction of ^{131}I atoms or fractional activity remaining after 6 days?

Since $T^b = 8.1$ days, the elapsed half-lives $= 6/8.1 = 0.741$. Thus, $f = 0.598$ (Part Two, Fig. 2.11).

Example 3.4   How many atoms of ^{131}I in the sample described in Example 3.2 decay during the 6-day period?

The number of atoms decayed $=$ initial number $-$ number left.

$$N_d = N - fN = N(1 - f)$$

The initial number of atoms is 5×10^6. At 6 days, the fraction 0.598 remains, or $0.598 \times 5 \times 10^6 = 2.99 \times 10^6$ atoms. Atoms decayed $= 5 \times 10^6 - 2.99 \times 10^6 = 2.01 \times 10^6$. An alternate calculation is $5 \times 10^6 \times (1 - 0.598) = 2.01 \times 10^6$.

Since the activity is the quantity that is known rather than the number of atoms, it is more useful to write N as AT^a and the equation for the number of atoms decayed in a given time period becomes $N_d = AT^a (1 - f)$.

1.4   Relationships Involving Both Radioactive Decay and Biological Elimination

The loss of atoms of a radionuclide from a region may be due not only to physical decay but also to biological elimination of the nuclide from the region. The total rate of loss of atoms is the sum of the rates of the loss from physical decay and biological elimination.

We introduce a biological decay constant given by λ_b and call the total fractional rate of loss by all mechanisms, the effective decay constant, λ_e.

$$\begin{matrix} \text{Effective fractional} \\ \text{rate of loss} \end{matrix} = \begin{matrix} \text{Fractional rate from} \\ \text{physical decay} \end{matrix} + \begin{matrix} \text{Fractional rate from} \\ \text{biological elimination} \end{matrix}$$

$$\lambda_e \quad = \quad \lambda_p \quad + \quad \lambda_b$$

Example 3.5   Phosphorous-32 undergoes physical decay at the rate of 4.85 percent per day. In addition, it is eliminated biologically at the rate of 1.44 percent per day. What is the effective decay constant?
   $\lambda_e = 0.0485 + 0.0144 = 0.0629/\text{day}$. The total loss rate is 6.29 percent per day.

As a result of biological elimination, the effective average life of the radioactive atoms in the tissue is less than the physical average life. Just as the physical average life $T_p^{a} = 1/\lambda_p$, we can write:

$$\text{Effective average life: } T_e^{a} = \frac{1}{\lambda_e} = \frac{1}{\lambda_p + \lambda_b}$$

$$\text{Effective half-life: } T_e^{h} = \frac{0.693}{\lambda_p + \lambda_b} = \frac{T_p^{h} T_b^{h}}{T_p^{h} + T_b^{h}}$$

For phosphorus-32:

$$T_p^{a} = \frac{1}{0.0485} = 20.6 \text{ days}$$

$$T_b^{a} = \frac{1}{0.0144} = 69.5 \text{ days}$$

$$T_e^{a} = \frac{1}{0.0629} = 15.9 \text{ days}$$

$$T_e^{h} = 0.693 \times 15.9 = 11.0 \text{ days}$$

We can now calculate the total number of disintegrations in a time period if there are both physical decay and biological elimination. We calculate it in the same manner as we did when we had only physical decay, but use the effective average life instead of the physical average life. Thus the total number of disintegrations that can occur in a region is given by the product of the initial activity and the effective average life of the activity in the region.

Example 3.6   The initial disintegration rate of a phosphorous-32 sample is 10,000 per minute. Calculate the total number of disintegrations expected.

Initial disintegration rate, expressed as disintegrations per day, = $1{,}440 \times 10{,}000 = 1.44 \times 10^7$. Total disintegrations = $15.9 \times 1.44 \times 10^7 = 22.9 \times 10^7$.

The number of disintegrations over a specific time interval is given by the product of the total disintegrations for complete decay times the fraction of the total that occurs during the time interval. The fraction that occurs is determined by subtracting the fraction of atoms remaining from 1. That is, disintegrations = initial activity $\times T_e^a \times$ (1 − fraction remaining).

The fraction remaining is calculated using the effective decay constant, or the number of effective half-lives elapsed during the period of interest.

Example 3.7   If the initial activity of ^{32}P is 10,000 dis/min, how many disintegrations would occur in 31 days?

$T_e^b = 11.0$ days, so the number of effective half-lives = 31/11.0 = 2.82. From Figure 2.11, the fraction remaining = 0.14. Therefore the number of disintegrations in 31 days = $22.9 \times 10^7(1 - 0.14) = 19.7 \times 10^7$.

1.5   Absorbed Beta Dose over a Period of Time

The disintegrations evaluated in section 1.4 are accompanied by radiations that are responsible for the dose to the patient over the period of interest. Since the dose and disintegrations are directly related to each other, the same formula can be used to calculate total dose as was used to calculate total disintegrations merely by substituting initial dose rate for initial disintegration rate.

Example 3.8   Suppose the initial dose rate to the bone marrow from phosphorous-32 is calculated to be 1 mGy/hr. What is the dose imparted in 31 days?

The initial dose rate = $1 \times 24 = 24$ mGy/day. The total dose = initial dose rate $\times T_e^a = 24$ mGy/day $\times 15.9$ days = 382 mGy. The dose in 31 days = $382(1 - 0.14) = 329$ mGy.

The following formula for the beta dose in a given time is readily derived:[6]

$$D = 19,900 C E_\beta T_e^{h} (1 - f) \tag{3.1}$$

where

D = absorbed dose in mGy
C = concentration in MBq/g
E_β = average beta energy per disintegration in MeV
T_e^{h} = effective half-life in days
f = fraction remaining at end of time period

When radioactive substances follow complex pathways in the body, it may not be possible to describe the variations in regions of interest in terms of a single half-life. Sometimes the variation can be approximated as the sum of two or more components, each decaying with a characteristic half-life, and it is still possible to use the equations given in this section (see section 5.5). If the simple decay relationships do not apply, however, it may be necessary to obtain detailed data on the accumulation and elimination as a function of time, and from these data to determine the total number of disintegrations during the period of interest and hence the dose (see section 5.2).

2 A CLOSER LOOK AT THE DOSE FROM BETA PARTICLES

Our treatment of beta dose from a source distributed in tissue was based on local absorption of the beta particle energy averaged over the tissue in which the beta particles were emitted. We did not need to concern ourselves with the specific details of the distribution of the source and the penetration of the beta particles after emission. However, there are instances where the manner in which the beta particles impart their dose after emission from a source must be considered for a proper evaluation of the dose. Examples of cases requiring such analysis include the use of beta plaques for radiotherapy and the evaluation of injury from irradiation of the skin by beta contamination deposited on it.

2.1 Beta Particle Point Source Dose-Rate Functions

The dose rate from beta particle sources may be evaluated with the use of basic equations known as beta particle point source dose-rate functions.

---

6. The factor 19,900 comes from multiplying 86,400 × 10[6] (dis/day per MBq) × 1.44 (conversion from half-life to average life) and dividing the product by 6.24 × 10[6] (conversion from MeV/g to mGy).

These functions give the dose rate as a function of distance from a point source of beta particles. They are derived either by experimental or by theoretical methods. The general equations are quite complex and are given by Hine and Brownell (1956). With these dose-rate functions, one can determine the dose distribution from any arbitrary source configuration.

2.2 Evaluation of Beta Particle Dose from the Fluence and Stopping Power

Dose calculations are often made by first evaluating a quantity known as the fluence. The fluence at a point irradiated by a beam of particles is calculated by dividing the total number of particles in the beam by its cross-sectional area. If, at the point of interest, beams of particles are incident from many directions, then the fluence must be evaluated for each direction separately, and the results summed.

Example 3.9   Particles are incident upon a point from three different directions, as illustrated in Figure 3.1. Beam A has a cross-sectional area of 5 cm^2 and consists of a total of 1,000 particles. Beam B has a cross-sectional area of 2 cm^2 and consists of a total of 500 particles. Beam C has a cross-sectional area of 1 cm^2 and consists of a total of 100 particles. What is the fluence at point *P*?

The fluence is 1,000/5 contributed by A, 500/2 contributed by B, 100 contributed by C, for a total of 550 particles/cm^2.

Note that the fluence refers to an accounting of particles over a period of time. Frequently, we are concerned with irradiation per unit time, in which case we use a quantity known as the flux. The flux at a point is the number of particles passing per unit time, per unit area of the beam. For example, if the 1,000 particles carried by beam A pass a point in 2 seconds,

---

3.1   Fluence at a point irradiated by multiple beams.

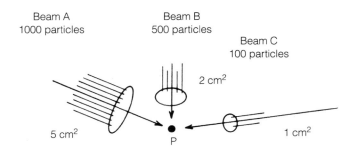

Beam A
1000 particles

Beam B
500 particles

Beam C
100 particles

2 cm^2

5 cm^2

1 cm^2

P

then the flux is 1,000/(5 × 2), or 100/cm²-sec. The calculation of the flux at various distances from a point source is described in section 3.2.2.

If the fluence is multiplied by the rate at which the particles impart energy per centimeter of travel through the medium, the result is the energy imparted per unit volume. If the energy imparted per unit volume is divided by the density, the result is the energy imparted per unit mass, or the absorbed dose. The rate at which beta particles lose energy per unit distance in a medium is called the stopping power. For the energies emitted by beta particle sources, almost all of the stopping power is due to ionization and excitation and the energy is imparted close to the track. Thus the rate of energy loss along the track, the stopping power, is also essentially equal to the energy absorbed along the track, the linear energy transfer. The stopping power may also include a contribution from the loss of energy by radiation (bremsstrahlung). The radiation is not absorbed locally and hence does not add to the stopping power. However, this occurs to an appreciable degree only for electrons at higher energies, such as are attained in accelerators.

Example 3.10   The linear stopping power for 1 MeV electrons in water is 1.89 MeV/cm.[7] What is the dose at the point where a beam of 1 MeV electrons enters a water medium, if the beam contains 10,000 electrons/cm² (that is, fluence = 10,000/cm²)?

The initial rate of loss by the electrons in 1 cm² of the beam is 10,000 × 1.89 = 18,900 MeV/cm. The energy lost per unit volume at the point of incidence is 18,900 MeV/cm³. The energy imparted per gram (obtained by dividing by the density of 1 g/cm³) = 18,900/1 = 18,900 MeV/g. The absorbed dose in mGy = 18,900/6.24 × 10⁶ = 0.00303 mGy.

For practical reasons, the evaluation of radiation hazards is based on the absorbed dose. However, the absorbed dose is determined by averaging the energy transferred over gross volumes or masses, while the actual damage-producing mechanisms involve the transfer of energy to individual molecules. Much of the research in radiation biology is concerned with determining the actual patterns of ionizations and excitations and interpreting their significance. As a consequence of this work, the analysis of radia-

---

7. At a rate of loss of 1.89 MeV/cm, a 1 MeV particle would travel a distance of 1/1.89 cm before coming to rest if the value of the stopping power remained constant as the energy of the particle decreased. The distance traveled is less than this value, since the stopping power increases as the particle slows down.

tion effects may be based ultimately on consideration of ion clusters, delta rays, time sequences of ionizations, and so on (Rossi, 1968; Zaider, 1996; Kellerer, 1996), although operational surveys probably will retain, as their basis, determination of the mean energy imparted per gram of tissue.

3   Calculation of the Absorbed Dose from Gamma Emitters in the Body

In considering the dose from beta emitters distributed uniformly throughout an organ, we were able to assume that the energy carried by the beta particles was locally absorbed, because of their short range. As a result, if we knew the energy released per gram, we could equate this to the energy absorbed per gram, and thus calculate the absorbed dose.

When we deal with gamma photons, the situation is different. The photons are uncharged in contrast to the charged beta particles. Upon leaving a nucleus, they travel in a straight line without any interaction until there is a collision with an atom or one of its external electrons. No matter what distance is considered, there is a finite probability that a photon can travel that distance without making a collision, but the probability is less as the amount of material in the path of the photon increases. On the average, the distance traveled by gamma photons in tissue before their first collision is measured in centimeters.

The collision of the photon gives rise to an energetic electron that actually imparts the dose to the tissue and that, because of its short range, is considered as locally absorbed.

3.1   Dose Rate from a Point Source of Photons—The Specific Dose-Rate Constant for Tissue

Calculation of the dose rate within the body from gamma emission throughout the body is more complicated than that for beta particles because the dose is imparted over an extended distance from the source. The dose to tissue at a given point depends on both the distance to the source and the attenuation in the intervening medium. If we exclude the attenuation factor, we can write for the dose rate from a point source of gamma photons an expression of the form

$$DR = \Gamma_{tis} A / r^2 \tag{3.2}$$

where Γ_{tis} is the specific dose-rate constant for tissue.[8] A is the activity of the source, and r is the distance.

The point source, to which the equation above applies, is, of course, a mathematical idealization. All actual sources have finite dimensions. The

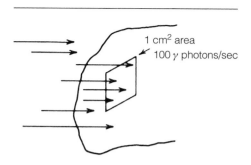

3.2 Photon flux incident on tissue.

"point source" equation applies to a source when the relative distances to various parts of the source from the dose point of interest do not vary appreciably, and attenuation within the source is negligible. A few milliliters of radioactive liquid in a test tube may for all practical purposes be considered a point source if the dose is to be evaluated 10 cm from the test tube. On the other hand, if the test tube is being held in the hand, the contributions to the dose to the fingers vary greatly for different parts of the source. When it is possible to represent sources of radiation as point sources, the calculations are simplified considerably.

3.2 Evaluation of the Specific Dose-Rate Constant

The constant Γ for use in the equation above is determined by calculating the absorbed dose rate for the flux of photons produced at a distance of 1 cm from a source which has an activity of 1 MBq.

3.2.1 Calculation of the Absorbed Dose Rate from the Flux

Consider a beam of gamma photons incident on tissue, as in Figure 3.2. Let us mark out at the surface an area of 1 cm^2 perpendicular to the beam and assume 100 photons of energy equal to 1 MeV are crossing the area per second.

These photons will impart to the medium a fraction of their energy per unit distance as they travel through, about 3 percent per centimeter in soft tissue. Thus, the energy imparted in a region of cross-section 1 cm^2 and 1 cm deep—that is, to a volume of 1 cm^3—is 100/cm^2-sec $\times$ 1.0 MeV $\times$ 0.03/cm $=$ 3.0 MeV/cm^3-sec. Since the density is 1 g/cm^3, the energy imparted per gram is

$$\frac{3 \text{ MeV/cm}^2\text{-sec}}{1 \text{ g/cm}^3} = 3 \text{ MeV/g-sec}$$

The absorbed dose rate is

$$\frac{3 \text{ MeV/g-sec} \times 3{,}600 \text{ sec/hr}}{6.24 \times 10^6 \text{ MeV/g-mGy}} = 0.00173 \text{ mGy/hr}$$

If the energy per photon had been twice as great, the dose rate from the 100 photons/cm^2-sec would also have been twice as great, or 0.00346

8. The specific dose-rate constant, when dose units are expressed in cGy to soft tissue, is only 3 percent less than the specific gamma ray constant (Γ), which gives the exposure in roentgens. The data in the literature are generally for the specific gamma ray constant, since this is defined for interactions in air, which can be more precisely defined than tissue. For purposes of numerical calculations in radiation protection, however, both constants may be used interchangeably. Elementary examples on the use of Γ were given in Part Two, section 21.2.

mGy/hr. The dose rate due to 100 photons/cm^2-sec of energy E MeV per photon is $0.00173E$ mGy/hr.

Let us designate the number of photons per square centimeter per second as Φ. The dose rate is equal to $(1.73 \times 10^{-5})\Phi E$ mGy/hr.

The quantity Φ is the flux (introduced in section 2.2). The fraction of the incident energy locally absorbed per centimeter is called the energy-absorption coefficient, μ_{en}.[9] The mass energy-absorption coefficient, μ_{en}/ρ, is obtained by dividing the energy-absorption coefficient by the density.[10] The product of μ_{en}/ρ, Φ, and E gives the energy locally absorbed per unit mass per unit time. Detailed tables have been prepared giving the values of μ_{en}/ρ in various media (Evans, 1968; Jaeger et al., 1968; Chilton et al., 1984; Kahn, 1997). Some values are given in Table 3.1.

Example 3.11   What is the energy absorbed per gram per second and the resultant dose rate in muscle from a flux density of 100 gamma photons/cm^2-sec with photon energy equal to 2 MeV? Note that μ_{en}/ρ = 0.0257 cm^2/g.

The rate of energy absorption = $0.0257 \times 100 \times 2 = 5.14$ MeV/g-sec. Therefore, the absorbed dose rate = $(5.14 \times 3{,}600)/(6.24 \times 10^6)$ = 0.00296 mGy/hr.

While the example here is for photons all traveling in the same direction, regions exposed to radiation are usually irradiated from many directions (as discussed in section 2.2). One way of determining the flux in a region is to imagine a sphere is present with a cross-sectional area of 1 cm^2. Then the flux is equal to the number of particles crossing the surface of the sphere per second. (If the sphere were drawn in Figure 3.1 in section 2.2, the unit areas of the beams shown would correspond to sections of the sphere that passed through its center and were oriented perpendicular to the directions of the beams.)

3.2.2   Calculation of Flux from the Activity and Distance

The flux at a distance from a point source can be calculated very simply in the absence of attenuation, since the photons are emitted with equal in-

---

9. Compare to the attenuation coefficient, which gives the fraction (or probability) of photons interacting per unit distance (see Part Two, section 8.4).

10. The *mass energy-absorption* coefficient excludes energy carried by Compton-scattered, fluorescence, annihilation, and bremsstrahlung photons as locally absorbed. The *mass energy-transfer* coefficient excludes all of these except bremsstrahlung photons. The *mass absorption* coefficient excludes Compton-scattered photons only.

Table 3.1 Values of the mass energy-absorption coefficient (cm²/g).

Photon energy (MeV)	Medium				
	Air	Water	Aluminum	Iron	Lead
0.01	4.61	4.84	25.5	1.42	1.31
0.05	0.0403	0.0416	0.1816	1.64	6.54
0.10	0.0232	0.0254	0.0377	0.219	2.28
0.50	0.0296	0.0210	0.0286	0.0293	0.0951
1.0	0.0278	0.0309	0.0269	0.0262	0.0377
1.5	0.0254	0.0282	0.0245	0.0237	0.0271
3.0	0.0205	0.0227	0.0202	0.0204	0.0234
6.0	0.0164	0.0180	0.0172	0.0199	0.0272

Sources: Chilton et al., 1984, after Hubbell and Berger, 1966; Kahn, 1997, after Hubbell, 1982.

tensity in all directions and travel in straight lines from the source. To evaluate the flux at a distance r, we imagine a spherical surface of radius r around the source (Fig. 3.3). This surface has an area of $4\pi r^2$, which is irradiated uniformly by the photons emitted from the source. Thus, if the source strength S is defined as the number of photons emitted per second, the photons penetrating through unit area per second is $S/4\pi r^2$. When the activity of the source (disintegrations per unit time) is given, the number of photons resulting from one disintegration must be determined before the strength can be calculated.

3.3 Illustration of inverse square law. Spherical surface of radius r around source has area of $4\pi r2$. Number of photons passing through 1 cm² of surface per second (flux density) equals number emitted per second divided by $4\pi r2$ and thus varies as $1/r^2$.

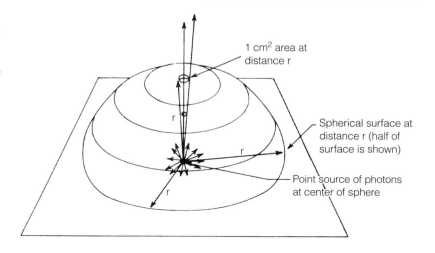

1 cm² area at distance r

Spherical surface at distance r (half of surface is shown)

Point source of photons at center of sphere

Example 3.12   What is the flux at 1 and 5 cm from a 1 MBq source of ^{131}I?

The photon emission from iodine-131 appears as 0.364 MeV gammas in 80 percent of the disintegrations and 0.638 MeV gammas in 8 percent of the disintegrations.

The source strength is equal to 10^6/sec-MBq $\times$ 0.8 = 8.0×10^5 photons/sec of 0.364 MeV energy, and $10^6 \times 0.08 = 8.0 \times 10^4$ photons/sec of 0.638 MeV energy.

The flux at 1 cm is equal to $(8.0 \times 10^5)/4\pi(1)^2 = 63{,}662$ photons/cm^2-sec at 0.364 MeV and $(8.0 \times 10^4)/4\pi(1)^2 = 6{,}366$ photons/cm^2-sec at 0.638 MeV

The flux at 5 cm is 1/25 the flux at 1 cm and is equal to 2,546/cm^2-sec at 0.364 MeV and 254.6/cm^2-sec at 0.638 MeV.

3.2.3   *Conversion from the Flux for Unit Activity and Unit Distance to the Specific Dose-Rate Constant*

To calculate the specific dose-rate constant for tissue, Γ_{tis}, we need to determine the flux at a distance of 1 cm from a 1 MBq source and then convert to dose rate by the method shown in section 3.2.1.

Example 3.13   Calculate Γ_{tis} for ^{131}I.

In the preceding section, we calculated the flux at 1 cm for a 1 MBq source of ^{131}I for the two photon energies emitted by the radionuclide. The mass energy-absorption coefficients are 0.032 cm^2/g for both energies. At 0.364 MeV, $DR = 0.032 \times 63{,}662 \times 0.364 \times 3{,}600/(6.24 \times 10^6) = 0.4278$ mGy/hr. At 0.638 MeV, DR = $0.032 \times 6{,}366 \times 0.638 \times 3{,}600/(6.24 \times 10^6) = 0.075$ mGy/hr. The total dose rate = 0.428 + 0.075 = 0.503 mGy/hr.

This differs somewhat from the actual value given in the literature because ^{131}I emits additional photons that were not included in the calculation.

The specific dose-rate constant gives the dose rate from 1 MBq of a specific radionuclide at any distance r, in the absence of attenuation, by dividing by r^2. In any specific instance, attenuation of the radiation in the medium is usually significant and can be evaluated by the methods presented in Part Two, section 8.5.

3.3 Dose Rate from Distributed Gamma Sources

When we cannot consider the gamma source as a single point source—for example, when we must calculate the dose from a gamma emitter distributed throughout the body—we must divide the larger source into smaller sources, each of which can be considered as a point source. Ideally, the source should be divided into infinitesimally small elements and the dose evaluated by means of the calculus. For all but the simplest geometries, the resultant equations become too difficult to solve. As an alternative, subdivisions are established on a larger scale and the problem is solved numerically with the use of computers.

3.4 The Absorbed-Fraction Method—Dose within the Source Volume

Gamma-dose equations are more complex than beta-dose equations because most of the emitted gamma energy is absorbed at some distance from the source. If the dose points of interest are within the source volume, however, it is possible to adapt the methods and equations for beta dose to evaluate gamma dose by introducing a correction factor known as the absorbed fraction, ϕ. This factor gives the fraction of the gamma energy emitted by the source that is absorbed within the source. It is specific for the source geometry, photon energy, and source material. A calculation of gamma dose by the absorbed fraction method proceeds by first assuming that all the gamma energy emitted by the source is absorbed within the source. The dose is calculated as for beta particles, and the results are multiplied by the absorbed fraction to correct for the gamma energy that is not absorbed within the source. The concept of absorbed fraction was developed to provide a simple method for gamma-dose calculations. Values of absorbed fractions have been calculated for a large variety of radionuclides and organs (Snyder et al., 1969).

Referring to equation 3.1 for beta dose in section 1.5, we can convert it to an equation for gamma dose by replacing E_β by $E_\gamma n_\gamma \phi$, where n_γ is the mean number of photons of energy E_γ emitted per disintegration and ϕ is the absorbed fraction. When photons of several energies are emitted, the contributions are summed (see Example 3.14). Thus,

$$D = 19{,}900 \cdot C \cdot E_\gamma \cdot n_\gamma \cdot \phi \cdot T_e^b \cdot (1-f) \quad (3.3)$$

$$\text{mGy} \qquad \text{MBq/g} \quad \text{MeV} \qquad \text{days}$$

Example 3.14 Calculate the gamma dose imparted in 7 days from an intravenous injection of 500 MBq ^{24}Na in a 70 kg man.

The dose contributions must be calculated separately for the 1.37 and 2.75 MeV photons emitted for each disintegration. By linear extrapolation of the data of Snyder et al. (1969), we determine absorbed fractions of 0.307 for 1.37 MeV photons and 0.268 for 2.75 MeV photons. The effective half-life is 0.625 days, and the term $(1 - f)$ is close to 1 for a 7-day dose.

The dose in mGy is:

$$\overset{C}{19,900} \times \frac{\overset{}{500}}{70,000} \times [(\overset{E_{\gamma 1}}{1.37} \times \overset{\phi_1}{0.307}) + (\overset{E_{\gamma 2}}{2.75} \times \overset{\phi_2}{0.268})] \times \overset{T_e^h}{0.625} \times \overset{f}{(1 - .000425)}$$

$$= 103 \text{ mGy}$$

Some values of the absorbed fraction are given in Table 3.2.

3.5   Dose to Targets outside the Source Volume by the Absorbed-Fraction Method

The absorbed-fraction concept may also be applied to evaluation of the dose to organs external to the region where the source is localized. In this case the absorbed fraction is defined as

$$\phi = \frac{\text{photon energy absorbed by target}}{\text{photon energy emitted by source}}$$

To calculate the dose in a target removed from the source, we multiply the photon energy emitted from the source by the absorbed fraction and divide by the mass of the target. The dose equation then becomes:

$$D = 19,900 \cdot \frac{\text{source activity}}{\text{target mass}} \cdot E_\gamma \cdot n_\gamma \cdot \phi \cdot T_e^h \cdot (1 - f) \quad (3.4)$$

$$\text{mGy} \qquad\qquad \frac{\text{MBq}}{\text{g}} \quad \text{MeV} \qquad\quad \text{days}$$

The contributions at each photon energy must be summed.

Example 3.15   Calculate the dose to the vertebral marrow from 0.37 MBq of ^{131}I in the thyroid.

We use values of absorbed fractions calculated for a mathematical model of the spine, which is treated as an elliptical cylinder, length 56.5 cm, volume 887.5 cm^3, density 1.5 g/cm^3, mass 1,331 g (Snyder et al., 1969).

The calculation for the various photon energies (Dillman, 1969) emitted from ^{131}I atoms proceeds as follows:

Photon energy	Mean number/ disintegration	Absorbed fraction[11]	$E_\gamma \cdot n_\gamma \cdot \phi \cdot 10^7$
0.03	0.047	0.00255	36
0.080	0.017	0.0099	135
0.284	0.047	0.0060	802
0.364	0.833	0.00578	17,500
0.637	0.069	0.0052	2,310
0.723	0.016	0.0052	603
			21,386

The total absorbed energy is 0.00214 MeV. The gamma dose to the spine in mGy $= 19,900 \times (0.37/1,331) \times 0.00214 \times 7.2 = 0.0852$ mGy.

Table 3.2 Absorbed fractions for gamma radiation sources uniformly distributed in various organs in the adult male.

Organ	Mass (g)	Energy (MeV)				
		0.03	0.1	0.5	1.0	1.5
Thyroid	20	0.15	0.029	0.033	0.031	0.029
Lung (inc. blood)	1,000	0.23	0.051	0.050	0.046	0.043
Liver	1,800	0.53	0.16	0.16	0.15	0.13
Kidney	310	0.32	0.073	0.078	0.070	0.066
Whole body	70,000	0.80	0.38	0.35	0.33	0.31

Source: Derived from specific absorbed fractions in ICRP, 1975, Publication 23. See also Snyder et al., 1969.

3.6 The Specific Absorbed Fraction—Sparing the Need to Divide by the Target Mass

The computer code that goes through the complicated mathematics of calculating the fraction of the energy emitted by the source that is absorbed by the target can just as well divide the result by the target mass and spare the need for this extra step in determining the dose. The resulting quantity is called the *specific absorbed fraction.* Extensive compilations of the specific absorbed fraction as a function of energy, source organ, and

11. Obtained by plotting and drawing curve through values of absorbed fractions given by Snyder et al., 1969.

Table 3.3 Equilibrium dose constants (Δ_i) and mean energies of radiation ($\bar{E}_i$) emitted from technetium-99m (half-life = 6 hr).

Input data		
Radiation	%/disintegration	Transition energy (MeV)
Gamma-1	98.6	0.0022
Gamma-2	98.6	0.1405
Gamma-3	1.4	0.1427

Output data			
Radiation (i)	Mean number/ disintegration (n_i)	($\bar{E}_i$) (MeV)	(Δ_i) g-rad/μCi-hr
Gamma-1	0.000	0.0021	0.0000
M int. con. electron, Gamma-1	0.986	0.0017	0.0036
Gamma-2	0.883	0.1405	0.2643
K int. con. electron, gamma-2	0.0883	0.1195	0.0225
L int. con. electron, gamma-2	0.0109	0.1377	0.0032
M int. con. electron, gamma-2	0.0036	0.1401	0.0011
Gamma-3	0.0003	0.1427	0.0001
K int. con. electron, gamma-3	0.0096	0.1217	0.0025
L int. con. electron, gamma-3	0.0030	0.1399	0.0009
M int. con. electron, gamma-3	0.0010	0.1423	0.0003
K α-1 x rays	0.0431	0.0184	0.0017
K α-2 x rays	0.0216	0.183	0.0008
K β-I x rays	0.0103	0.0206	0.0005
K β-2 x rays	0.0018	0.0210	0.0001
L x rays	0.0081	0.0024	0.0000
KLL Auger electron	0.0149	0.0155	0.0005
KLX Auger electron	0.0055	0.0178	0.0002
KXY Auger electron	0.0007	0.0202	0.0000
LMM Auger electron	0.106	0.0019	0.0004
MXY Auger electron	1.23	0.0004	0.0010

Source: Dillman, 1969, supplement no. 2. To convert to g-mGy/MBq-hr, multiply by 270.27.

target organ are available (ICRP, 1975; Cember, 1996). The calculations proceed along the same lines as given in the previous section for the absorbed fraction, except that the ratio *absorbed fraction/mass* is replaced by the *specific absorbed fraction*.

3.7   Use of the Equilibrium Dose Constant—Computer-Generated Source Output Data

When sources emit primarily beta particles and high-energy gamma photons, it is usually possible to calculate doses directly from the data given in decay schemes. However, in the case of radionuclides that emit low-energy gamma rays or decay by electron capture, the radiations emitted in significant numbers from the source may result from several complex processes, including internal conversion, x-ray emission, the production of Auger electrons, and so on. The contributions from these sources can be evaluated only by extensive calculations. Fortunately, the required calculations have been made for most of the radionuclides of interest in nuclear medicine (Dillman, 1969). The output data obtained with digital computers include a listing of each particle emitted from the atom, the mean number per disintegration, the mean energy per particle, and an equilibrium dose constant, Δ. The equilibrium dose constant gives the absorbed dose rate on the assumption that all the emitted energy is locally absorbed. The traditional units, rad/hr for a concentration of 1 μCi/g, are given in Dillman (1969). To convert to SI units, multiply by 270.27 (10 mGy/rad divided by 0.037 MBq/μCi). An example of the input and output data for technetium-99m and the equilibrium dose constants is given in Table 3.3.

It is convenient to divide the particles listed in the output data into nonpenetrating and penetrating radiations. The doses from the nonpenetrating radiations are evaluated as for beta particles, while the doses from the penetrating radiations are determined from the absorbed fractions.

The dose equation for nonpenetrating radiation is

$$\underset{\text{mGy}}{D_{NP}} = 1.44 \cdot \underset{\frac{\text{MBq}}{\text{g}}}{C} \cdot \underset{\frac{\text{g-mGy}}{\text{MBq-hr}}}{\Delta} \cdot T_e^b \cdot (1-f) \qquad (3.5)$$

The equation for doses within a source of penetrating radiation is

$$\underset{\text{mGy}}{D} = 1.44 \cdot \underset{\frac{\text{MBq}}{\text{g}}}{C} \cdot \underset{\frac{\text{g-mGy}}{\text{MBq-hr}}}{\Delta} \cdot \phi \cdot \underset{\text{hr}}{T_e^b} \cdot (1-f) \qquad (3.6)$$

The equation for doses external to a source of penetrating radiation is

$$\underset{\text{mGy}}{D} = 1.44 \cdot \underset{\frac{\text{MBq}}{\text{g}}}{\frac{\text{source activity}}{\text{target mass}}} \cdot \underset{\frac{\text{g-mGy}}{\text{MBq-hr}}}{\Delta} \cdot \phi \cdot \underset{\text{hr}}{T_e^b} \cdot (1-f) \qquad (3.7)$$

Example 3.16   Repeat the calculation in Example 3.15 of the dose to the spine from 0.37 MBq of ^{131}I in the thyroid using the equilibrium dose constant.

The equilibrium dose constant (Dillman, 1969) is multiplied by the appropriate absorbed fraction (Snyder et al., 1969) for each photon energy and the products are summed to give $\Sigma \Delta_i \phi_i = 0.0046$ g-rad/μCi-hr or 1.243 g-mGy/MBq-hr. The dose to the spine is calculated from equation 3.7 (note the half-life is expressed in units of hours);

$$D = 1.44 \times \frac{0.37}{1331} \times 1.243 \times 7.2 \times 24$$

$$= 0.0860 \text{ mGy}$$

The result is within one percent of the value calculated in Example 3.15.

3.8   The *S* Factor—Doses from Cumulated Activity

The dose rate (milligrays/day) at a point results mainly from the local distribution of activity (megabecquerels). The total dose over a period of time is given by the cumulative effect of dose rate and time, and thus can be related to an exposure expressed in units of cumulated activity, $\bar{A}$ (megabecquerel-days). The average absorbed dose per unit cumulated activity is known as the *S* factor. The introduction of the *S* factor spares considerable numerical work in making dose calculations, and values of *S* have been tabulated for selected sources and target organs (Snyder et al., 1975). In applying this concept to dose calculations, it is necessary to plot activity, *A,* versus time and then to obtain the area under the curve for the time period of interest ($\int A dt$). The result (in MBq-days) is then multiplied by the *S* factor (in mGy/MBq-day) to give the dose. When the activity falls off with a constant average life, the cumulated activity is merely the initial activity times the average life times (1 − fraction remaining).

S values for a target organ are tabulated in handbooks both for uniform distributions of activity within the organ and for activity in remote organs. The organ masses used in the tables are typical for a 70 kg adult. The results can be scaled approximately for other cases (Synder et al., 1975).

4 SUMMARY OF FORMULAS

4.1 Radioactive Decay

$A = \lambda N$ A = activity

$T^a = 1/\lambda$ λ = decay constant

$T^b = 0.693/\lambda$ N = number of radioactive atoms, and also total number of disintegrations

$T^a = 1.44T^b$ T^a = average life

$N = AT^a$ T^b = half-life

$N = 1.44AT^b$ f = fraction remaining; λ and t (time) must be expressed in the same units of time—that is, if t is in sec, λ is in 1/sec

$f = e^{-\lambda t}$ n = number of half-lives

$f = (1/2)^n$

4.2 Physical Decay and Biological Elimination

$\lambda_e = \lambda_p + \lambda_b$ λ_e = effective decay constant

$T_e^a = 1/(\lambda_p + \lambda_b)$ λ_p = physical decay constant

$T_e^b = 0.693/(\lambda_p + \lambda_b)$ λ_b = biological decay constant

$T_e^b = 0.693T_e^a$ T_e^a = effective average life

$T_e^b = T_p^b T_b^b / T_p^b + T_b^b$ T_e^b = effective half-life

$A = \lambda_p N$ T_p^b = physical half-life

$N' = AT_e^a$ T_b^b = biological half-life

$f = e^{-\lambda_e t}$ A = activity

$f = (1/2)^n$ N = number of atoms

     N' = total no. of disintegrations

     f = fraction remaining

     n = number of effective half-lives

4.3 Dose from Nonpenetrating Radiation from Internal Emitters

$D_{NP} = 19,900 CET_e^b(1 - f)$ D_{NP} = dose in mGy from nonpenetrating radiation

     C = activity concentration of emitter in MBq/g

     E = average energy in MeV per disintegration carried by particles

     T_e^b = effective half-life in days

4.4   Dose from Penetrating Radiation from Internal Emitters

$$D_P = 19{,}900 C E_\gamma n_\gamma \phi T_e^h (1 - f)$$

D_P = dose in mGy within source of penetrating radiation

C = activity concentration in MBq/g

ϕ = absorbed fraction

T_e^h = effective half-life in days

E_γ = photon energy

n_γ = mean number of photons of energy E_γ per disintegration

$$D_P = (1.44)\Delta \phi T_e^h (1 - f)$$

Δ = equilibrium dose constant in g-mGy/MBq-hr

T_e^h = effective half-life in hr

$$D_{PX} = (1.44)(A_s/M_t)\Delta T_e^h \phi (1 - f)$$

D_{PX} = dose in mGy to organ external to source

A_s = activity of source in MBq

M_t = mass of target organ in g

Equations for penetrating radiation may be used for nonpenetrating radiation (beta particles and low-energy photons, say below 0.011 MeV) by setting $\phi = 1$.

4.5   Inverse Square Law

$$\Phi = S/4\pi r^2$$
$$= 0.080 S/r^2$$

Φ = flux in particles/cm^2-sec

S = particles emitted from source per sec

r = distance from source in cm

Approximate flux to dose rate conversion factors:

100 β particles/cm^2-sec = 0.1 mGy/hr to skin

100 γ photons/cm^2-sec, energy E per photon = 0.00172E mGy/hr to tissue

4.6   Dose Rates at a Distance from Gamma Sources

$$DR = (6/37)A E_\gamma n_\gamma / r^2$$

DR = approximate exposure rate in mR/hr

A = activity in MBq

$$DR = \Gamma A / r^2$$

$E_\gamma n_\gamma =$ mean photon energy per disintegration in MeV

$r =$ distance in ft

$DR =$ exposure rate in mGy per hr

$\Gamma =$ specific dose rate (or gamma ray) constant in mGy per MBq-hr at 1 cm

$r =$ distance in cm

$A =$ activity in MBq

4.7 Attenuation of Radiation

First, determine the dose rate or flux by neglecting attenuation. Then, multiply by the attenuation factor (AF).

$$AF = e^{-\mu x}$$

$$\mu = 0.693/\text{HVL}$$

$$\text{HVL} = 0.693/\mu$$

$AF =$ attenuation factor

$\mu =$ attenuation coefficient in cm^{-1}

$x =$ thickness of attenuating medium in cm

$\text{HVL} =$ half-value layer

AF by half-value layers method. Determine the number of half-value layers, *n,* corresponding to the thickness of the medium between the source and dose point. Obtain the attenuation factor AF from $(\frac{1}{2})^n$ or from the HVL curve.

4.8 Equivalent Dose

$$ED = QD$$

$$ED = w_R D$$

$ED =$ equivalent dose

$Q =$ quality factor

$w_R =$ radiation weighting factor

$D =$ absorbed dose

5 DOSE CALCULATIONS FOR SPECIFIC RADIONUCLIDES

We have seen that the calculation of dose is relatively simple in a region that is uniformly irradiated and in which the effective average life is constant over the period of interest. In practice, the distribution of radionuclides is quite complex, and the effective average life changes as the material is acted on through physiological and biochemical processes. It is the need to acquire accurate metabolic data rather than the performance of subsequent mathematical analyses that provides most of the difficulties in evaluating the dose from internal emitters. Simplifying assumptions made in the absence or in place of accurate data can sometimes lead to significant errors.

Two of the major potential sources of error produced by simplifications in dose calculations are: failure to consider localization of the radionuclides in special parts of an organ, or even within special parts of a cell, and failure to investigate the possibilities of long-term retention of a fraction of the ingested material, when most of the radioactivity appears to be eliminated with a short half-life.

In this section, we shall investigate these and other factors that influence the evaluation of radiation dose from internal emitters, with reference to some of the more important radionuclides that are now used in nuclear medicine or in other ways that result in the exposure of human beings. Detailed metabolic models and dose assessments for radionuclides are published periodically by the International Commission on Radiological Protection (ICRP, 1979–1989, 1993b, 1995a, 1995b, 1998).

5.1   Hydrogen-3 (Tritium, Half-life 12.3 yr)

Beta particles—maximum energy and percent of disintegrations	0.018 MeV (100)%
Average beta particle energy per disintegration	0.006 MeV

Tritium emits only very low energy beta particles and has a long half-life. When released to the environment as a gas (HT), it is slowly converted to tritiated water (HTO). Since tritiated water is approximately 10,000 times more hazardous than the gas, it is important to know the dynamics of the conversion from HT to HTO in any assessment of personnel exposure resulting from a release of the gas. In one study, no evidence was found for the rapid conversion of HT to HTO in the atmosphere; the HTO observed in air, during and after release, arose mainly from HT oxidation in the soil, followed by the emission of HTO (Brown et al., 1990).

Tritium is introduced into the environment as a result of cosmic ray interactions in the atmosphere, the continuous release from nuclear power plants and tritium production facilities, and from consumer products. An important source will be nuclear fusion facilities, if nuclear fusion is developed successfully as a source of power. Thus, knowledge of the dosimetry, radiation biology, and environmental transport of tritium is an important area in radiation protection (Straume, 1993).

5.1.1 Tritiated Water

Tritiated water is administered to patients in tests for the determination of body water. It appears in the atmosphere as a consequence of the release and oxidation of tritium from nuclear reactors. It has also been produced by nuclear explosions in the atmosphere.

The general assumption is that tritiated water, whether ingested or inhaled, is completely absorbed and mixes freely with the body's water. It permeates all the tissues within a few hours and irradiates the body in a fairly uniform manner. Thus body tissue is the critical tissue for intakes of tritiated water. The percentage of body water varies in different tissues. The average is about 60 percent. It is about 80 percent in such important tissues as bone marrow and testes (Vennart, 1969).

We can obtain an estimate of the average time the tritiated water molecules remain in the body from the following reasoning. Under normal conditions the body maintains its water content at a constant level. This means that over a period of time, as much water is eliminated as is taken in. If an adult male takes in 2,500 ml of water per day (2,200 by drinking, 300 as water of oxidation of foodstuffs), he also eliminates 2,500 ml/day. Assume a total water pool of 43,000 ml in the adult male. Thus the fraction of the body water that is lost per day is 2,500/43,000, or 0.058 per day. This is the biological elimination rate. The average life is 1/0.058, or 17.2 days, and the biological half-life is 0.693 × 17.2, or 12 days.

Example 3.17   Calculate the dose to tissues in the testes from the administration of 37 MBq of tritiated water.

Physical half-life, 4,480 days
$\lambda_p = 0.693/4,480 = 0.000155/\text{day}$
$\lambda_b = 0.058/\text{day}$
$\lambda_e = 0.058/\text{day}$
Effective average life, $1/\lambda_e = 17.2$ days
Effective half-life, $0.693T_e^a = 12$ days
Activity/g testes (0.8 of activity in body water),
  $0.80 \times 37 \text{ MBq}/43,000 \text{ g} = 0.000688 \text{ MBq/g}$
Average beta particle energy, 0.006 MeV
Dose $= 19,900 \times 0.000688 \times 0.006 \times 12 = 0.99$ mGy

The injection or ingestion of 37 MBq of tritiated water in a test of body water in an adult male results in a dose of 0.99 mGy to the testes.

The assumption used in the dose calculation that the tritium atoms remain in the body water and are eliminated at a constant and relatively rapid rate is only an approximation. Experiments with animals and studies of humans who accidentally ingested tritiated water have shown that a small fraction of the tritium is excreted at a much slower rate (Snyder et al., 1968). The concentration of tritium in the urine of a worker who had accidentally taken into his body 1,700 MBq of tritiated water was followed for over 400 days (Sanders and Reinig, 1968) and the concentration on the 415th day was still significantly above the concentration of tritium in urine from unexposed employees. The excretion curve could be interpreted as due to excretion of almost all of the tritium with a half-life of 6.14 days (the unusually rapid elimination was produced through administration of a diuretic) and excretion of the remainder in two fractions with half-lives of 23 days and 344 days. The complex excretion curve indicates that some of the tritium exchanges with organically bound hydrogen. The effect is to increase the dose calculated on the basis of a single short half-life, but by only a small percentage.

Example 3.18   What is the dose imparted to the testes as a result of breathing air containing tritiated water vapor at a concentration of 0.185 Bq per cubic centimeter of air?

Assume an exposure of 40 hours in a week. Also assume all inhaled water vapor is assimilated into body water. (Use physical data given in preceding problem.)

Air intake during working day (8 hr), 10^7 cc
Air intake during working week (40 hr), 5×10^7 cc
Becquerels inhaled during week, $0.185 \times 5 \times 10^7 = 9.25$ MBq
Dose to the testes from 9.25 MBq (from preceding problem, 37 MBq delivers 0.99 mGy), $(9.25/37)0.99 = 0.25$ mGy.

If exposure to the radioactive water vapor continues after the week at the same rate, the dose will accumulate, and the weekly dose rate will increase to a maximum value, which will be reached after several effective half-lives. This equilibrium weekly dose rate will be equal numerically to the total dose from a week's exposure,[12] and thus the maximum dose rate from inhalation will be 0.25 mGy/week.

12. Assume the total dose from a week's inhalation is essentially imparted in 4 weeks and the inhalation of the activity goes on for 40 weeks. Then the total dose imparted is 40 × 0.25 mGy = 10 mGy and the maximum time over which this is imparted is 44 weeks. Under these conditions there is little error in assuming the dose from the inhaled activity is im-

The actual dose rate to the individual immersed in the tritiated water atmosphere will be considerably larger because a significant amount of HTO is absorbed through the skin, even when the individual is clothed. The actual buildup of tritium in the body is about 80 percent greater than that deduced on the basis of inhalation alone (Morgan and Turner, 1967, p. 336).

We thus estimate the total dose rate from 0.185 Bq/cc at 0.25 mGy per week. We may perhaps increase this by another 10 percent to account for the dose from long-term retention. An RBE of 1.7 was used by ICRP (1960) in deriving concentration limits for tritium. A quality factor of 1.0 is currently used in radiation protection. However, theoretical estimates for the RBE for tritium tend to be greater than one because its LET (5.5 keV/μm) is higher than that of more energetic beta-gamma emitters (less than 3.5 keV/μm) (NCRP, 1979b, Report 62; NCRP, 1979a, Report 63; Till et al., 1980).

5.1.2 Tritiated Thymidine

The calculation of the dose from tritiated water in the preceding section was relatively simple because of the uncomplicated history of the water molecules in the body. The problem becomes much more difficult when the tritium is carried by a molecule with a specific metabolic pathway. As an example, let us consider the dose from thymidine labeled with tritium.

As a precursor of DNA, thymidine is used in studies of the synthesis of DNA molecules by cells. Such synthesis occurs in the nuclei of cells, and most of the thymidine is concentrated selectively in cell nuclei that are undergoing DNA synthesis at the time the thymidine is present.

How do we evaluate the dose in a situation like this? How do we assess the consequences? Since the tritiated precursors of DNA are extremely valuable in studies of cell function, can we establish permissible activities of tritiated DNA precursors for administration to human beings?

Animal experiments tell us that when tritium-labeled thymidine is administered intravenously, it is initially uniformly distributed throughout the body. It is then either promptly incorporated into DNA in cells that are actively proliferating or degraded to nonlabeling materials, primarily water. The result is a "flash" labeling.

The following data were obtained by Bond and Feinendegen (1966) following the administration of tritiated thymidine intravenously to rats. Radioautograph and counting techniques were used.

---

parted over the 40-week inhalation period, and the average dose rate during this period is 10 mGy/40 wk, or 0.25 mGy/wk. The error in neglecting the delay in imparting the dose decreases as the inhalation time increases.

Activity administered: 1 μCi/g

Average activity per 10^6 nucleated bone marrow cells, 1–2 hr after administration (by counting): 6,000 dis/min per 10^6 cells

Fraction of nucleated bone marrow cells labeled (by radioautography): 0.28

Average activity per labeled bone marrow cell: $6,000/(0.28 \times 10^6)$ = 0.022 dis/min

Because of its very low energy, the tritium beta particle does not travel very far. The maximum range in water is 6 μm and the average range is 0.8–1 μm. The diameter of a nucleus may be of the order of 8 μm. Thus most of the energy of the beta particles emitted in the nucleus is deposited in the nucleus. For a nuclear diameter of 8 μm, 80 percent of the emitted energy remains in the nucleus.

The tritium in a given nucleus will irradiate the nucleus until the cell is destroyed and breaks up, if the dose is high enough, or until the cell divides. If the cell undergoes division, some of the tritium will be transferred through chromosome exchanges to the daughter cells. Thus, successive divisions of cells containing tritiated DNA will tend to lower the tritium activity in individual nuclei over successive generations, with a consequent lowering of exposure to successive generations of cells. Eventually the doses to individual cells will become very small.

An estimate of the time it takes a cell to divide after uptake of thymidine may require special experimentation, using autoradiography to determine the activity in individual nuclei. The generation time of most rat bone marrow cells is 10–15 hr.

Example 3.19   Using uptake data given in the paper by Bond and Feinendegen, calculate the dose from administration of 1 μCi/g of tritiated thymidine.

Disintegrations during generation time of of 15 hr, 0.022/min $\times$ 900 min = 20

Mass of nucleus, diameter of 8 μ, 2.7×10^{-10} g

Fraction of emitted energy absorbed in nucleus, 0.8

Average dose over 15 hr:

$$\frac{20 \text{ dis} \times 0.006 \text{ MeV/dis} \times 0.8}{2.7 \times 10^{-10} \text{ g} \times 6.24 \times 10^6 \text{ MeV/g-mGy}} = 57.0 \text{ mGy}$$

Some cells receive as much as four times the average dose, or 240 mGy.

The level of 1 μCi/g was the minimum found to produce a definite physiological effect in experiments with rats, the effect involving diminution in the turnover rate of the cells. Occasional bi- and tetranucleated bone marrow cells were seen 3 days after injection.

The effects from tritiated thymidine were compared with the effects from various exposures to x rays. The authors reported that the dose levels to the nuclei producing effects such as mitotic delay, cell abnormalities, and some cell deaths in their experiments were comparable to x-ray dose levels to the whole cell needed to produce similar results.

The following conclusions were drawn by Bond and Feinendegen with regard to the administration of radioactive precursors of DNA:

> The present studies indicate that early somatic effects can be predicted on the basis of the absorbed dose. The evidence is consistent with long-term somatic and genetic effects not exceeding those expected on the basis of absorbed dose, although further work is needed for adequate evaluation. Thus guides for "allowable" levels of tritiated thymidine probably can be safely related to the amount of the compound that will deliver a dose to the cell nuclei of bone marrow, gonads, or other proliferating tissues that will not exceed the dose of external x or gamma radiation "allowed" for these same tissues. Until further data become available, however, it remains prudent not to use ^{3}H DNA-precursors in young individuals, particularly those in the childbearing age.   (1966, p. 1019)

5.2   Iodine-131 (Half-life 8.05 days) and Iodine-125 (Half-life 60.14 days)

	^{131}I		^{125}I	
	MeV	%	MeV	%
Beta particles—maximum energies and percent of disintegrations	.61	89.9		
	.33	7.3		
	.25	2.1		
Gamma photons—energies and percent of disintegrations	.72	1.8	0.027(x ray)[13]	114
	.64	7.2	.031-x	25
	.36	81.7	.035	6.7
	.28	6.1		
	.08	2.6		
Average beta particle or electron energy per disintegration	.187			
Specific gamma ray constant (R-cm²/MBq-hr)	0.059		0.0073	

Source: National Nuclear Data Center. www.nndc.bnl.gov/nndc/, MIRD and NUDAT links, accessed February 2002.

13. Energy of x radiation following electron capture (see Part Two, Fig. 2.8).

Iodine-131 emits approximately equal numbers of medium-energy beta particles and gamma rays. Iodine-125, because it decays by electron capture, gives off a mixture of low-energy gamma rays, conversion electrons, and x rays. Both isotopes have fairly short half-lives.

Radioiodines have been among the most important contributors to the environmental hazard of fission products from past nuclear weapons tests. They play a major role in the evaluation of the consequences of radioactivity releases resulting from reactor accidents. Their significance is due to their high yield in fission and selective uptake by the thyroid. In nuclear medicine, radioiodine is used primarily for medical examination and treatment of thyroid conditions. For this use it is administered as an inorganic salt, such as NaI. Its use stems from its remarkable property of concentrating in the thyroid gland. The radioactivity can be used either to trace the uptake of iodine in the gland and throughout the body or, in larger amounts, to destroy diseased tissue.

If NaI is injected intravenously, it will distribute in a short period throughout an "iodide space," comprising about 30 percent of the body weight (Spiers, 1968, pp. 55–57, 111–116). It will be eliminated from this space with a biological half-life of about 2 hr. It will reach a concentration in the red blood cells of about 0.56 that of plasma. As the bloodstream passes through the thyroid, perhaps 20 percent of the plasma iodide is removed per passage. In normal patients, 0.5–6.8 percent in the circulating pool is extracted per hour and the gland may accumulate 30 percent of the total injected or ingested activity. In the gland, the iodine becomes bound rapidly to the protein hormone. It is released from the thyroid only very slowly, in accordance with control mechanisms that exist throughout the pituitary.

The biological half-life may be deduced from the following data. A typical value of the iodine content of the normal thyroid gland is approximately 8 mg, though it varies greatly in individuals. The body tissues have an iodine concentration of about 1 μg/g. The thyroid releases about 0.08 mg/day, or 0.01 of its content per day (λ_b). The biological average life is thus 1/0.01, or 100 days, and the biological half-life is 69 days. Since the physical half-life of ^{131}I is much shorter, that is, 8.05 days, the effective half-life is 7.2 days, only slightly less than the physical half-life. The biological half-life varies in individuals from 21 to 200 days.

In addition to concentrating in the thyroid, iodine also concentrates in the salivary glands and gastric mucosa. Iodine released by the thyroid into the bloodstream circulates as protein-bound iodine (PBI), which is degraded in the peripheral regions. As a result, the iodine is set free as inorganic iodide. This again enters the body iodide pool and is available for use in the same manner as ingested iodine. Excretion of the iodines is almost entirely by way of urine.

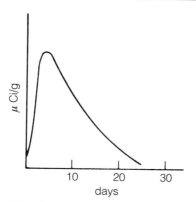

3.4 Retention curve for activity in blood.

Example 3.20 Evaluate the dose to the thyroid from the intravenous administration of 0.37 MBq of ^{131}I.

Initial activity (uptake of 30 percent) = 0.11 MBq
Mass of thyroid = 20 g
Effective half-life = 7.2 days
Absorbed fractions in thyroid (calculated from interpolation in Snyder, 1969):

E_i (MeV)	n_i	ϕ
0.72	0.018	0.032
0.64	0.072	0.033
0.36	0.817	0.032
0.28	0.061	0.032
0.08	0.026	0.037

$\sum E_i n_i \phi$ = 0.0120 MeV
Beta dose = 19,900 × 0.11 MBq/20 g × 0.187 MeV × 7.2 days = 147 mGy
Gamma dose = 19,900 × 0.11 MBq/20 g × 0.0120 MeV × 7.2 days = 9.46 mGy

---

The dose to the blood and bone marrow comes primarily from thyroxine and other ^{131}I-labeled organic molecules released slowly to the blood by the thyroid. The retention history is complicated, and it cannot be described by an essentially instantaneous uptake and a constant half-life. Experimental data on the activity in the blood as a function of time are shown in Figure 3.4.

Example 3.21 Calculate the dose to the blood from the retention curve for iodine.

The total number of disintegrations is given by the area under the curve in Figure 3.4. For the conditions of the problem, the total number of disintegrations per gram of blood is 5.5×10^6. The beta dose then equals

$$\frac{5.5 \times 10^6 \text{ dis/g} \times 0.187 \text{ MeV/dis}}{6.24 \times 10^6 \text{ MeV/g-mGy}} = 0.165 \text{ mGy}$$

The gamma photons emitted by iodine in the thyroid also contribute to irradiation of the blood. The calculation is complicated. Basic data may be found in Spiers (1968) or in Medical Internal Radiation Dose Committee pamphlets. According to Spiers, the gamma dose to the blood from ^{131}I in the thyroid contributes another 24 percent to the dose from ^{131}I in the blood. The dose to the bone marrow is probably more critical than the dose to the blood. Compartmentation of the bone marrow by partitions known as trabeculae serves to reduce the dose through beta shielding by the trabeculae. The reader is referred to Spiers for details. The gamma contribution to the dose is less affected by the trabeculae, and the total dose to bone marrow is about 80 percent of the total dose to blood.

The iodine problem illustrates the many factors that can be involved in a thorough analysis of dose from administration of a radioisotope. The calculation considers selective dilution of the isotope in body tissues, selective shielding from body structures, and complex retention curves. The combination of high uptake in the thyroid and low mass available for irradiation produces a high dose to the thyroid for a small amount of ingested activity. It is because of this physiological peculiarity that radioiodine is considered a very hazardous isotope, particularly for infants, and maximum levels in air and water have been set very low. On the other hand, iodine is administered in large quantities to destroy the thyroid. In this case the dose to the bone marrow becomes significant and may have to be considered in treatment planning.

Example 3.22   What is the dose to the thyroid if 0.37 MBq of ^{125}I is given instead of ^{131}I?

Initial activity (uptake 30 percent) = 0.11 MBq
Mass of thyroid = 20 g
Physical half-life = 60 days
Biological half-life = 69 days
Effective half-life = 32 days
Absorbed fractions in thyroid (from interpolation in Snyder, 1969):

E_i (MeV)	n_i	ϕ
0.027	1.14	0.184
.031	.25	.144
.035	.067	.124

$\sum E_i n_i \phi$ = .0071
Electron dose = 19,900 × (0.11 MBq/20 g) × 0.022 × 32 days = 77.1 mGy
Gamma dose = 19,900 × 0.11 MBq/20 g × 0.0071 MeV × 32 days = 24.9 mGy

The local dose rate for ^{125}I is lower than that for ^{131}I because of the lower energy radiation. This difference is largely offset by the longer half-life. However, ^{125}I may be the radioisotope of choice in some thyroid tests because of the superiority of the imagery and localization data resulting from the lower photon energy.

5.2.1 Microdosimetry of Iodine-125: Irradiation by Auger Electrons

Iodine-125 decays by electron capture, the capture by the nucleus of an electron in an inner shell, usually the innermost K shell. This leaves an inner shell of the atom devoid of an electron, which results either in the emission of x rays or a number of very low energy electrons, called Auger electrons (see Fig. 2.8). Much of the radiation dose imparted to the thyroid is by the Auger electrons. Because they have very short ranges in tissue at their low energy, the evaluation of the dose is dependent upon the distribution of the iodine and the microstructure of the gland.

The thyroid gland is made up of a conglomeration of tiny units called follicles, more or less rounded in shape and varying from 50 to 500 μm in diameter. The functional cells are closely packed together as a shell, which encloses a viscous fluid called colloid. This fluid (which makes up about 50 percent of the total gland mass in a normal gland) contains 90 percent of the iodine in the thyroid. This is not particularly significant for ^{131}I, because the energy imparted by beta particles is more or less uniform over the whole gland, so it is valid to talk about a mean gland dose. However, because of the short range of the lower-energy electrons emitted by ^{125}I (typically 35 keV, range of 22 μm in tissue), there is a non-uniform energy distribution across the follicle. Much of the electron energy is absorbed within the lumen, and the interfollicular tissue in the gland neither contains activity nor gets the particulate radiation dose. Microdose calculations give a maximum at the center of the follicle, dropping off to 49 percent at the colloid cell interface, 25 percent over the nucleus, and 14 percent at the basal membrane. Calculations indicate follicular cell dose equal to 50 percent of the mean gland dose. Experiments on rat thyroid indicate that the mean cell dose for 50 percent cell survival is about twice as high for ^{125}I as for ^{131}I, which supports the hypothesis that, in this case, the follicular cell nucleus dose is the significant dose.

The dosimetry and some experimental results with the Auger electrons point to a quality factor significantly greater than one because of the low energies imparted, but not enough is known to change the current quality factor of 1 in the standards (Persson, 1994).

5.3   Strontium-90 (Half-life 28 yr) → Yttrium-90 (Half-life 64 hr) → Zirconium-90 (Stable)

	^{90}Sr		^{90}Y	
	MeV	%	MeV	%
Beta particles—maximum energy and percent of disintegrations	0.544	100	2.27	100
Average beta particle energy per disintegration	0.21		0.89	

We have here an example of a radioactive decay series. Strontium-90 atoms, with an average life of $1.44 \times 28 = 40$ years, decay into yttrium-90 atoms, with an average life of 92 hours, and these decay to stable zirconium-90. Every strontium decay is followed within a few days on the average by an yttrium decay with the emission of an yttrium beta particle. Most of the energy in this double decay comes from the yttrium. If we start out with fresh strontium, we will have no significant yttrium activity until the yttrium atoms have a chance to build up. In time, they will build up until they decay at the same rate at which they are produced. They will attain half their maximum level in one yttrium half-life (64 hr) and make up half the interval to the equilibrium activity in each succeeding half-life. In 4 half-lives they will be within 93 percent of their maximum level. By 7 half-lives, the yttrium is essentially in equilibrium with the strontium and has the same activity.[14] Note that the yttrium beta particles have a high maximum energy (2.27 MeV) for a radionuclide.

14. The fraction of equilibrium reached by the ^{90}Y at any time after starting from pure ^{90}Sr can be calculated by the following reasoning. Assume first that we have a source with ^{90}Sr and ^{90}Y in equilibrium. They are both disintegrating at the same rate, with the long half-life of the ^{90}Sr, so the ^{90}Y remains constant. Now imagine that the ^{90}Y is separated from the ^{90}Sr. The separated ^{90}Y decays with its half-life of 64 hr and new ^{90}Y is formed from the decay of the ^{90}Sr. The total ^{90}Y still remains constant and disintegrates at the equilibrium rate. Thus we can say for any time after separation:

Disintegration rate of all ^{90}Y (equilibrium rate)
  = Disintegration rate of separated ^{90}Y + disintegration rate of new ^{90}Y
Disintegration rate of separated ^{90}Y
  = Equilibrium rate × fraction remaining at time t
Disintegration rate of new ^{90}Y
  = Equilibrium rate − disintegration rate of separated ^{90}Y
Disintegration rate of new ^{90}Y
  = Equilibrium rate − (equilibrium rate × f)
  = Equilibrium rate × $(1 - f)$

where f (fraction remaining) is calculated from the ^{90}Y half-lives elapsed after starting with pure ^{90}Sr. Alternatively, we may write the exact mathematical expression for the fractional buildup to the equilibrium activity as $(1 - e^{-0.693t/T^b})$, where T^b is the half-life of the ^{90}Y.

Strontium-90 is produced with a high yield in the fission of ^{235}U (5.8 atoms/100 fissions). It is one of the most important constituents of fallout from nuclear weapons tests, producing population exposure primarily from ingestion of contaminated milk and milk products. It is the most important and most hazardous constituent of aged radioactive wastes from nuclear power plants.

Almost all the strontium (99 percent) retained by the body ends up in the skeleton (NCRP, 1991c). The chemistry and metabolism of strontium are very much like that of calcium, since these nuclides are in the same family in the periodic table. One group of experiments gave the following data following intravenous injections of $^{85}SrCl_2$ in humans: 73 percent was eliminated with a half-life of 3 days; 10 percent was eliminated with a half-life of 44 days. The remaining 17 percent became fixed in the body—at least over the time period studied. A nominal value for the biological half-life of ^{90}Sr in bone is 1.8×10^4 days. The effective half-life is thus 6.4×10^3 days (ICRP, 1968b, Publication 10).

A measure of the dose from ^{90}Sr can be obtained by determining the total energy emitted in the skeleton over a period of time and dividing it by the mass of the skeleton. More refined calculations evaluate the dose only to the radiation-sensitive parts of the skeleton, such as the bone marrow and bone-forming cells. These are contained within a complex matrix in the skeleton. The dose to the marrow may be one-third of the actual absorbed dose averaged over the skeleton. In addition, bone is a dynamic structure and is continuously being remodeled. Calcium, accompanied by strontium, is released from certain sections and incorporated into newly formed bone in other sections. The strontium that is ingested is not deposited uniformly over the skeleton but is incorporated into those regions where bone is being produced. As a result, local dose rates may be several times higher than those calculated by determining an average value over the skeleton. Certain models for dosimetry consider the strontium-calcium ratio rather than the ^{90}Sr activity alone. The analysis takes into account the discrimination in favor of calcium over strontium retention in the body and makes use of an observed ratio, the Sr/Ca ratio, in bone relative to the Sr/Ca ratio in the diet (UNSCEAR, 1962). This ratio is of the order of 0.25.

Example 3.23   What is the expected dose to the skeleton in 13 weeks and in 1 year from ingesting 0.37 MBq of ^{90}Sr, assuming 9 percent becomes fixed in bone?

Fraction reaching bone = 0.09
Activity initially in bone = 0.0333 MBq

Mass of skeleton = 7,000 g
Effective half-life = 17.5 yr (6,400 days)
Effective average life = 25.2 yr (9,200 days)
Effective decay constant = 0.0397/yr
Average beta energy per decay (from Sr-90 + Y-90) = 1.1 MeV

$$\text{Beta dose} = 19,900 \times \frac{0.0333}{7,000} \times 1.1 \times 6,400 = 666 \text{ mGy}$$

The total dose of 666 mGy would be delivered over several effective half-lives, with a duration longer than the lifetime of the exposed individual. Because of the long effective half-life, the dose rate may be considered essentially constant for periods of several months. Thus the total dose in the first 13 weeks (91 days) equals 0.072 mGy/day × 91 days = 6.55 mGy. By the end of the first year, the activity is 0.961 of the original activity. The average activity during any year is approximately 0.98 the activity at the beginning of the year. The dose during the first year is then 0.98 × 0.072 × 365 = 25.75 mGy.

We have neglected the dose from pools with rapid turnover in this calculation. We have also not considered the possible variations in dose throughout the regions that take up strontium. The analysis is complicated and beyond the scope of this treatment. A discussion of the detailed dosimetry of bone-seekers is given by Spiers (1968).

An original limit for internal radioactivity from bone-seekers was the Maximum Permissible Body Burden (MPBB), the activity accumulated in the body of a Standard Man after exposure to the Maximum Permissible Concentration of the radionuclide for the whole of a working lifetime of 50 years that resulted in the emission of energy that was equivalent (with respect to the dose imparted) to the energy emitted from 0.1 μCi of radium-226 fixed in the body and the radium decay products. This is the maximum permissible body burden for radium. Many individuals with radium body burdens were studied, and no cases of bone cancer or other fatal diseases were found in individuals with several times the 0.1 μCi level.

The decay of radium atoms in the body is followed by the successive decay of a series of decay products with relatively short half-lives. The first decay product is the noble gas radon, with a half-life of 3.83 days. Because of the average time for decay of the radon atoms (5.5 days), about 70 percent of them leave the body before decaying and giving rise to several more short-lived emitters of both alpha and beta particles. The effective energy deposited in bone by the radium series is 220 MeV. This result takes into account the loss of 70 percent of the radon and also includes a multiplication of the energies of the alpha particles by 20 to weight their increased hazard over that from beta particles. However, animal experiments on the toxicity of some of the beta-emitting bone-seekers indicated that the energy they released should also be increased by some modifying factor because of their deposition patterns. For ^{90}Sr-Y, a modifying factor of 5 was used.

The maximum body burden of a bone-seeker was then determined as that activity which gave off the same amount of energy absorbed in the body, multiplied by appropriate modifying factors, as the energy absorbed from a body burden of 0.1 μCi (0.0037 MBq) of radium.

Example 3.24   Estimate a maximum permissible body burden for ^{90}Sr-Y, on the basis of energy release compared to 0.1 μCi of ^{226}Ra.

Average beta energy of ^{90}Sr-Y per disintegration = 1.1 MeV
Modifying factor = 5
Effective energy per disintegration = 5.5 MeV
Energy imparted locally per radium disintegration (based on quality factor of 20 for alpha particles) = 220 MeV
Maximum permissible body burden for ^{90}Sr-Y = 0.1 × 220/5.5
    = 4 μCi (0.148 MBq)

Detailed analyses of the deposition of strontium-90 in bone and related dosimetry have been performed by the ICRP (ICRP, 1979, Publication 30; 1994a, Publication 68) and Spiers (1968).

5.4   Xenon-133 (Half-life 5.27 days) and Krypton-85 (Half-life 10.3 yr)

	^{133}Xe		^{85}Kr	
	MeV	%	MeV	%
Beta particles—maximum energy and percent of disintegrations	0.34	100	0.15	0.4
			0.67	99.6
Average beta particle energy per disintegration	0.13		0.22	
Gamma photons—energy and percent of disintegrations	0.029 (x ray)[15]	60		
	0.081	40	0.51	0.4
	0.160	0.1		
Gamma constant (R/mBq-hr at 1 cm)	0.0119		0.00054	

---

15. Energy of x radiation following internal conversion in 60 percent of transitions as compared to 40 percent resulting in emission of 0.081 MeV gamma photons from nucleus.

Krypton-85 is one of the waste products from fission reactors. Because it is a noble gas, it is difficult to remove, and it is discharged to the atmosphere when it is released from spent reactor fuel elements. It has caused some concern as a potential environmental pollutant if a large fraction of the nation's power requirements come from nuclear power.

Both ^{85}Kr and ^{133}Xe are very useful in nuclear medicine in the diagnosis of lung function and cardiac shunts. (Similar uses are also being found for cyclotron-produced radioactive gases such as ^{15}O and ^{13}N.) Because it produces a much lower beta dose in the subject for an equivalent external gamma intensity, ^{133}Xe is favored over ^{85}Kr as a tracer.

5.4.1 Calculation of Beta and Gamma Dose to Surface of Body from Krypton in Air

Consider first exposure from a radioactive gas in the air. The body is irradiated externally by both beta particles and gamma rays and is irradiated internally from gas that is brought into the lungs and dissolved in body tissues.

The maximum range of the beta particles in air is 180 cm. Since an individual will always be surrounded by a volume of air greater than this maximum range, the body receives the same exposure as if it were immersed in an infinite source of beta radiation. Because of the low attenuation of the gamma rays in air, the magnitude of the gamma exposure will depend on the extent of the source of radioactive gas. It will begin to approach that from an infinite source if all the air within a distance of about 200 m is polluted with the gas. The gamma dose will be much lower than that from an infinite source if the volume is limited, for example, to the size of a laboratory. The geometry for the dose calculation is different from any we have encountered previously. The irradiated object is immersed in the source. Although the solution may appear complex, with a little reasoning we can find a simple way to solve this problem.

Consider the dose at a point inside a large volume of air, much larger than the half-value layer of gamma rays. We have then a situation where essentially all the energy produced in the volume is absorbed in the volume, except for boundary effects, which are minor in such a large volume. The energy absorbed per gram of air is equal to the energy produced per gram of air, and the dose to the air is readily determined, as was done previously for beta sources distributed uniformly throughout a region.

Now if we introduce a small mass of tissue into the region, the energy absorption per gram of tissue will be approximately equal to the energy absorption per gram of air because of the similarity in tissue and air absorption for the energies under consideration. Accordingly, the gamma dose to a small mass of tissue within an infinite volume of air will be equal to the dose to the air itself.

If the tissue is of significant size, as in the case of a human being, then the dose will depend on the attenuation of the incident radiation to the dose point of interest. In particular, the dose on a surface would be reduced by a factor of 2, if the back side of the surface effectively shielded the radiation incident upon it and the front of the surface saw essentially an entire half-plane. Also, if tissue were next to the ground, only half the dose for an infinite medium would be imparted at the most, because the incident gamma radiation was again limited to the upper half-space.

In the case of beta particles, the dose on the surface would be expected to be half the dose in an infinite medium, as betas cannot penetrate through the body. However, beta particles impart energy to tissue at about a 10 percent greater rate per centimeter than in air, and thus the surface dose to tissue from beta particles originating in air would have to be increased by the same amount, relative to the dose to air.

Example 3.25   Calculate the dose rate to the surface of the body from exposure to air containing 0.37 Bq/cc of krypton-85.

Mass of 1 cc air at 20° C, 760 mm = 0.0012 g
Beta energy emitted per gram air per hour

$$= \frac{0.37 \text{ dis}}{\text{sec-cc}} \times \frac{1}{0.0012 \text{g/cc}} \times \frac{0.22 \text{ MeV (av)}}{\text{dis}} \times \frac{3,600 \text{ sec}}{\text{hr}}$$

$$= 2.44 \times 10^5 \text{ MeV/g-hr}$$

$$\text{Beta dose rate to air} = \frac{2.44 \times 10^5 \text{ MeV/g-hr}}{6.24 \times 10^6 \text{ MeV/g-mGy}} = 0.039 \text{ mGy/hr}$$

Mass stopping power tissue relative to air = 1.13

$$\text{Beta dose rate to tissue} = \frac{1.13 \times 0.039}{2} = 0.022 \text{ mGy/hr}$$

Gamma energy emitted per disintegration = $0.51 \times 0.004 = 2.04 \times 10^{-3}$ MeV

$$\text{Gamma dose rate to air} = \frac{0.37 \times 0.00204 \times 3,600}{0.0012 \times 6.24 \times 10^6}$$

$$= 0.00036 \text{ mGy/hr}$$

Gamma dose rate to tissue (assuming roughly that tissue receives half the air dose) = 0.00018 mGy/hr

Total dose rate from beta and gamma radiation = .0222 mGy/hr

This gives 0.89 mGy per 40-hr week.

5.4.2   Calculation of Dose from Krypton Inhaled into Lungs

Let us now evaluate the dose to the lungs when air containing ^{85}Kr is inhaled. We shall use the following model for the calculations.

The volume of inspired air in a single breath (called the tidal volume) is 600 cc, and 450 cc of this reaches the alveolar region of the lungs, where it mixes with 2,500 cc of air available for exchange with the blood. This air is distributed through a lung mass of 1,000 g. The remaining 150 cc is confined to the "dead space," consisting of the nasal passages, trachea, and larger bronchi preceding the alveolar region.

There are several ways to evaluate the dose to the lungs from breathing krypton, ranging from simple algebraic considerations to the use of calculus. As an example, consider the following simplified breathing pattern applied to a single breath of air containing krypton followed by a breath of nonradioactive air:

- Instantaneous inhalation of tidal volume with concentration of krypton $= C$ and retention of gas in lung for time τ.
- Instantaneous exhalation after time τ.
- Immediate inhalation of nonradioactive air and retention for period τ.
- Continued washout of krypton by repeating the same breathing pattern.

Note that when air containing krypton is expired from the lungs, a portion is retained in the dead space and inspired in the next cycle along with an amount of nonradioactive air given by the tidal volume less the dead space volume. Let V_A = alveolar air volume (that is, excluding dead space) following exhalation, V_T = tidal volume, and V_D = dead space volume. In the first breath the alveolar region takes in an amount of krypton $C(V_T - V_D)$ and receives an exposure proportional to $C(V_T - V_D)\tau$. After exhalation and inspiration of nonradioactive air, the krypton activity is

$$C(V_T - V_D)\frac{V_A + V_D}{V_A + V_T}$$

and the exposure is proportional to

$$C(V_T - V_D)\frac{V_A + V_D}{V_A + V_T}\tau$$

The exposure accumulates in successive breaths according to the relationship

$$C(V_T - V_D)\tau[1 + \left(\frac{V_A + V_D}{V_A + V_T}\right) + \left(\frac{V_A + V_D}{V_A + V_T}\right)^2 + \left(\frac{V_A + V_D}{V_A + V_T}\right)^3 + \dots]$$

$$= C(V_T - V_D)\tau\left(\frac{V_T + V_D}{V_A + V_T}\right)$$

$$= C\tau(V_A + V_T)$$

Another approach is to consider the fractional loss per unit time as given by

$$\left(\frac{V_T + V_D}{V_A + V_T}\right)\frac{1}{\tau}$$

This is equivalent to the decay constant, λ, so the average time in the lungs is given by $1/\lambda$ or

$$\tau\left(\frac{V_A + V_T}{V_T + V_D}\right)$$

Thus, the total exposure is proportional to the original activity in the lungs times the average life, or

$$C(V_T - V_D)\left(\frac{V_A + V_T}{V_T + V_D}\right)\tau = C\tau(V_A + V_T)$$

The same result was obtained by the method described above.

This result may be generalized to any type of breathing pattern if $V_A + V_T$ is replaced by the average volume of air in the alveolar region during a breathing cycle and τ is the time interval between successive breaths. In this treatment, we have assumed that the physical average life is much longer than the biological average life. If the physical and biological average lives are comparable, the effective average life must be used.

Example 3.26   Calculate the dose to the alveolar region from a single breath of air containing a concentration of 1 MBq/cc.

At end of first inspiration,

Activity in alveolar region = 450 MBq
Activity in dead space = 150 MBq
Volume of air in alveolar region = 2,500 + 600 = 3,100 cc

Concentration of activity in alveolar region = 450/3,100 = 0.145
MBq/cc

At end of first expiration,

Fractional activity exhaled from alveolar region = 600/3,100 =
0.19
Fractional activity retained in alveolar region = 2,500/3,100 = 0.81
Activity in alveolar region = 0.81 × 450 = 364 MBq
Activity in dead space = 0.145 × 150 = 21.8 MBq

In succeeding breaths, we shall assume the inhaled air does not contain ^{85}Kr. However, a small amount of ^{85}Kr is carried into the alveolar
region by the air in the dead space of the lungs.

At end of second inspiration (occurs at essentially same time as end
of first expiration in our model),

Activity in alveolar region = 364 + 21.8 = 386 MBq, or 0.855 of
initial activity
Activity in dead space = 0

At end of second expiration,

Activity in alveolar region = 386 × 2,500/3,100 = 311 MBq
Activity in dead space = 386 × 150/3,100 = 18.7 MBq

At end of third inspiration,

Activity in alveolar region = 330, or 0.855 of previous activity

We see the activity drops to 0.855 of its previous activity in each
breath. If we assume the subject is taking 16 breaths per minute, the
duration of a single breath is 1/16 min and the dose during the first
breath is

$$\frac{450 \text{ MBq} \times 10^6 \text{ /sec-MBq} \times 0.22 \text{ MeV} \times 1/16 \text{ min} \times 60 \text{ sec/min}}{1,000 \text{ g} \times 6.24 \times 10^6 \text{ MeV/g-mGy}}$$

$$= 0.059 \text{ mGy}$$

Total dose = $0.059 + 0.855 \times 0.059 + (0.855)^2 \times 0.059 + \ldots$

$$= 0.059 \left(\frac{1}{1 - 0.855} \right) = 0.407 \text{ mGy}$$

The same result would be obtained with the aid of the formulas preceding the example.

The calculation of the dose from a single breath is of interest when krypton is administered to a patient in a medical test. The dose from taking several breaths is obtained by adding the contributions of the individual breaths.

When one is exposed to krypton in the environment, it is, of course, being breathed in continuously, and the levels in the lungs will rise until a maximum concentration equal to the concentration in the inspired air is reached. The time to reach essential equilibrium with the concentration in the air is practically a few effective half-lives. For inhaling any inert gas, we see that this would take just a minute or so. Under these conditions, the calculation of the dose rate is very simple. The concentration in the lungs is equal to the concentration in the air, and for the conditions we have been using, it is thus 1 MBq/cc. The dose rate to the lungs is

$$\frac{1 \text{ MBq}}{\text{cc}} \times \frac{10^6}{\text{sec-MBq}} \times \frac{3,100 \text{ cc}}{1,000 \text{ g}} \times \frac{0.22 \text{ MeV}}{6.24 \times 10^6 \text{ MeV/g-mGy}} = 0.109 \text{ mGy/sec}$$

In our calculations, we have neglected the gamma dose from krypton in the lungs. Because gamma photons are given off in only 0.4 percent of the disintegrations, the gamma dose is negligible in comparison with the beta dose.

5.4.3 Dose to Tissues from Inhalation of a Radioactive Gas

Radioactive gas is transferred from the lungs to the tissues by the blood. The blood leaving the lungs contains gas in equilibrium with the gas in the lungs, and, as the blood flows through the tissues, the dissolved gas transfers to the tissues. The concentration builds up to a value given by the relative solubility in the tissues versus the solubility in blood. The speed with which the equilibrium value is reached is given by the rate of flow through the tissues and the solubility. Intuition tells us that for insoluble radioactive gases, the doses to the tissues will be much less than the dose to the lungs. The degree of difference can be determined by the following analysis.

Assume that as the blood leaves the lungs, the concentration of gas in the blood (C_b^L) is in equilibrium with the gas concentration in the air in the alveolar region (C_g^A). Thus, relation is expressed as

$$C_b^L = \alpha_b \times C_g^A \tag{3.8}$$

where α_b is the solubility coefficient for blood.

The gas diffuses from the blood into the body tissues permeated by the blood. If we know the concentration of gas in the blood as it enters the tissue (C_b^L) and the reduced concentration upon leaving (C_b^v), then the difference represents the flow of gas into the tissues. If we multiply the difference by the blood flow per unit volume of tissue (f_b^v), we obtain the rate of increase in gas concentration in tissue (C_v):

$$\frac{dC_v}{dt} = f_b^v(C_b^L - C_b^v) \qquad (3.9)$$

The concentration of gas in the blood leaving the tissue may be considered to be in equilibrium with the concentration of gas in the tissue. Thus

$$C_v = \frac{\alpha_v}{\alpha_b} C_b^v$$

where α_v/α_b is the partition coefficient between tissue and blood. Substituting for C_b^v in equation 3.9,

$$\frac{dC_v}{dt} = f_b^v(C_b^L - \frac{\alpha_b}{\alpha_v} C_v)$$

The concentration in the tissues will increase until there is no net flow between blood and tissue, that is, until $C_b^v = C_b^L$ and $dC_v/dt = 0$. Under these conditions, $C_v = (\alpha_v/\alpha_b)C_b^L$. Substituting for C_b^L (see equation 3.8), $C_v = \alpha_v C_g^A$.

Example 3.27   A patient breathes air containing 1 MBq/cc ^{133}Xe for 2 minutes in a lung-function test. Calculate the beta dose to the lungs and other tissues in the body resulting from the solubility of xenon in these tissues.

Mass of lungs = 1,000 g
Average beta energy = 0.13 MeV
Solubility coefficient (α_v):
    fatty tissue–air 1.7
    nonfat tissue–air 0.13

Assume lung and tissue exposure results from equilibration with an air concentration of 1 MBq/cc for 2 minutes. (This assumption may be shown to be valid by a mathematical treatment in which the gas equilibrates in an exponential manner (Lassen, 1964). The activity/cc of nonfat tissue $(C_g^A \times \alpha_v)$ is $1 \times 0.13 = 0.13$ MBq/cc. Assuming a density of 1 g/cc, the nonfat tissue is exposed to an activity of 0.13 MBq/g for 2 min. The dose is 0.13 MBq/g $\times$ 60 $\times$ 10^6/min-MBq $\times$ 0.13 MeV $\times$ 2 min $\div$ 6.24 $\times$ 10^6 MeV/g-mGy = 0.32 mGy.

> The dose to fatty tissue would be greater by the ratio of the solubility coefficients, or $(1.7/0.13) \times 0.32 = 4.24$ mGy.
>
> The average beta dose to the lungs is
>
> $$\frac{1 \text{ MBq/cc} \times 3{,}100 \text{ cc} \times 60 \times 10^6/\text{min-MBq} \times 0.13 \text{ Mev} \times 2 \text{ min}}{1{,}000 \text{ g} \times 6.24 \times 10^6 \text{ MeV/g-mGy}}$$
>
> $$= 7.75 \text{ mGy}$$

In diagnostic tests, instead of the gas being administered by inhalation, it may be dissolved in physiological solution and injected intravenously over a brief period of time. For this procedure, we assume that all the gas is released to the alveolar region in the lungs, on passing through, and that the alveolar gas maintains equilibrium with the gas in the blood in the pulmonary capillaries. The level of xenon in the blood then falls off in the same way as the concentration in the lungs as the lung-air exchanges with clean inspired air. Dosimetry for this and other conditions of administration is treated by Lassen (1964).

5.5 Uranium-238 and Its Decay Products

Uranium is a naturally occurring primordial radionuclide; that is, it was one of the elements created after the Big Bang (Part One). It is the last naturally occurring element in the periodic table. Natural uranium consists of ^{238}U (99.27%), ^{235}U (0.72%), and ^{234}U (0.0054%—proportion based on equilibrium with ^{238}U). The ICRP definition of a special curie of natural uranium is 3.7×10^{10} dis/sec of ^{238}U, 3.7×10^{10} dis/sec of ^{234}U, and 1.7×10^9 dis/sec of ^{235}U. The decay of the ^{238}U nucleus by alpha emission is followed after two beta decays by the decay of uranium, thorium, and radium nuclei in succession, all by alpha emission (see Table 3.4):

$$^{238}\text{U} \rightarrow {}^{234}\text{Th} \rightarrow {}^{234m}\text{Pa} \rightarrow {}^{234}\text{U} \rightarrow {}^{230}\text{Th} \rightarrow {}^{226}\text{Ra} \rightarrow {}^{222}\text{Rn} \rightarrow$$

Radium-226 decays to radon-222, an alpha-emitting noble gas with a half-life of 3.82 days. The radon may decay in the ground or it may diffuse out to the air before it undergoes an additional series of decays (discussed in the next section), including three by alpha emission. Thus, the uranium series is responsible for significant contamination of the environment with alpha radioactivity.

The normal daily intake of uranium by an adult male is 1.9 μg, with a resultant equilibrium body content of 90 μg. The distribution is 66 percent in the skeleton and 34 percent in soft tissue (7.8 percent in the kidneys) (ICRP, 1975, Publication 23). The uptake to blood following ingestion is 0.05 for water-soluble inorganic compounds (hexavalent uranium)

Table 3.4 Alpha radiations from uranium-238 and its decay products.

Radionuclide	Half-life	Radiation energies (MeV) and percent of disintegrations	Specific activity (MBq/mg)
Uranium-238 (^{238}U)	4.5×10^9 yr	4.20(75%)	0.124×10^{-4}
		4.15 (25%)	
Thorium-234 (234Th)	24 day		
Protactinium-234m (^{234m}Pa)	1 min		
Uranium-234 (^{234}U)	2.47×10^5 yr	4.77 (72%)	0.228
		4.72 (28%)	
Thorium-230 (^{230}Th)	8×10^4 yr	4.68 (76%)	0.718
		4.62 (24%)	
Radium-226 (^{226}Ra)	1,602 yr	4.78 (95%)	37
		4.60 (5%)	
Radon-222 (^{222}Rn)	3.823 day	[See Table 3.5]	

Notes: Thorium-234 emits low-energy beta particles and ^{234m}Pa emits a high-energy (2.28 MeV, 98.6%) beta particle. Some gamma radiation is also emitted (Slaback, Birky, and Shleien, 1997). Uranium-235, which also occurs naturally, is the progenitor of a separate decay series.

and 0.002 for insoluble compounds (UF_4, UO_2, U_3O_8, usually tetravalent uranium) (ICRP, 1976b, 1979, 1988).

The long-term retention in bone and kidney following unit uptake to blood is given by the following equations, where t is given in days (ICRP, 1979). (The first one says that 20 percent is retained with an effective half-life of 20 days and 2.3 percent with an effective half-life of 5000 days.)

$$R_{bone} = 0.20e^{-0.693t/20} + 0.023e^{-0.693t/5000}$$

$$R_{kidney} = 0.12e^{-0.693t/6} + 0.00052e^{-0.693t/1500}$$

The thorium-232 series is the other major source of alpha radioactivity in humans. The normal daily intake is 3 μg, and the content of mineral bone is 30 μg. The uptake to blood following ingestion is 0.001. Retention in bone following unit uptake to blood is given by the equation

$$R_{bone} = 0.70e^{-0.693t/8000}$$

where t is given in days.

Example 3.28 Calculate the annual alpha dose to bone (7,000 g) from ingestion of 10 mg per week of soluble natural uranium. Assume 5 percent of the ingested soluble uranium reaches the blood.

The activity of a daily intake of 2 mg is:

For ^{238}U, 2 mg × 0.9927 × 1.24 × 10^{-5} MBq/mg = 0.0000246 MBq, 4.19 MeV α

For ^{234}U, 2 mg × 5.4 × 10^{-5} × 0.229 MBq/mg = 0.0000246 MBq, 4.76 MeV α

For ^{235}U, 2 mg × 0.0072 × 7.9 ×10^{-5} MBq/mg = 1.14 × 10^{-6} MBq, 4.4 MeV α

The total activity is 5.03 × 10^{-5} MBq and the average alpha particle energy is 4.47 MeV.

The contribution of the short-lived component to the annual dose from daily ingestion for 50 weeks, 5 days per week, will be 19,900 × (5.03 × 10^{-5} MBq × 0.05 × 0.20)/7,000 g × 4.47 MeV × 20 day × 250 day = 0.032 mGy or 0.64 mSv (Q = 20). If a distribution factor (DF) of 5 is assumed, the calculated dose equivalent is 3.2 mSv. The total committed dose from the long-lived component following intake for one year is = 0.92 mGy and the actual amount imparted in the first following year would be approximately 0.92(1 − 0.5$^{365/5000}$) = 0.045 mGy (4.5 mSv), or a dose about 40 percent greater than the dose from the short-lived component. The total committed dose from both long- and short-lived components is 9.52 mSv. If the uranium is in insoluble form (that is, UO$_2$, U$_3$O$_8$), the uptake from the GI tract is only 4 percent of the soluble form and the dose will be correspondingly less.

5.5.1 Radium-226

The normal daily intake of radium by an adult male is 0.0851 Bq, with a resultant equilibrium body content of 1.147 Bq (31 pg). The distribution is about 85 percent in the skeleton and 15 percent in soft tissue (4.5 percent in muscle). The uptake to blood following ingestion is 0.2. The long-term retention in bone following unit uptake to blood is given by the equation

$$R_s = 0.54e^{-0.693t/0.4} + 0.29e^{-0.693t/5} + 0.11e^{-0.693t/60} + 0.04e^{-0.693t/700} + 0.02e^{-0.693t/5000}$$

where t is counted in days (ICRP, 1988, Publication 54).

Radium is a decay product of uranium and is in radioactive equilibrium with it on a global basis. It is much more water soluble than uranium is, however, and thus is leached out of the soil by groundwater and makes its way to drinking water and food. The actual concentrations vary greatly in different locations and different foods. Radium in the body becomes incorporated in bone, where it remains virtually indefinitely.

The protection standard for radium was originally based on a maximum level in the body (maximum body burden). From studies of hundreds of persons who had accumulated radium in their bones occupationally or medically, it was determined that an acceptable radium level was 0.1 μg fixed in the body. This limit then served as a model for setting limits for other bone-seekers. One approach was to determine levels of other bone-seekers that gave the same effects in animals as does radium. The other was to make comparisons of doses. The International Commission on Radiological Protection recommended that "the effective RBE dose delivered to the bone from internal or external radiation during any 13-week period averaged over the entire skeleton shall not exceed the average RBE dose to the skeleton due to a body burden of 0.1 μCi (3700 Bq) of ^{226}Ra" (ICRP, 1960, p. 3). The associated dose rate was 0.6 mGy/week.

The current protection standard for radium is associated with an effective committed dose, determined on the basis of a detailed metabolic model and expressed as an annual limit on intake.

5.6　Radon-222 and Its Decay Products

The radioactive noble gas radon-222 is produced continuously from the decay of radium in the ground. It dissolves in groundwater, which often carries it in high concentrations and releases it to areas inhabited by humans. It diffuses readily through soil and into the atmosphere. Thus it is always present in the air at levels which are determined by local geology and meteorology. It imparts the highest organ dose (to the lungs) of any radioactive environmental contaminant.

Large quantities of radon are emitted to the environment from radium-containing wastes produced at uranium mills. The radium is left behind in the wastes, known as mill tailings, after the uranium is extracted from the ore. These wastes are often piled up in huge mounds, tens of meters high. Mill tailings have also been used as fill and in the construction of foundations for residential and commercial buildings, producing dramatic increases in radiation and radon levels.

The hazards of radon were first suspected in the 1930s, when a high incidence of lung cancer was discovered among miners working in the radium mines of Schneeberg, Germany, and Joachimstal, Czechoslovakia. Over half the deaths of the miners were from lung cancers and most occurred before the miners had reached 50 years of age.

Because high levels of radon gas occurred in these mines, attention was

Table 3.5 Radiations from radon and its decay products.

Radionuclide	Half-life	Major radiation energies (MeV) and percent of disintegrations		
		α	β	γ
Radon-222 (^{222}Rn)	3.823 day	5.49 (100%)		
Polonium-218 (^{218}Po)	3.05 min	6.00 (100%)		
Lead-214 (^{214}Pb)	26.8 min		0.65–0.98	0.295 (19%)
				0.352 (36%)
Bismuth-214 (^{214}Bi)	19.7 min		1.0–3.26	0.609 (17%)
				1.120 (17%)
				1.764 (17%)
Polonium-214 (^{214}Po)	164 μsec	7.6 (100%)		
Lead-210 (^{210}Pb)	21 yr		0.016–0.061	0.047 (4%)
Bismuth-210 (^{210}Bi)	5 day		1.161 (100%)	
Polonium-210 (^{210}Po)	138 day	5.31 (100%)		
Lead-206 (^{206}Pb)	Stable			

Source: BRH, 1970.

turned to the alpha dose to the lungs from inhaled radon and its short-lived decay products as the possible causes. The decay products and the radiations emitted are shown in Table 3.5.

Early dose calculations (Evans and Goodman, 1940) were based on the amount of energy imparted to the lung by the alpha particles from radon itself and from the decay products of the radon molecules decaying in the lung. The doses did not seem high enough to produce the lung cancers. In 1951 Bale pointed out that the "radiation dosage due to the disintegration products of radon present in the air under most conditions where radon itself is present conceivably and likely will far exceed the radiation dosage due to radon itself and to disintegration products formed while the radon is in the bronchi. This additional dosage is associated with the fact that disintegration products of radon remain suspended in the air for a long time; tend to collect on any suspended dust particles in the air; and that the human respiratory apparatus probably clears dust from air, and the attached radon disintegration products, with reasonable efficiency" (Bale, 1951). Bale calculated an average dose to the lungs by dividing the alpha energy released by the decay products deposited on lung tissue by the mass of the lung. He then assigned the problem to me, as his graduate student, and I spent the next four years determining the dose imparted to the lungs and to various portions of the respiratory tract and developing simplified sampling procedures for radon and its decay products (Shapiro, 1954; Bale and Shapiro, 1955; Shapiro, 1956a,b;

Burgess and Shapiro, 1968). The maximum permissible concentration (MPC) for radon and its decay products was based "quite considerably on these data and on the work of Chamberlain and Dyson (1956)" (ICRP, 1966).

Consider an enclosed volume containing radon at a typical atmospheric concentration of 3.7 Bq/m^3. All of the short-lived decay products will be in equilibrium with the radon and thus will have the same activity as the radon in the volume.[16] A fraction of them will be attached to the surface of the enclosure so the airborne activity concentrations will be somewhat less than 3.7 Bq/m^3, depending on the dust loading, air motions, and so on. (Clean air vigorously stirred would favor deposition on the surface rather than suspension in air.)

It is a good first approximation to assume that any atom of the short-lived decay products that is deposited in the lung will decay in the lung. The energies of the ^{218}Po and ^{214}Po alpha particles are 6.00 MeV and 7.69 MeV, respectively. Every ^{218}Po atom deposited will lead to the release of 6 + 7.69 = 13.69 MeV of alpha energy. Every ^{214}Pb, ^{214}Bi, and ^{214}Po atom deposited will lead to the release of 7.69 MeV, of alpha energy. The total alpha energy emitted in the lungs by the decay product per cubic meter of air inhaled is equal to the number of atoms of each of the decay radionuclides of concern per cubic meter times the fraction of those atoms deposited in the lungs times the alpha energy associated with the decay of each of the radionuclides.

Example 3.29   Calculate the energy released in the lungs from deposition of the short-lived decay products in equilibrium with 3.7 Bq of ^{222}Rn.

The number of ^{218}Po atoms (with T^b = 3.05 min) = 3.7 Bq × 60/min-Bq × 1.44 × 3.05 min = 975. Similar calculations for ^{214}Pb, ^{214}Bi, and ^{214}Po give 8,567, 6,298, and 0.0009 atoms, respectively. Therefore, the total alpha energy associated with the decay of 3.7 Bq of the short-lived radionuclides = 975 × 13.69 + (8,567 + 6,298) × 7.69 = 1.3 × 10^5 MeV.

Example 3.30   Calculate the average dose to the lungs from one year's exposure to the short-lived decay products in equilibrium with 3.7 Bq (100 pCi)/m₃ of ^{222}Rn.

---

16. See section 5.3 on ^{90}Sr-^{90}Y and note 14 for discussion of equilibrium.

Assume a total daily inhalation volume (adult male) of 23 m³ (ICRP, 1975, Publication 23, p. 346), a lung mass of 1,000 g, and an average deposition (and retention) of 25 percent (Shapiro, 1956a). The total alpha energy absorbed in the lungs per year is

$$\frac{365 \text{ days}}{\text{yr}} \times \frac{23 \text{m}^3}{\text{day}} \times 0.25 \times \frac{1.3 \times 10^5 \text{ MeV}}{\text{m}^3} = 2.73 \times 10^8 \text{ MeV}$$

The average annual absorbed dose to the lungs is

$$\frac{2.73 \times 10^8 \text{ MeV}}{1000 \text{ g} \times 6.24 \times 10^6 \text{ Mev/g-mGy}} = 0.0437 \text{ mGy}$$

The Working Level (WL) is defined as any combination of short-lived radon decay products (through ²¹⁴Po) per liter of air that will result in the emission of 1.3×10^5 MeV of alpha energy. From the preceding example, this is the alpha energy released when all the short-lived decay products in equilibrium with 3.7 Bq of ²²²Rn undergo decay. The Working Level is characteristic of radon levels in uranium mines (3.7 Bq/l) and is 1,000 times as high as the value used in the previous example as characteristic of naturally occuring radon levels. A person exposed to 1 WL for 170 hr is said to have acquired an exposure of one Working Level Month (WLM). The Working Level and Working Level Month are the units used for expressing occupational exposure to radon, and most risk assessments are based on the direct relationship between lung cancer incidence among miners and exposure as experienced in WLM. Original exposure standards allowed for an annual exposure of 12 WLM. Current standards call for a maximum of 4 WLM per year.

Example 3.31   Calculate the mean dose to the lungs from exposure to 1 WLM of radon decay products.

The same approach is used as in example 3.31. Assume 9,600 liters of air are breathed during an 8-hour working day and 25 percent of the decay products are retained.

$$\frac{1.3 \times 10^5 \text{ MeV}}{\text{WL-l}} \times 0.25 \times \frac{1}{1,000 \text{ g}} \times \frac{\text{g-mGy}}{6.24 \times 10^6 \text{ MeV}}$$

$$= 5.2 \times 10^{-6} \text{ mGy/WL-l}$$

$$\frac{5.2 \times 10^{-6} \text{ mGy}}{\text{WL-l}} \times \frac{9,000 \text{ l}}{8 \text{ hr}} \times \frac{170 \text{ hr}}{\text{month}} = 1.00 \text{ mGy/WLM}$$

3.5 Front view of the trachea and bronchi. (*Source:* Gray, 1977.)

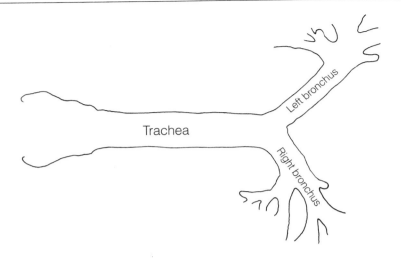

Doses to individual regions of the lung vary greatly from the mean dose because of the anatomy of the respiratory tract (Fig. 3.5). The lung passageways make up a complex labyrinth of tubes that branch out from the trachea (windpipe) to hundreds of thousands of successively smaller tubes (bronchioles) to millions of tiny ducts at the end that terminate in tiny sacs, or alveoli, with walls one cell thick. These serve as partitions between air and blood through which oxygen, carbon dioxide, and other gases pass in and out of the body. The largest particles in the inspired air do not get very far; many are filtered by the hairs lining the nasal passageways. In fact, particles with dimensions larger than 5 microns are considered nonrespirable. The smaller particles are deposited in varying degrees as they proceed down the tract, some by settling under the influence of gravity, some by colliding with the sides as the air makes sharp turns from one branch to another, and the smallest by diffusing to the walls and sticking. A fraction of those of appropriate size will actually reach the alveolar region. Replicas of the respiratory passages have been cast from lung specimens and detailed measurements made of passageway dimensions. The data have been used to develop idealized models of the respiratory tract that have been used to calculate deposition and dose distributions for various types of radioactive aerosols and breathing patterns (Findeisen, 1935; Landahl, 1950; Weibel, 1963).

Dose calculations have been performed with increasing degrees of sophistication. Early approaches assumed that the deposited particles remained at the point of deposition and that all the emitted alpha energy was absorbed by the tissue at risk (Shapiro, 1954). Subsequently, account was taken of absorption of some of the alpha energy by the mucous lining the passageways, and the clearance of particles through transport by the mucus back up the respiratory tree to the mouth, where they were swallowed and largely excreted (Jacobi and Eisfeld, 1981; Altshuler et al., 1964; Jacobi, 1964; James et al., 1980; Harley and Pasternack, 1972; Burgess and Shapiro, 1968).

There are no cilia in the alveoli; other means are used to remove particles deposited there. Some particles will dissolve in regional fluids and be transferred to the blood, from which they may be taken up by cells or excreted. Others may be absorbed by mobile cells known as macrophages and carried away, generally to lymph nodes. Others may resist all methods that can be mobilized for clearance and remain in place for very long periods of time. While alveolar clearance is very important in determining the dose from radioactivity of long half-life, the significance is minor when considering the radon decay products with half-lives of just minutes.

The different calculation methods result in different values for doses to the various portions of the respiratory tract. Since lung cancer among uranium miners appears primarily in the area of the large bronchi and presumably originates in the basal cells of the upper bronchial epithelium, there is particular interest in the dose to this region. Calculations by various investigators of the dose vary from 2 to 100 mGy/WLM. A nominal value for the dose from a typical mining atmosphere is 10 mGy per WLM. Detailed calculations of doses to the various portions of the respiratory tract are also of interest (Harley, 1996; Burgess and Shapiro, 1968; NCRP, 1984a; ICRP, 1981; UNSCEAR, 2000, Annex B, Table 26). UNSCEAR (2000) lists various assessments of lung dose published between 1956 and 1988. Most take as the target region basal cells in the upper airways. The dose rates from inhalation and deposition of radon decay products in equilibrium with a radon concentration in air of 1 Bq/m^3 range from 5.7 to 71 nanograys per hour, with a median value of 21 mGy/hr.

5.7 Plutonium-239 (Half-life 24,390 yr) and Plutonium-240 (Half-life 6,600 yr)

	239PU		240PU	
	MeV	%	MeV	%
Alpha particles—energies and percent of disintegrations	5.16	88	5.17	76
	5.11	11	5.12	24
Gamma photons	[see Radiological Health Handbook]			

Plutonium-239 is used as fuel in nuclear power reactors and as the explosive for nuclear weapons. Reactor-grade plutonium is roughly 70 percent ^{239}Pu and 30 percent ^{240}Pu. Weapons-grade plutonium is roughly 93 percent ^{239}Pu and 7 percent ^{240}Pu. A liquid-metal fueled fast-breed reactor could contain 3,000 kg of plutonium.

Only 0.003 percent of ingested plutonium is considered to be transferred from the GI tract to the blood. The resultant fractional retention in organs is (ICRP, 1979): $0.45e^{-0.69t/37,000}$ in bone; $0.45e^{-0.693t/37,000}$ in liver; $0.1e^{-0.693t/0.3}$ excreted; and 0.00035 in testes (t in days).

5.7.1   Dose to Lungs

The main hazard to humans from plutonium in the environment is from inhalation. An example of a lung model used for dose calculations is shown in Figure 3.6 (ICRP, 1966a). The lung model gives values for purposes of radiation-protection calculations of the fraction of inhaled dusts that are deposited in various regions of the respiratory tract, the biological half-lives, and the fractions that become translocated through the bloodstream to other organs in the body. The values shown in the figure apply to plutonium dioxide, a highly insoluble form to which humans are likely to be exposed in releases from nuclear power plants. Following inhalation, 10

3.6 Lung model applicable to data for dose calculations for plutonium. *N-P, T-B,* and *P* stand for nasopharynx, tracheobronchial, and pulmonary regions respectively. Parameters apply to PuO$_2$ particles, size 0.4 μm. Fraction in box gives deposition in compartment. Flow lines show percentages removed from compartment and clearance rate (expressed roughly as a biological half-life). (*Source:* Bennett, 1974.)

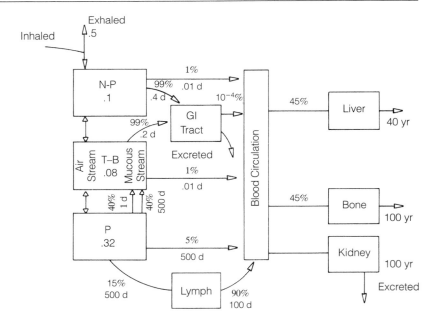

percent is deposited in the nasopharynx region, 8 percent in the tracheo-bronchial region, and 32 percent in the pulmonary region.

Sixty percent of the activity in the pulmonary region is eliminated with a half-life of 500 days—15 percent goes to the lymph nodes (10 percent of which is retained permanently), 5 percent goes to the blood, and 40 percent is carried back along the respiratory tract to the mouth and then swallowed. Another 40 percent is carried back along the respiratory tract with a half-life of 1 day. The plutonium in the blood is divided equally between liver and bone. The fractional transfer from the GI tract to the blood is extremely low, about 10^{-6}. (For a more recent model, see ICRP, 1994b.)

Example 3.32   Calculate the equilibrium annual average dose to the pulmonary region of the lungs from continuous exposure to the environmental limit of 7.4×10^{-4} Bq/m^3 (0.02 pCi/m^3)of plutonium-239 in insoluble form (10CFR20, Table II, col. 1).

The daily intake (in 23 m^3 of air inhaled) is 0.017 Bq. Pulmonary deposition is 32 percent, 40 percent with a half-life of 1 day (1.44 average) and 60 percent with a half-life of 500 days (720 average) (Fig. 3.7).

$$7.4 \times 10^{-4}\,\text{Bq/m}^3 \times 23\,\text{m}^3/\text{day} \times 365\,\text{day/yr} \times 0.32$$
$$\times\, 86{,}400/\text{day-Bq} \times 5.1\,\text{MeV} \times (0.4 \times 1.44\,\text{day}$$
$$+\, 0.6 \times 720\,\text{day}) \div (1{,}000\,\text{g} \times 6.24 \times 10^6\,\text{MeV/g-mGy})$$
$$= 0.0607\,\text{mGy/yr}$$

$$0.0607\,\text{mGy/yr} \times 20 = 1.21\,\text{mSv/yr}$$

Tables of committed equivalent dose from plutonium and other radionuclides to various organs per unit intake via ingestion and inhalation are given in Slaback, Birky, and Shleien, 1997.

The calculation of the dose from highly radioactive plutonium particles by averaging over the whole lung is defended as radiologically sound by Bair, Richmond, and Wachholz (1974).

5.7.2   Determination of a Plutonium Standard for Bone

A protection standard was originally set for plutonium by deriving it from an existing standard/standard limit for radium (Healy, 1975). The limit for radium, set well below the level at which no cases of bone cancer attributable to radium were found in extensive epidemiological studies, was 0.1 μg (0.1 μCi) for radium fixed in the body (that is, bone). Comparison of bone tumors in rats, mice, and rabbits caused by plutonium and radium indicated that plutonium was 15 times more toxic as measured by activity injected. When corrected for retention (75 percent of plutonium and 25 percent of radium are retained in rodents), and twice the ratio of radium to plutonium energy deposition in man as in the rat (80–85 percent of radon escapes from rats, whereas only 50 percent escapes from the human body), plutonium was deduced to be 2.5 times as effective as radium. Thus, allowable limits for plutonium in the body were set at 0.1 μCi/2.5 or 0.04 μCi.

The average dose to bone from 0.04 μCi in the body (bone fraction = 0. 9) is

$$\frac{0.04\mu Ci \times 0.9 \times 1.17 \times 10^{12} \ yr^{-1} \text{-}\mu \ Ci^{-1} \times 5.1 \ MeV}{7000 \ g \times 6.24 \times 10^{6} \ MeV/g\text{-}mGy} = 4.9 \ mGy/yr$$

or 4.9 × 5 × 20 = 490 mSv/yr. This is comparable to the dose from 0.1 μCi radium (600 mSv/yr).

Considering the judgments that must be made in the application of available data, it is not surprising that experts disagreed on permissible levels of exposure and that proposed values could disagree by orders of magnitude. Morgan (1975) suggested that the maximum permissible body burden based on bone be reduced by at least a factor of 200.

The current standards for exposure to plutonium are based on extensive computer models (NCRP, 1985a; ICRP, 1989, 1993a, 1991b, 1995a).

6 DOSE RATES FROM SMALL, HIGHLY RADIOACTIVE PARTICLES

Contamination produced as a result of operations with radioactive materials of high specific activity—that is, of high radioactivity per unit mass—consists of particles that are individually significant radioactive sources, down to very small particle sizes. Examples of sources of highly radioactive particulate contamination include the alpha emitters polonium-210, plutonium-238, plutonium-239, and americium-241; high specific activity compounds prepared by radiochemical companies for research purposes; fission products produced in the fuel of nuclear power reactors or as a result of nuclear explosions; and corrosion products from the coolant systems of nuclear power plants that reach the reactor core and are activated in the intense neutron field.

A spherical hot particle 100 μm in diameter originating from reactor fuel immediately following 500 days of core irradiation has typically a specific activity of 518 GBq/g (14 Ci/g) and an activity of 2.6 MBq (70.3 μCi). Most of the fuel hot particles, however, have activities less than 11 kBq and follow a log-normal distribution. Hot particles arising from corrosion products consist primarily of cobalt-60 with activities of a few kBq. Hot particles with activities up to 40 GBq have been found.

The activity of a hot particle is not diluted by distance from the point of release. Following a series of above-ground nuclear weapons tests by China prior to and during 1970, hot particles were found in the atmosphere near Sweden with specific activities up to 370 GBq/g. Hot particles found in the atmosphere near Norway following the Chernobyl accident had specific activities up to 520 GBq/g.

A researcher can purchase high specific activity phosphoric acid (H_3PO_4) with an activity of 5.55 GBq (150 mCi) of phosphorous-32 per milliliter from a radiopharmaceutical company. A 0.05 ml droplet has an activity of 277.5 MBq (7.5 mCi). The spread of a few droplets can contaminate an entire building and require a very costly cleanup operation, not to mention an accounting with the institution staff, the media, the public, and the regulatory authorities (NCRP, 1999a).

6.1 Evaluation of the Dose from Beta Particles

The dose rates from particles that emit beta particles can be evaluated in various ways. A practical, straightforward approach was conceived 50 years ago (Loevinger, 1956) and still has relevance today. Dose rates are calculated from formulas for the dose rate in tissue for point sources of beta particles (Hine and Brownell, 1956), which are derived from actual measurements of dose rates in water. This method was the basis of such classic work as the Los Alamos report "Surface Contamination: Decision Levels" (Healy, 1971), which provided extensive data on beta dose rates associated with skin and surface contamination. The VARSKIN computer code is used by the NRC for evaluating the dose to contaminated skin and comparing it to regulatory limits. The National Institute of Standards and Technology (NIST) developed a code (NISTKIN) that refined the dose calculations from point sources (point-kernel-integration approach). This code starts with dose calculations from point sources of monoenergetic electrons, follows the electrons through their range by Monte Carlo calculations, accounting for the effects of electron energy-loss straggling, and then integrates over beta spectra to get doses from specific radionuclides (NCRP, 1999a). The code includes a parallel routine for photons. Some results obtained with this code are given in Table 3.6.

6.2 Biological Effects of Hot Particles

The NCRP classifies highly radioactive particles greater than 10 μm but less than 3,000 μm in any dimension that emit beta or gamma radiation as "hot particles," and particles less then 10 μm as general contamination

Table 3.6   Dose rates in water near spherical beta-particle sources with diameters of 10 and 100 μm.

Distance from surface (μm)	Beta dose rate in mGy/h-kBq			
	^{60}Co in alloy, $\rho = 8$ g/cm^3		^{90}Sr/^{90}Y in "fuel," $\rho = 11$ g/cm^3	
	10 μm	100 μm	10 μm	100 μm
50	5,534	569	4,116	984
100	1,153	176	1,079	405
300	39.2	6.80	109	61.6
500	2.19	0.295	33.6	21.2
2,000			1.05	0.944
5,000			0.0717	0.0588

Source: NCRP, 1999a.

(NCRP, 1999a, Report 130). As sensitive personnel and area detection instrumentation is developed that can monitor large areas efficiently and rapidly, these particles are easily detected. As a result, they are found with increasing frequency on workers and surfaces at facilities, such as nuclear power plants, that handle large amounts of radioactivity, and they are therefore a source of considerable interest and concern with respect to their health effects.

Because the volume of tissue that receives high doses from hot particles is very small, the NCRP considers risks of cancer to be insignificant—the purpose of their recommendations is to limit damage to tissue. Even when the radiation dose is damaging to the skin around a hot particle, the small area of injury is repaired by replacement with healthy cells from surrounding tissue. Thus, the limit recommended by NCRP is intended to prevent the type of exposure that would compromise the barrier function of the skin, including the skin in the ear, and cause ulceration. The eye is another organ given special consideration. Here, the limits are designed to prevent loss of visual function, damage to the cornea, and injury to the skin around the eye, such as the eyelid, with the possibility of infection. Limits for the skin can also be applied to the eye.

According to the NCRP, there has, in fact, been only one reported clinically observed human injury due to hot-particle exposures in the workplace. A worker at a nuclear reactor suffered radiation injury from a hot particle in his ear. The particle had a diameter of 70 μm and was found to produce a dose rate of one roentgen per hour at 3 cm from aged fission products. The particle was apparently blown into the ear and lodged on the tympanic membrane by a mild blast of air from a contaminated heat exchanger. The particle was not discovered and removed until three days after the worker complained of a draining ear and initial efforts at treating it as an ear infection failed. The duration between the initial introduction of the particle into the ear and its removal was approximately 10 days. A hearing loss of 10 percent was claimed as a result of the accident.

Quantitative data on the effects of hot particles on the skin have, by necessity, to be obtained from experiments with animals, normally pigs. A characteristic area that receives most of the dose from hot particles is 1.1 mm^2. It takes about 110 Gy of beta rays of intermediate energy (0.5 MeV < E_{max} < 1.5 MeV) and 340–540 Gy of low-energy beta rays (E_{max} < 0.5 MeV) imparted at a depth of 16 μm and an area of 1.1 mm^2 to produce a lesion in 10 percent of the exposures. Limits in the regulations, as well as those in previous reports of the NCRP, refer to an average dose to 1 cm^2 of the skin at a depth of 70 μm. Averaged in this way, the lesion-producing dose would be 1.3–2.2 Gy for intermediate-energy radionuclides and in the range of 3.4–4.1 Gy for low-energy beta-emitting radionuclides.

6.3 Risk of Cancer from Hot Particles

Hot particles do pose a risk of inducing skin cancer, although the risk is low and is confined primarily to basal cell cancer and squamous cell cancer, which are rarely fatal. There is no significant data to indicate the risk of melanoma, though if it exists, it must be low. The risk is proportional to the area of skin irradiated. Because ultraviolet radiation (UV) acts to promote the induction of cancer initiated by ionizing radiation, skin pigmentation appears to be a factor in protecting against the incidence of the disease. Risks for nonmelanoma skin cancer, calculated by combining the results of several studies, are 0.18 per million per year per sievert per cm^2 skin irradiated (absolute risk) for skin exposed to UV and about one-fifth as much in skin shielded from UV. Expressed as excess relative risks, compared to the normal incidence of skin cancer, they are 61 percent per sievert for UV-exposed skin and 1.4 per sievert for protected skin.

The lung dose from hot particles can be evaluated as the average dose to the lung for purposes of assessing the risk of inducing cancer, which is also the position presented in previous reports pertaining to nonuniform distribution of doses in the lungs, including the dose from alpha emitters. However, the NCRP recommends in its Report 130 that the averaging should be performed separately for the different regions (nose, pharynx, larynx, trachea, lung, etc.) because of the very different sensitivities in these regions and the weighting factors should be apportioned appropriately. In particular, the doses to the thoracic portions of the lung should be determined separately for the bronchial, bronchiolar, and alveolar-interstitial portions, each of which is assigned 0.333 of the 0.12 weighting factor assigned to the lung, and for the lymphatics, with 0.001 of the 0.12 weighting factor. The dose to the extrathoracic portion should be determined separately, with the 0.025 remainder weighting factor assigned to the combination of posterior nasal passage, larynx, pharynx, and mouth and 0.001 of the 0.025 (that is, 0.000025) applied to the anterior nasal cavity and to the lymphatics. This gives an extremely low risk for cancer of the anterior nasal cavity and lymphatics. The concern for health is focused on the tissue damage from high doses, rather than the risk of inducing cancer.

6.4   Highly Radioactive Particles in Fallout

Of particular significance from a public health point of view are the radioactive particles released to the environment as a result of past or potential nuclear power plant accidents and releases in the past from the testing of nuclear weapons in the atmosphere. The particles are carried to high altitudes and circulate around the earth to be deposited on the ground, mainly by rain and snow. Some of the particles in the air at ground level are inhaled by human beings and are deposited in the lungs. Their activity is not diminished by the inverse square law, even though they have traveled thousands of miles from their origin. They have the same activity anywhere in the world as they had at the place they were produced, decreasing only in time with the half-life of the radionuclide.

Most of the particles that make their way into the lungs are eliminated by normal clearance processes, but a few remain for months or longer. Many of the retained particles are carried to the lymph nodes by scavenger cells, where they accumulate and produce unusually high local doses.

The activities of the particles depend on the particle size, time after the detonation, and the nature of the explosion. A characteristic activity found for 1-micron diameter particles in fallout 100 days old was 6 disintegrations per minute (Sisefsky, 1960). For other times in days *(T)* and diameters in microns *(D)*, the activity of particles of similar composition would be given, in disintegrations per minute, by $6D^3(100^{1.15}/T^{1.15})$. Note that the activity does not follow a single half-life mode of decay characteristic of single radionuclides, since fallout contains a mixture of radioactive fission products of different half-lives.

6.5   Recommendations of the NCRP on Limits of Exposure to Hot Particles

The following recommendations are quoted from NCRP Report 130 (1999a):

Skin and ear. It is recommended that the dose to skin at a depth of 70 μm from hot particles on skin (including ear), hair, or clothing be limited to no more than 0.5 Gy averaged over the most highly exposed 10 cm² of skin. [In the event that the areas of skin exposed by two or more hot-particle exposure events overlap, then the limit applies to the calendar year, rather than to the individual events.]

Eye. It is recommended that the dose at a depth of 70 μm to any ocular tissue from hot particles be limited to 5 Gy averaged over the most highly exposed 1 cm² of ocular tissue. [The doses imparted by separated instances of exposure to hot particles should be added, and the sum must meet 5 Gy as an annual limit, even if the particles were at different places on the eye and different times, because of the small size of the eye.]

Respiratory system. It is recommended that limitation be based on currently applicable effective dose limits, with the effective dose determined using general respiratory system models and residence times for insoluble material. [The method of evaluating the effective dose is given in section 6.3.] For the special case of hot particle sequestration in the anterior nasal compartment, the dose at a depth of 70 μm to nasal tissue from hot particles be limited to 5 Gy averaged over the most highly exposed 1 cm². [Because of the small size of the anterior nasal compartment, this should be viewed as an annual limit, even if particles are believed to have been present at different locations and/or times.]

Gastrointestinal system. It is recommended that limitation be based on currently applicable effective dose limits, with the effective dose determined using general GI system models and residence times for insoluble material. [Adjustments should be made if bioassay data give exposure conditions that vary significantly from the default model values.]

6.6 NRC Enforcement Policy for Exposures to Hot Particles

The NRC defines a hot particle as a discrete radioactive fragment that is insoluble in water and is less than 1 mm in any dimension. Its limits for exposure to the skin, as averaged over 1 cm², are relaxed if the hot particle is in contact with the skin (NRC, 1990). For this case, the limit is a total beta emission value of 10^{10} beta particles (75 microcurie-hours). If it can be determined that the particle was never in contact with the skin, the annual limit for the dose to the skin applies, that is, the limit is 500 mSv at a depth of 7 mg/cm² averaged over an area of 1 cm² in the region of the highest dose. This policy follows the recommendations in NCRP Report 106 (NCRP 1989d), where the limit was set to prevent deep ulceration.

7 THE RADIOACTIVE PATIENT AS A SOURCE OF EXPOSURE

Every year, 8 to 9 million mobile radiation sources are released into public areas. That is about how many people are given radioactive pharmaceuticals for diagnosis or therapy or are implanted with a radioactive source in the treatment of cancer in a single year. The regulatory agencies have the problem here of placing controls on the release of patients from hospitals and the subsequent activities of the patients to limit the dose they could impart to others.

The NRC allows a hospital or any other authorized licensee to release a radioactive patient from its control only if the estimated effective dose the person might impart to another individual is not likely to exceed 5 mSv (500 mrem) in any one year. This is higher than the general limit in 10CFR20 (NRC, 1991) of 100 mrem per year for members of the public but consistent with the provision allowing the higher limit for limited periods of time. The dose imparted in unrestricted areas by the patient is also exempt from the requirement imposed on other external sources that it not exceed 0.02 mSv (2 mrem) in any one hour. If the dose imparted to any individual by the patient is likely to exceed 1 mSv (100 mrem) in a year from a single administration, the licensee must provide the patient before release with radiation safety guidance and written instructions on how to minimize exposures of others; it must also maintain for three years a record of the released patient and the calculated total effective dose to the individual likely to receive the highest dose. This control procedure replaced a previous NRC directive, which permitted the release of patients only if the measured dose rate from the patient was less than 0.05 mSv (5 mrem) per hour at a distance of 1 meter or the radiopharmaceutical content of the patent was less than 1,110 MBq (30 mCi). The new criteria replaced an activity-based limit with a dose-based limit and were consistent with recommendations of the International Commission on Radiological Protection and the National Council on Radiation Protection and Measurements. They permitted earlier release of the patient and reduced both the cost of hospitalization and risks of infection sometimes incurred by patients as a result of extended hospitalization.

An NRC Regulatory Guide, No. 8.39 (NRC, 1997), contains values of activities in a patient of commonly used radionuclides that would meet the requirements for release. Special instructions are given for patients who could be breast-feeding after release.

Absorbed doses to family members of patients released after ^{131}I therapy for thyroid cancer and for hyperthyroidism were measured using thermoluminescent dosimeters worn on the chest (Mathieu et al., 1999). The thyroid cancer patients were treated with 3,700–7,400 MBq and hospitalized for two days. The hyperthyroid patients were treated on an outpatient basis with 200–600 MBq. The activity decayed with a mean half-life of 2.2 days in the thyroid patients and 6.2 days in the hyperthyroid patients. The dose was less than 0.5 mSv in all children and spouses in the cancer group. The doses in the hyperthyroid group ranged from 0.05 to 5.2 mSv (median 1.04) to the partners and 0.04 to 3.1 mSv (median 0.13) to the children. Mean doses to family members were, for different levels of administered activity to the patient, 0.31 mSv (200 MBq), 0.92 mSv (400 MBq), and 1.50 mSv (600 MBq).

The highest dose of 5.2 mSv was received as a result of circumstances accompanying the treatment. The patient was a woman treated with 600 MBq who went on vacation with her family soon after treatment. The family of four stayed in a little flat. The father received a dose of 5.2 mSv and the two daughters, respectively 16 and 21 years old, received more than 3 mSv. Since dosimeters were worn, the father was probably aware of the policy to exercise special care to limit exposure. A benefit-risk decision was made in which the benefit of taking a family vacation in housing at close quarters outweighed the risk imposed by the radiation exposure. Rescheduling of either the vacation or the treatment would have significantly reduced the exposures received.

The doses to the families of patients treated for hyperthyroidism were significantly higher than those to families of patients treated for cancer, even though the cancer patients were administered higher activities in the thyroid. The reason was the longer effective half-life of the ^{131}I in the hyperthyroid patients than in the cancer patients, whose thyroids were no longer functional. Furthermore, the cancer patients were hospitalized for a couple of days so that contamination by perspiration and saliva could be reduced and their urine could be collected for disposal as radioactive waste.

8 RADIATION DOSES IN NUCLEAR MEDICINE

Radiopharmaceuticals now rank with x rays in the diagnosis and treatment of disease. The most frequently used radionuclide is technetium-99m, used in 65 percent of all tests in 1987, with runner-up honors going to iodine-131, used in 12 percent of the tests (UNSCEAR, 1993). Gold-198 was phased out in 1977. Iodine-125 is losing market share in diagnostic tests, giving place to the short-lived iodine-123 where practical, though it has found a place in the radiotherapy of cancer. The use of thallium-201 is increasing, though it is being replaced with antimyosine immune scintigraphy, radionuclide ventriculography, and other methods that give lower patient doses. Indium-111, iodine-131, and technetium-99m are used to tag monoclonal antibodies to locate tumors and metastases through radioimmunoscintigrapy. Other evolving diagnostic tools are single photon emission tomography (SPET), positron emission tomography (PET), and whole-body imaging for use in tumor localization, functional brain studies, cardiac studies, bone imaging and abdominal imaging.

Examples of absorbed doses imparted to organs in the body and the ef-

Table 3.7 Organ doses and effective doses from technetium-99m (mGy/MBq).

Organ	Pertechnetate in test for thyroid uptake	Macroaggregated albumin used for lung scan	Methylene diphosphonate used for bone scan
Adrenals	0.0034	0.010*	0.0019
Bladder wall	0.014*	0.0011	0.050*
Bone surfaces	0.0033	0.0064	0.063
Breast	0.0018	0.0027	0.00088
GI tract			
Stomach wall	0.050*	0.0062	0.0012
Small intest	0.030*	0.0043	0.0023*
Upper large intestine wall	0.074*	0.0056	0.0020*
Lower large intestine wall	0.024*	0.0018	0.0038*
Kidneys	0.0054	0.0097*	0.0073*
Liver	0.0040	0.074*	0.0013
Lungs	0.0022	0.0055	0.0013
Ovaries	0.012	0.0022	0.0035
Pancreas	0.0092	0.012*	0.0016
Salivary glands	0.0061		
Red marrow	0.0062	0.011	0.0096
Spleen	0.0060	0.077*	0.0014
Testes	0.0019	0.00062	0.0024
Thyroid	0.015	0.00079	0.0014
Uterus	0.0087	0.0019	0.0061
Other tissues	0.0032	0.0028	0.0019
EDE (mSv/MBQ)	0.015	0.014	0.0080

Source: ICRP, 1988b. Another source for radiopharmaceutical doses, using slightly different modeling, is Stabin et al., 1996.

Notes: The dose to organs or tissues not mentioned in the table can usually be approximated with the value given for "Other tissues." The dose to the embryo, and to the fetus when diaplacental transfer is known not to occur, can be approximated by the dose to the uterus. In the calculation of effective dose equivalent (EDE), doses to the gonads, breast, red bone marrow, lung, thyroid, and bone surfaces are always considered, with their specific weighting factors. In addition, the five remaining organs and tissues receiving the highest dose are also included and marked with an asterisk in the table. Multiply doses in mGy/MBq by 3.7 to get rad/mCi.

fective dose per unit activity administered for three radiopharmaceuticals tagged with ^{99m}Tc are given in Table 3.7.

Doses imparted to the patient for the activity used in some of the diagnostic procedures in nuclear medicine are given in Table 3.8. Here the whole-body dose and dose to the organ receiving the highest dose are listed. The data are given for an adult. Doses to children are several times higher. The doses are only rough estimates of the doses actually imparted to a particular patient. Metabolic data are limited and must be inferred from whatever data are available. Adjustments are required for age, sex, or

Table 3.8 Doses imparted by radiopharmaceuticals in routine nuclear medicine procedures.

Function or organ examined	Technetium-99m labeled radiopharmaceuticals	Activity given (mCi)	Organ dose (rad)	Effective dose equivalent (rem)
Bone	Pyrophosphate	19–27	4.4–6.3	0.56–0.80
Brain	Gluconate	20–27	4.2–5.6 (bladder)	0.68–0.90
Liver	Sulfur colloid	2.7–5.0	0.75–1.40	0.14–0.26
Lung	Microaggregated albumin	2.7–5.4	0.67–1.34	0.12–0.24
Heart (cardiac output)	Erythrocytes	20–24.3	1.7–2.07	0.63–0.77
Renal	Gluconate	10–16.2	2.07–3.36 (bladder)	0.33–0.54
Thyroid (scan)	Pertechnetate	2.7–10	0.23–0.85 0.62–2.30 (intestine)	0.13–0.48

Function or organ examined	Other radionuclides	Activity given (mCi)	Organ dose (rad)	Effective dose equivalent (rem)
Thyroid (uptake, 35 percent)	Sodium iodide (^{131}I)	0.0051–.054	6.8–72	0.21–2.2
Thyroid (scan, 35 percent uptake)	Sodium iodide (^{123}I)	0.203	3.4	0.11
Inflammation	Gallium citrate (^{67}Ga)	2.7–6.75	5.9–14.8 (bone surface)	1.2–3.0
Heart	Thallous chloride (^{201}Th)	2.0–2.16	4.1–4.5 (testes)	1.7–1.84

Notes: I millicurie = 37 megabecquerels. I rem = 10 millisieverts. Activities administered taken from Huda and Gordon (1989), who give the upper values generally used in administrations to adults. Data on doses are taken from ICRP (1987a, b), Publications 53 and 54, and refer to adults. Doses to a 5-year-old child relative to adults are 3–5 times as high. Organ doses are to organ examined unless otherwise indicated.

condition of patient. Sometimes the metabolic data must be inferred from animal experimentation or from compounds or radionuclides that differ from the substance administered. There are just too many compounds being used in nuclear medicine to permit the detailed studies that have been done, for example, on the most significant nuclides in radioactive fallout. The dose problems are much simpler in the nuclear medicine field, however, because of the short half-lives of the compounds generally used.

As a first approximation, it may be assumed that percentage uptake and time dependence of activity in an organ are independent of age. The extrapolation of doses from nonpenetrating radiation then depends only on knowledge of the organ mass. Organ mass may be assumed proportional to body mass as a first approximation, but more representative values are available (ICRP, 1975). The extrapolation of doses from penetrating radia-

Table 3.9 Fraction of activity ($\times$ 10^6) in the embryo/fetus (excluding the thyroid) and in the thyroid initially and every 30 days after uptake of ^{125}I and ^{131}I into bloodstream of the mother.

Age (days) of fetus at uptake	Fraction $\times$ 10^6 in fetus / Fraction $\times$ 10^6 in fetal thyroid at age (days) of fetus						
	90	120	150	180	210	240	270
Iodine-125							
0		34/13[a]	1/33	2.9/44	4.3/47	5.2/43	5.3/35
30		—/21	2.1/52	4.6/71	6.9/75	8.3/69	8.5/57
60		—/32	3.3/80	7.2/112	11/119	13/109	14/90
90	19/166	2.9/157	8.0/192	14/217	20/213	23/187	23/151
120		91/4010	112/2850	137/1930	130/1290	111/838	88/535
150			235/7820	334/5260	338/3340	279/2070	211/1250
180				420/11600	658/7410	596/4470	446/2630
210					600/15300	1070/9310	895/5360
240						736/19000	1560/11000
270							802/23000
Iodine-131							
90	24/158	—/17	—/2.2				
120		120/3810	12/312	1.6/22	—/1.6		
150			318/7420	37/574	3.9/38	–/2.5	
180				584/11000	72/808	6.8/51	—/3.2
210					863/14500	117/1020	10/62
240						1110/18000	170/1200
270							1270/21500

Source: Sikov et al., 1992.

a. For example, these numbers indicate that the activity was administered to the mother at or before conception of the fetus (age = 0 day) and that, when the fetus was age = 120 days, the fraction of activity in the fetus (exclusive of thyroid) was 34 $\times$ 10^{-6} and the fraction of activity in the fetal thyroid was 13 $\times$ 10^{-6}.

tion to age groups that differ from those listed is affected by anatomical factors in addition to masses. As a first approximation, body dimensions can be estimated as proportional to the one-third power of the mass (Webster et al., 1974).

The ICRP recommends that the activity administered to patients should be the minimum consistent with adequate information for the diagnosis or investigation concerned (ICRP, 1971, Publication 17). When pregnant patients are being treated, consideration must be given to the quantity of activity transmitted across the placenta and to the resulting fetal uptake. Examinations of women of reproductive capacity should be restricted to the first 14 days following onset of menses. Activities administered to children should be reduced according to weight or other reasonable basis. When repeat and serial tests are being considered, the overall dose received during each series of tests should be considered rather than the dose in any one investigation. Certainly, unnecessary repeat investigations should be avoided. Blocking agents, when they are available, should always be considered before administering a radioactive nuclide. An estimate must be made of the dose to the organ that then is likely to receive the maximum dose. A review of the factors influencing the choice and use of radionuclides in medical diagnosis and therapy has been issued by NCRP (1980). The report includes examples of dose calculations in nuclear medicine.

8.1 Dose to the Fetus from Uptake of Radionuclides from the Mother

An intake of radioactivity by a pregnant worker calls for as accurate an assessment as possible of the dose to the fetus and any health consequences that might result. The dose calculations are very difficult, entailing knowledge of the source term in the mother and deducing from that the transfer across the placenta and the evaluation in the fetus of an uptake that varies greatly as the organs develop. Analyses have been made for a number of radionuclides that are of particular interest with respect to occupational exposure (Sikov et al., 1992).

Results are presented in Tables 3.9 and 3.10 for an uptake of iodine-125 and iodine-131 into the bloodstream of the mother. Table 3.9 gives the fractions of the activity initially appearing in the bloodstream of the mother that are deposited in the fetus (excluding the thyroid) and in the fetal thyroid over the course of the gestation period. Table 3.10 gives the

Table 3.10 Dose (mGy) to the fetus for remainder of gestation period following uptake of 1 μCi (0.037 MBq) of ^{125}I and ^{131}I into bloodstream of mother.

Age (days) of fetus at uptake	Dose from ^{125} (mGyI)		Dose from ^{131}I (mGy)	
	Fetus	Fetal thyroid	Fetus	Fetal thyroid
0	0.00083	0.198	0.00062	0.000060
30	0.00090	0.314	0.0010	0.00090
60	0.0011	0.489	0.00099	0.013
90	0.0029	2.2	0.0067	4.6
120	0.014	14	0.036	39
150	0.013	13	0.025	31
180	0.011	9.5	0.030	24
210	0.0090	6.6	0.021	20
240	0.0053	3.9	0.010	14

Source: Sikov et al., 1992.

doses imparted to the fetus and to the fetal thyroid as a result of the introduction of 0.037 MBq (1 μCi) into the bloodstream of the mother.

9 EVALUATION OF DOSES WITHIN THE BODY FROM X RAYS

The radiation incident on a patient in an x-ray examination is absorbed
strongly as it enters the body. As a result, the dose falls off rapidly with
depth. Figure 3.7 shows the depth dose in water. The fall-off in soft tissue
is very similar and the data may be applied to determine the dose imparted
to points inside the body if the only shielding material encountered by the
incident beam is soft tissue.

The attenuation curves of Fig. 3.7 include the effect of distance as well
as attenuation by the medium. Thus they apply strictly only to the 30-inch
source-skin distance under which the data were obtained. However, they

3.7 Central axis depth dose in water: single-phase system, 30-inch source-to-skin distance, 40-inch source-to-film distance, 14 × 17-inch field at 40 inches, 3 mm aluminum filtration. (*Source:* Kelley and Trout, 1971.)

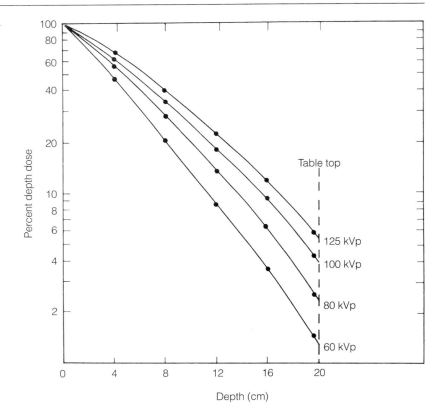

may be used for a limited range of other distances between the source and surface and, in particular, to surface doses imparted to patients x-rayed at the typical target-film distance of 40 inches.

There is also some scattered radiation dose outside the direct (collimated) beam, and while it is only a small fraction of the direct radiation, it is often of interest. For a given energy, the scattered dose at different depths depends primarily on the distance from the edge of the field as defined by the collimator. Measurements in a water phantom with 100 kV$_p$ x rays give a scattered dose of 10 percent of the central beam dose at 1.3 cm from the edge and 1 percent at 8.8 cm (Trout and Kelley, 1965).

Figure 3.8 gives a model of the adult human torso developed for calculating the dose to organs within the body. An anterior view of the principal organs is given in Figure 3.9 and their dimensions are shown on an isometric drawing in Figure 3.10. The organs are represented mathematically primarily as ellipsoids, and the figure shows the principal axes. Because the dose through an organ is not uniform, the value at the midline may be used as an average value for radiation-protection purposes and thus may be calculated with the use of the depth dose data for water given in Figure 3.7. In radiotherapy, a detailed dose plot is necessary, and the location and geometry of the organs of patients undergoing treatment must be accurately determined, either by taking a series of x rays or by computerized axial tomography.

Dose distributions may also be determined from basic data. One calculation (Rosenstein, 1976b) considers monoenergetic parallel collimated beams normally incident on grid elements, 4 cm × 4 cm, that cover the vertical midplane of the phantom. The photons are followed through successive interactions in the body (by a mathematical technique known as the Monte Carlo method) until they are absorbed or leave the body. The results are determined for several photon energies that adequately represent the energy range of diagnostic x rays. Any diagnostic x-ray spectrum is simulated by a weighted combination of these energies. The calculations are very lengthy and tedious and must be made with the use of high-speed computers. Doses to organs in a reference adult patient have been obtained by this method for a variety of common x-ray projections using the mathematical models of Figures 3.8 and 3.9 (Rosenstein, 1976b). Corresponding data are also available for children (Beck and Rosenstein, 1979; Rosenstein, Beck, and Warner, 1979). The basic data are presented in terms of the *tissue-air ratio,* which is the average absorbed dose (grays) in the organ per unit exposure *(R)* at the organ reference plane in the absence of the attenuating medium (that is, *free-in-air).* The organs selected for the calculations included the testes, ovaries, thyroid, active bone marrow, and the uterus (embryo). To facilitate use of the data by practitioners, organ

3.8 Dimensions and coordinate system of adult human phantom. (*Source:* Snyder et al., 1969. The trunk is an elliptical cylinder given by the equation $(x/20)^2 + (y/10)^2 \leq 1, 0 \leq z \leq 70$. The head is also an elliptical cylinder, $(x/7)^2 + (y/10)^2 \leq 1, 70 < z \leq 94$. The legs are considered together to be a truncated elliptical cone, $(x/20)^2 + (y/10)^2 \leq [(100 + z)/100]^2$, $-80 \leq z < 0$.

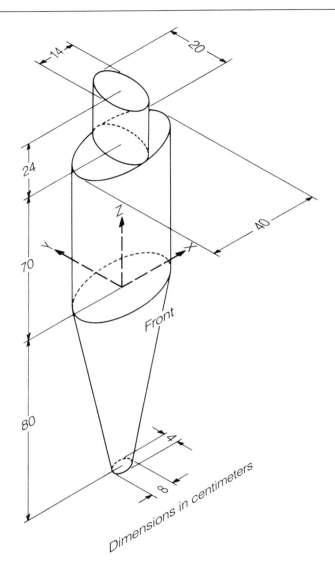

doses in terms of the entrance skin exposure (free-in-air) have also been compiled in handbook form (Rosenstein, 1976a; Suleiman et al., 1999). An example of the data for a lumbar spine x ray, expressed in millirads per 1,000 mR entrance skin exposure is given in Table 3.11.

The beam quality (HVL, mm Al) is either measured directly or esti-

3.9 Anterior view of principal organs in head and trunk of phantom. (*Source:* Snyder et al., 1969.

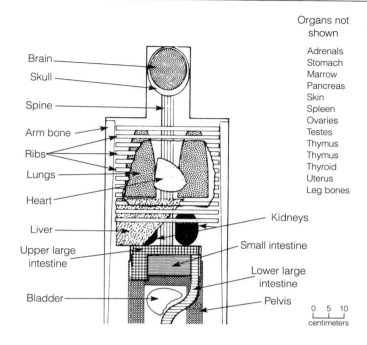

Brain
Skull
Spine
Arm bone
Ribs
Lungs
Heart
Liver
Upper large intestine
Bladder

Organs not shown

Adrenals
Stomach
Marrow
Pancreas
Skin
Spleen
Ovaries
Testes
Thymus
Thymus
Thyroid
Uterus
Leg bones

Kidneys
Small intestine
Lower large intestine
Pelvis

0 5 10
centimeters

mated from the peak potential (kV_p), total filtration, and machine waveform. Some data are given in Table 3.12. A characteristic value for total filtration on diagnostic x-ray machines is 3 mm Al. Table 3.13 gives a compilation of data from the handbook of organ doses for a 3 mm Al HVL beam. Note that the skin exposures are considerably higher for a lateral than for an anterioposterior (AP) projection, and much less for a thoracic examination than for an abdominal examination.

9.1 Patient Doses in Mammography

Organ doses are often evaluated from knowledge of the entrance dose at the surface of the body, but the situation is much more complicated at the energies used in mammography. The radiation dose throughout the breast varies significantly with peak potential (kV_p), beam filtration, breast thickness, and other factors (Hendee, 1995). As a result, dose to the breast is determined as the absorbed dose to glandular tissue, the mean glandular dose (Rosenstein et al., 1985).

Recommendations on limits for patient dose in mammography need to be made for specific examination conditions. For example, restrictions

3.10 Isometric drawing of torso showing locations of organs, which are idealized mathematically as ellipsoids. The dimensions are taken from Snyder et al., 1969. The solid lines are the principal axes of the ellipsoid. The y axis is drawn at 45° to x and z and serves to locate the organ with respect to the midline (that is, the plane that divides the body in half, front to back). To obtain organ location and dimensions: (1) follow the dashed line leading from the y axis to its terminal point in the x-z plane, shown as a dot. This dot gives the z and x coordinates of the centroid of the organ. (2) Determine the distance from the dot to the center of the organ (as given by the intersection of the three axes). (3) Determine the distance from the surface of the body through the y principal axis in the midline by following from the dot up along the z axis to $z = 70$ and then measuring out along the y axis to the anterior surface. (4) Calculate the depth of the organ, which is the difference between the surface y coordinate and the organ centroid y coordinate.

Example: Calculate the depth of the kidney below the surface in an AP exposure. The kidney centroid (right side of the body) is 6 cm behind the dot, which is at $z = 32$ cm and $x = 5$ cm as read in the x-z plane. The distance from the midline to the surface at $x = 5$ cm is 9.5 cm as read at $z = 70$. Therefore the depth of the kidney from the anterior surface is 9.5 + 6 = 15.5 cm.

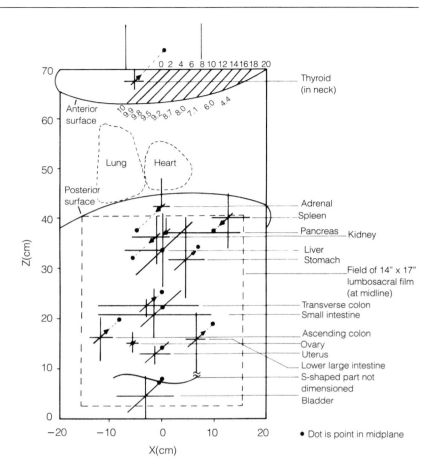

have been stated in terms of the average glandular dose per view to a compressed breast of 4.5 cm thickness and composition of 50 percent adipose and 50 percent glandular tissue (Rothenberg, 1993).

A summary of doses to the breast compiled by the American College of Radiology (ACR) as part of its Mammography Accreditation Program (Barnes and Hendrick, 1994) is given in Table 3.14.

9.2   Evaluation of Doses in CT Examinations

The nature of the radiation exposure in a CT examination is very different from that in other examinations in diagnostic radiology. Surveys

Table 3.11 Example of organ dose data for diagnostic x rays: lumbar spine organ dose (mrad) for 1,000 mR entrance skin exposure (free-in-air).

Organ	View	Beam HVL (mm Al)					
		1.5	2.0	2.5	3.0	3.5	4.0
		Dose (mrad)					
Testes	AP	1.1	2.2	3.7	5.6	7.8	10.0
	LAT	0.2	0.4	0.7	1.1	1.6	2.3
Ovaries	AP	91	139	188	238	288	336
	LAT	15	27	41	58	76	96
Thyroid	AP	0.05	0.2	0.3	0.5	0.8	1.1
	LAT	a	a	a	a	a	a
Active bone	AP	13	21	32	46	62	81
marrow	LAT	8.2	13	19	27	37	48
Uterus	AP	128	189	250	309	366	419
(embryo)	LAT	9.4	17	27	39	53	68

Source: Rosenstein, 1976a.

Conditions: SID-102 cm (40 in.). Film size = field size, 35.6 cm × 43.2 cm (14 in. × 17 in.). Entrance exposure (free-in-air), 1,000 mR. Projection: Lumbar spine.

Note: Divide by 100 to convert to mGy.

a. <0.01 mrad.

Table 3.12 Half-value layers as a function of filtration and the tube potential for diagnostic units[a].

Total filtration (mm Al)	Peak potential (kVp)					
	70	80	90	100	110	120
	Typical half-value layers in millimeters of aluminum					
0.5	0.76	0.84	0.92	1.00	1.08	1.16
1.0	1.21	1.33	1.46	1.58	1.70	1.82
1.5	1.59[b]	1.75	1.90	2.08	2.25	2.42
2.0	1.90	2.10	2.28	2.48	2.70	2.90
2.5	2.16	2.37[b,c]	2.58[b,c]	2.82[b,c]	3.06[b,c]	3.30[b,c]
3.0	2.40	2.62	2.86	3.12	3.38	3.65
3.5	2.60	2.86	3.12	3.40	3.68	3.95

Source: NCRP, 1968, Report 33.

a. For full-wave rectified potential.

b. Recommended minimum HVL for radiographic units.

c. Recommended minimum HVL for fluoroscopes.

Table 3.13 Selected organ doses for projections common in diagnostic radiology.

Projection	View	Exposure at skin (mR)[c]	Organ dose (mrad)[a]				
			Active bone marrow	Uterus (embryo)	Ovaries	Testes	Thyroid
Retrograde pyelogram,	AP	470	23	155	121	10	X[b]
KUB, barium enema,	PA	250	33	44	50	3	X
lumbosacral spine, IVP	LAT	1000	31	53	70	4	X
Lumbar spine	AP	736	34	227	175	4	X
	LAT	2670	72	104	155	3	X
Pelvis, lumbopelvic	AP	1100	52	388	288	103	X
	LAT	3500	116	196	256	105	X
Upper GI	AP	256	8	9	11	X	0.6
Chest (72 in. SID)	PA	14	1.4	0.03	0.02	X	0.6
	LAT	45	1.9	0.04	0.04	X	5.2

Source: BRH, 1976.

Note: Divide by 100 to convert mrad to mGy.

a. For source to image (SID) distance of 102 cm; 14 in. x 17 in. film size; 3.0 mm Al HVL, unless otherwise indicated.

b. X = insignificant.

c. Measured in air in absence of patient. Doses for other exposures may be determined by ratio provided beam quality (HVL) is the same. (See Table 2.16 for recent exposure data.)

Table 3.14 Values of mean glandular dose in institutions participating in the ACR Accreditation Program

Image receptor	Number of facilities	Mean glandular dose (mGy)	
		Average	Range
Screen-film, nongrid	164	0.77	0.12–2.48
Screen-film, grid	5,054	1.28	0.15–7.45
Xeroradiography	420	2.90	0.56–8.90

Source: Hendee, 1995, after Rothenberg, 1993.

of CT doses therefore utilize different protocols and analyses of the data. The basic exposure is from a beam of radiation that is incident on the body in a 180 degree arc or larger. The beams are delivered in "slices" and the machine settings are adjusted by the radiologist to give the desired image. The adjustments include the slice width, the intervals between slices, and electrical settings such as high voltage and tube current. The dose at the central slice in a multiple scan is then made up of contributions from radiation incident directly on the slice; from radiation imparted to an adjacent slice that may overlap it; from radiation scattered from the other slices that contribute to the examination; and from divergence of the beam, which occurs if the beam is not limited precisely to the boundaries of the slice.

As a result, a mapping of the dose across the slice in a single scan looks something like the dose profile in Figure 3.11, instead of a constant value. The contribution to the dose at the central slice by a contiguous slice is also shown in the figure. The profile can be determined for either single or multiple scans by taking radiation measurements with a series of TLD chips placed across the slice or producing an image on a strip of film. The results are obtained with the use of phantoms to simulate the head or the abdomen. The most commonly used phantoms are acrylic cylinders in which holes are drilled for the insertion of dosimeters.

3.11 Radiation dosimetry in computed tomography. *(a)* Collimated x-ray beam incident on body, producing a slice of width *T*. *(b)* Profile of radiation dose through a slice. The center of the slice is at point z_0 and successive slices are generated in the *z* direction. *(c)* Cylindrical phantom for body scans. Pencil chambers, 10 cm long, are placed in holes drilled in the phantom. The dose produced in the chamber from a single slice gives the dose produced at the center slice from exposures from the multiple scans in a CT examination.

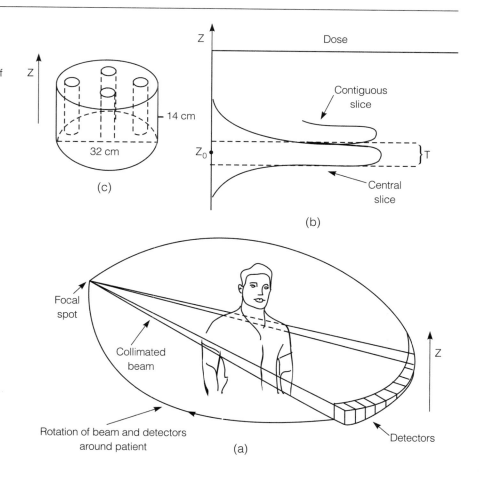

It is not necessary, however, to determine the dose profile to assess the dose from a multiple scan. The line integral of the dose profile for a single scan is equal to the dose at the central slice for a multiple scan, provided the contribution of the outermost slices to the central slice is insignificant and the distance between scans is equal to the slice thickness (Rothenberg, 1993). The single scan line integral—that is, the area under the curve for the single scan in Figure 3.11—can be obtained with a pencil ionization chamber. In practice, pencil ionization chambers with an active length of about 10 cm are inserted into the phantoms and then exposed to a single scan, with the slice going through the center of the dosimeter. The pencil chamber receives the same total radiation that is imparted to the chips or film and the charge produced in the chamber is converted to roentgens or milligrays with the appropriate calibration factor. The result is the computed tomography dose index (CTDI), defined as

$$\text{CTDI} = \frac{1}{T} \int_{-\infty}^{\infty} D(z)\,\mathrm{d}z$$

where $D(z)$ is the dose at point z, as measured from the midpoint of the central slice, and T, as before, is the thickness of the slice.

Another dosimetric quantity used in computed tomography is the multiple scan average dose (MSAD). It is more generally applicable, as it is based on the number of slices in a given examination. It is equal to the CTDI if the number of slices is large enough that the contributions of the end slices to the central slice are negligible and the slice interval is equal to the slice width. If the increment (I) between scans is less than the slice thickness (T)—that is, if the multiple scans overlap—the MSAD may be obtained from the single scan CTDI by the relationship MSAD = (T/I)(CTDI).

Head phantoms are typically 16 cm in diameter by 14 cm long and body phantoms are 32 cm in diameter by 14 cm long to comply with federal regulations applicable to CT manufacturers.

The pencil ionization chambers are calibrated in a uniform radiation field. However, measurement of a slice exposes only a portion of the chamber for the width of the slice. Thus, if a 10 cm long chamber is exposed to 1 R in a 10 mm slice, it will produce only 10 mm/100 mm or 1/10 the charge of a chamber uniformly exposed, and its reading will have to be increased by a factor of 10. The reading would have to be multiplied by 8.7 to convert the exposure to air kerma in milligrays. CTDI values published by manufacturers are given in Table 3.15. Since the values of absorbed dose were given for acrylic (7.8 mGy/R), they were multiplied by 9.7/7.8 to give the dose to tissue reported in the table.

Table 3.15 Manufacturer's published CTDI values.

Unit	kVp	mA	Scan time (sec)	Slice thickness (mm)	Head CTDI (mGy)	Body CTDI (mGy)
GE 9800 (Xenon)	120	170	2.0	10	50	14
Picker PQ 2000	130	65	4.0	10	42	15
Toshiba Xpress	120	200	2.0	10	53	21
Siemens	120	500	2.0	5	38.6	
Somatom+	120	290	1.0	10		12.0

Source: Seibert et al., 1994.

Note: Values are for standard head and body techniques measured at the center of the standard CT dose phantoms.

10 SURVEY RESULTS, HANDBOOKS, AND THE INTERNET

It's important to solve selected problems in dosimetry to appreciate the elements that go into an assessment of radiation dose. However, much of the time it is not necessary to go through a complete dose calculation to evaluate exposure to a radiation source. Chances are that there are publications in technical journals, tables of data in handbooks, computer programs for customized solutions, and sites on the Internet that provide ready solutions to radiation dose problems. It should be borne in mind, however, that the data are more likely to be used appropriately and effectively if the user has had training in dose calculations as provided in the preceding sections. The user should also look for information on the accuracy of the data and analyses on which the dose estimate is based, including assessments of the uncertainty in the results.

Excellent sources for internal doses from radiopharmaceuticals are the publications of the Oak Ridge Radiation Internal Dose Information Center (RIDIC), the International Commission on Radiological Protection (ICRP), and the Medical Internal Radiation Dose (MIRD) Committee of the Society of Nuclear Medicine. A comprehensive collection of links to information on the Internet is provided by the Oak Ridge Associated Universities at www.orau.com/ptp/infores.htm. Details are given in the Selected Bibliography.

10.1 Surveys of Doses in X-Ray Examinations

Surveys of radiographic techniques in different countries and at different institutions reveal differences in techniques and equipment used that can result in large variations in doses imparted, variations that exist even within an institutional facility, as from one x-ray tube to another. Publication of the results and analysis of the reasons often lead to changes that not only reduce the dose but improve the quality of the examinations.

Doses to patients may be based either on measurements of the dose free-in-air (the dose in the absence of the patient) or the entrance surface dose (ESD) (the dose on the surface of an attenuating medium, or phantom, designed to simulate the effect of the presence of the patient on the measurement). Doses to organs and tissues inside the body are calculated from the measurements. They may then be used to assess the detriment to the health of the individual, for example, as measured by the effective dose or the effective dose equivalent. The values given for the effective dose and the effective dose equivalent (developed earlier) are generally used interchangeably, although they are conceptually different (sections 12 and 13 of Part Two). Reported ratios for the two quantities may differ by 20 percent and more (UNSCEAR, 1993) but because of the indeterminacy in the effective dose concept, the difference is not particularly relevant in radiation protection.

The United Nations conducts a continuing review of doses incurred in diagnostic examinations (UNSCEAR, 1993, 2000).

10.1.1　Organ Doses in Complex X-Ray Examinations

Doses imparted in barium meal, barium enema, and intravenous urography contribute a large fraction of the collective effective dose from diagnostic radiology. Fluoroscopy can also contribute a significant fraction of the dose, which makes it difficult to simulate. Evaluations of organ dose in actual patient examinations are given in Table 3.16 (Calzado et al., 1992).

11　Producing an Optimum Radiation Field for Treating a Tumor

Strategies for attacking a tumor by radiation are based on delivering a dose high enough to defeat the tumor while at the same time limiting collateral damage to healthy tissue. The planning and execution of a campaign of this type require the following major steps: identify the extent of the tumor; specify the dose required to treat the tumor; select from available options the type of radiation and radiation source that can most effectively attack the specific tumor targeted; specify an irradiation protocol that will impart the optimum dose configuration to treat the tumor within particular clinical constraints, including limiting the dose to healthy tissue.

Table 3.16 Mean patient organ doses (mGy) and effective doses.

Organ	Barium meal		Barium enema		Intravenous urography	
	Mean	Effective	Mean	Effective	Mean	Effective
Breasts	2.83	0.62	1.34	0.49	0.19	0.06
Stomach	22.1	3.0	18.2	6.2	18.6	3.7
Upper large intestine	8.02	1.14	19.3	5.5	15.1	3.7
Kidneys	3.64	0.53	28.2	6.2	2.89	0.62
Liver	15.1	2.0	13.6	4.0	10.9	1.9
Lungs	7.57	1.23	2.26	0.61	0.51	0.12
Ovaries	1.49	0.53	17.4	5.4	6.8	1.93
Pancreas	10.5	1.6	13.6	3.5	7.76	1.68
Testes	0.85	0.21	5.66	2.42	5.55	1.22
Thyroid	3.13	0.73	0.15	0.07	Negligible	
Urinary bladder	1.66	0.51	14.1	2.6	21.0	4.3
Effective equivalent dose	5.66	0.68	10.7	2.4	6.7	1.38

Source: Calzado et al., 1992.

Irradiation of the tumor can be done by external beams from radiation therapy x-ray machines operating in the megavolt range or, less frequently, from cobalt-60 units; by the placement of sealed radioactive sources within the tumor (brachytherapy); and by the administration of short-lived radioactive drugs that localize within the tumor. The radiation field may be provided by x-ray photons or by charged particles, including electrons and protons. These different sources have their special advantages for treatment of different types of tumors.

The placement of a radioactive source within a tumor has the advantage of imparting a greater dose to the tumor relative to healthy tissue. On the other hand, external beam therapy can better define the volume to be irradiated, cover volumes occupied by larger tumors, or impart high doses outside the tumor boundaries to treat infiltrations by cancer cells. The tumor can be irradiated by beams from several directions (multiple fields) to better concentrate the radiation in the region of the tumor and produce less radiation damage in healthy tissue.

The extent of the tumor is determined by appropriate imaging techniques, including diagnostic x rays and, in particular, computer-assisted tomography (CAT scans). The determination of the target volume and the dose to be delivered to that volume is followed by design of the actual treatment protocol to produce the required dose distribution to the tumor.

The dose distribution over the *treatment volume,* the volume that will actually be irradiated for the purpose of treating the tumor, as well as the distribution over all the tissues outside the tumor that are irradiated in the

3.12 Isodose curves (in grays) in treatment of prostate cancer. *(a)* Characteristic form of dose distribution produced by external irradiation with a linear accelerator to treat both the prostate gland and the pelvic lymph nodes. The dose is delivered to the tumor in fractions through anterior/posterior and lateral portals to limit damage to surrounding healthy tissue.

(b) Dose distribution from irradiation by permanent interstitial implants using iodine-125 seeds. Bulges along the isodose curves appear in the vicinity of the seeds, which produce fields similar to those of point sources, so the dose rate increases sharply close to the seeds. Since the prostate gland surrounds the urethra, the placement of the seeds must be adjusted to limit the dose to the urethra (as indicated by the dip in the isodose curve at the center of the prostate) so as not to cause urethral damage. (*Source:* H. Ragde, Brachytherapy for clinically localized prostate cancer, *J. Surgical Oncology* 64:79–81, 1997. Reprinted by permission of Wiley-Liss, Inc., a subsidiary of John Wiley & Sons, Inc.)

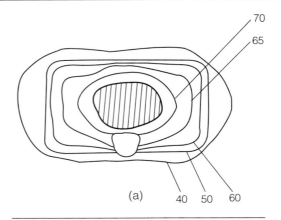

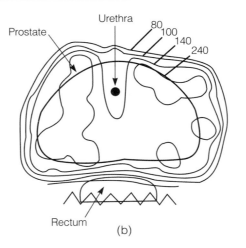

course of the treatment (the *irradiated volume*), is determined by calculation using high-speed computers. The calculation follows paradigms similar to those used in dose calculations in other applications, such as in the design of radiation shielding or in the evaluation of occupational and environmental radiation doses. The accuracy required is much greater in radiation therapy, however. The ICRU recommends that the dose to the target volume should be delivered to within 5 percent (1 standard deviation) of the prescribed dose. This may be compared to the accuracy of doses determined and reported for occupational exposures, which are not normally associated with accuracies greater than 30 percent.

If the radiation is imparted by a radiation therapy machine, the treatment can be simulated with a radiation therapy simulator, a diagnostic x-ray machine mounted on a rotating gantry that provides beam geometries identical to those used by the therapy machine. In this way, the simulated beams can mimic the actual beams proposed to produce the therapy field with respect to the anatomy exposed and to verify the coverage of multiple beams. The simulation can also be used to modify and optimize the proposed treatment technique.

The results of dose calculations are presented schematically as isodose distributions, contours that show regions of equal dose. Isodose distributions are shown in Figure 3.12 for fields that are generated in the treatment of prostate cancer by external beam therapy and by brachytherapy with iodine-125 seeds. A treatment plan for prostate cancer by external beam radiotherapy may call for irradiating the tumor with x rays produced by a linear accelerator at 18 MV and administering the dose in daily fractions of 1.8–2 Gy, five fractions per week, for a total dose to the prostate of 65 Gy (Perez and Brady, 1998). A treatment plan for prostate cancer by iodine-125 seeds may place the seeds to produce a minimum dose of 160 Gy to the periphery of the prostate (resulting in doses several times greater close to the seeds), with adjustments to limit the dose to the urethra (Radge, 1997).

PART FOUR

Radiation Measurements

The most widely used radiation detectors are devices that respond to ionizing particles by producing electrical pulses. The pulses are initiated by the imparting of energy by the ionizing particles to electrons in the sensitive volume of the counter.

Two major modes of signal production are utilized in radiation counter designs. In one mode, the deposited energy serves merely as a trigger to produce an output electrical pulse of constant form every time an interaction occurs in the detector. The output pulse is constant regardless of the amount of energy deposited in the detector or the nature of the particle. This type of behavior is exhibited by the Geiger-Mueller counter. In the other mode, the magnitude of the output pulse is proportional to the amount of energy deposited in the detector; that is, the greater the energy deposited, the larger the output pulse. This type of behavior is exhibited by scintillation counters, gas proportional counters, and semiconductor detectors.

We shall now examine various kinds of radiation detectors, the types of signals they produce, and means of analyzing these signals, with particular attention to Geiger-Mueller counters and scintillation detectors as examples of the constant-output and proportional-output detectors, respectively. Detailed treatments of these and other types of radiation measuring instruments are given in Knoll, 1999, and Tsoulfanidis, 1995.

1 Radiation Counting with a Geiger-Mueller Counter

The Geiger Mueller (G-M) counter is the best-known radiation detector. It is popular because it is simple in principle, inexpensive to construct, easy

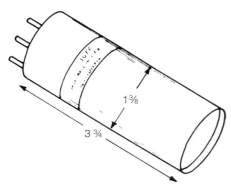

4.1 End-window Geiger-Mueller tube. Specifications (Anton Electronic Laboratories, Model 220): Fill, organic admixture with helium; Life, 10^9 counts; Mica window thickness, 1.4–2.0 mg/cm^2; Operating voltage, 1,200–1,250 V; Operating plateau slope, approx. 1%/100 V; Operating plateau length, approx. 300 V; Starting voltage, 1,090–1,160 V.

to operate, sensitive, reliable, and very versatile as a detector of ionizing particles. It is particularly suitable for radiation-protection surveys. It is used with a scaler if counts are to be tallied or with a ratemeter if the counting rate (generally specified as counts per minute) is desired.

1.1 A G-M Counter Described

Put a gas whose molecules have a very low affinity for electrons (for example, helium, neon, or argon) into a conducting shell, mount at the center a fine wire that is insulated from the shell, connect a positive high-voltage source between the wire and the shell, and you will have a Geiger counter.

Any incident particle that ionizes at least one molecule of the gas will institute a succession of ionizations and discharges in the counter that causes the center wire to collect a multitude of additional electrons. This tremendous multiplication of charge, consisting of perhaps 10^9 electrons, will produce, in a typical G-M circuit, a signal of about 1 volt, which is then used to activate a counting circuit.

Details and specifications for a commonly used G-M tube are given in Figure 4.1. The tube is called an end-window detector because the end of the cylindrical shell is provided with a very thin covering to allow low-energy beta particles to penetrate into the counter. A window thickness equivalent to 30 microns of unit-density material is thin enough to allow about 65 percent of the beta particles emitted by carbon-14 to pass through. Windows in tubes used to detect alpha particles, which are less penetrating, should be thinner, that is, less than 15 microns at unit density. Gamma rays do not require a special window and can penetrate the counter from any direction.

Practically every beta particle that reaches the counter gas will cause a discharge and register a count on the counting equipment. On the other hand, when gamma photons are incident upon the counter, only a small fraction will interact with the walls, and a much smaller fraction with the gas, to liberate electrons that will penetrate into the gas and produce a discharge. The other photons will pass through without any interaction and thus will not be recorded. Clearly, G-M counters are much more efficient in detecting the charged beta particles than the uncharged photons. Of course, the window must be thin enough to let the charged particles through.

The signals from a G-M counter that is working properly are all of constant size, independent of the kinds of particles detected or their energies. The G-M tube is thus purely a particle counter, and its output signal cannot be used to provide information on the particles that triggered it.

1.2   Adjusting the High Voltage on a G-M Counter and Obtaining a Plateau

The operation of a G-M counter requires an applied high voltage, and this must be adjusted properly if the counter is to give reliable results. If the voltage is too low, the counter will not function. If the voltage is too high, the gas in the counter will break down, and the counter will go into continuous discharge, which will damage it.

The high voltage is adjusted by exposing the counter to a source of radioactivity and recording the counts accumulated in a given time period as the voltage is increased. At low voltages, the signals from the counter will be very small and of varying sizes and will not activate the counting circuit or scaler. As the voltage is raised, a level will be reached at which some pulses from the detector are just large enough to activate the scaler and produce a low counting rate. This is called the starting voltage. As the voltage is increased further, the pulses will increase in height, and when the Geiger region is reached, all the pulses will become uniform, regardless of the type of particle or energy deposited in the detector. The voltage at which the pulses become uniform is known as the threshold voltage, and it can be recognized from the counting rate, because once the threshold voltage is reached, the counting rate increases only slowly as the voltage is increased. The region over which the counting rate varies slowly with increased voltage is known as the plateau. The plateau slope is usually expressed as percent change in counting rate per 100 V change in high voltage. An example of a plateau is shown in Figure 4.2. The slope is approximately 10 percent/100 V at 900 V. This is adequate for most counting purposes, but much flatter plateaus can be obtained.

1.3   How a G-M Counter Can Lose Counts and Even Become Paralyzed

A G-M tube requires a certain recovery time after each pulse. If a succeeding event is initiated by an incident particle before the tube recovers, the discharge will not occur and the event will not be recorded. The recovery time generally varies between 10^{-4} and 10^{-3} seconds. For low counting rates, essentially all the intervals between pulses are larger than this recovery time, and all the events are recorded. With higher counting rates, the intervals between successive events in the counter decrease, and more events occur at intervals shorter than the recovery time and fail to produce counts. Corrections for lost counts must be made to the observed counting rate.

A useful method for increasing the observed counting rate to correct for

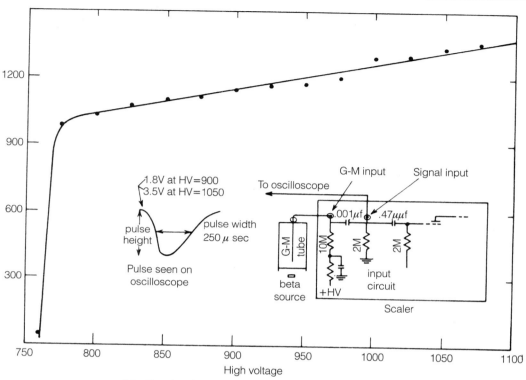

4.2 Results of measurement of plateau of G-M counting tube. The tube was connected to a scaler that provided the high voltage and counted the events initiated in the detector. The shape of the pulse was observed by connecting an oscilloscope to the "signal" input to the scaler. The input circuit for the scaler used is shown in the graph. Note that the signal input is isolated from the high voltage and is used when the high voltage is not required. G-M tube: Nuclear Chicago Model T108, 900 V halogen-quenched, end window. Caution! When determining a plateau, do not cover the full plateau range because of the risk of applying excessive voltage, which will drive the tube into continuous discharge and damage it.

counting losses is to add $R^2\tau/(1 - R\tau)$ to the count rate. R is the observed counting rate and τ is the recovery time. (Alternate terms, with slightly different meanings, are the *dead time* and the *resolving time*.) R and τ must of course be expressed in the same time units (see Evans, 1955, pp. 785–789, for detailed treatment).

Example 4.1 The recovery time of a G-M counter is given as 3.5 × 10^{-4} seconds. What is the counting rate corrected for dead-time losses, if the counting rate measurement of a sample is given by the detector as 26,000 counts per minute (c/min)?

The recovery time is 5.83×10^{-6} minutes.

$$\text{Correction} = \frac{(2.6 \times 10^4)^2 \times 5.83 \times 10^{-6}}{1 - 2.6 \times 10^4 \times 5.83 \times 10^{-6}}$$

$$= 4{,}645 \text{ c/min}$$

Corrected rate $= 26{,}000 + 4{,}645 = 30{,}645$ c/min.

In most G-M survey meters, if the rate of triggering by incident particles becomes too high, the counter becomes paralyzed and records no counts. This is a serious deficiency for G-M counters. Some G-M meters have special circuitry to produce a full-scale reading in very high fields. Since a G-M counter always gives some response to the radiation background, a condition of absolutely no counts can be recognized as a sign of either a defective counter or a seriously high radiation level. G-M counters that can become paralyzed must not be relied upon as the sole monitor for levels potentially high enough to paralyze the counter; they must be replaced or supplemented with other instruments, as described in later sections.

1.4 How to Distinguish between Beta and Gamma Radiation with a G-M Counter

As noted above, once a G-M counter is triggered it is impossible to tell from the resulting signal the type of radiation responsible or the amount of energy deposited in the counter. However, beta particles and gamma photons can be readily distinguished by the use of absorbers. If a thin absorber is placed in front of the window, it will stop the beta particles but will have relatively little effect on the gamma photons. Thus the counting rate with and without the absorber can be used to distinguish beta particles from gamma photons. A suitable absorber thickness for most applications is the equivalent of 5 mm of unit-density material. Thus, 5 mm of lucite, which has a density fairly close to 1 g/cc, or 1.85 mm of aluminum (density 2.7), would be equally effective in stopping beta particles.

Example 4.2   Using an end-window G-M tube to survey his laboratory bench after an experiment, an investigator found a counting rate of 15,000 c/min. When he covered the window with a beta shield, the counting rate was reduced to 300 c/min. The radiation background away from the contaminated area was 35 c/min.

The contamination was reported as follows:

Beta counting rate: $15,000 - 300 = 14,700$ c/min
Gamma counting rate: $300 - 35 = 265$ c/min

If one is dealing with very low energy gamma photons, a correction may have to be made for attenuation of the gamma photons in the beta absorber, but generally this is not necessary.

1.5 How to Determine Source Strength of a Beta Emitter with a G-M Counter

Suppose one needs to determine the strength of a source of beta radioactivity, that is, the number of particles it emits per unit time. The source may be material contained in a planchet or contamination on a surface. The counter is positioned over the source, and a reading is taken in counts per minute. How is this converted to the rate of emission of radiation from the source?

The simplest approach is to compare the counting rate for the unknown source with that of a reference source of known emission rate. If the unknown and reference sources are identical except for activity, they are detected with the same efficiency. In this case, all we have to do is compare the measured counts per minute, since the ratio of the two counting rates is the ratio of the source strengths. By multiplying the known emission rate of the reference source by the ratio of the counting rates of the sample and reference source, we obtain the emission rate of the sample.

Example 4.3   A surface is surveyed with a G-M counter for contamination and a small spot of contamination is found which reads 15,000 c/min. The background reading is 45 c/min. The counting rate becomes negligible when a 5 mm plastic absorber is placed between the source and the counter, showing the counts are due to beta particles. The counting rate of a bismuth-210 beta reference source counted in the same manner as the contaminated surface is 9,400 c/min. The reference source emits 32,600 beta particles/min. What is the rate of emission of beta particles from the surface?

The beta particles from ^{210}Bi have a maximum energy of 1.17 MeV, which is fairly energetic for a beta-particle source. Their attenuation in a thin end-window G-M counter is low and can be neglected for purposes of this example. If the beta particles emitted from the contaminated spot are also energetic and attenuation in the counter window can be neglected, the strength of the source of contamination is

$$14{,}955 \times \frac{32{,}600}{9{,}355} = 52{,}115 \text{ beta particles/min}$$

In converting the measured counting rate to disintegrations per minute, it was assumed that the counter detected the same fraction of beta particles emitted from the reference source and the contaminated surface. However, there are many factors that affect the actual counting rate for a sample of a given emission rate, and if the sample and reference source differ in any way, corrections may have to be made.

1.6   Factors Affecting Efficiency of Detection of Beta Particles

A detector intercepts and registers only a fraction of the particles emitted by a radioactive source. The ratio of the detector counts per minute to the number of particles emitted per minute by the source is called the detector efficiency. If the efficiency is known, the source strength can be determined from the counting rate of the detector:

$$\frac{\text{Counts/min}}{\text{Efficiency}} = \text{Particles emitted/min from source}$$

The major factors determining the fraction of particles emitted by a source that actuate a detector are depicted in Figure 4.3 (Price, 1964). They include:

- f_w, the fraction of particles emitted by the source traveling in the direction of the detector window. For a point source located on the axis of an end-window G-M tube, a distance d from the circular detector window of radius r,

$$f_w = \frac{1}{2}\left(1 - \frac{d}{\sqrt{d^2 + r^2}}\right) \qquad (4.1)$$

- f_b, the fraction emitted in the direction of the detector window that actually reach the window. The beta particles may be prevented from reaching the window by absorption in the source itself, in any material covering the source, or in the air or other media between the source and the detector; or by deflection away from the source

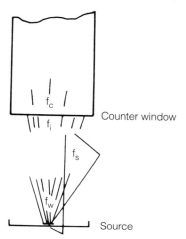

4.3 Factors affecting the efficiency of detection of beta particles by a G-M counter.

window in the intervening media. An approximate formula for the attenuation of beta particles is given by equation 2.4 in Part Two.

- f_o, the fraction of particles incident on the window that actually pass through the window, into the sensitive volume of the counter, and produce an ionization. The sensitive volume of the counter is the volume in the gas within which a particle must penetrate if any ionization it produces is to result in a discharge. There is a narrow region behind the window and also at the extreme corners of the tube where G-M discharges will not be initiated, even if an ionization is produced.

- f_s the fraction of particles that leave the source in a direction other than toward the detector but are scattered into the detector window. This factor increases the detector response, in contrast to f_i and f_o which attenuate the particles. The additional particles may be backscattered into the detector from backing material for the source, or from the source material itself, or they may be scattered into the detector by the medium between the source and the detector.

The backscattering from the support of the source depends strongly on the energy of the beta particles and the material of the scatterer, increasing with the energy and the atomic number of the scatterer. Materials with low atomic numbers, such as plastics, give minimal backscattering. Scattering from aluminum is low, while backscattering from lead is high. Curves giving the backscatter factor as a function of atomic number of the sample planchet are given in Figure 4.4 for several beta sources.

The four factors contributing to the efficiency of detection are multiplied to give an overall efficiency factor relating source emission rate and

4.4 Backscattering factors versus atomic number (Zumwalt, 1950). (RaE, the classical designation for one of the decay products of radium, is ^{210}Bi.)

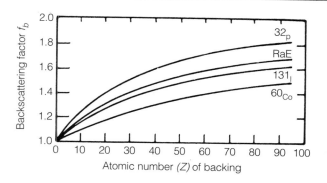

detector counting rate. Thus, letting R and S equal counting rate and emission rate, respectively:[1]

$$R = S f_w f_i f_i f_c$$

Example 4.4   Calculate the efficiency of detection of an end-window G-M counter for beta particles from "point" sources of ^{14}C and ^{32}P deposited on a copper planchet ($Z = 29$), for the following specifications:

Distance from source to detector window, 1 cm
Diameter of detector window, 2.54 cm
Window thickness, expressed for unit-density material, 30 μm

	^{14}C	^{32}P
$f_w = \frac{1}{2}(1 - [1/\sqrt{1^2 + 1.27^2}])$	0.19	0.19
Air + water thickness, unit density	0.0042 cm	0.0042 cm
Half-value layer (from eq. 2.4)	0.0049 cm	0.076 cm
Number of half-value layers in air + window	0.86	0.055
Attenuation factor in air and window ($f_i f_c$)	0.55	0.96
Backscatter factor	1.2	1.5
Overall efficiency ($f_w f_i f_c f_s$)	0.13	0.27

Note that low-energy beta particles from ^{14}C are attenuated significantly, even by the air. Combined with the attenuation in the window, the efficiency of detection is about half that calculated for the high-energy beta particles emitted by ^{32}P. These undergo very little attenuation, either in the air or in the window of the detector.

1.7   Correcting for Attenuation of Beta Particles by Determining Absorption Curves

A correction for the absorption of beta particles in the medium between the source and detector can be made experimentally by obtaining an absorption curve. Absorbers of known thickness are added between the detector and the source, and the counting rate is plotted as a function of total absorber thickness.

The total absorber thickness includes not only the added absorbers but also the intrinsic thickness associated with the counting geometry, such as the air and counter window, all expressed in terms of unit-density equivalent. The measured count rate is plotted against total thickness and extrap-

---

1. For a detailed treatment, see Price, 1964, pp. 127–137.

olated to zero thickness for the case of no attenuation. Results obtained with this method are shown in the insert in Figure 4.5.

If the sample emits gamma photons also, the contribution from the photons must be subtracted to obtain the beta counting rate. This is done by adding absorbers until the counting rate falls off in a slow manner, indicating that only gamma radiation remains. The gamma tail to the absorption curve is then extrapolated to zero absorber thickness, and its value is subtracted from the total count rate to give the beta contribution. An example of the gamma correction in counts of bets particles from ^{131}I is shown in Figure 4.5.

1.8 Counting Gamma Photons with a G-M Counter

Gamma photons, because they are more penetrating than beta particles, undergo very little interaction in the gas of the G-M counter. Most of the gamma photons that make a collision in a detector interact in the detector wall. As a result of the collisions, electrons are liberated. Some of these electrons enter the sensitive volume of the detector and initiate a discharge. Generally, only a small fraction of the gamma photons incident on a G-M detector are counted.

The actual efficiency of detection of a G-M counter may be determined experimentally by using a source of known gamma emission. The efficiency depends strongly on the energy of the gamma photons; thus efficiencies measured with one type of nuclide cannot be used for other nuclides emitting gamma photons with different energies, unless corrections are made.

Measurements of the energies of gamma photons may be made in principle with G-M counters by making attenuation measurements with calibrated absorbers and estimating the energy from the half-value layer. The accuracy is poor, however, and attenuation methods are not used for any but the grossest determinations. An example of a half-value layer measurement is given in Figure 2.11.

1.9 Standardization of Radionuclides with G-M Counters

The standardization of a radionuclide requires an accurate determination of its activity. The procedures to be followed are essentially the same as those required for the determination of the source strength. Whereas source strength refers to the particle emission per unit time, activity is defined as the number of nuclear transformations per unit time. It is the nuclear transformations that result in the emission of particles—sometimes in a one-to-one correspondence, sometimes in more complex relationships. Thus, the activity may be determined from a measurement of the source strength, provided the decay scheme of the nuclide is known.

4.5 Absorption curve for correcting for window absorption and gamma counts in counting beta particles from ^{131}I. The total absorber thickness includes the counting tube window (0.017 mm). The insert includes, for comparison, results for ^{14}C and ^{204}Tl. The ^{14}C beta curve extrapolates to 15,000 c/min for zero window thickness. The ^{131}I curve extrapolates to 9,900 c/min. The gamma contribution to this counting rate is only 36 c/min. Note that the absorber is placed as close to the window as possible when an extrapolation to give window thickness is made.

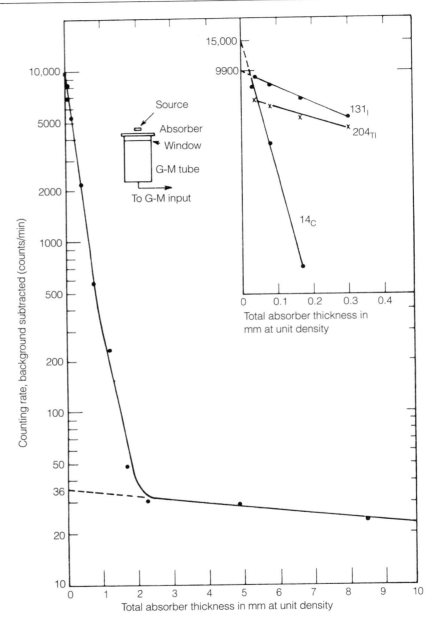

As with the determination of source strengths for beta sources discussed previously, the simplest method of standardization is to compare the counting rate of the unknown with a standard of identical composition and geometry. When sample and standard are identical except for activity, the ratio of counting rates is also the ratio of activities. Any detector, including a G-M detector, is satisfactory for determining the ratio of counting rates. The only precaution to be taken is that, if the counting rates for the sample and the standard vary widely and are excessively high, appropriate corrections must be made for "dead time" losses.

Generally, there are differences of size and construction between sample and standard, and the effort to be made in correcting for the differences will depend on the degree of accuracy required. Experience shows that as the accuracy requirements become stringent—say, below 15 percent—detailed corrections for all differences between sample and standard must be made.

Example 4.5   It is desired to standardize an aliquot of ^{131}I on a planchet. The sample is counted on September 30 at 11 A.M., and its counting rate is 2,580 c/min. The background is 15 c/min. The sample activity is to be determined with the use of a standardized solution of ^{131}I whose activity is given as 9,250 Bq/ml as of September 26 at 9 A.M. Describe the procedures for standardizing the sample.

The period between the time for which the activity of the standardized solution is known and the time of the measurement is 4 days and 2 hours, or 4.083 days. The half-life of ^{131}I is 8.1 days. The elapsed time is 0.504 half-lives, and the corresponding fraction remaining is $0.5^{.504}$ = 0.705. Thus, the activity of the standardized solution is 6,521 Bq/ml at the time of the measurement, or 391,275 dis/min-ml.

A standard prepared from the solution should not give more than about 10,000 c/min to minimize dead-time losses. If it is assumed (from experience) that the counting rate is approximately 20 percent of the sample activity, a reference source should be prepared with an activity of about 50,000 dis/min. A suitable reference source may be obtained with 0.1 ml of the solution. This has an activity of 39,128 dis/min at the time of measurement.

The standard solution is counted on the same kind of planchet as used for the sample. If we assume the reference source gives a counting rate of 7,280 c/min, the net counting rate is 7,280 − 15, or 7,265 c/min. The activity of the sample is thus (2,565/7,265) × 39,128, or 13,815 dis/min.

1.10   Interpreting Counts on a G-M Counter

The G-M counter is a very sensitive detector and gives high counting rates at low radiation levels, particularly when monitoring for contamination from beta particles. The radiation is normally detected through clicks on a speaker that is incorporated in a G-M survey meter. The detection of high clicking rates when hands or work surfaces are monitored after working with radioactivity can be very traumatic to a worker who does not know how to interpret the significance of those clicks. It is recommended, as part of a training program, to have the participants listen to a high clicking rate produced by a source of beta particles, and then to explain the significance of the clicks. A demonstration of the significance of high clicking rates as given in a training program in radiation safety goes as follows.

The students first listen to a clicking rate of 1,000 clicks per minute obtained by exposing a pancake G-M tube to a small beta source. They then listen to the background count rate and the instructor explains the source of the background counts, pointing out that background radiation penetrates throughout the body whereas the beta particles penetrate only a few millimeters into the skin. This is followed by a discussion of the significance of detecting 1,000 clicks per minute for two cases of monitoring for beta particles: *(a)* monitoring for beta radiation in the environment and getting a high count rate from beta contamination on a surface and *(b)* monitoring for radioactivity on the skin and finding a high count rate, even after thorough washing of the skin. The discussion takes the form of the following script, more or less:

> Essentially every beta particle incident on the window of a thin end-window G-M tube gives a count, so the number of clicks per second from a beta source represents the number of beta particles per second impinging on the end window. About 3 percent of incident gamma photons give a count.
>
> First, listen to the background of a pancake G-M detector. This is produced by natural background radiation in the environment (from cosmic rays and radioactive materials, mainly the uranium and thorium decay series, in the earth).
>
> If there are no artificial sources present, the background will be about 70 clicks per minute. This adds up to 36,792,000 clicks in one year.
>
> The background is composed of penetrating radiation. Each one of those clicks represents a ray that is imparting energy somewhere within the body, and with large numbers of clicks, the exposure is uniform throughout the body. The clicks are produced by a nominal dose from background radiation of 0.87 mSv in a year. Thus 36,792,000 clicks = 0.87 mSv = 1 year background penetrating radiation.
>
> Now, listen to 1,000 clicks per minute. (A small beta source is positioned

next to the window to give a counting rate as indicated on the ratemeter of 1,000 clicks per minute.) What is the significance of this reading?

Assume that this represents radiation incident on the skin externally from radioactivity on the floor or a laboratory bench contaminated as the result of a spill, in particular a spill of phosphorus-32. Thus, the count rate is caused by beta particles emitted from P-32. We can calculate that those 1,000 clicks per minute from beta particles on the pancake detector represent an exposure rate from beta particles that would produce a dose rate to the skin, if it were at the location of the G-M tube, of 0.0012 mSv/hr. The limit to the skin is 500 mSv/yr or 27 billion clicks. It would take a long time, 417,000 hours, to reach the dose limit. Also, the clicks represent particles that enter only a very short distance into the skin (most, less than a few millimeters), whereas the background clicks represent radiation that produces a dose in all the organs of the body.

Finally, suppose the reading of 1,000 clicks per minute is obtained by placing the counter about 1 cm above a spot of contamination on skin, consisting of radioactivity from P-32. The counter window has a diameter of 4.45 cm and an efficiency of 30 percent. That is, 30 percent of the beta particles emitted from the radioactive contamination are incident on the G-M tube and produce clicks when it is placed close to the skin. Thus, the activity on the skin is 1,000/0.3 or 3,333 disintegrations per minute. Averaging over 1 cm^2, this gives an activity of 55.5 disintegrations per second per square centimeter. The associated dose rate to the skin for a spot of radioactive contamination from P-32 is 0.138 mSv/hr.

It would take 500/0.138 = 3,623 hours to receive a dose of 500 mSv. The number of clicks would be 60,000/hr × 3,623 = 217 million.

Our experience in one case of contamination of a finger was that, as the radioactivity decayed and was washed off the skin, the average time it remained on the skin was 5.76 days. The total accumulated counts would be 8,294,400 for complete decay. This produces a total dose of about 19.1 mSv, a small fraction of the limit.

This demonstration has been very helpful in providing workers with perspective on the significance of the results of surveys with G-M counters.

2 ENERGY MEASUREMENTS WITH A SCINTILLATION DETECTOR

We have seen how a G-M counter can be used to count ionizing particles that trigger it and noted that, once the counter is triggered, the signal is independent of the characteristics of the particle that initiated the discharge. We now turn to detectors that produce signals whose magnitudes are proportional to the energy deposited in the detectors. Such detectors are generally much more useful than G-M counters in tracer experiments or in analytical measurements. We shall consider the scintillation counter as an example of energy-sensitive devices.

2.1   Description of Scintillation Detectors and Photomultipliers

When an energetic charged particle, such as a beta particle, slows down in a scintillator, a fraction of the energy it imparts to the atoms is converted into light photons. The greater the amount of energy imparted by a beta particle to the scintillator, the greater is the number of light photons produced and the more intense is the light signal produced in the scintillator. When gamma photons pass through a scintillator, they impart energy to electrons, which also cause the atoms in the crystal to emit light photons as they slow down. The more energetic the gamma photons, the more energetic (on the average) are the electrons they liberate, and the more intense are the pulses of light produced. The amount of light in each pulse, which is determined with a photomultiplier tube, represents a measure of the energy deposited in the scintillator. The ability to evaluate these energies provides a means of sorting radiations from different sources and of identifying and determining the magnitudes of the radiation sources. The operation of a scintillator-photomultiplier combination is shown in Figure 4.6.

Scintillators of all kinds of materials—gaseous, liquid, and solid—and in all shapes and sizes are available. Some are of plastic; others are of dense inorganic material such as sodium iodide. The larger and heavier scintillators are used to detect gamma photons, since their greater mass gives them a higher detection efficiency.

2.2   Pulse Counting with a Scintillation Counter and Scaler

Figure 4.6 shows how an electrical charge proportional in size to the energy imparted by the radiation in the scintillator is produced in a photomultiplier tube.

Increasing the voltage applied to the tube accelerates the electrons to higher energies between dynodes and increases the number of secondary electrons produced per stage. Hence, the gain of the tube and the output signal are increased.

When a source of beta particles or gamma rays is counted with a scintillator, a variety of magnitudes of energy transfer occurs in the scintillator and a wide distribution of pulse heights is produced by the photomultiplier tube. The pulses also include spurious pulses not produced by radiation interactions in the scintillator. Some are very small and are associated with so-called noise in the amplifier or in the thermionic emission

4.6 Scintillation crystal-photomultiplier tube radiation detector. The light emitted from the scintillator as a result of absorption of energy from the ionizing particle is converted to an electrical signal in a photomultiplier tube, which consists of a photosensitive cathode and a sequence of electrodes called dynodes. The cathode emits electrons when irradiated by the light released in the scintillator. The number of electrons emitted by the photocathode is proportional to the amount of light incident on it. The electrons are accelerated within the tube and strike the first dynode, which emits several additional electrons for each electron that strikes it. The multiplied electrons from the first dynode are accelerated in turn to collide with successive dynodes. In this way, a large multiplication is obtained. Photomultiplier tubes of 10 dynodes, with accelerating voltages of about 100 V between dynodes (that is, high voltage of about 1,000 V), have multiplication factors of the order of 10^6. The electrons are collected at the anode and deposited on a capacitor, where they produce a voltage signal for subsequent processing and analysis by electronic circuits.

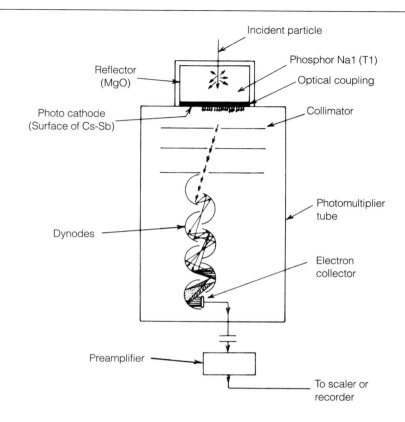

of electrons from the cathode of the photomultiplier. Some are large, usually the result of electrical interference.

In the simplest application of a scintillation detector, one merely counts the pulses produced by means of a scaler. The scaler counts all pulses above a certain voltage level determined by its sensitivity. Many scalers for general use respond only to signals of 0.25 V or larger.

As mentioned earlier, the sizes of the pulses from the photomultiplier can be increased by increasing the high voltage. There is usually an optimum setting or range of settings depending on the objectives of the investigator. If the voltage is too low, many of the pulses supplied by the detector to the scaler will be too small to produce counts. As the voltage on the photomultiplier tube is increased, more pulses have heights above the sensitivity threshold of the scaler and are counted. Obviously, the detection efficiency of the system is increased at the higher voltage setting.

If the voltage gets too high, low-energy electrons emitted by thermionic emission from the cathode of the photomultiplier tube will acquire enough energy to produce pulses large enough to be counted. The counting rate increases greatly, and the system is useless for counting particles.

The best setting of the high voltage is determined by counting a source over a range of high-voltage settings. Care must be taken not to exceed the maximum voltage that can be applied to the tube as specified by the manufacturer. A convenient source gives about 3,000 counts for a 10-second counting time. A background count without the source is also taken at each high-voltage setting for a time sufficient to accumulate at least a few hundred counts. The net counts per minute and the background counts per minute are plotted on a graph as a function of the high voltage. Statistical considerations show that the best value for the high-voltage setting is given by the highest ratio of the square of the net counting rate to the background. A graph of count rates as a function of high voltage is given in Figure 4.7.

2.3 Pulse-Height Distributions from Scintillation Detectors

In the previous section we noted that the detection of beta particles or gamma rays by scintillators produced a wide range of pulse heights in the

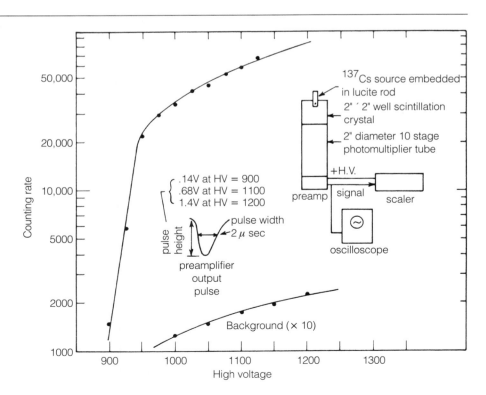

4.7 Net count rate of a simple scintillation counting system as a function of high voltage. At each voltage setting, the sample was counted for 10 sec, and the background for 1 min. The values given for the pulse heights are the maximum observed for ^{137}Cs and are due to complete absorption of the energy of the 0.667 MeV photon.

output, in contrast to the uniformly large pulses from a G-M tube. The scintillation detector becomes a much more powerful tool for radiation measurements if we can perform detailed analyses on the pulses it produces. In the next section we shall describe electronic methods of analyzing the pulses, but first let us examine the general form of pulse distributions that are produced. We shall consider the two most frequent methods of using scintillation detectors—beta counting with a liquid scintillation system and gamma counting with a sodium-iodide solid crystal.

2.3.1 *Pulse-Height Distributions from Beta Emitters in a Liquid Scintillator*

Two radionuclides widely used in research in the life sciences are the beta emitters carbon-14 and hydrogen-3. The beta rays from ^{3}H are emitted with energies up to 0.018 MeV, and those from ^{14}C with energies up to 0.154 MeV. Because of the low energies and short ranges in matter, the most feasible way to detect these radionuclides is to dissolve them in a liquid scintillator. The beta particles are released directly into the scintillating medium, producing light pulses of intensity proportional to their energies (Fig. 4.8). This technique eliminates problems posed by the existence of attenuating matter between the source and the detecting medium in external counting. It is because of the development of liquid scintillation counting that ^{14}C and ^{3}H play the wide role they do in medical experimentation.

Theoretically, the distribution of pulse heights produced in a liquid scintillation detector should have the same shape as the distribution of energies of the beta particles emitted by the radioactive source, since the amount of light emitted is proportional to the energy imparted to the scintillator by the particle. However, an actual measurement of the distribution of pulse heights produced by a source will differ from the true energy spectrum. The main reason for the difference is that the liquid scintillation detector has a relatively poor energy resolution. By this, we mean that the absorption of charged particles of a single energy within the scintillator will not produce a unique light or voltage signal but instead a distribution of pulse heights about a mean value characteristic of the energy of the particle. The spread of the pulse about the mean value determines the energy resolution of the detector.

Another cause of a discrepancy between the observed pulse-height distribution and the distribution of energy absorption events in the detector is that the mean amount of light produced by the absorbed energy may not be strictly proportional to the energy imparted. This lack of proportionality is most marked for low-energy electrons in organic scintillators.

Even with these limitations, useful estimates of the energies of the parti-

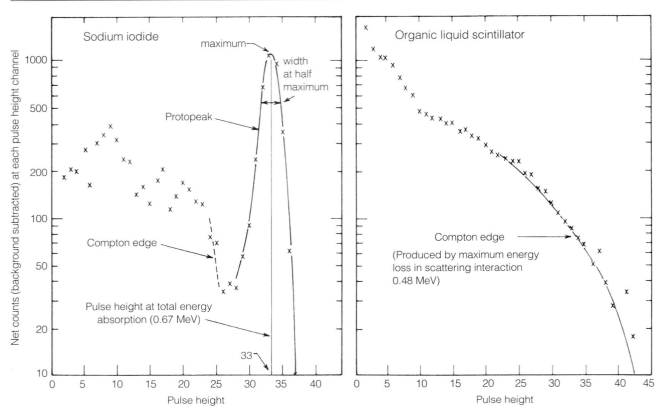

4.8 Comparison of pulse-height distributions from interactions of the monoenergetic gamma photons from cesium-137 (0.667 MeV) in sodium iodide and organic liquid scintillators. The interactions of the 0.667 MeV gamma photons from ^{137}Cs in the sodium-iodide crystal include a large number of interactions with the relatively high atomic number iodine atoms that result in complete absorption of the gamma photons and produce the peak located at pulse height 33. This peak is generally referred to as the *photopeak* or *full-energy peak*. Note that although all the pulses in the photopeak are produced by absorption of the same amount of energy, the pulse heights vary somewhat. The spread of pulse heights about the most probable value is given by the resolution of the system, which is defined as the ratio of the difference in heights at which the counting rate is half the peak to the pulse height at the peak. The resolution of the measuring system is (3/33) × 100 = 9.1 percent. The pulse-height distribution for the organic scintillator does not show a photopeak. Because of the low atomic numbers of the atoms constituting the scintillator, the number of photon interactions resulting in complete absorption of the photon is negligible, and most of the interactions are photon scatterings involving small photon energy losses. The electrons liberated as a result are relatively low in energy and are responsible for the steep rise of the curve at the lower pulse heights. The interaction accounting for this distribution is known as Compton scattering. The maximum energy imparted in Compton scattering is less than the photopeak and is known as the Compton edge. Although, theoretically, the Compton edge should show a vertical drop at the upper limit, the finite resolution of the detection system produces a more gradual slope. The relative pulse-height values for the NaI and liquid scintillators are not significant and were obtained by adjusting the amplifiers in each case to distribute the pulses over a convenient number of channels.

cles can be obtained from the measured pulse heights. However, the main use made of the pulse-height distributions produced in liquid scintillators is to separate the individual contributions from coexisting beta emitters. Use of the method for distinguishing between the signals from different nuclides is discussed in section 2.4.2 and illustrated in Figure 4.12.

2.3.2 Pulse-Height Distributions from a Sodium-Iodide Solid Scintillation Detector Exposed to Gamma Emitters

One of the best scintillators for detecting gamma photons is a single crystal of sodium iodide to which has been added a small amount of thallium. The advantages of this detector are: (a) It is dense (specific gravity 3.67); therefore, the probability of interaction per centimeter is higher. (b) It has a high light yield from deposited energy. (c) It has a high atomic number (because of the iodine). The gamma interactions are more likely to result in photon absorption (with all the photon energies imparted to the scintillator), rather than photon scattering (with only part of the energy imparted). In contrast, in organic scintillators, which consist primarily of carbon and hydrogen, both of low atomic number, most of the interactions are scatterings with relatively small energy transfers; that is, the photons escape from the detector, carrying most of the initial energy with them.

An example of the distribution of pulse heights corresponding to the interactions of the 0.66 MeV gamma photons from cesium-137 in a 3″ by 3″ sodium-iodide crystal is given in Figure 4.8. (Pulse-height distributions obtained with a germanium detector, which gives much better discrimination between energies, are shown in Figure 4.17b, section 3.2.) For a comparison, a pulse-height spectrum obtained with a liquid scintillator (volume 10 ml) is also presented. As noted earlier, because of the low atomic number of the plastic crystal, only an insignificant fraction of the gamma photons are totally absorbed.

The distribution of pulse heights produced by the gamma scintillation detector and its photomultiplier will not reproduce exactly the events produced in the scintillator because of the finite energy resolution of the system. The effect of finite resolution may be observed in Figure 4.8 by examining the part of the spectrum labeled *photopeak*. The pulse heights near the photopeak are all due to the absorption of the same amount of energy, 0.66 MeV from ^{137}Cs. A quantitive measure of the resolution is given by the ratio of the energy pulse-height span over which the pulse count-rate drops by a factor of 2 to the value of the pulse height at the peak count-rate. The resolution for the crystal shown is 9 percent at 0.66 MeV, which is characteristic of a high performance NaI detector.

2.4 Electronic Processing of Pulses Produced by Scintillation Detectors

The output electrical charge from a photomultiplier tube (Fig. 4.6) is processed by a variety of electronic circuits whose functions include amplification, discrimination, and counting. A block diagram of the instrumentation is given in Figure 4.9, that also shows pulse shapes as determined with an oscilloscope at various points. The output of the radiation detector is connected first to a preamplifier, which is located physically as close as possible to the detector. The function of the preamplifier is to couple the detector to the other electronic circuits, which may be located at any convenient location.

The most common type of preamplifier currently used is the charge-sensitive type, shown schematically in Figure 4.9. The charge-sensitive preamplifier provides a signal proportional to the charge produced in the detector. Exceptionally low noise operation, important with semiconductor detectors, is obtained if the input is provided with a field effect transistor (FET). Voltage-sensitive preamplifiers were used before charge-

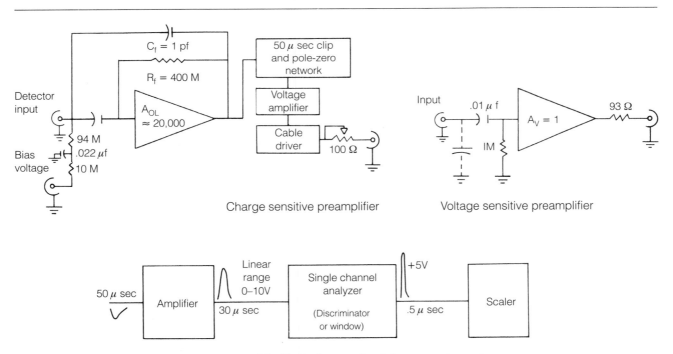

4.9 Block diagram of radiation measuring system.

sensitive preamplifiers but are no longer used in spectroscopy or with semi-conductor detectors. (The circuit is shown in Fig. 4.9.)

Let us estimate the magnitude of the electrical charge produced by the interaction of radiation with a scintillator. About 500 eV absorbed in a high-quality scintillation detector will result in the release of one photo-electron at the cathode of the photomultiplier. Multiplication factors of 10^6 are readily achieved in the photomultiplier tube. Thus, the charge developed from a 0.018 MeV tritium beta particle is (18,000 eV/500 eV) $\times$ 1.6×10^{-19} C $\times 10^6 = 5.76 \times 10^{-12}$ C. The charge from a 0.154 MeV carbon-14 beta particle would be 4.93×10^{-11} C. The noise generated in a commercial preamplifier for scintillation detectors is $<10^{-15}$ coulombs at 0 picofarads input capacitance (*www.canberra.com,* Model 2005). The noise increases with increased input capacitance, and since the leads connecting the preamplifier to the detector add capacitance in proportion to their length, they should be as short as possible.

Not all inputs would be exactly the same charge. Ionization is a statistical process and the exact number of ion pairs produced will vary from one particle to the next of exactly the same energy. The spread is of the order of the square root of the number of ion pairs (see section 6.5). For example, the absorption of a 0.05 MeV beta particle (average energy for ^{14}C) will release 100 photoelectrons on the average with a spread of $\pm\sqrt{100}$ or 10 electrons. This is 10 percent of the average. The absorption of 0.006 MeV, average energy for tritium, gives only 12 photoelectrons with a spread of $\pm\sqrt{12}$ or 29 percent. The spread of pulse heights is further increased a small amount by the noise of the signal-processing electronics.

The pulses pass from the preamplifier through the amplifier, which clips and shapes them to produce maximum signal-to-noise ratio and minimum spectrum distortion at high counting rates. The amplifier changes the size of each input pulse by a constant factor (as set by the operator) to bring the pulse heights into the working range of the analyzer (usually 0–10 V). The amplifier can also change the polarity of the pulse, as shown in Figure 4.9.

A hypothetical sequence of pulse heights appearing at the output of an amplifier is shown in Figure 4.10. The pulses are labeled according to origin—radiation background, radiation source, and noise. Noise pulses arise from erratic and random behavior of various electrical and electronic elements and from outside electrical interference. Almost all the larger noise pulses can be eliminated by proper design, including effective shielding and grounding of the system. Low-level noise is inherent to the system, and although by selection of specially designed high-quality components its magnitude can be minimized, it cannot be eliminated completely.

The pulses that pass through the analyzer are counted by a scaler. The

4.10 Examples of discriminate counting of sequence of pulses from single-channel analyzer. The dots represent counts from pulses greater than the baseline setting. The histogram at the bottom gives registered counts from pulses that pass through a window between the baseline setting and 1 V above the baseline.

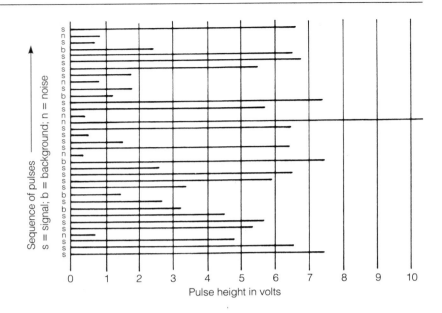

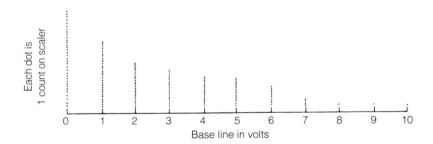

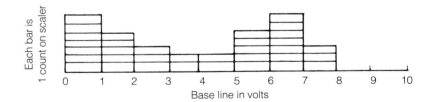

analyzer determines which pulses will get through for counting. It may allow through all pulses above a certain level, or only those pulses that fall within a certain voltage interval, known as a window. Let us examine the various kinds of pulse-selection methods and their application to counting problems.

2.4.1   Integral Counting and Integral Bias Curves

The simplest selection of pulses is through integral counting. All pulses above a certain value are counted. Those below that value are not counted. The level at which pulses are counted is set by a control known as a discriminator, which acts as an electrical gate. It is used mainly to reject low-level pulses from spurious sources such as electrical noise. It will, of course, also reject pulses from the lower-energy radiation interactions. The effect of a discriminator on the transmission of pulses is shown in Figure 4.10.

For any particular counting problem, a determination of the optimum discriminator setting may be made by obtaining an integral bias curve. Counts in a fixed time interval are determined as the discriminator is varied systematically from a minimum value to a maximum value. The measurements are repeated with the source removed to determine the effect of the background. An example of an integral bias curve and its use in obtaining the best discriminator setting is given in Figure 4.11.

2.4.2   Counting with the Use of a Window

Additional discrimination in analyzing pulses is obtained by the imposition of an upper limit to size of the pulses that will be counted. This is done by an upper-level discriminator. An upper-level and lower-level discriminator can be used together to selectively pass pulses in a specified energy range to a counter. A combination of upper and lower discriminators is called a window. The effect of a window on the transmission of pulses is shown in Figure 4.10. Usually two or three windows are used in liquid scintillation counters to separate contributions from several radionuclides counted simultaneously, such as ^{14}C and ^{3}H. A block diagram of the major components of a liquid scintillation counter is given in Figure 4.12. The figure also presents measurements on the efficiencies of detection of disintegrations from ^{3}H and ^{14}C samples in two separate channels, each adjusted for optimum detection of one of the radionuclides. Data such as those presented in the figure allow adjustment of the analyzer to distinguish between the activities of ^{3}H and ^{14}C in a sample containing both radionuclides.

4.11 Determination of integral bias curve. Note that while higher net counting rates from the source are obtained as the discriminator setting is lowered, they are obtained only at the cost of higher background counting rates. There are times when a higher counting rate from the source may be preferred and other times when a lower background is desired, even at the expense of lower net counts. Normally, one sets the discriminator to give the highest value of the ratio (square of net counting rate)/background.

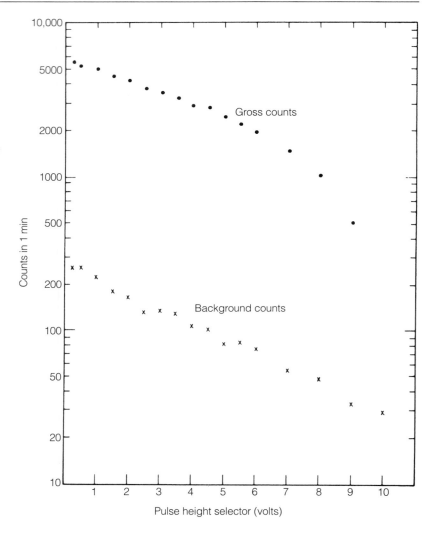

Example 4.6   The liquid scintillation counter used to derive the curves in Figure 4.12 is used to count a sample containing both ^{14}C and ^{3}H. The lower-level discriminators are set at 0.5 for channel A and 1 for channel B. The sample is counted for 20 minutes and gives a count of 7,920 in channel A and 8,340 in channel B. The background counting rate for both channels is 20 c/min. Calculate the ^{14}C and ^{3}H activities in the sample.

4.12 Use of windows to distinguish between ^{3}H (E_{max} = 0.0186 MeV) and ^{14}C (E_{max} = 0.156 MeV) counts in a double-labeled isotope experiment. The plotted points give the efficiencies of detection for ^{3}H and ^{14}C liquid scintillation standards as a function of the lower-level settings of the windows in a two-channel system. The upper-level setting remained constant at 10 V and the lower-level settings were set at 0.5 V. The gain of Amplifier B was increased until the counting rate of a ^{14}C standard solution approached a maximum level. At this point, the window was passing almost all the ^{14}C signals. The largest pulses, produced from absorption of the ^{14}C beta particles of maximum energy (0.156 MeV), were at a voltage approximately equal to the upper-level setting of 10 V. The lower level of 0.5 V passed pulses greater than 1/20 of the maximum energy pulses, or 0.008 MeV. Since the tritium beta particles have energies up to 0.018 MeV, a significant number were counted in the ^{14}C channel at these settings. The gain of Amplifier A was adjusted to maximize the net count rate of a tritium standard solution. After the adjustments were made, backgrounds were taken in counts per minute for each channel, with the use of a blank scintillation solution. The ratio (percent efficiency)^2/background, (E^2/B), is evaluated as an index of performance of the system. Values obtained for E^2/B were 150 for tritium and 308 for ^{14}C. These values are indicative of a high-performance system.

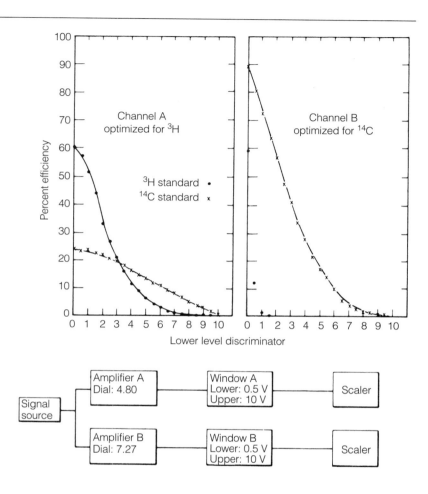

Let e represent the detection efficiency for a channel; use a superscript to indicate the nuclide and a subscript to denote the channel. Thus e_A^C is the efficiency of detection of ^{14}C in channel A. From Figure 4.12, e_A^H = 0.57; e_A^C = 0.24; e_B^H = 0.013; e_B^C = 0.72. The net counting rates in channels A and B are 376/min and 397/min, respectively. If we denote the activities of ^{3}H and ^{14}C by X and Y, respectively, the equations to be solved are 376 = 0.57X + 0.24Y and 397 = 0.013X + 0.72Y. The solutions are X = 430 ^{3}H dis/min and Y = 545 ^{14}C dis/min.

If we had assumed the tritium contribution to channel B was negligible, we would have calculated Y = 551 ^{14}C dis/min and X = (376 − 0.24 × 551)/0.57 = 426 ^{3}H dis/min.

2.4.3   Differential Pulse-Height Analysis with a Single-Channel Analyzer

If, instead of counting only the pulses in a specific energy range, we wish to determine the distribution of pulses over the whole energy spectrum, we have to use a differential analyzer. In a single-channel analyzer, a window is provided that allows through only those pulses above a continuously variable lower-level setting, known as the baseline, and below an upper level that is always a fixed voltage above the lower level. The difference between the upper and lower levels is determined by the "window setting."

If the window setting is at 0.1 V, it will pass only those pulses between the lower-level setting and a level 0.1 V above that setting. A curve can be plotted of counts obtained in a fixed counting time as we increase the level of the baseline stepwise. Such a curve is known as a differential spectrum. An example of a spectrum obtained with a single-channel analyzer for ^{137}Cs is shown in Figure 4.13. The amplifier was adjusted to give an output pulse of 3.3 V when all the energy of the 0.667 MeV photon from ^{137}Cs was absorbed in the crystal (that is, photopeak is at pulse height of 3.3 V).

The integral bias curve, obtained with the same analyzer, is shown along with the differential spectrum for comparison. Note that the value for the integral count above any baseline may be obtained in principle from the differential count by adding up the counts obtained at each setting above the baseline. On the other hand, the differential spectrum can be obtained from the integral curve by determining the difference in integral counts at the settings bracketing each differential reading. However, an accurate sorting of pulses into consecutive voltage increments can be obtained only through the use of a multichannel analyzer.

2.4.4   Use of Multichannel Analyzers in Pulse-Height Analysis

A multichannel analyzer sorts pulses into a large number of intervals, known as channels. Transistorized analyzers have a working range of about 10 V. A 1,000-channel analyzer would thus separate the incoming pulses into 0.01 V intervals.

Data obtained from analysis of ^{137}Cs and ^{60}Co sources with a multichannel analyzer are presented in Figure 4.14. The amplifier of the analyzer was adjusted so that the 0.667 MeV ^{137}Cs photopeak fell in channel 33. This caused the ^{60}Co 1.17 and 1.33 MeV photopeaks to fall in channels 58 and 66, respectively. The number of pulses with heights that fall between the limits set for the photopeak (referred to as counts under the photopeak) is generally used in evaluating the data obtained with the scintillator. The simplest approach is to add the counts in a fixed number

4.13 Differential and integral spectra obtained with a single-channel pulse-height analyzer. Source: [137]Cs; Detector: 3" × 3" solid sodium-iodide crystal; Window: 0.2 V; Interval between successive settings on baseline: 0.2 V. The ^{137}Cs photon energy of 0.667 MeV is identified with the "photopeak" on the differential curve and with the maximum rate of change of counts on the integral curve. In reading the differential curve, the baseline value should be increased by half the window width since it represents the lower level of the window.

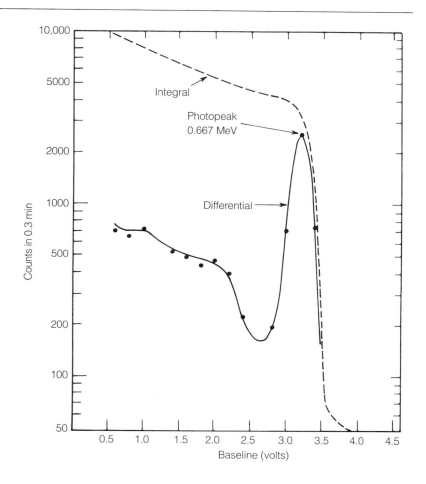

of channels on each side of the channel containing the greatest number of counts. It is preferable to sum up the counts in a number of channels than to use only the counts in a single channel. This procedure minimizes the effect of drift, that is, the shifting of the channel containing the peak associated with a particular energy as a result of slight changes in amplification. Also, the larger number of counts obtained with several channels provides improved counting statistics for the same measuring time.

When we discussed pulse-height distributions in section 2.3, we noted that the interaction of gamma photons with a NaI detector resulted in a large range of energy absorption, giving a distribution of pulse heights with a maximum pulse-height value corresponding to absorption of all the energy of the photon. Thus, when photons of several energies are counted,

4.14   Pulse-height distributions obtained with a multichannel analyzer. The distributions are given for counts of a 137Cs source, a ^{60}Co source, and the ^{137}Cs and ^{60}Co sources counted together. The channel counts include the background, and the results of a separate background count are plotted as horizontal bars. The contribution from ^{137}Cs to the counts in the channels assigned to its photopeak (29–37) is determined approximately by extending the curve from the right to the left side of the photopeak in a smooth manner and subtracting it from the photopeak, as shown in the insert. The extended curve represents the contributions from scattering interactions of the higher-energy photons, in this case ^{60}Co. Note how the ^{137}Cs peak, which is very prominent when the source is counted alone, appears suppressed when plotted on semilog paper on top of the ^{60}Co distribution.

The following counts were obtained for the sources and channels indicated:

	Channels 63–71		Channels 29–37	
	(^{60}Co photopeak)		(^{137}Cs photopeak)	
Source	Gross	Net	Gross	Net
^{137}Cs	49	0	1,925	1,735
^{60}Co	9504	9437	10,411	10,221
Background	67		190	

The counts in channels 29–37 contributed by ^{60}Co equal 1.08 times the counts in channels 63–71. To determine the ^{137}Cs contribution in any other combination of 137Cs and ^{60}Co, one would multiply the net counts in channels 63–71 by 1.08 and subtract from the net counts in channels 29–37. The same approach is used for other pairs of energies. For more than two energies, the procedure is to start with the two highest energies and work down in succession. Of course, it is necessary to have standard pulse-height distributions for each of the photon energies contributing to the composite spectrum. References for more sophisticated analytical methods with the use of computers are given in the text.

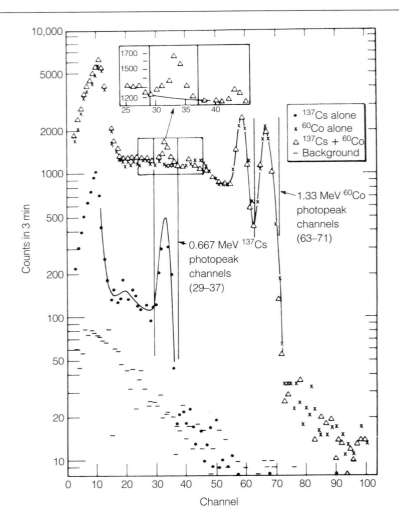

any peak below that corresponding to the maximum photon energy contains, in addition to contributions from complete absorption of the photons of energy associated with the peak, contributions from higher-energy photons that are only partially absorbed. An approximate way to subtract out the counts that are not attributable to photons at the photopeak energy is to draw a line between the left and right sides of the peak and subtract the counts under the line that fall in the channels under consideration. An example of the calculation of the photopeak area and the photopeak center is given in Figure 4.15. The procedure for correcting for con-

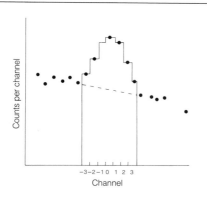

4.15 Determination of peak area. Call peak channel $i = 0$ and select n channels on each side ($n = 3$ in the figure). Draw a straight line between the channel counts outside the left- and right-hand boundaries of the peak. Let a_i refer to the counts in the channels inside the peak and b_R and b_L equal the backgrounds at the left and right boundaries. Then the peak area is given by $A = \sum_{i=n}^{+n} a_i - (n + \frac{1}{2})(b_L + b_R)$. To determine the peak center, $\bar{X}$, weight each channel in the symmetrical portion of the peak (for example, above the half-maximum height) by the net counts, add, and divide the sum by the total counts in the channels weighted: $\bar{X} = \Sigma N_X X / \Sigma N_X$ where X is the channel number and N_X is the net counts in channel X (Baedecker, 1971).

tributions from higher-energy photons is known as "spectrum stripping." For complex spectra, accurate identification and evaluation of peaks by spectrum stripping can be a laborious process if done by hand computations, and the data are best processed by computer techniques (Heath, 1964; DeBeeck, 1975; Quittner, 1972; DeVoe, 1969). The processing can include background subtraction, correction for interference effects between channels, and even calculations of source strengths from calibration data stored in the computer. The computer circuitry may be built into the multichannel analyzer, or the data may be fed to an external computer facility.

Multichannel analyzers cannot process incoming pulses instantaneously, so the live counting time (that is, the time during which the analyzer is actually counting photons) is less than the clock counting time. Normally the analyzer timer is designed to operate on "live" time. Its accuracy can be checked by counting the source along with a pulser of constant known rate. The pulser signal is adjusted to appear in a channel that is approximately 10 percent above the gamma-ray peak. The live time can be computed from the number of pulses accumulated during the counting period.

3   DETECTORS FOR SPECIAL COUNTING PROBLEMS

We have examined two of the most frequently used detectors for radiation counting—the G-M tube and the scintillator. Other detectors are preferred for special applications. Two of the most important are gas-filled proportional counters and semiconductors.

3.1   Gas-Filled Proportional Counters

Gas-filled counters can be used to measure the energy of particles but their use is confined to particles with very short range because of the low density of the gas (for example, alpha particles, very low energy beta particles, very low energy x rays). The signal is produced by the electrical charge or current resulting from ionization of the gas by the radiation. Only part of the imparted energy goes into ionization, the rest produces excitation, which does not contribute any charge. On the average, one ion pair is produced for every 34 eV of energy absorbed in air; similar values apply to other gases commonly used in counters, such as helium, argon, and nitrogen. The total number of ion pairs produced is a measure of the total energy absorbed in the gas.

The production of ion pairs is a statistical process, and the exact number of charges produced from absorption of a given amount of energy varies. The variation is measured by the square root of the average number of charges produced (see section 6.5). This variation is one of the factors affecting detector resolution, discussed previously in connection with the photopeak produced by scintillation detectors (sections 2.3.2, 2.4).

The negative electrons and positive ions produced by ionization are collected by the imposition of an electrical voltage between the central wire and the outer shell of the counting tube (see section 1.1 on G-M counters). The central wire is made positive to attract the electrons. As they drift toward it, they make many collisions with the gas molecules in the detector. In between collisions, they are accelerated and given energy that may be sufficient to ionize the molecules with which they collide. These in turn produce additional ionization that constitutes an amplification process. There is a range of operating voltages over which the amplified charge remains directly proportional to the energy absorbed in the detector. A counter operating in this range is called a proportional counter. Proportional counter amplification of the initial charge is generally a thousand times or more.

Gas-filled proportional detectors can easily distinguish between alpha and beta particles through pulse-height discrimination. A 5.49 MeV alpha particle from americium-241 has a range at standard temperature and pressure of 4 cm. Since alpha particles ionize at an average energy expenditure of 35 eV per ion pair, a ^{241}Am alpha particle produces $5.49 \times 10^6/35 = 1.56 \times 10^5$ ion pairs. A beta particle would lose 0.01 MeV in the same 4 cm, producing (at 34 eV/ion pair for β particles) 294 ion pairs. Thus in a gas counter with dimensions of 4 cm, the ratio of the heights of α to β pulses would be $1.56 \times 10^5/294 = 531$, and they would be easily separable by pulse-height discrimination. Proportional counters do not work well with air; the oxygen has a strong affinity for electrons and prevents multiplication. A 90 percent argon, 10 percent methane mixture is popular.[2] The counters are often operated as flow counters, that is, the gas flows through at a slow rate at atmospheric pressure. This avoids the buildup of impurities in the gas that can occur through outgassing in a sealed counter and that degrades the counter performance.

Gas-filled proportional counters are particularly suited for low-level alpha measurements because they can be built with a large detection area and very low background. A typical proportional counting system and operating characteristics are shown in Figure 4.16.

2. The range of an alpha particle in a medium other than air is given approximately by the Bragg-Kleeman rule (Evans, 1955, p. 652), $R = (\rho_a/R_a)\sqrt{A/A_a}$ where R_a, ρ_a, and A_a are the range, density, and effective atomic number ($= 14.6$) for air, respectively. The range in argon is about 20 percent greater than the range in air, and the average energy expended in producing an ion pair is 26 eV (Attix and Roesch, 1968, p. 320).

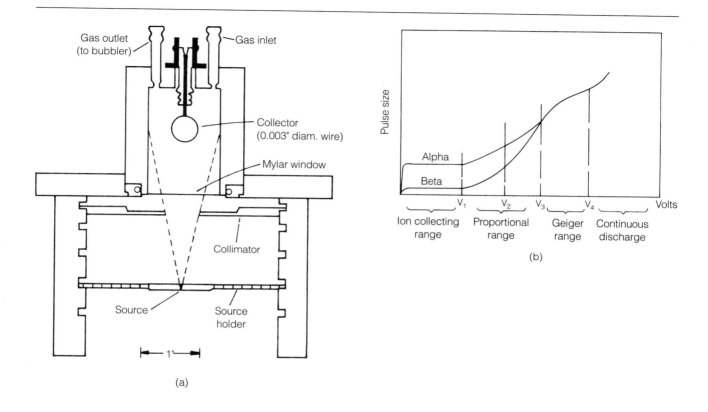

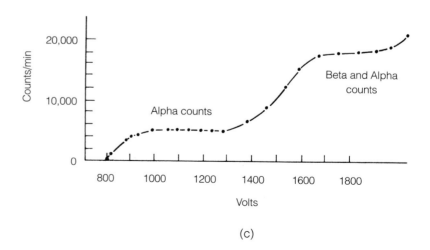

4.16 Operation of gas counters. (*a*) Typical setup for flow-type counter. (*b*) Variation of pulse size with voltage. (*c*) Counting rate versus counter voltage for a flow-type proportional counter.

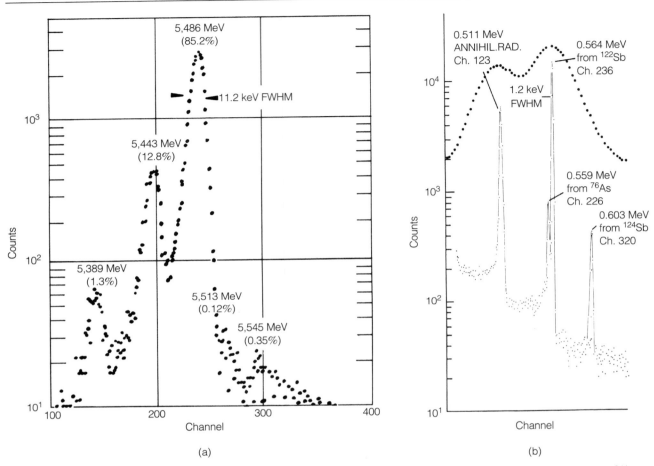

4.17 Spectra obtained with high-resolution detectors. (*a*) Alpha spectrum (^{241}Am) obtained with silicon surface barrier detector. (*b*) Gamma spectrum obtained with germanium (Li) detector. Plotted above it is the expanded portion of a spectrum taken with a 3" × 3" NaI(Tl) scintillation detector (Courtesy ORTEC).

3.2   Semiconductor Detectors

Semiconductor detectors are the detectors of choice for very high resolution energy measurements. The basic detection medium is silicon (specific gravity = 2.42) or germanium (specific gravity = 5.36). The reason for the superior performance of semiconductors is that much less energy is required to produce an ion pair than is required in gases or scintillators. The average energy needed to produce an ion pair in silicon is 3.62 eV

(300 K) and in germanium is 2.95 eV (80 K), compared to 34 eV in air and 500 eV in a scintillator. Thus, many more ion pairs are produced per unit energy absorbed. This produces a smaller spread in the distribution of pulses and improved resolution.

Silicon detectors are generally used for alpha and beta particles. The sensitive volume is made thicker than the maximum range of the particle in the medium. A standard thickness (depletion depth) for the sensitive volume of a silicon detector is 100 μm. This is equal to the range of 12 MeV alpha particles and 0.14 MeV beta particles. Detectors with active areas up to 900 mm^2 are standard catalog items. Silicon detectors with smaller areas are offered to 500 μm thickness (range of 2.5 MeV β particles).

Germanium detectors operated at liquid nitrogen temperature (80 K) are generally used for gamma radiation. They must be thick enough to give adequate sensitivity for photons in the energy range of interest. The needed thickness can be evaluated from the half-value layer (1.9 cm for the 0.67 MeV ^{137}Cs photons). The improvement in resolution of a gamma spectrum obtained with a semiconductor compared to NaI is shown in Figure 4.17. Notice the additional energies that can be distinguished with a germanium crystal—these overlap and cannot be distinguished in a scintillation detector. It costs much more to make semiconductor than NaI detectors of comparable detection efficiency for gamma rays, so NaI detectors are the detector of choice when cost is a factor or high-efficiency detection of photons of known energy is required.

4 MEASURING RADIATION DOSE RATES

The pulse detectors and associated instrumentation discussed in the previous sections are used primarily for counting particles and determining their energies. We shall now consider radiation measuring devices which give readings closely representative of the absorbed dose rate.

4.1 Measuring X and Gamma Dose Rates with Ionization-Type Survey Meters

The simplest type of detector that responds to the absorbed energy in the detector medium is the ionization chamber. This consists of a container filled with gas. If the gas is air, each ion pair is associated (on the average) with the expenditure in the air of 33.73 eV of energy. The ion pairs produced by the absorbed energy are collected by maintaining a suitable voltage difference between the wall of the chamber and an inner electrode. However, the operating voltage is much lower than in proportional and G-M counters because there is no multiplication. Continuous exposure to radiation produces an ionization current that is measured with a sensitive electrometer. Thus the electrometer current is a measure of the dose rate to the gas. If the gas is air, and the radiation consists of x and gamma rays, the ionization is expressed in terms of a quantity known as the exposure. The exposure is defined as the total negative or positive charge liberated by photons as a result of interactions in unit mass of air.

The charge is that produced in the process of the complete slowing down in air of all the electrons liberated by the photons. However, the dimensions of actual chambers are generally much smaller than the range in air of the electrons. As a result, the chamber gas gets only a fraction of the ionization (and resultant current) from the slowing down of the electrons liberated by photon interactions in air. Most of the ionization is produced by electrons released from photon interactions in the chamber walls. If the walls of the chamber are made of air-equivalent material, the thickness of which approximates the maximum range of the electrons released by the x or gamma radiation, electron equilibrium exists. This means that the ionization produced in the chamber gas by electrons liberated from the walls compensates exactly for the ionization that would have been produced in the air by the electrons liberated in the gas. In other words, the ionization produced in the chamber air is the same as if it were caused by the complete absorption in the air of all the electrons liberated in the air. The special unit of exposure is called the roentgen, R (defined in Part Two, section 14). Since 1 R produces a dose of 0.93–0.97 rad to muscle, depending on the photon energy, we may equate an air exposure of 1 R to a soft-tissue dose of 1 rad or 0.01 Gy for all practical purposes.

Air-filled ionization chamber-type survey instruments for measuring exposure rates of the order of 1 milliroentgen per hour (mR/hr) can be built simply and are very useful for monitoring work spaces. The ionization chamber is usually in the form of a cylinder. An ion-chamber survey meter is shown in Figure 4.18. The sensitivity of the ion-chamber meter can be increased greatly by pressurizing the gas in the detector. Ionization chambers used to measure radiation at low levels comparable to the natural background must be larger than those used in work areas to achieve the required sensitivity. A spherical chamber with a volume of 16 liters has been used extensively for background measurements.

Example 4.7 Calculate the current produced in a cylindrical air-ionization chamber 8 cm in diameter by 10 cm in length, exposed in a field of 1 mR/hr. The temperature and pressure are 25° C and 770 mm Hg, respectively. Assume that radiation equilibrium exists.

4.18 Count rate and dose rate monitoring instruments. *(a)* General purpose count rate meter for G-M and scintillation detectors shown in *(b)* through *(f)*. *(b)* Energy compensated G-M tube, energy response between 50 keV and 1.25 MeV. *(c)* Low-energy Gamma Scintillator with 2.5 cm × 1 mm thick sodium iodine crystal. *(d)* 3" × 3" NaI gamma scintillation detector. *(e)* 15 cm^2 pancake G-M detector for alpha, beta, gamma surveys. *(f)* 76 cm^2 alpha scintillator probe with ZnS(Ag) scintillator. *(g)* Ion chamber survey meter for measuring exposure rates around x-ray machines and gamma sources. *(h)* Pressurized (6 atmospheres) ion chamber radiation dose meter for measuring exposure rates around gamma sources. (*Sources: (a)–(f)*, Ludlum Measurements, Inc., *www.ludlums.com; (g)–(h)*, Inòvision Radiation Measurements, *www.surveymeters.com*.)

(a) (b) (c) (d) (e) (f) (g) (h)

The chamber volume is 503 cc. The mass of air in the chamber equals:

$$503 \text{ cc} \times \frac{0.001293 \text{ g}}{\text{cc}} \times \frac{273 \text{ K}}{298 \text{ K}} \times \frac{770 \text{ mm Hg}}{760 \text{ mm Hg}} = 0.604 \text{ g}.$$

The current is then:

$$\frac{1\text{mR}}{\text{hr}} \times \frac{2.58 \times 10^{-10} \text{ C}}{\text{mR-g}} \times 0.60 \text{ g} \times \frac{1\text{hr}}{3,600 \text{ sec}} = 4.3 \times$$

$$\frac{10^{-14} \text{ C}}{\text{sec}} = 4.3 \times 10^{-14} \text{ amperes}$$

This is a very small current and is about 10 times the limit detectable by portable survey instruments. The air in one commercial ionization chamber survey meter is pressurized to six atmospheres for increased sensitivity (*www.inovision.com,* model 451P).

4.2   Use of Scintillation Detectors to Measure Dose Rates

Scintillators may be used to indicate the rate of energy absorption if current from the photomultiplier is measured. Used in this way, the scintillation detector is a dose-rate measuring device. If an organic scintillator is used, the energy absorption, and consequently the dose, are very similar to that produced in tissue.

The output current from scintillation dose-rate instruments is hundreds of times greater than that from ionization chambers of the same volume, but scintillation detectors are not used as widely for radiation monitoring. Some of the reasons are: a strong dependence of the gain of the photomultiplier tube on temperature and applied voltage, a high inherent background current from thermionic emission from the dynodes of the photomultiplier, fragility, and high cost.

4.3   Use of G-M Counters to Monitor Dose Rates

The type of radiation survey instrument used most often for monitoring of beta and gamma radiation is a G-M counter connected to a count rate meter with a scale reading in terms of mR/hr (Fig. 4.18). These instruments are used because of their low cost, simplicity, reliability, and high sensitivity. From our previous discussion, it is apparent, however, that the G-M counter does not actually measure exposure rate or dose rate. All it does is register counts for whatever incident radiation happens to produce an ionization in the sensitive volume. It turns out, however, that the G-M survey meter can be made into a fairly accurate indicator of actual exposure rate over a wide range of x and gamma ray energies. This is done through the use of appropriate filters around the detector. The energy response of a specially filtered detector is shown in Figure 4.19. When the dose rate is very high, G-M counters cannot be used. Ionization-type survey instruments should be used instead.

4.3.1   *Use of Accumulated Counts on a G-M Counter*

While the G-M counter is normally used to monitor for anomalous levels of radiation in the environment by measuring the count rate, it is also useful as a monitor of radiation levels over an extended period of time. For this application, the accumulated counts are registered on a scaler, which

4.19 Energy response of G-M tube. *(a)* Unmodified and *(b)* filtered, 0.053 in. tin +0.010 in. lead on sides; 0.022 in. tin + 0.004 in. lead on end (Wagner and Hurst, 1961; see also Jones, 1962).

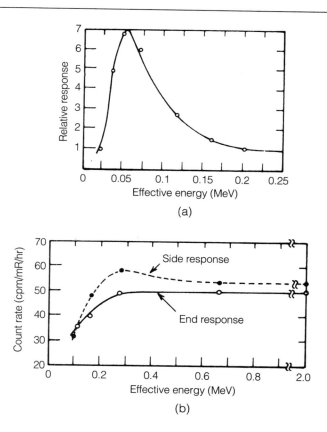

(a)

(b)

may be incorporated in the instrument or attached to the earphone connection on a portable unit. Devices that provide only count rate outputs are readily modified to indicate individual counts by adding a connector attached to the speaker outlet. Instruments with this capability can be used to note counts over an extended period—for example, a day or a week—and thus keep track of changes in the background radiation with high precision. This capability also provides the user with a good feeling for the radiation background and its variations from both natural and artificial radiation sources in the environment. The counts can be converted to dose with suitable calibrations.

4.4 Routine Performance Checks of Survey Meters

Survey meters should be checked for signs of malfunction whenever they are taken off the shelf for use. The following procedures apply specifically to instruments with needle indicators, but the general approach is also applicable to digital readouts.

1. Look over instrument carefully. Are there indications that it may have been dropped or otherwise mistreated? Is anything damaged on it? If the detector is used as a probe, does the cable show signs of breakage? Was the instrument used to monitor loose radioactive contamination? If so, check that it is free of significant radioactive contamination. Is the needle at its zero position or at the point specified by the manufacturer? Does it bounce around excessively when the meter is moved, indicating poor damping?

2. Turn on the instrument and allow it to warm up. You may find that the instrument was already on, that the person who used it before you forgot to turn it off. This means that the batteries are probably weak. If there is no battery check, replace the batteries.

3. Turn to the battery check position and see that the batteries are good. Record the battery check reading.

4. Turn the knob gently to the first scale and let the needle stabilize. Continue turning to more sensitive scales until you get a response. (Give the needle time to settle down at each new position. It generally takes longer on the more sensitive scales.) The needle will fluctuate, more so at the lower readings, because of the random nature of the detected events. Record the background, noting the highest and lowest reading at which the needle remains for a second or so; that is, do not record momentary swings. Always record the scale along with the meter reading.

5. Check the response of the meter on each operational scale with a radiation check source that gives a reading close to midscale. Repeat three times to test reproducibility. The readings obtained should not deviate from the mean value by more than 10 percent.

6. Note the meter readings with the check source held in place while the instrument is oriented in three mutually perpendicular planes. If there are significant differences, be careful to use the instrument in the same orientation as during calibration.

7. Compare the check source reading with the reading obtained when the meter was given a complete calibration. If the two readings differ by more than 20 percent, the instrument should be recalibrated. Use the check source on the instrument prior to each use if the instrument is used intermittently and several times a day if it is used continuously.

4.5   Calibration of Survey Meters

Meters are calibrated by exposing them to a radiation source with a known output. The calibration should be performed under conditions of minimum scatter, by placing the source and detector far from walls, floor, and other scattering surfaces or by collimating the beam so no radiation hits scatter surfaces. (If a collimator is used, the value for the source output must include the radiation scattered from the collimator.) Otherwise, scatter corrections should be made. Some experimental data showing the effect of scatter on readings with a G-M counter are given in Figure 4.20.

At least two points should be calibrated on each scale to determine the degree of linearity of response of the instrument. Readings also should be taken on each operational scale with a check source at the time of the calibration and repeated routinely prior to use. Calibrations should be performed on a regular schedule and after maintenance, after change of batteries, and when check-source readings indicate departure from calibration greater than 20 percent. It is important to keep records of all calibration, maintenance, repair, and modification data for each instrument. A label should be affixed to the instrument giving the date of calibration, check-source reading, any calibration curves or correction factors, special energy or specific use correction factors, and the identity of the calibrator. Extensive treatments of calibration may be found in NCRP, 1991a; ANSI, 1975; ICRU, 1971a, Report 20; IAEA, 1971a, Technical Reports Series No. 133.

4.5.1 Calibration Sources

Ideally, a meter should be calibrated in a radiation field identical to the field to be monitored; for example, a meter that will be used to measure the exposure rate around a cesium irradiator should be calibrated with a cesium source attenuated to give an energy spectrum similar to that from the irradiator, and an x-ray survey meter should be calibrated with an x-ray machine, and so on. Calibrated in this way, any radiation detector could be used to monitor exposure rates. In practice, one must rely on the *energy independence* of the meter; that is, one calibrated to read accurately in roentgens with a source of a specific energy, such as ^{137}Cs, will be satisfactory for monitoring fields at other energies. Most instruments, however, are accurate only over a limited energy range. It is important to know the energy response and the corrections that should be made for sources that differ from the calibration source. The information should be supplied by the manufacturer or, if necessary, obtained by the use of calibration sources with different energies.

Calibration sources should have reasonably long half-lives so they do not have to be corrected for decay or replaced frequently. Some of the more popular sources are listed in Table 4.1.

Commercial calibration sources should be checked independently. They may be calibrated by the National Institute of Standards and Tech-

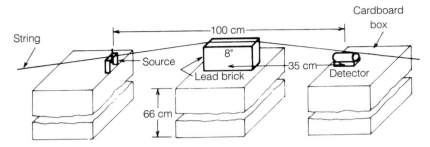

4.20 Setup for calibration of gamma survey meters with a standard source. The dose rate due to the primary photons at a given distance from the source is evaluated by the inverse square law (air attenuation can usually be neglected). The contribution from the additional photons scattered into the detector by the floor, walls, and other structures is evaluated experimentally by shielding out the direct photons with a 2" × 4" × 8" lead brick. The shield cross section should be as small as practical and the shield should be positioned very precisely or part of the detector will see the source and the reading for the scattered radiation will be erroneously high. In the figure, the equipment is shown lined up with the use of a taut string. The source, shield and detector are positioned on top of cardboard cartons, which scatter a minimum amount of radiation because of their small mass.

The following calibration data were taken at 1 and 2 m. The purpose of taking the readings at two points was to check that the inverse square law applied and that the correction for scattered radiation was being made properly. (*Source:* ^{137}Cs, 5 mCi; dose rate, 1.6 mrad/hr at 1 m.)

Source-detector distance (m)	Unshielded		Shielded		Counts/min from direct beam
	Counting time (min)	Counts/min	Counting time (min)	Counts/min	
1	1	4,778[a]	5	373[a]	4,405
2	1	1,215	10	187	1,028

a. Corrected for coincidence loss (G-M resolving time 3.5×10^{-4} sec) and background rate of 42 c/min.

The measured ratio of 4,405/1,028 = 4.3 follows the inverse square law approximately. The use of longer counting times to reduce the statistical variation in the counts and a shadow shield contoured to shield the diect radiation with a minimum effect on the scattered component would improve the agreement.

Using the measurements at 1 m, the calibration constant for the G-M tube becomes 4,405 c/min = 1.6 mrad/hr, or 1,000 c/min = 0.36 mrad/hr. In calibrating a survey meter, measurements should be made at enough distances to cover all ranges of the instrument. The problems connected with reporting G-M readings in mrad/hr or mR/hr should be understood and the possibilities of making large errors at low energies recognized.

Table 4.1 Some photon-emitting radionuclides suitable for instrument calibration.

Radionuclide	Effective energy (keV)	Half-life (yr)	Specific exposure rate constant (mR/hr-MBq at 1m)[a]
^{241}Am	60	433	0.00349
^{137}Cs	662	30.1	0.00873
^{226}Ra	836[b]	1600	0.0223[c]
^{60}Co	1250	5.27	0.0351

Source: ANSI, 1975.

a. Assume negligible self-absorption, scattering, and bremsstrahlung.

b. ^{226}Ra calibration sources emits gamma rays of many energies, from 19 to 2,448 keV.

c. In equilibrium with its decay products and with 0.5 mm platinum filtration.

nology (NIST) or checked against other sources that were standardized by NIST.

4.5.2 Determination of Exposure Rates for Calibration

Since there is some attenuation of the source materials by the encapsulating materials, calibration sources generally are specified not by the activity contained but by the equivalent activity of an uncapsulated source. When this is the case, the exposure rate produced by the direct radiation from the source can be determined readily with the specific exposure rate constant.

Example 4.8   A cesium-137 source is certified to have an equivalent activity of 185 MBq on September 1, 1977. What is the exposure rate at a point 50 cm from the source on March 15, 1980?

From Table 4.1, the specific exposure rate constant is 0.00873 (mR-m^2)/(hr-MBq) or 0.00873 mR/hr from 1 MBq at 1 m. The elapsed number of half-lives equals 0.084 and the decay factor equals 0.94. Therefore, the exposure rate at 50 cm on March 15, 1980, equals $0.00873 \times 185 \times (100/50)^2 \times 0.94 = 6.07$ mR/hr.

The value of the specific exposure rate constant may be due to photons of a single energy or of a range of energies. It usually does not include very low energy photons emitted by the source that ordinarily do not penetrate through encapsulating materials, or if they did, would probably not get through the detector walls. On the other hand, if a source had only a very thin covering and was monitored with a thin-walled detector (one designed for low-energy x rays), the meter might respond to the low-energy photons and these would have to be included (with appropriate corrections for attenuation) in the specific exposure rate constant. Generally, this is not a problem, but the possibility of complicating source emissions under special conditions should be kept in mind. For example, the specific exposure rate constant for ^{137}Cs is calculated for the 0.662 MeV photons. Barium x rays in the 30 keV range are also emitted, but their contribution to the dose rate if unattenuated is only about 1 percent. Their actual contribution is even less because of attenuation in the source's encapsulating material. Beta particles emitted from the source are generally absorbed by the encapsulating material and detector walls. However, high-energy beta particles may emerge and contribute to the detector reading. This can lead to significant calibration errors, particularly in the case of ^{226}Ra sources, which emit 3.26 MeV (max) beta particles from the decay product ^{214}Bi. The contribution of beta particles can be checked by taking an absorption curve. The results will also indicate how thick an absorber is necessary to eliminate their effect on a given detector. Of course, the use of added absorber requires a correction in the stated output of the calibration source. No one ever said that careful calibration was easy.

4.5.3   *Effects of Calibration Geometry*

The basic calibration exercise involves determining the distance from the source to give a desired exposure rate by use of the inverse square law; setting up the detector at this distance; and adjusting the detector to read properly. This procedure is quite adequate for radiation-protection purposes most of the time, but there are pitfalls. The calculated exposure rate refers to a point that is located at a precise distance, while the detector has a finite size and the distance is somewhat indefinite. Radiation scattered from the structures near the source and the detector—ground, walls, or other surfaces in the calibration area—adds to the direct emission from the source. Absorption in air can be significant if the distance between source and detector is large enough or the energy is low enough. Source containers may also introduce some attenuation. Fortunately, these factors often tend to cancel out each other or be insignificant in the first place.

The calibration procedure can be designed to minimize some of these effects. The uncertainty in the effective distance due to the detector or source dimensions is minimized by making the separation distance large enough, preferably greater than seven times the maximum dimension of the source or detector, whichever is larger. On the other hand, the relative effect of scattering from external surfaces is reduced by decreasing the distance between source and detector, since the direct radiation increases inversely as the square of the distance while the scattered radiation changes more slowly. The distance to scattering objects from the source and from the detector should be greater than the distance between the source and the detector. The structure holding the source and detector should be as light as possible.

It is well worthwhile to check the radiation levels at the calibration points with a standard instrument that is highly accurate and reliable, such as a cavity ion chamber (R meter). Preferably, the instrument should have been calibrated at the National Bureau of Standards or, alternatively, with sources traceable to NBS. Close agreement between the measurements with the standard instrument and the calculations should give confidence in the accuracy of the calibration values.

4.5.4 Corrections for Scattered Radiation

If there is any question about the significance of scattered radiation, it can be evaluated in several ways. One is a shadow shield technique, whereby the direct radiation is blocked out by a shield just large enough to cut out the source from the field of the detector (Fig. 4.20). Another way is to compare the readings in the calibration facility with readings in as scatter-free a situation as possible, say 2 m above the roof of the building. The differences in readings will indicate whether scattering is important. A third check is made by testing whether measurements at different source-detector distances follow the inverse square law. Significant scatter will result in a drop-off considerably slower than the inverse square.

4.5.5 Directionality Checks

A survey meter should give a reading that does not depend on the angle of incidence or directionality of the radiation. Differences due to directionality should be checked by reading the meter at various orientations to the source. If there is a directional dependence, this must be taken into account in the radiation survey.

4.5.6 Linearity Checks

The linearity of an instrument's response can be checked over its entire operating range by a "two-source" procedure. Each of the two sources should have sufficient output to produce maximum dose rates desired for calibration when the sources are placed close to the detector. Choose a scale, position source A to give 40 percent of the upper limit, and mark the position. This is the reference position. Remove source A and place source B in a different location that also gives 40 percent of the upper limit. Record the reading, return source A to its original location, and record the reading for the two sources together. The ratio of the sum of the two separate readings to the combined reading is the correction factor for the combined reading. Repeat at positions where the two sources together produce a reading that is 40 percent of full scale. The ratio of the combined reading to the sum of the two readings for the sources used separately is the correction factor for the single source reading. Repeat for each range and plot correction factors as a separate curve for each range. Then obtain between-range corrections by positioning each source to give 80 percent of full scale. The sum should be 16 percent of full scale on the next higher range if the ranges differ by a factor of 10. The overall correction factor is the product of all within-scale and between-scale correction factors determined in reaching a particular value.

In this method, the sources may be placed very close to the detector; thus, readings on the high ranges with sources of relatively low activity may be obtained. Errors may be introduced, however, when the source-detector distance is much shorter than the detector dimensions, because of the nonuniform production rate of ions. At high exposure rates, this could create high local concentrations of ions, resulting in excessive recombination and erroneously low meter readings.

4.6   Beta Dose-Rate Measurements

Monitoring for beta radiation is usually done with the same types of instruments used for gamma radiation (ionization chambers and G-M counters), except that the instruments have a thin window for admitting the beta particles. In principle, a gamma calibration of an ionization chamber–type instrument should also hold for beta measurements, since the ionization is accomplished by electrons in each case. However, a survey meter is usually calibrated with a radiation field that is uniform through the sensitive volume of the detector, while the field is not likely to be uniform in monitoring beta radiation. The reason is that the beta measurements are usually made very close to the source of beta particles and there is strong inverse-square-law attenuation over the dimensions of the detector. The effect can be minimized by measuring the energy absorption in a thin tissue-equivalent detector, such as a plastic scintillator. One should determine a correction factor for converting the instrument reading to actual beta dose rate, where the actual beta dose rate is determined with an extrapolation chamber. Because of the strong attenuation of the lower-energy beta particles in the dead layer of the skin, it may be desirable to use the detector with an absorber to include this effect. A good nominal value for absorber thickness is 7 mg/cm^2 (0.007 cm at unit density).

When G-M survey meters are used to monitor for beta contamination, the results are usually reported in terms of the meter reading, which is in milliroentgens per hour (based on a gamma calibration). There is no reason why the gamma calibration should apply, since the instrument is actually giving the number of beta particles crossing the window of the detector per unit time and the survey instrument must be calibrated separately for beta radiation. Accurate ways to evaluate beta dose rates are to determine the energy distribution and flux of the beta particles, to measure the energy absorption in a thin tissue-equivalent detector, such as a plastic scintillator, or to calibrate the detector with a source that corresponds in geometry and radiation energy spectrum to the radiation field being monitored. For general monitoring, a uranium slab is often used as the calibration source (2.33 mGy/hr with 7 mg/cm^2 absorber), but the calibration applies only to monitoring higher-energy beta emitters.

4.7   Neutron Monitoring

Neutrons are not normally encountered in the research or medical environment. They are produced incidentally in the medical environment in the operation of high-energy accelerators now used in radiation therapy (above 10 MeV), through interactions of x-ray photons with nuclei in structural and equipment materials, and around cyclotrons used for the production of short-lived radionuclides for radiation diagnosis and therapy. They can also be found in the environment as leakage through the shielding around neutron sources used for educational purposes or as spurious radiation around synchrocyclotrons that produce high-energy protons for therapy.

The conversion factor for neutron absorbed dose to equivalent dose is strongly dependent on neutron energy (Table 2.8), which makes neutron dosimetry more complicated than gamma dosimetry. Therefore, a neutron dose measuring device should be able to discriminate neutrons from gamma rays, determine the energies of the neutrons, and weight them appropriately. Alternatively, an instrument could weight signals produced in a tissue-equivalent medium from their linear energy transfer without consideration of the types of radiation or their energies; this approach is the basis of a dose-equivalent meter that uses a tissue-equivalent gas proportional counter as the detector (Baum et al., 1970; Kuehner et al., 1973). Perhaps the simplest and most rugged neutron dose measuring instruments are those based on a thermal neutron detector surrounded by a moderator. The thermal neutron detector responds only to neutrons, and the moderator is shaped to produce a count from thermal neutrons proportional to the neutron dose equivalent. One of the most successful designs utilizes a small BF$_3$ tube enclosed in a polyethylene cylinder. The desired response is obtained by sizing the moderator, drilling holes in it, and incorporating a thin cadmium absorber (Andersson and Braun, 1963). This instrument is very sensitive, giving about 700,000 c/mSv, with a background of only a few counts per hour. There are also designs of moderated thermal neutron detectors based on spherical geometry (Hankins, 1968; Nachtigall and Burger, 1972). Other thermal neutron detectors used with moderators include ^{6}LiI(Eu) scintillators (Tsoulfanidis, 1995), indium activation foils (McGinley, 1992), and boron loaded track plates (Vives and Shapiro, 1966; Shapiro, 1970). An Am-Be neutron source is often used for calibration (Nachtigall, 1967; ICRU, 1969, Report 13).

5 Measuring Accumulated Doses over Extended Periods—Personnel and Environmental Monitoring

In long-term monitoring of low levels of radiation, the dose rate is not particularly interesting. We are more interested in knowing the dose accumulated by personnel or in the environment over an extended period. Several different types of detectors are suitable for this purpose.

The nuclear emulsion monitor. Nuclear emulsions are unique detectors in that they are actually a composite of tiny individual radiation detectors, silver halide microcrystals or grains, primarily AgBr, dispersed in gelatin. Charged particles passing through the grains raise electrons from the valence into the conduction band in the AgBr lattice (requiring radiation particles with energy exceeding 2.5 eV). This initiates a chain of events leading to the formation of silver specks that render the grain developable; that is, under the action of a developer, the grain is reduced to free silver. After the emulsion is developed, it is fixed to produce a negative as with regular photographic film. Depending on the size and the spacing of the grains, the result is a general darkening, which can be evaluated in terms of the energy absorbed in the emulsion, or a pattern of spots revealing the track of charged particles passing through the emulsion. The larger the grains, the greater is the sensitivity (speed) of the emulsion. High-speed x-ray emulsions have grain diameters of about 2 μm. Emulsions for observing tracks of charged particles have smaller, well-separated grains (0.1–0.6 μm).

Emulsions similar to those used in dental x-ray film are widely used as personnel monitors for x, gamma, and beta radiation. For such use, the film is inserted in a special holder which can be clipped to the clothes and is therefore called a "film badge." The amount of darkening is read with a densitometer and is related, through appropriate calibration, to the absorbed dose to the film. Because the nuclear emulsion is composed largely of grains of silver bromide (in contrast to tissue, which is largely carbon, hydrogen, and oxygen), the response to radiations of different energy is different from that of tissue, and the dose to the film as indicated by its darkening is not representative of the dose to a human being. However, with the use of selective radiation filters over the film, such as copper, lead, and plastic, the resultant density distribution produced by the radiation can be used to identify the general energy range of the radiation and allows conversion of the film dose to tissue dose.

The emulsion can be a source of much information in addition to the dose. Unusual exposure patterns are produced by contamination on the badge and thus reveal its presence, or patterns can provide information on the source of the exposure as well as the direction of the incident radiation. Not to be ignored is the low cost of film monitoring compared to other methods if the monitor must be changed at frequent intervals.

Since suitably prepared emulsions will record the tracks of charged particles, they may be used to identify and quantify exposures from protons and alpha particles (Barkas, 1963; Becker, 1966, 1973; Fleischer et al., 1975). Personnel monitoring films for neutrons are based on the effects of neutron collisions with the hydrogen atoms in the emulsion, propelling them as protons that leave tracks in the emulsion when developed. The dose is determined by counting the tracks, but the analysis is quite complicated (Dudley, 1966). Membrane filters, used for sampling alpha-emitting aerosols, produce interesting radioautographs when placed in contact with nuclear-track emulsions, such as neutron NTA personnel monitoring film, ranging from one or two tracks from radon decay products in the environment to bursts of tracks from hot plutonium particles sampled in a glove box (Shapiro, 1970, 1968).

The disadvantage of the nuclear emulsion is that the information stored as silver specks (the latent image) is degraded at high temperatures and humidity and fades over a few weeks, even under temperate conditions. Thus, the emulsion is not useful for long-term monitoring, although the lifetime can be extended by special packaging.

Thermoluminescent dosimeters. Thermoluminescent (TLD) detectors are well-suited to general personnel and environmental monitoring of x and gamma radiation. The principle of operation is that energy absorbed from the radiation raises the molecules of the detector material to metastable states. They remain in these excited states until they are heated to a temperature high enough to cause the material to return to its normal state with the emission of light. The amount of light emitted is proportional to the energy absorbed, hence is proportional also to the dose to the detector. The emitted light is measured with a photomultiplier tube.

Table 4.2 Properties of some thermoluminescent materials used as dosimeters.

Property	LiF	$Li_2B_4O_7(Mn)$	$CaF_2(Mn)$	$CaF_2(Dy)$
Density (g/cc)	~2.6 (solid) ~1.3 (powder)	~2.4 (solid) ~ 1.2 (powder)	3.18	3.18
Temperature of main TL glow peak	195° C	200° C	260° C	180° C
Efficiency at ^{60}Co relative to LiF	1.0	0.15	10	30
Energy response (30 keV/^{60}Co)	1.25	.9	~13	12.5
Useful range	mR to 3×10^5 R	50 mR to 10^6 + R	mR to 3×10^5 R	10 μR to 10^6 R
Fading	Negligible (5%/yr at 20° C)	<5%/in 3 mo.	10% first 16 hr; 15% in 2 wk	10% first 24 hr; 16% in 2 wk

Source: Harshaw Chemical Co.

Note: Thermoluminescent dosimeters are available from Bicron Radiation Measurement Products, Solon, Ohio.

The most commonly used thermoluminescent material is lithium fluoride activated with magnesium and titanium. Calcium fluoride is much more sensitive but it is not tissue-equivalent (as LiF is) and has poor low-energy response. This, however, can be largely corrected with the use of appropriate filters (that is, 2 mm steel; Shambon, 1974). Both materials performed well in international field-monitoring tests that were conducted for six weeks in a hot and humid climate (Gesell et al., 1976). Properties of various TLD materials commercially available are given in Table 4.2.

In the evaluation of any long-term dose registration device, the possible loss of information before read-out is important. The retention of the information can be affected by such environmental variables as temperature and humidity. TLD materials are less affected by environmental changes and hold the information longer than photographic emulsions do. Emulsions (before development) are particularly affected by humidity, and much of the information can be lost after a few days if the humidity is high. However, the emulsion equipped with filters contains more information than the TLD detector and the developed emulsion provides a permanent record for future reference. We could continue to propound the relative advantages of TLD and nuclear emulsion for personnel dosimetry, and the choice of one system over another may not be clearly indicated in many applications.

Pocket ion chamber dosimeters. The direct-reading pocket ionization chamber consists of a small capacitor, charged prior to use and connected to a glass fiber electroscope. The unit is mounted in a pen-type housing which can be clipped into the pocket of a shirt or laboratory coat. Exposure of the chamber to ionizing radiation results in loss of charge proportional to the amount of exposure and a corresponding deflection of the fiber. The deflection can be viewed directly by means of a lens and a scale built into the instrument. Simpler versions of the pocket ionization chamber are read not directly but by means of auxiliary electrometers.

Although pocket ionization chambers are convenient to use, they must be handled carefully. If they are exposed to excessive moisture, leakage across the insulator will result and cause deflection of the fiber and erroneous readings. Rough handling can also produce spurious results. The direct-reading chambers, however, are the best available monitors for following significant exposure levels directly.

Electronic dosimeters. Developments in the miniaturization of electronic circuitry along with almost unlimited capabilities in processing and displaying data have resulted in the commercial development of personal electronic dosimeters incorporating direct-reading electronic detectors. The detectors include both energy-compensated miniature G-M tubes and semiconductor detectors. The dosimeters are small enough to clip onto a shirt pocket. They measure dose and dose rate, can alarm at pre-set doses and dose-rates or after a specified period in a high-dose area, provide a dose history—and they also chirp. Their many features make them the dosimeter of choice in situations requiring close control of radiation exposure or a detailed accounting of exposure history.

Optically stimulated luminescence dosimetry (OSL). Aluminum oxide activated with carbon (Al_2O_3:C) is the most recent addition to detector materials developed for use in dosimetry. As with TLD materials, the absorbed energy raises the molecules of the detector material to metastable states. However, unlike TLD materials, which require heating to cause them to return to their normal state with the emission of light, OSL materials are stimulated to release the stored energy resulting from exposure to ionizing radiation by the absorption of light (McKeever et al., 1995; Akselrod et al., 1996). The method, which found early application in dating materials in archeological and geological research, has several features which make it particularly advantageous for personnel radiation dosimetry, including a wide dynamic range, high sensitivity and precision, capability of assignment for an extended period, and readout without destruction of the signal, so the dosimeter can be reanalyzed if necessary.

Miscellaneous dosimeters. A variety of other devices are useful for special applications. Silver-activated phosphate glass on exposure to radiation undergoes two effects that can be used for dosimetry: an increase in optical density and the formation of stable fluorescing centers that continuously emit orange light under ultraviolet excitation. The detector generally consists of a small glass rod (for example, 1 mm diameter by 6 mm long). Commercially available systems offer a reliable, economical, and fast method for gamma-radiation dosimetry that is insensitive to neutrons and approaches TLD or emulsion in sensitivity.

Certain plastics when exposed to highly ionizing particles—such as fission products, alpha particles, or protons—will show pitting or tracks at the site of deposition of energy from the ionizing particle after being etched with suitable caustic solutions. The etched track method is useful for monitoring neutrons present near high-energy accelerators.

A personal neutron monitor that is noteworthy for simplicity and ease of use is the bubble dosimeter. The detection medium is a gel throughout which are dispersed a high concentration of microscopic droplets of a superheated liquid. When a neutron collides with a droplet, the potential energy is released and transforms it into a visible bubble. The number of bubbles that appear in the gel are proportional to the neutron dose. The gel is contained in a transparent glass tube in the form of a fountain pen so it can be clipped in a pocket.

Good reviews of the types of detectors available for dosimetry may be found in Knoll, 1999; Tsoulfanidis, 1995; Attix and Roesch, 1966; and ICRU, 1971a, Report 20.

5.1   Use of Biodosimetry in Reconstructing Radiation Exposures

Persistent effects of radiation in the body can give an indication of the magnitude of high past radiation exposures incurred in radiation accidents. These include biological effects, such as chromosome aberrations in blood lymphocytes, and subtle "solid state" effects, such as detection of free electrons in dental enamel by electron spin resonance (ESR). A comparison of doses given by the two methods was made on a victim of the Chernobyl nuclear reactor accident who received high exposures in the course of fighting the fire from the roof of the building. He received relatively uniform gamma radiation from the major sources of radiation, the reactor core and the radioactive plume. Although he received supportive care and a bone marrow transplant, he died 86 days after the accident from pneumonia and "graft-vs-host disease" (Baranov et al., 1989). No dose information was available from personal dosimetry. Biological dosimetry consisted of assays for the presence of dicentric chromosomes in blood and bone marrow (cytogenetics) and making serial determinations of the levels of granulocytes and lymphocytes following the accident. The weighted average of the dose from these two techniques was 5.2 Gy. Enamel dosimetry by ESR gave values of 11.0 Gy and 8.2 Gy from Russian and Canadian laboratories, respectively (Pass et al., 1997).

In another accidental exposure, the victim was exposed for 1–2 min at 0.5 m from a 3×10^7 GBq (0.81 MCi) ^{60}Co source used for sterilizing medical supplies that had not retracted properly. Dose calculations by Russian researchers based on simulation of the accident gave a dose of 12.5 Gy. Analysis of chromosome aberrations in cultured blood lymphocytes indicated a dose range of 9.6 to 11.7 Gy. The dose estimate from blood granulocyte kinetics was 9 to 11 Gy and ESR in clothing material gave an exposure range of 12 to 18 Gy. ESR studies of dental enamel by the Canadian laboratory gave a dose of 13.7 Gy. The victim was treated with supportive measures, transfusions, and hematopoietic growth factor (but no transplants) in a Moscow hematology ward (Baranov et al., 1994). The patient died 113 days following the accident from radiation pneumonitis infection secondary to diffuse and focal fibrosis of the lungs.

The yield of chromosome aberrations (dicentric and ring chromosomes) in cultured lymphocytes was used to provide an estimate of the equivalent whole-body dose to patients who had been extensively exposed to diagnostic x rays, fluoroscopies, and some computed tomography scans (Weber et al., 1995). Most of the procedures were performed between one and several years prior to the blood sampling. None of the subjects had radiation therapy. The equivalent whole-body dose was the dose which, if received homogeneously by the whole body, would produce the same yield of chromosome aberrations as the dose actually absorbed over a period of time. It was calculated from a dose-effect curve for dicentrics obtained in vitro with ^{60}Co gamma rays, $Y = \alpha D + \beta D^2$, with $\alpha = 0.0168$ and $\beta = 0.0583$. D is the dose (Gy). A background yield of dicentrics of 0.0005 per cell was used. Results, expressed in terms of the number of dicentrics/number of cells examined, and the equivalent whole-body dose following subtraction of the background were 11/500, 0.48 Gy; 5/95, 0.25 Gy; 16/1,615, 0.13 Gy; 11/1,042, 0.30 Gy; and 6/1,201, 0.17 Gy. A small number of cells in the pooled studies had more than one dicentric. One difficulty in the assignment of the whole-body dose was that the reference curve was obtained with ^{60}Co gamma rays, which have a much higher energy than the diagnostic x rays. In addition, some of the x-ray examinations employed iodized contrast media, which contributed a "dose-enhancement factor." The yields of chromosome aberrations from diagnostic x rays were much higher than in a study of workers in two German nuclear energy plants (0.0028) and in a study of pilots and stewardesses (0.0024). It was pointed out that the subjects incurred doses much greater than those usually involved in diagnostic x rays. One weakness was that there were no control values for aberrations in the subjects prior to their radiation exposure. Thus there were a number of reasons to consider the equivalent whole-body doses as rough estimates, at best, of the real doses. Nevertheless, the authors emphasized that the finding of an unexpectedly high yield of chromosome aberrations supported the need to exercise caution in the prescription of x rays.

Another study was performed following the claims of a worker that the official dosimetry records substantially underestimated his actual dose (0.56 Sv, compared to his estimate of 2.5 Sv over 36 years). Results showed that the frequencies of chromosome translocations and glycophorin A (GPA) gene mutations, chosen as stable markers of accumulated dose, were significantly elevated when compared with those from unexposed controls, but that the worker's estimate of his dose seemed unlikely (Straume, 1993).

The assay for reciprocal translocations in lymphocytes in blood by chromosome painting is of particular interest in evaluating a history of occupational exposures, because reciprocal translocations are reported to be totally stable with time after exposure, in contrast to the behavior of dicentric chromosomes. Dose-response curves are obtained by exposing human blood samples to a range of doses. The results are applied directly to the evaluation of blood samples drawn from a subject exposed occupationally or from other sources. The curve has the same form as the curve given previously for dicentrics, a linear slope, with a coefficient α and a quadratic term with coefficient β. One published relationship, obtained by calibration with ^{60}Co gamma rays, is $Y = 0.023D + 0.053D^2$, where D is the dose in Gy and Y is the yield in translocations per cell. The background is 0.005 translocations per cell (Lucas et al., 1995).

Biodosimetry has many shortcomings—because of technical difficulties, variability in response to dose, and low sensitivity—but results can be improved by the use of more than one technique. It is useful in providing information on exposure when personnel dosimetry data are not available or in question or in complementing modeling and other methods of assessing a dose in the past.

6 SPECIFYING STATISTICAL VARIATIONS IN COUNTING RESULTS

If you are given a dozen apples to count, the chances are your answer will be 12 apples. If you repeat the count, your answer probably will be the same, and you will always come up with the same result unless you become fatigued or bored. On the other hand, when you repeat a measurement with a radiation counter, you will very likely come up with a different result. A count of 12 may be followed by 11, or 13, or 14, and by an occasional 6 or 18 or even larger discrepancies. These variations are due not to any inherent malfunction of the counter or its readout system but to the radiation and radiation-absorption process. Since the probability of penetration of the radiation and interaction in the detector is strictly the result of random processes, the variability of repeat counts with a properly functioning detector can be described with mathematical rigor, and a good estimate of its magnitude can be made, based solely on the magnitude of the count.

6.1 Nature of Counting Distributions

To appreciate the variability among many successive counts on the same sample, let us examine some actual data. The results of a hundred 10-second counts and of a hundred 100-second counts on the same sample are given in Table 4.3. If we should make an additional count on the sample, we would probably obtain one of the values presented in the table. There would be a possibility of obtaining a value outside the limits of the data obtained, but the probability of this would be very small.

Which values would we be most likely to find in repeat counts? We can find out by presenting the data in a manner that shows the frequency with which different values of the count were obtained. A histogram of the distributions for the data in Table 4.3 is presented in Figure 4.21. We see that the distribution may be characterized by a most probable value, and that the greater the deviation of any particular value from the most probable value, the less is the chance that it will be obtained.

Table 4.3 Results of repeat counts on a radioactive sample.

10-second counts					100-second counts				
7	9	11	13	9	120	92	108	107	117
13	13	9	11	11	98	146	117	112	92
11	6	11	15	15	131	86	90	109	112
8	9	13	12	19	119	123	88	101	112
18	11	9	12	9	127	104	85	111	99
12	10	10	12	10	114	118	133	127	107
11	17	8	15	11	96	115	118	113	120
15	24	7	8	14	119	109	97	110	94
12	11	15	10	8	114	123	114	123	97
14	11	10	4	11	121	80	98	108	126
8	8	9	13	8	97	131	97	105	125
15	11	6	11	17	93	120	112	115	118
14	14	14	8	12	130	121	111	110	114
8	10	10	9	15	114	101	117	109	122
12	12	9	14	19	113	108	106	128	122
9	14	6	6	13	100	90	126	111	94
11	11	8	14	10	115	104	119	105	102
10	10	10	14	7	103	98	105	120	108
7	9	11	13	11	116	123	130	109	110
14	10	15	12	10	107	112	122	109	131

4.21 Histogram of frequency distribution of 100 repeat counts for two counting periods.

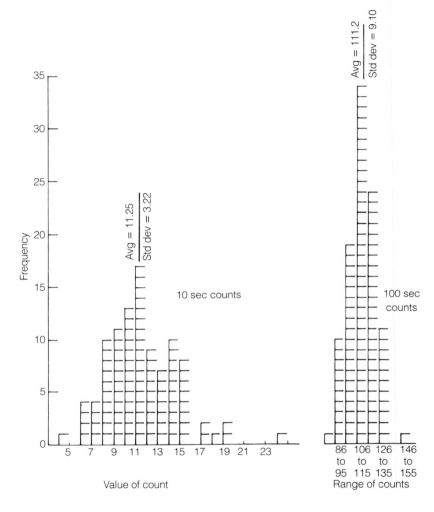

6.2 Sample Average and Confidence Limits

If we make a number of repeat measurements and the results differ, we ordinarily average the results to obtain a "best" value. The averages of the values of the counts presented in Table 4.3 are recorded in the histogram in Figure 4.21. As the number of repeat counts increases, the average of the values approaches a limit that we shall call the true mean. Ideally, it is this limit that we should like to determine when we make a radiation measurement. However, radiation counts generally consist of a single measurement over a limited time interval and provide only an estimate of the true mean for the sample. Accordingly, the result of a measurement should specify the degree of confidence in the estimate.

One way of expressing the degree of confidence is in terms of confidence limits. These are upper and lower values between which it is highly probable that the true mean lies. The greater the confidence interval, the greater the probability that the true mean lies within the limits. Referring again to the measurements in Table 4.3, let us determine within what ranges 95 percent of the counts lie. For the 10-second measurements, 95 percent of the values (that is, 95 determinations in this example) lie between 6 and 17 counts, or within 51 percent of the average of 11.25. For the 100-second measurements, 95 percent lie within 25 counts, or within 23 percent of the average of 111.2. We refer to the intervals containing 95 percent of the values as "95-percent confidence limits." Other confidence intervals for expressing the result of an investigation may be preferred. The limits the investigator chooses, however, are usually expressed as a multiple of a quantity known as the standard deviation.

6.3   Standard Deviation

The standard deviation is the most useful measure of the spread of a distribution of values about the average. It is defined as the square root of the arithmetic average of the squares of the deviations from the mean. The calculation of an unbiased estimate of the standard deviation of a frequency distribution from a sampling of observations proceeds as follows.

Call the number of measurements made n, call an individual measurement X, and call the sum of the measurements ΣX.

a.   Calculate the average value, $\bar{x}$, of the n measurements: $\bar{x} = \Sigma X / n$.
b.   Subtract each measurement from the mean, square the resulting difference, and sum the squares: $\Sigma(\bar{x} - X)^2$.
c.   Divide this result by one less than the number of measurements, to obtain the variance s^2: $s^2 = \Sigma(\bar{x} - X)^2/(n - 1)$.
d.   Take the square root of the variance to obtain the standard deviation, s: $s = \sqrt{[\Sigma(\bar{x} - X)^2/(n - 1)]}$.

Example 4.9   Calculate the standard deviation from the following sample of five readings: 8, 13, 7, 4, 9.

$n = 5$
$\bar{x} = 8.2$
$\Sigma(\bar{x} - X)^2 = 42.8$

$s^2 = 10.7$
$s = 3.271$

An alternate method is to enter the data into a scientific calculator and press the "s" key.

The calculation of confidence limits from the value of the standard deviation is based on the close resemblance between the shape of the distribution curve for successive radiation counts and the normal curve of error.

6.4 The Normal Error Curve—A Good Fit for Count Distributions

A set of measurements that are the result of numerous random contributing factors—and this includes most counting determinations—corresponds approximately to the normal error curve.[3] This curve may be drawn from knowledge of just two parameters—the average and the standard deviation. The distribution and its properties are illustrated in Figure 4.22. It is characteristic of the normal distribution that 68 percent of the values fall within one standard deviation from the mean, 95 percent within 1.96 standard deviations, and 99 percent within 2.58 standard deviations.

The histogram in Figure 4.21 for the repeated counts with an arithmetic mean of 111.2 can be fit nicely with a normal error curve. The histogram for the distribution with a mean of only 11.25 counts shows a distinct asymmetry, and some error is introduced in replacing it with the symmetrical normal curve. The asymmetry becomes more pronounced as the mean value decreases, because the accumulation of counts from a radioactive source results from random processes whose probability of occurrence is very small and constant. As a result, the variability in repeat counts follows the Poisson rather than the normal distribution.[4] While the Poisson distribution is asymmetrical for small numbers of observed events (that is, less than 16), it rapidly approaches the shape of the normal distribution as the number of events increases. It is a property of the Poisson distribution that the mean value is equal to the square of the standard deviation. Thus the normal curve fitted to the data can be drawn if only one parameter, the arithmetic mean, is given. The standard deviation is obtained simply by taking the square root of the mean. Referring to the mea-

---

3. Theoretically, repeat counts follow a Poisson distribution (see Table 4.5), but when more than just a few counts are accumulated—that is, more than 16—the normal distribution is a good approximation.

4. The Poisson distribution gives the probability of obtaining a count x when the mean is m, as $e^{-m}m^x/x!$ (Evans, 1955, chap. 26).

4.22 Properties of the normal distribution. The abscissa y is equal to $(x - \mu/\sigma)$, where x is the value of a specific measurement, μ is the limit approached by the average value as the number of measurements increases, and σ is the limit approached by the standard deviation as the number of measurements increases. The area under the probability-density function between any two points, y_1 and y_2, gives the probability that a measured value will fall between y_1 and y_2. The area under the tails of the curve, bounded by $y = \pm 1.96$, and the area under the curve to the right of $y = 1.645$ are both 5 percent of the total area under the curve. The value of the ordinate at any point y_1 on the cumulative distribution curve gives the probability that the value of a measurement will be less than y_1. The difference between values of the ordinates at y_1 and y_2 on the cumulative distribution curve gives the probability that a measured value will fall between y_1 and y_2 and, therefore, the area under the probability-density function between y_1 and y_2.

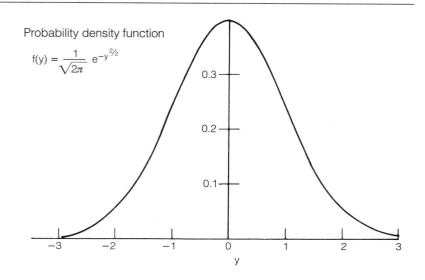

Probability density function

$$f(y) = \frac{1}{\sqrt{2\pi}} e^{-y^2/2}$$

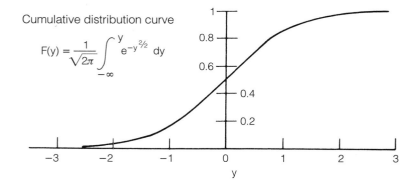

Cumulative distribution curve

$$F(y) = \frac{1}{\sqrt{2\pi}} \int_{-\infty}^{y} e^{-y^2/2} \, dy$$

surements presented in Figure 4.21, the standard deviation for the 10-second measurements as calculated for a normal distribution with a mean of 11.25 is 3.35, and the standard deviation for the 100-second measurements as calculated for a normal distribution with a mean of 111.2 is 10.5. These values are in good agreement with the values of 3.22 and 9.10 as calculated for the series of 100 measurements.

From the theory of the normal distribution, we expect 32 percent of the values to differ from the mean by more than one standard deviation. This expectation may be compared to the actual numbers found, which were 23 and 37 out of 100 determinations for the 10- and 100-second measurements, respectively.

6.5 Precision of a Single Radiation Measurement

In previous sections, we were concerned primarily with the properties of repeat determinations of radiation counts. Now we come to the problem of presenting and interpreting actual results. First let us consider the meaning of a single measurement. The number of counts determined in the measurement may be many thousands, if the sample is fairly active, or it may be only a few counts, such as that obtained in low-level counting. What is the significance of the number?

First, it is obviously the best estimate of the mean value we would obtain if we were to make many repeat determinations. It is best because we have no other measurements. Second, we know that the true mean is probably different from our measurement, and we can specify limits within which the true mean probably lies. The probability that the true mean lies within specified limits around the measured value is determined from the normal error curve, drawn with a mean value equal to the measured count (*N*) and standard deviation[5] (*s*) equal to the square root of the measured count: $s = \sqrt{N}$. There is a 68 percent probability that the true mean lies between $N + \sqrt{N}$ and $N - \sqrt{N}$. There is a 95 percent probability that the true mean lies between $N + 1.96\sqrt{N}$ and $N - 1.96\sqrt{N}$. There is a 99 percent probability that the true mean lies between $N + 2.58\sqrt{N}$ and $N - 2.58\sqrt{N}$.

Using statistical language, we may say: the 68 percent confidence interval is equal to the measured value $\pm$ 1*s;* the 95 percent confidence interval is equal to the measured value $\pm$ 1.96*s;* the 99 percent confidence interval is equal to the measured value $\pm$ 2.58*s.*

The spread of the limits we choose for the measurements will depend on the consequences of the error if the true sample mean should in fact fall outside the limits applied to the measurement, and the judgment of the investigator must determine what type of confidence limits should be chosen.

In practice, the 95 percent confidence level is usually chosen; that is, most investigators will set these limits so there is a 95 percent chance that the average is within the limits applied to the estimate. In very critical situations, 99 percent limits may be chosen; in others they may be set as low as 50 percent.

---

5. Common practice is to use the symbol σ to represent the true value of the standard deviation of a distribution and *s* to represent the estimate of the standard deviation as determined from the sample data.

Table 4.4 Limits of confidence intervals.

	Limits of confidence intervals expressed as counts and as fraction of measured value			
	68% C.I.		95% C. I.	
Measured counts, N	$\sqrt{N}$	$\sqrt{N}/N$	$1.96\sqrt{N}$	$1.96\sqrt{N}/N$
20	4.47	0.224	8.77	0.438
60	7.75	0.129	15.2	0.253
100	10.0	0.100	19.6	0.196
600	24.5	0.0408	48.0	0.0800
1,000	31.6	0.0316	62.0	0.0620
6,000	77.5	0.0129	152.0	0.0253
10,000	100.0	0.0100	196.0	0.0196

In Table 4.4, we show the 68 and 95 percent confidence intervals for various accumulated numbers of counts. We present the limits not only as absolute numbers of counts but as fractions of the measured values. Note how the fractional limits decrease as the number of accumulated counts increases.

When the number of counts accumulated falls to a very small number, say, less than 16, the distributions no longer follow the normal curve accurately and the analysis is based on the Poisson distribution. Some results are presented in Table 4.5.

It is important to reemphasize that the inferences of confidence intervals and standard deviations depend on the random nature of the counts. The counts do not have to come from a single nuclide or a single source. They may come from any number of radiation sources, including background, but they must not include nonrandom events. Such nonrandom events might come, for example, from electrical interference, such as operation of a welding machine in the vicinity of the counter; interference from a radar transmitter; and high-voltage discharges through the insulation of the detector.

6.6 The Effect of Background on the Precision of Radiation Measurements

In the preceding section we discussed the estimate of the true mean from a single measurement. When the measurement also includes a significant contribution from the radiation background, the background value must be subtracted. Values of repeated background determinations are also random and follow a normal distribution. As a result, the

Table 4.5 Probable limits (95% confidence level) of true mean based on observed count for low numbers of counts (Poisson distribution).

Observed count	Lower limit	Upper limit
0	0.000	3.69
1	0.0253	5.57
2	0.242	7.22
3	0.619	8.77
4	1.09	10.24
5	1.62	11.67
6	2.20	13.06
7	2.81	14.42
8	3.45	15.76
9	4.12	17.08
10	4.80	18.39
11	5.49	19.68
12	6.20	20.96
13	6.92	22.23
14	7.65	23.49
15	8.40	24.74
16	9.15	25.98

Source: Pearson and Hartley, 1966.

difference between sample and background values follows a normal distribution with true mean equal to the difference of true sample and background means, but with variability greater than the variability of sample or background counts alone. If the counting times of sample and background are equal, then the best estimate of the standard deviation of the difference is given by the square root of the sum of the two measurements, except when they consist of only a few counts. The statistics for differences of small numbers of counts depart somewhat from the normal distribution and have been analyzed in detail (Sterlinski, 1969).

Example 4.10   A sample counted for 10 min gives a measurement of 650 c. The background, also counted for a 10 min period, is 380 c. Give the net sample count rate and a measure of its precision.

The net count is $650 - 380$, or 270 in 10 min. The standard deviation of the difference is $\sqrt{(650 + 380)}$ or 32.1, and $1.96s = 62.9$.

The difference may be expressed as 270 ± 63, at the 95 percent confidence level. The count rate may be expressed as 27 ± 6.3 c/min, at the 95 percent confidence level.

When the counting times for the sample and background are different, the net count rate would be

$$R = \frac{S}{t_S} - \frac{B}{t_B}$$

S and B are total sample and background counts accumulated in times t_S and t_B, respectively. The standard deviation of the net count rate is estimated by

$$s = \sqrt{\left(\frac{S}{t_S^2} + \frac{B}{t_B^2} \right)}$$

Example 4.11 A sample counted for 10 min yields 3,300 c. A 1 min background count gives 45 c. Find the net counting rate of the sample and a measure of its precision.

The net counting rate is 330 − 45, or 285 c/min. The standard deviation is

$$\sqrt{\left(\frac{3,300}{10^2} + \frac{45}{1^2} \right)} = \sqrt{(33 + 45)} = 8.83 \qquad 1.96s = 17.3$$

As in example 4.10, the count rate may be expressed as 285 ± 17.3 c/min at the 95 percent confidence level.

When a finite time is available for counting both sample and background, statistical analysis provides a means of determining the most efficient way of partitioning that time. If R_S is the counting rate of the sample and R_B is that of the background, then

$$\frac{t_S}{t_B} = \sqrt{\frac{R_S}{R_B}}$$

Example 4.12 A 1 hr period is available for counting a sample which is approximately twice as active as the background. How can the counting time be best divided between sample and background?

$$\frac{t_S}{t_B} = \sqrt{\left(\frac{2R_B}{R_B} \right)} = \sqrt{2} \qquad t_S = 1.4\, t_B$$

The sample counting time should be 1.4 times the background counting time. Thus, $1.4t_B + t_B = 60$, $t_B = 25$ min. The background is counted for 25 min and the sample for 35 min.

6.7   The Precision of the Ratio of Two Measurements

Instead of calculating the difference between two measurements, it may be necessary to determine the ratio of one measurement to another, where each measurement is given with a degree of precision. For example, in medical uptake studies, a ratio is obtained of the activity retained in the patient to the activity administered to him. The standard deviation of the ratio is obtained by

$$\frac{M_1}{M_2} \sqrt{\left(\frac{s_1^2}{M_1^2} + \frac{s_2^2}{M_2^2} \right)}$$

where M_1 is the value of the measurement of the patient, with estimated standard deviation s_1, and M_2 is the value of the measurement of the reference activity with standard deviation s_2.

Example 4.13   The following data were obtained in a thyroid uptake study. The background measurement was 3,200 c in 10 min or 320 c/min

	Patient	*Reference*
Gross counts	4,200	7,200
Counting time (min)	5	5
Gross c/min	840	1,440
Net (gross − background) c/min	520	1,120

Calculate the percent uptake and the 95 percent confidence interval.

$$\text{For the patient, } s_{net} = \sqrt{\left(\frac{4{,}200}{(5)^2} + \frac{3{,}200}{(10)^2} \right)} = 14.1$$

$$\text{For the reference, } s_{net} = \sqrt{\left(\frac{7{,}200}{(5)^2} + \frac{3{,}200}{(10)^2} \right)} = 17.9$$

$$s_{ratio} = \frac{520}{1{,}120} \sqrt{\left(\frac{200}{(520)^2} + \frac{320}{(1{,}120)^2} \right)} = 1.46 \times 10^{-2}$$

Percent uptake is $(520/1{,}120)(100) = 46.4$ percent, and with 95-percent confidence limits it is $46.4 \pm 1.96 \times 1.46 = 46.4 \pm 2.86$.

6.8   Testing the Distribution of a Series of Counts—The Chi-Square Test

Counting statistics are generally based on a single count under the assumption that the counts follow a Poisson distribution and the standard deviation can be estimated as the square root of the count. This is much simpler than the use of multiple counts to estimate the standard deviation. However, external influences can affect the distribution of counts and it is prudent to take a number of counts and apply statistical methodology to the results. The standard deviation of multiple counts can be compared with the standard deviation from the square root of a single count as one check. A more rigorous approach is to compare the distribution of the repeat measurements (x_i) with a true Poisson distribution through the chi-square (χ^2) test:

$$\chi^2 = \sum_{i-1}^{n} \frac{(x_i - \bar{x})^2}{\bar{x}}$$

(Evans, 1955, 1963). The chi-square statistic is commonly used in statistical tests and its integrals are plotted in Figure 4.23 as a function of the *degrees of freedom, F.* In this case, $F = n - 1$, where n is the number of measurements. The number of measurements are reduced by 1 because they are used to determine one parameter, the sample mean, in the summation given. A series of measurements is considered to be suspect if the probability *(P)* of χ^2 is too high or too low relative to designated values. For example, we may say that if P lies between 0. 1 and 0.9, the system that generated the measurements would be accepted as working normally—that is, the distribution of the observed measurements is not very different from a Poisson distribution—while if P is less than 0.02 or more than 0.98, the system performance is suspect.

4.23 Integrals of the χ^2 distribution (shown in inset), where P is the probability that a χ^2 statistic is higher (as shown here) than the indicated reference value (or, if shading appeared at the lower tail of the distribution, the probability that a χ^2 statistic is lower than the reference value) and F is the number of degrees of freedom: $dPd\chi^2 = [(\chi^2)^{(F-2)/2}/2^{F/2} (F/2)]e^{-x^2/2}$ (Evans, 1955, 1963).

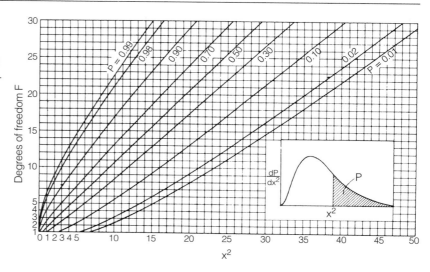

Example 4.14   Test the first 5 readings in the third column of Table 4.1 for conformance with the Poisson distribution. The numbers are 11, 9, 11, 13, 9.

$$\bar{x} = 10.6 \quad \Sigma(x_i - \bar{x})^2 = 11.2 \quad \chi^2 = 11.2/10.6 = 1.06 \quad F = 4$$

According to Figure 4.24, where $F = 4$ and $\chi^2 = 1$, $P = 0.9$. The data look somewhat more uniform than expected, but meet the criteria for acceptance. The next 5 numbers are 10, 8, 7, 15, 10.

$$\bar{x} = 10 \quad \Sigma(x_i - \bar{x})^2 = 38 \quad \chi^2 = 3.8 \quad F = 4$$

P is between 0.3 and 0.5. This indicates that the system is performing satisfactorily.

6.9   Measurements at the Limits of Sensitivity of Detectors

There are times when it is necessary to make measurements at the lowest levels possible with a given detector. In theory, one could get the degree of precision desired by accumulating as many counts as necessary, but at low counting rates the length of time needed would be impractical. Also, the longer the counting time, the greater the possibility of extraneous and nonrandom effects on the detector counts, both of which could nullify the advantages of longer counting times.

How do we determine the lowest practical limits of sensitivity of a detector and how do we express and interpret results obtained near these lowest limits? Obviously, the limits are determined by the magnitude of the background counts and their variability, and by the precision with which it is possible to accumulate data with the detector. To define the problem in concrete terms, consider a hypothetical measurement of a very low level of activity. Let us assume that in a 10 hr period, we accumulate 13 counts from the sample. A background count for the same counting period yields 10 counts (see Donn and Wolke, 1977, for treatment of different sample and background counting periods).

We are interested in evaluating the significance of the two measurements: the sample reading of 13 and the background reading of 10. Because of the variability in repeat determinations, we may not conclude that the sample radioactivity has a net value of 3 counts in 10 hr. How can we present and interpret the results?

1. We can assign upper and lower limits to the true values of sample and background measurements, and hence estimate upper and lower limits for the net sample count. Thus, the upper estimate of the true sample count (background included), using 95 percent confidence limits, is $13 + 1.96\sqrt{13}$. A lower limit to the true background count is $10 - 1.96\sqrt{10}$. We

may be tempted to conclude that the maximum value of the net counts could be as high as $20.1 - 3.8$ or 16.3 counts. However, it is a gross overestimate, from a statistical point of view, to report that the sample activity may be as high as 16 counts on the basis of a measured difference of 3 counts.

2. We can present the difference with confidence limits. If we assume the normal distribution still holds, even at these low levels, we can say the standard deviation of the difference is $\sqrt{(10 + 13)} = \sqrt{23}$ counts, and, at the 95 percent confidence level, the difference is $3 \pm 1.96\sqrt{23}$. This means that we are estimating the difference between sample and background as being between -6.4 and 12.4 and expect that 5 percent of the time the true difference will fall outside these limits. On the basis of this analysis, the precision of the difference is too poor to accept.

To this point, we have not used the information that, if the instrumentation is working properly, the difference between true sample and background means cannot be less than zero. With this restriction, only the upper limit of the difference is tested (one-tailed test) rather than both upper and lower limits (two-tailed test). If we select the probability of exceeding the upper limit as being no greater than 5 percent, the upper limit is $3 + 1.645\sqrt{23}$, or 10.9 counts. The value of 1.645 is chosen from the curve in Figure 4.22 so that the area under the curve to the right of $y' = 1.645$ is 0.05. Limits for other probability values are readily obtained from Figure 4.22.

3. We can use the results to test certain hypotheses, such as that the true mean count of the sample is no different from the background or that the true mean count of the sample is no greater than the background. Because this method uses more information than the first, it provides a more sensitive test. However, it is based on the assumption that the mean value of the background does not change.

4. We can compare the measurement with a minimum count specified (in advance) as significant. After the minimum significant count is specified, we can also determine the minimum true count that would give a positive reading with significant probability, say at the 95 percent confidence level.

Let us examine how we apply these statistical concepts to our counting results.

6.9.1   Test of the Hypothesis That the Sample Activity Is No Different from the Background

The measured net count is $13 - 10 = 3$. If the net difference of the true sample and background mean values were in fact 0, repeat net counts would be distributed about 0 with an estimated standard deviation of $\sqrt{(13 + 10)} = 4.8$.

The chance of obtaining net counts different from 0 as a function of the estimated number of standard deviations has already been discussed in section 6.5. At the 95 percent confidence level, a difference of 1.96 standard deviations, or 9.4 counts, will be needed to establish a significant difference. Since this is greater than the measured difference of 3 counts, the measured difference is not significant.

6.9.2   Test of the Hypothesis That the Sample Activity Is Not Greater Than the Background

If we feel that the mean background rate did not change during the measurements and, therefore, that the mean sample count must be at least equal to or greater than the mean background count, we may use the more sensitive one-tailed test of significance. Values of multiples of standard deviations for different confidence levels, as derived from the curves in Figure 4.22, are given in Table 4.6. The significant difference at the 95 percent confidence level is 1.645 standard deviations, or 7.9 counts. The smaller difference calculated on the basis of this hypothesis is still greater than the measured difference, which is therefore not considered significant. If there is no significant difference between the sample count and background count, then we would conclude that the sample that was measured is not radioactive.

6.9.3   Calculation of the Minimum Significant Difference between Sample and Background Counts

Rather than test the validity of individual measurements, we can calculate the minimum count above a given background count that could be considered significant. The net sample count that is significant must be

Table 4.6   Limits at various confidence levels in one-tailed tests of significance.

Confidence level (%)	α or β	Limit in terms of standard deviations (k_α or k_β)
99	0.01	2.326
95	0.05	1.645
90	0.10	1.282
84	0.16	1.000
75	0.25	0.674

equal to or greater than a specified multiple, k_α, of the estimated standard deviation of the difference between sample and background.

$$S - B \geq k_\alpha \sqrt{S + B} \qquad (4.2)$$

where S = counts measured from a sample (including background), k_α = factor associated with the assigned confidence level, and B = background counts. We rewrite equation 4.2 to allow solution in terms of $S - B$.

$$S - B \geq k_\alpha \sqrt{S - B + 2B} \qquad (4.3)$$

For the lowest value of $S - B$, we solve equation 4.3 as an equality and obtain

$$(S - B)_{min} = \frac{1}{2}\left(k_\alpha^2 + \sqrt{k_\alpha^4 + 8k_\alpha^2 B} \right) \qquad (4.4)$$

For a 95 percent confidence level

$$(S - B)_{min} = \frac{1}{2}(1.645^2 + \sqrt{(1.645)^4 + 8(1.645)^2 (10)} = 8.8$$

Thus, for a measured background count of 10, it would take a net difference of 9 counts for the sample count to be considered significantly different from the background count. The difference of 9 counts is significant at the 95 percent confidence level. A smaller difference would be significant if we used a smaller confidence level—that is, if we were willing to accept a higher probability of making the error of assuming that the counts were different when in fact they were the same. Thus, we could consider a difference of 1 standard deviation as significant—this would constitute a test at the 84 percent confidence level. On the average, in 16 cases out of 100 we would be saying a difference existed when in fact it did not. We tend to accept higher probabilities of making the wrong conclusion when it is important to accept a difference, if there is a chance that it exists.

6.9.4 On the Probability That a Given True Source Plus Background Count Will Provide a Positive Result

The elements involved in assigning confidence levels to counting results are illustrated in Figure 4.24. Two counting distributions are shown, one for a true background count of 10 and one for a true sample count (source + background) of 19 for the same counting time. The true counts were obtained from long counting times to give an accurate estimate of the true count during the actual counting time for which the values are given.

Normal curves of error are drawn with standard deviations estimated as the square root of the number of counts. The values of the ordinates are taken from tables of the normal curve, which give the ordinate as a func-

4.24 Frequency distributions of sample and background counts.

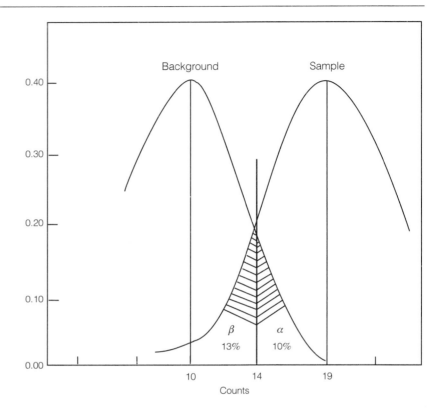

tion of the number of standard deviations from the mean. Once the curve is drawn, the probability of a count between any two values on the abscissa is given by the area under the curve. The area is also given in probability tables.

Example 4.15   An accurate measurement of the background count of a detector for a given counting time is 10 counts. What is the probability of error in detecting a source with a true count of 9 in the same counting time if a count of 14 is chosen as the boundary between calling the sample radioactive or attributing the count to background? Refer to Figure 4.24.

(*a*) The true mean count of the sample is 9 + 10 = 19. Determine the ordinates of the normal error curve for a true mean count of 10 at 10 counts and 14 counts, and the area under the curve between 10 and 14 counts.

The standard deviation, taken as the square root of 10, is 3.162 counts. Deviations from the mean, in terms of the standard deviation, are 0 for the 10 counts and $(14 - 10)/3.162 = 1.266$ at 14 counts. The ordinates (given, for example, in the *Handbook of Chemistry and Physics*) are 0.3989 and 0.1791 (by linear interpolation), respectively. The area under the curve between the mean and 1.266 standard deviations above or below the mean is given as 0.3972.

(b) Determine the probability that a single background count will be greater than 14.

From the symmetry of the normal curve, the area between ± 1.266 standard deviations is 2×0.3972 or 0.7944. The area under the two tails is $1 - 0.7944 = 0.2056$. Therefore the area in the tail to the right of 14 counts is $0.2056/2 = 0.1028$. There is thus a 10 percent probability that a count will exceed 14 if the background is 10.

(c) Determine the ordinates of the normal error curve for a true mean count of 19 at 19 counts and 14 counts, and the area under the curve between 19 and 14 counts.

The standard deviation, taken as the square root of 19, is 4.359 counts. Deviations from the mean in terms of the standard deviation are 0 for the 19 counts and $(19 - 14)/4.359 = 1.147$ at 14 counts. The ordinates (given, for example, in the *Handbook of Chemistry and Physics*) are 0.3989 and 0.2066 (by linear interpolation), respectively. The area under the curve between the mean and 1.147 standard deviations above or below the mean is given as 0.3743.

(d) Determine the probability that a single sample (source + background) count will be less than 14 and hence the sample considered background (in error).

From the symmetry of the normal curve, the area between ± 1.147 standard deviations is 2×0.3743 or 0.7486. The area under the two tails is $1 - 0.7486 = 0.2514$. Therefore the area in the tail to the left of 14 counts is $0.2514/2 = 0.1257$. There is thus a 13 percent probability that a count will be less than 14.

Thus, choosing a detection level of 14 counts for a true sample count (source + background) of 19 would leave a 10 percent probability that a background count would be considered positive (false positive) and a 13 percent probability that a sample count would be considered negative (false negative). If smaller errors were desired at a 10-count background, a higher detection count (and longer counting time) would be required. The empirical procedure employed here could be used to explore statistical errors associated with other true sample counts and the associated detection limit.

The analytical approach to the problem of sensitivity limits in counting is to derive an equation for the minimum true source count that will yield a significant sample count under the two error constraints considered above, designated in statistical parlance as type 1 (α) and type 2 (β) errors (Altshuler and Pasternack, 1963. The α error is the fraction of times a count is accepted as significant when in fact it is only due to background. The β error is the error in classifying a count as indicative of the absence of a radioactive source when in fact a radioactive source is present.

The analyses to this point were based on the assumption that the counts followed a normal distribution, which is an acceptable assumption for background counts as low as 4. The situation for counts below 4 has been studied by Sterlinski (1969). He concluded that the significant difference in counts of the sample and background taken in the same time interval (at the 0.01 level of significance) was 6 for average values of the background count between 2.96 and 4. Other values of significant differences for corresponding ranges of the mean background count (in parentheses) were: 5 (2.0–2.96), 4 (1.18–2.0), 3 (0.56–1.18), 2 (0.18–0.56), and 1 (0–0.18).

7   Comments on Making Accurate Measurements

The successful acquisition of experimental data does not follow automatically from good technical training and sophisticated measuring equipment. There are less tangible requirements revolving about the attitudes and habits of the experimenter or operator—the exercise of great care, preoccupation with accuracy, concern for details, use of controls, and so on. These habits, of course, all should have been developed during the technical education of the worker, and there should be no need to mention them here except that their breach is so often the cause of invalid results. Accordingly, we conclude this part with a list of practices for meaningful radiation measurements.

1. Test the detection instruments frequently with a test source to determine whether they are responding accurately and reproducibly.
2. Repeat measurements wherever possible to verify reproducibility of technique.
3. Make measurements of lengths of distances as accurately as possible. The ruler is used frequently in radiation measurements, and

in spite of its simplicity, it is often used incorrectly. Only a small error in reading a ruler or positioning a detector often will give a large error in the results obtained.

4. When using instruments in the field, check to see that the reading is not dependent on the orientation in space. If the reading depends on how the instrument is positioned, and this fault cannot be corrected, make sure that the instrument is always positioned the same way.

5. Ascertain the effect on the instrumentation of environmental factors, such as temperature, humidity, and atmospheric electricity.

6. Be sure the right instrument is used for the quantity being measured. (Do not use a thick-walled detector to measure alpha particles.)

7. Always examine instruments and equipment for modifications in the original construction that may have been introduced by previous users. This is particularly important if several users are sharing the same equipment.

In summary, always execute measurements with great care and never take the performance of the instrumentation for granted. Continually check operation of the measuring equipment with test equipment or test runs and scrutinize the results critically for consistency and reasonableness.

PART FIVE

Practical Aspects of the Use of Radionuclides

Users of sources of radiation find their lives complicated by governmental and institutional regulations, legal requirements, inspections, committee reviews, record keeping, public relations, and so on. All result from the efforts of society to protect the worker, the public, and the environment, to prevent radiation accidents, and to institute proper corrective action if they occur. The extent of the regulations to which a radiation user is committed is generally but not always commensurate with the degree of hazard involved.

The technical and administrative measures are relatively innocuous for users of radionuclides in tracer amounts in the research laboratory. They become more stringent in the medical use of radiation, with the complexity increasing as the doses increase from radiation for diagnostic purposes to radiation in medical therapy. Special consideration must be given to the administration of radionuclides to volunteers in medical research connected with the development of new drugs. The most complex and costly regulations are promulgated for operators of nuclear power plants or nuclear fuel processing facilities. Obviously the control of strong radiation sources that can significantly affect the worker, the patient, the public or the environment must be in the hands of highly trained professionals. Here we shall limit ourselves to discussing the responsibilities and working procedures of interest to individuals who work with radioactive chemicals in support of their normal routines in research or medicine. Material on the protection of individuals who work with radiation machines and sealed radionuclide sources is given in Part Two, sections 31 and 32. Material relating to regulations for the protection of the public from radiation sources, whether as patients or as users of consumer products, is given in Part Two, section 34. Material on the impact of the use of radioactive materials on the public health is given in Part Six for ionizing particles and

Part Seven for nonionizing particles. More technical references developed primarily for specialists in radiation protection are listed in the selective bibliography.

Our approach will be to consider, in turn, the various items that are likely to be of concern to an individual from the time he decides to use radionuclides to the time he terminates such work. Our presentation will be consistent with the standards for protection against radiation of the U.S. Nuclear Regulatory Commission as promulgated in the Code of Federal Regulations, Title 10, Part 20 (NRC, 1991), as they form the basis of most regulations pertaining to the use of radioactive materials in the United States. While this part is designed to familiarize readers with the regulatory process and help them to understand the regulations, it is by no means meant to be a comprehensive treatment of radiation regulations. The reader will have to become familiar with whatever current regulations apply to a specific use of radioactive materials.

Since the NRC regulations still express activity in terms of the traditional units (microcuries, millicuries, etc.), these units will generally be used along with the SI units in citations from the regulations in this part.

1 OBTAINING AUTHORIZATION TO USE RADIONUCLIDES

The chances are that an individual initiating a program of work with radionuclides will have to obtain authorization from a government agency on the federal or state level, either the U.S. Nuclear Regulatory Commission (NRC) or a state health department. Anyone working at an institution will need to obtain this authorization through the institution. It is instructive to consider some features of the licensing of the use of radionuclides by the U.S. Nuclear Regulatory Commission.[1]

---

1. Federal regulations pertaining to radiation control are issued in the Code of Federal Regulations. The regulations of the Nuclear Regulatory Commission are issued as Title 10 of the Code (Energy). They may be accessed at the NRC website, *www.nrc.gov*. The most important sections for users of radionuclides are: *Standards for Protection against Radiation*, Part 20 (10CFR20); *Rules of General Applicability to Licensing of Byproduct Material*, Part 30 (10CFR30); *Human Uses of Byproduct Material*, Part 35 (10CFR35); and Notices, Instructions, and Reports to Workers; Inspections (10CFR19). Users of exempt quantities are exempt from the regulations in parts 30–34 and 39 (10CFR30.18a) but not from the regulations in part 20. Material obtained under this exemption may not be used for commercial distribution.

The NRC also issues Regulatory Guides that describe methods acceptable to its regulatory staff for implementing specific parts of the commission's regulations. Regulatory Guides are useful in preparing the applications and complying with regulations. Some of the available guides are listed in the Selected Bibliography. They may be obtained upon request from

The commission is authorized by Congress through the Atomic Energy Act of 1954 as amended to regulate the distribution of all radionuclides produced in nuclear reactors.[2] Under this authorization, it has decreed that reactor-produced radionuclides (referred to as by-product material) can be obtained and transferred only under specific or general licenses,[3] except for certain exempt items, concentrations, and quantities. Some exempt quantities and concentrations are listed in Table 5.1. These exempt quantities are still covered under a general license and the commission has the authority to regulate their use. The exemption applies only to the need to make actual application for a license.

A specific license is issued to a named individual, corporation, institution, or government agency. Applicants must fill out a special material license form, NRC Form 313, which requests information on the radionuclides to be obtained, the intended use, maximum quantities to be possessed at any one time, training and experience of the person responsible for the radiation safety program and the individual users, radiation detection instruments available, personnel monitoring procedures to be used (including bioassay), laboratory facilities, handling equipment, and waste management procedures. In addition, a description of the radiation safety program is required. The license authorizes possession and use only at the locations specified in the application, and radioactive material may not be transferred except as authorized by the regulations and the license.

Compliance with the terms of the NRC regulations and the conditions of the license is the responsibility of the management of the institution. The radiation safety program must include, to the extent practicable, procedures and engineering controls to ensure that doses to workers and members of the public are kept as low as reasonably achievable (ALARA). The management must review the content and implementation of the radiation-protection program at least annually (10CFR20.1101).

The radiation safety program is implemented by the Radiation Safety Officer (RSO). The duties of the RSO include preparing regulations, providing advice and training on matters of radiation protection, maintaining a system of accountability for all radioactive material from procurement to

---

the Nuclear Regulatory Commission, Washington, DC 20555; Attention: Director, Division of Document Control.

2. Radionuclides not produced in a reactor (such as radium), as well as cyclotron-produced nuclides, are not under the control of the NRC. However, if a user is working with both reactor-produced nuclides and other sources of radiation, the NRC assumes the authority to review all radiation work under the reasoning that it has the responsibility to determine that the user has complied with exposure standards for the use of reactor-produced isotopes *when added to* exposure from all other radiation sources.

3. Title 10, parts 30, 31, and 35 of the Code of Federal Regulations relate to licensing of users of radioactivity. The basic standards of radiation control for licensed users are given in part 20.

Table 5.1   Quantities and concentrations exempt from a specific byproduct material license.

Radionuclide	(a) Quantity (μCi)	(b) Concentration in liquids and solids (μCi/1)
Calcium-45	10	0.09
Carbon-14	100	8
Cesium-137	10	—
Chlorine-36	10	—
Chromium-51	1,000	20
Cobalt-60	1	0.5
Copper-64	100	3
Gold-198	100	0.5
Hydrogen-3	1,000	30
Iodine-123	10	
Iodine-125	1	—
Iodine-131	1	0.02
Iron-55	100	8
Iron-59	10	0.6
Molybdenum-99	100	2
Phosphorous-32	10	0.2
Potassium-42	10	3
Sodium-24	10	2
Strontium-90	0.1	—
Sulfur-35	100	0.6
Zinc-65	10	1

Sources: 10CFR30.14, 30.70 Schedule A; 10CFR30.18, 30.71 Schedule B (November 1988). Multiply μCi by 37,000 to convert to Bq.

disposal, inspecting work spaces and handling procedures, determining personnel radiation exposures, monitoring environmental radiation levels, and instituting corrective action in the event of accidents or emergencies. The main points that should be covered in radiation surveys of research laboratories are listed in the survey checklist (Fig. 5.1).

Institutions engaging in research and meeting requirements on staffing, experience, facilities, and controls may obtain a license of broad scope (10CFR33,35). This license does not require the naming of the individual users and does not limit radionuclides to specific uses. It allows the institution to review applications from individual staff members and to authorize use of radionuclides to qualified individuals without special application to the regulatory agency for each individual user. Authority resides in a Radiation Safety Committee (RSC), composed of the Radiation Safety Officer, a representative of management, and persons trained and experienced in the safe use of radioactive materials. The RSC also establishes policies, pro-

5.1 Radiation survey check list.

Environmental Health and Safety
Radiation Protection Office

Survey Check List

Date _____ Location _____
Licensee _____ Surveyed by _____

Measurement
- Meter check for external dose rates and contamination of surfaces (hot sinks, hoods, storage areas, refrigerators).
- Wipe tests of all suspected contamination (sink ledge, hood ledge, bench top) plus check of "clean areas".
- Air sampling (where required).

Inspection
- NRC Form 3 posted.
- Institutional regulations posted.
- Proper signs (radiation area, radioactive material) posted.
- Storage and laboratory areas containing radionuclides controlled, posted, and secured.
- Waste disposal area controlled and posted.
- Hood flow satisfactory.
- Sources, waste solutions, etc., properly labeled and secured.
- Sink disposal records posted and up to date.

Monitoring Instrumentation
- Available.
- Performance check, calibrated.

Review of Working Procedures
- No evidence of eating, drinking, smoking in laboratory.
- Personnel monitoring devices worn (whole body and hands).
- Records of monitoring saved.
- No pipetting by mouth.
- Protective clothing utilized, including gloves, coats.

vides overall guidance and supervision of the radiation-protection program, and enforces compliance with the program. The ultimate responsibility for compliance rests with the institution's management. *The key to an effective program, an uncomplicated coexistence with regulatory agencies, and enjoyment of the confidence of the media and the public is a close oversight and "hands-on" involvement by the Radiation Safety Committee and the support of management in resources and in implementation.*[4]

4. The National Council on Radiation Protection and Measurements has issued a handbook on the organization of radiation-protection programs (NCRP, 1998, Report 127). Professional journals have much useful information on experience in radiation protection (Schiager et al., 1996; Classic and Vetter, 1999; Shapiro and Ring, 1999). Procedures acceptable to the Nuclear Regulatory Commission are given in regulatory guides, which are issued periodically. These and other sources on practices and procedures are given in the Selective Bibliography.

1.1   Administration of Radioactive Material to Humans

The NRC issues specific licenses to medical institutions for the use of radioactive material within the institution in the practice of medicine. Only the institution's management may apply and the use is limited to physicians named in the license. These physicians must satisfy the commission's requirements regarding training, experience, access to hospital facilities, and monitoring equipment. They may use only the specific radionuclides and perform only the clinical procedures specifically mentioned in the license. Physicians may apply directly for a license only if the application is for medical use and will be administered outside a medical institution. A single by-product material license is issued to cover an entire radioisotope program except teletherapy, nuclear-powered pacemakers, and irradiators. Requirements and procedures for obtaining a license are given in the Code of Federal Regulations, Title 10, Part 35 (10CFR35) and in NRC Regulatory Guide 10.8 (NRC, 1987a).

The institution must have a Radiation Safety Committee to evaluate all proposals for clinical research, diagnostic, and therapeutic uses of radionuclides within the institution and to ensure that licensed material is used safely and in compliance with NRC regulations and the institutional license. The physicians named on the institution's license cannot conduct their programs without the approval of the Radiation Safety Committee. Institutional licenses provide a means whereby nonapproved physicians under the supervision of physicians named on the license may obtain basic and clinical training and experience that may enable them to qualify as individual users.

The NRC also issues specific licenses of broad scope for medical use, which do not name individual users nor limit the use of radionuclides to specific procedures. Instead, individual users and proposed methods of use are authorized by the institution's Radiation Safety Committee. To qualify for this type of license, an institution must be engaged in medical research using radionuclides under a specific license and must satisfy NRC requirements on personnel, equipment, and facilities (see NRC Regulatory Guide 10.8)

The NRC also issues a general license to any physician, clinical laboratory, or hospital for use of small quantities of specific radionuclides in prepackaged units for *in vitro* clinical or laboratory tests (10CFR31.11). These include ^{125}I and ^{131}I (maximum 0.37 MBq, 10 μCi); ^{14}C (0.37 MBq, 10 μCi); ^{3}H (1.85 MBq, 50 μCi); ^{59}Fe (0.74 MBq, 20 μCi); ^{75}Se (0.37 MBq, 10 μCi); and mock-^{125}I reference or calibration sources in units not exceeding 1,850 Bq (0.05 μCi) ^{129}I and 18.5 Bq (0.0005 μCi) ^{241}Am. The use of these materials is also exempt from the requirements of part 20, except for mock-^{125}I, which is still subject to waste disposal restrictions (10CFR20.1301).

1.2 Requirements for Obtaining Approval to Use New Radioactive Drugs

A new radiopharmaceutical drug must be approved by the Food and Drug Administration (FDA) as safe and effective before it can be administered clinically to humans.[5] This authority of the FDA is limited to the approval of new drugs. The NRC retains jurisdiction in the areas of licensing for possession, laboratory safety, and routine administration of drugs to patients by physicians.[6]

The developer of a new drug must submit a Notice of Claimed Investigational Exemption for a New Drug (IND) to the FDA. The IND presents the protocol for the investigation to determine its safety and effectiveness. It describes the drug, including radiochemical and radionuclidic purity, and presents the results of all preclinical investigations, a protocol of the planned investigation, and the qualifications of the investigators. The preclinical investigations generally include studies of the drug's pharmacology, toxicity, and biodistribution, as well as the acquisition of sufficient animal data to establish reasonable safety. The IND also provides for notification of adverse effects and annual progress reports.

Certain studies may be done without the filing of an IND if they are conducted under the auspices and approval of an FDA-approved Radioactive Drug Research Committee (RDRC). These studies are limited to the use of radioactive drugs in human research subjects during the course of a research project intended to obtain basic information regarding the metabolism (including kinetics, distribution, and localization) of a radioactively labeled drug in humans or regarding its effects on human physiology, pathophysiology, or biochemistry. They are not intended to serve immediate therapeutic, diagnostic, or similar purposes, or to determine the safety and effectiveness of the drug in humans for such purposes, that is, to be used as a clinical trial. Certain basic research studies, such as studies to determine whether a drug accumulates in a particular organ or fluid space and to describe the kinetics, may have eventual therapeutic or diagnostic implications, but the actual studies are considered basic research prior to established use.

The RDRC must consist of at least five individuals, including a physi-

---

5. Regulations of the FDA are given in Title 21, Part 361 of the Code of Federal Regulations (21CFR361).

6. *Guide for the Preparation of Applications for Medical Use Programs*, NRC Regulatory Guide 10.8, Rev. 2 (NRC, 1987a).

cian recognized as a specialist in nuclear medicine, a person qualified by training and experience to formulate radioactive drugs, and a person with special competence in radiation safety and radiation dosimetry. The remainder of the members of the committee must be qualified in various disciplines pertinent to the field of nuclear medicine. Details on the composition of the committee and its functions are given in 21CFR361.1 (FDA, 1999).

1.3   Protection of the Patient in Nuclear Medicine

Radiation protection in the administration of radiopharmaceuticals to patients in clinical medicine is normally concerned with minimizing the dose to the patient while achieving the desired effect. Although the International Commission on Radiological Protection (ICRP) recommends that "no practice shall be adopted unless its introduction produces a positive net benefit" (referred to as justification), the professional judgment of the nuclear medicine physician and of the referring physician that a proposed use of radiation will be of net benefit to a patient is all that is normally required as justification of the patient's exposure. Therefore, referring physicians should be aware of the biological effects of ionizing radiation and of the doses and associated risks for the prescribed tests.

Nuclear medicine studies on volunteer patients, which do not benefit the individual on whom they are performed, play an important role in the advancement of medicine through research. Most carry negligible risk, but they can be implemented only with the approval of the Radiation Safety Committee after a review has been made of the doses and considerations of age and state of health of the individuals. Prospective subjects must be fully informed of the estimated risks of the irradiation and of the significance of the exposures with respect to regulatory limits and normal background limits so that they are competent to give their "free and informed consent."

The FDA sets limits on radiation doses to adults from a single study or cumulatively from a number of studies conducted within 1 year (FDA, 1999, 21CFR361.1). They are: 30 mSv (3 rem) for a single dose and 50 mSv (5 rem) for an annual and total dose commitment to the whole body, active blood-forming organs, lens of the eye, and gonads; 50 mSv (5 rem) for a single dose and 150 mSv (15 rem) for an annual and total dose commitment to other organs. These limits are ten times the limits allowed for research subjects under 18 years of age.

Example 5.1   A proposal is presented before a Radioactive Drug Research Committee to determine under an IND whether a tumor-target-ing protein tagged with indium-111 will locate and stick to cancer tissue in the body. Using data from biopsy samples and gamma camera images, the manufacturer of the radiopharmaceutical has supplied values for organ doses for an administered activity of 0.185 Bq (5 mCi).

Evaluation of the effective dose proceeds as follows (for regulatory purposes, the radiation weighting factors given by the NRC in 10CFR20 are used rather than the latest recommendations of ICRP/ NCRP presented in other parts):

Part of body	Organ dose, D (mGy)	Weighting factor, W_T	$W_T \times D$ (mSv)
Basic organs			
Gonads (male or female)	20	0.25	5.0
Breast	20	0.15	3.0
Red marrow	30	0.12	3.6
Lung	20	0.12	2.4
Thyroid	20	0.03	0.6
Bone surfaces	20	0.03	0.6
Remaining organs (maximum of 5)			
Liver	160	0.06	9.6
Spleen	120	0.06	7.2
Kidney	100	0.06	6.0
Additional organ (a)	20	0.06	1.2
Additional organ (b)	20	0.06	1.2
	Effective dose equivalent		40 mSv

Because the study is being done under an IND to determine safety and effectiveness for administration clinically to humans, it is not subject to dose restrictions for investigations processed through the RDRC. If the proposal were submitted to the RDRC as a study to obtain basic information, it would not be approved because the effective whole-body dose and three single organ doses are above acceptable limits for a single administration. All radioactive materials included in a drug, either as essential material or as a significant contaminant or impurity, are considered in determining the total radiation doses and dose commitments. Radiation doses from x-ray procedures that are part of the research study (that is, would not have occurred but for the study) are also included. The possibility of follow-up studies must be considered when doses are calculated.

The radiation exposure must be justified by the quality of the study being undertaken and the importance of the information it can be expected to provide. Requirements are also specified regarding qualifications of the investigator, proper licensure for handling radioactive materials, selection and consent of research subjects, quality of radioactive drugs used, research protocol design, reporting of adverse reactions, and approval by an appropriate Institutional Review Committee. The RDRC must also submit an annual report to the FDA on the research use of radioactive drugs.

The radioactive drug chosen for the study must have the combination of half-life, types of radiations, radiation energy, metabolism, chemical properties, and so on that results in the lowest dose to the whole body or specific organs from which it is possible to obtain the necessary information.

The RDRC must determine that radioactive materials for parenteral use are prepared in sterile and pyrogen-free form. Each female research subject of childbearing potential must state in writing that she is not pregnant or, on the basis of a pregnancy test, be confirmed as not pregnant before she may participate in any study.

1.3.1 Consent Forms for Potential Research Subjects in Studies Involving Exposure to Radiation

Much thought has been given to the content of the statement of radiation risk in consent forms for potential research subjects. Such statements must be in accordance with standards set by the Food and Drug Administration.

An evaluation by Institutional Review Boards at 14 large medical-research institutions of various methods of expressing radiation risk provides useful guidance in the preparation of such statements (Castronovo, 1993). The most popular statement of radiation risk compared subject radiation dose with background radiation levels. The most common quantity for expressing radiation dose was the *effective dose*. The lowest approval rating was for consent form statements that expressed risk in terms of a fractional increased risk of cancer at low doses, except for effective doses to the subject comparable to occupational dose limits.

One institution developed three expressions of risk depending on the dose. For effective doses up to 10 mSv, the dose was presented as a ratio to the dose from natural environmental radiation that the average person receives in the United States with the added statement that there was no evidence that this level of radiation will be harmful. For effective doses between 10 and 50 mSv, the dose was expressed as the ratio to the annual radiation exposure limit allowed for a radiation worker and the risk was considered to be comparable to other everyday risks. For effective doses greater than 50 mSv, the risk was expressed as the ratio to the annual exposure limit allowed for a radiation worker along with an estimate of the excess cancer risk as derived from recommendations of the BEIR committee of the National Research Council. The benefits to the patient from the exposure were also given.

2 TRAINING REQUIRED FOR WORKING WITH RADIONUCLIDES

NRC training requirements vary depending on the magnitude of the radiation hazards to users and whether the purposeful irradiation of human beings for medical purposes is involved. A basic training program for authorized users and other persons who supervise and are responsible for the safety of technicians and other users of radioactive materials should cover five major areas (which also form the basis for the organization of this book): *(a)* principles of radiation protection; *(b)* radiation calculations; *(c)* radiation measurements and monitoring; *(d)* practical aspects of the use of radionuclides, including proper working procedures and compliance with regulations; and *(e)* biological effects, radiation risks, and exposure limits. Medical programs should also include training in radiopharmaceutical chemistry (Exhibit 2, Supplement A of Regulatory Guide 10.8; NRC, 1987a). The NRC also provides guidance for developing a training program for technicians working with radioactive materials acceptable to it in applications for medical use programs (Appendix A of Regulatory Guide 10.8). Appropriate instruction is required before any staff member assumes duties with, or in the vicinity of, radioactive materials, and should be supplemented by annual refresher training. The topics covered in training courses for new workers should include applicable regulations and license conditions, areas where radioactive material is used or stored, potential hazards, radiation safety procedures, work rules, each individual's obligation to report unsafe conditions to the Radiation Safety Officer, emergency response, worker's right to personnel monitoring results, and availability of license documentation. In addition, the training session should include a question and answer period followed by an examination. Refresher training will emphasize different topics each year, depending on the findings of inspections, new regulatory requirements, and working practices that need special attention, such as security of radioactive material or controls in the receipt of radioactive packages. The NRC has also issued a regulatory guide for the training of personnel at nuclear power plants (NRC Regulatory Guide 8.27; NRC, 1981a).

Training and experience requirements set by the NRC for physicians desiring licenses authorizing the use of radionuclides in humans are much more stringent. The details are given in NRC rules and regulations pertaining to medical use of by-product material (10CFR35, Subpart J, 1988). The training required, including appropriate certification in nuclear medicine or radiology, depends on the types of studies the applicant wishes to pursue. Training for uptake, dilution, and excretion studies requires 40 hours of classroom and laboratory training in basic radionuclide handling techniques applicable to the use of prepared radiopharmaceuticals and 20 hours of clinical experience under the supervision of an authorized user. Training for imaging and localization studies requires 200 hours of classroom and laboratory training in basic radioisotope handling techniques applicable to the use of prepared radiopharmaceuticals, generators, and reagent kits and 500 hours of supervised clinical experience.

2.1 Implementation of a Training Program

No element of a radiation protection program is more critical to its success than the quality and effectiveness of the training in radiation safety given to the workers. No matter how comprehensive the monitoring and inspection program, how technically impressive the measurements and analyses, how generous the resources provided in staff and facilities, the goals of the program will not be met if the object of all this effort, the radiation worker, is not committed to being knowledgeable in the practice of radiation safety and to compliance with the institutional regulations. Training in radiation safety cannot be limited to presenting the requirements of the program and the information for meeting the requirements, but it must inculcate a spirit of cooperation and teamwork between the users and the staff of the radiation safety office. Thus, the introduction of the worker to radiation safety should be in a session presented personally by the radiation safety officer or another staff member qualified to teach the fundamentals of radiation safety and to lay the groundwork for a harmonious working relationship between the worker and the radiation safety staff. Following the introductory session on fundamentals, a variety of training methods can be used to impart additional information and skills. In a second session, survey personnel might simulate a laboratory inspection and come upon the kinds of infractions and violations that are found. This rivets the attention of the participants if the script is prepared with dramatic embellishments on finding violations and spiced with touches of humor. For example, evidence of eating in the laboratory might be the discovery of a candy wrapper in a laboratory waste basket, and potential contamination of food might be illustrated by having the inspector find a milk or cream carton in the refrigerator where radioactive materials are stored. The scenario might include an interchange between the inspector and a renowned investigator or a recalcitrant graduate student regarding the finding of a number of violations. The exercise presents an opportunity for using one's playwrighting skills. Alternatively, one can resort to self-study material, relying heavily on interactive computer programs. The basic training program should be followed up by bringing the attention of the worker to various aspects of working safely on a regular basis. The use of posters reminding workers of the need to attend to safety; the placement on bulletin boards in laboratories of signs displaying the main elements of the rules for effective radiation protection and the basic regulations; the promulgation of notices on incidents that caused injury or jeopardized the health of workers, in a format designed to gain the attention of the reader with bold headings, such as "Radiation Protection Bulletin" or "Lessons Learned from Radiation Incidents"—all serve as reminders to workers to practice radiation safety. In short, the main objective of the radiation program is not to create a large bureaucratic framework of services to spare the worker the need to devote time and energy to complying with safety requirements, but to recruit the worker as an active partner in producing a safe working environment and to provide the training to accomplish this.

Mentoring in the laboratory by experienced co-workers is a most effective way of giving practical training. Where a laboratory has compliance problems, participation by the radiation safety officer in one of the regular laboratory meetings can be an informal way of initiating discussion of compliance and corrective measures. The laboratory meetings also present a favorable environment for providing refresher training, which can be directed toward the specific work performed by the group.

2.2 Radiation Safety within a Comprehensive Institutional Program in Laboratory Safety

An institution's training program in radiation safety operates at maximum effectiveness and efficiency when it is treated as part of a comprehensive program in laboratory safety which recruits all management and technical personnel to work cooperatively toward the highest levels of safety and regulatory compliance. Training is not then limited to radiation users, but special presentations are designed to cover all classes of employees in the institution. At the operational level, radiation safety technicians are trained to recognize instances of unsafe conditions or working habits in a variety of disciplines in the laboratory, including biological, chemical, electrical, or fire safety, in addition to their duties with regard to radiation safety. Every member in an institution's safety organization is trained to perform initial responses to all classes of emergencies as prescribed by a comprehensive emergency management protocol and to bring qualified personnel to deal with them. Managers are informed about the policies, regulations, and requirements pertaining to all areas of occupational and environmental health and safety that apply at their institution and are aware of the roles and responsibilities of all the personnel who are involved with the program.

The control of work with radiation sources originally developed as a relatively autonomous function in the organization of departments responsible for occupational and environmental health and safety. Radiation was an esoteric hazard understood only by the professionals in the field, its control was mandated by strong legislative and regulatory underpinning, and there was a high level of anxiety among all persons exposed or at risk of exposure to radiation. As a result, the discipline of radiation protection enjoyed the luxury of ample funding and resources, and a strong technical and experience resource base enabled a maximum level of safety. Over the years, however, legislation was introduced at the federal and state levels mandating safe practices in all areas of occupational health and granting strong enforcement powers to the Occupational Health and Safety Administration and to the Environmental Protection Agency, originally enjoyed only by the Nuclear Regulatory Commission. Where radiation protection commanded a large share of the health and safety budget in the past, more recently it has had to compete for funds with the other disciplines. At the same time, the organization and practice of radiation protection served as a model in many instances for the provision of health and safety services in other areas.

The result has been to erode the autonomy of the radiation-protection program and increasingly to integrate it with the other health and safety disciplines. Managers becoming familiar with the details of radiation protection discovered that, except for the extreme levels of radioactivity in the nuclear power industry, radiation protection was not so esoteric. There was much overlap with chemical and biological safety, and industrial hygiene, in the approaches and methods used in radiation control to promote health and safety. Except in the most extreme conditions, it was not necessary to have a specialist for every health and safety discipline, and technicians could be trained to perform inspections and necessary monitoring in more than one field. The resulting economies and increased effectiveness presented compelling arguments for integrating the fields. Currently, the momentum continues to increase in the direction of integrating activities in environmental health and safety at the operational level, and this trend in turn highlights the importance of developing interdisciplinary training programs at all levels (Christman and Gandsman, 1994; Emery et al., 1995; Classic and Vetter, 1999).

3 Responsibilities of Radionuclide Users

A user who receives authorization to work with radionuclides becomes directly responsible for (1) compliance with all regulations governing the use of radionuclides in his possession and (2) the safe use of his radionuclides by other investigators or technicians who work with the material under his supervision. He must limit the possession and use of radionuclides to the quantities and for the purposes specified in the authorization. He has the obligation to:

(a) Ensure that individuals working with radionuclides under his control are properly supervised and have obtained the training and indoctrination required to ensure safe working habits, security of licensed materials while in use, compliance with the regulations, and prevention of exposure to others or contamination of the surroundings. In addition, workers should be instructed in the health-protection problems associated with exposure to pertinent radioactive materials or radiation, and female workers should be given specific instruction about prenatal exposure risks to the developing embryo and fetus.[7] (Inadequate supervision and lack of training have been cited in radiation lawsuits as indicative of negligence.)

(b) Avoid any unnecessary exposure, either to himself or to others working under him.

(c) Limit the use of radionuclides under his charge to specified locations and to individuals authorized to use them and secure stored licensed materials from unauthorized removal or access.

(d) Keep current working records of the receipt and disposition of radionuclides in his possession, including details of use in research, waste disposal, transfer, storage, and so on.

(e) Notify the appropriate administrative departments of any personnel changes and changes in rooms or areas in which radioactive materials may be used or stored.

(f) Keep an inventory of the amount of radioactive material on hand and be prepared to submit this inventory to inspectors upon request.

(g) Ensure that functional calibrated survey instrumentation is available

---

7. See 10CFR19 for details on notices, instructions, and reports to workers and their rights regarding inspections. NRC Regulatory Guide 8.13 describes the instructions the NRC wants provided to female workers concerning biological risks to embryos or fetuses resulting from prenatal exposure.

to enable personnel to monitor for radiation exposure and surface contamination.

(h) Inform the radiation-protection office at his institution when he cannot fulfill his responsibilities because of absence and designate another qualified individual to supervise the work.

(i) Inform the radiation protection office when a female worker has voluntarily declared her pregnancy in writing (declared pregnant woman).

The importance of proper record keeping by the individual users as well as by the institution under whose auspices the work is being performed cannot be overemphasized. Records of personnel exposure, radiation surveys, instrument calibration, waste disposal, radiation incidents, and all the other activities discussed in this part represent the main proof of compliance with protection regulations. They are important for legal purposes as well as for effective administration of the radiation-protection program.

4 STANDARDS FOR PROTECTION AGAINST RADIATION

The basic radiation-protection standards formulated by the NRC for radionuclide users are published in the Code of Federal Regulations, Title 10, Part 20,[8] and can be reviewed at the NRC site on the internet, *www.nrc.gov.* Topics covered include permissible doses, permissible levels, permissible concentrations, precautionary procedures, waste disposal, records, reports, and notification of the NRC in the event of radiation accidents. Some of the radiation exposure limits defined in the regulations are given in Tables 2.10 and 2.11.

It should be emphasized again that regardless of limits that are set for allowable radiation exposures, the general policy is to avoid all unnecessary exposure to ionizing radiation.[9]

8. The standards are prepared from the recommendations of such advisory bodies as the National Council on Radiation Protection and Measurements (see NCRP, 1987b, Report 91).

9. This policy is referred to as the "as low as reasonably achievable" (ALARA) principle by the Nuclear Regulatory Commission (10CFR20.1C). The commission expects that its licensees will make every reasonable effort to maintain exposures to radiation as far below the limits as is reasonably achievable. Specific recommendations for implementing this policy are contained in NRC Regulatory Guides 8.8 (primarily for nuclear power stations), 8.1 (and an associated detailed report, NUREG-0267, for medical institutions), and 8.10. Some of the measures listed as indicators of a commitment by management to an ALARA policy include promulgation of the policy in statements and instructions to personnel; review of exposures and operating procedures to examine compliance with ALARA; and training programs including periodic reviews or testing of the understanding of workers on how radiation protection relates to their jobs.

5 P**ERSONNEL** M**ONITORING FOR** E**XTERNAL** R**ADIATION** E**XPOSURE**

Personnel monitoring devices are required by law, and records must be kept if workers receive or are liable to receive from sources external to the body an effective dose in one year in excess of 10 percent of the occupational dose limits. Thus, they are required if the external annual effective dose from whole-body radiation is likely to be in excess of 5 mSv. Exposure records are reviewed periodically by government inspectors. Exposure histories of workers are also often requested when employees leave one job and report to a new employer. The employment of minors is accepted, but the limits for minors are 10 percent those for adults—that is, 5 mSv per year—and accordingly, monitoring is required if the dose is likely to be 0.5 mSv in a year. The limit to the embryo or fetus of a declared pregnant woman is also 5 mSv in a year for the duration of the pregnancy, with monitoring required at one-tenth that level. If the dose is greater than the limit at the time pregnancy is declared, the dose must be limited to 0.5 mSv for the remainder of the pregnancy.

Monitoring devices are worn on the trunk, between waist and shoulder level. They should be located at the site of the highest exposure rates to which the body is subjected. If the hands are exposed to levels significantly higher than the rest of the body because of close work with localized sources, a separate monitoring device should be worn on the wrist or finger.

The personnel monitoring devices generally used for long-term monitoring are the film badge and thermoluminescent dosimeter (TLD), but these are beginning to be replaced by optically stimulated luminescence dosimeters (OSL, see Part Four, section 5). Film badges should not be worn for longer than a month between changes because of fading of the latent image. TLD and OSL dosimeters may be worn for longer periods between changes if the risk of excessive exposure is low. Direct-reading pocket ionization chambers and personnel electronic dosimeters are worn in high-risk areas where it is desirable to have immediate knowledge of integrated exposure. Pocket ionization chambers which cannot be read without a special readout system are also used for monitoring brief exposures.

Personnel monitoring devices are very useful in radiation control even when the possibility of significant exposure is small. Their use ensures that unexpected exposures will not go undetected and may help point out situations where controls are inadequate. They serve as a constant reminder to maintain safe working habits. They provide the best legal evidence of the actual exposure or lack of exposure of the worker.

Personnel monitoring devices, to be effective, must always be worn by the worker when he is exposed to radiation and must not be exposed to radiation when they are not being worn. It may seem unnecessary to make these obvious statements, but a significant percentage of high readings on personnel monitoring devices, upon investigation, are found to be due to the storage of the monitoring device in a high-radiation field to which the wearer was never exposed.

5.1 Ambiguities in Using the Personnel Dosimeter Dose as a Surrogate for Personnel Dose

The personnel dosimeter does not measure the dose to the worker. It measures the dose to the dosimeter. It is this reading that has to be interpreted in terms of the dose or the effective dose to the worker and to its relationship to regulatory standards of exposure. This leads to different questions in the design and implementation of programs of personnel dosimetry. Is provision of a single dosimeter to monitor whole-body exposure or the addition of a ring dosimeter for the hands enough? How long should a dosimeter be worn? At what frequency should they be changed? If the worker wears a lead apron (typically 0.5 mm lead equivalent in fluoroscopy), should the dosimeter be worn under the apron or over the apron? How does one process the results if a dosimeter is worn on the collar in addition to the waist? One formula devised for combining results when one dosimeter is worn on the collar (unshielded) and one is worn at waist level under the apron gives the effective dose as equal to 1.5 times the waist value plus 0.04 times the collar value (Rosenstein and Webster, 1994). Clearly, both judgment and intuition are involved in interpreting measurements from personnel dosimeters.

6 MONITORING PERSONNEL SUBJECT TO INTAKES OF RADIOACTIVE MATERIAL

Monitoring of intakes is required if the worker is likely to receive, in 1 year, an intake in excess of 10 percent of the applicable Annual Limit on Intake (ALI). Minors and declared pregnant women require monitoring of intakes if they are likely to receive, in 1 year, a committed effective dose in excess of 0.5 mSv.

Radioactive substances enter the body through inhalation, ingestion, or penetration through the skin. Methods acceptable to the NRC for monitoring intakes include determination of concentrations of radioactive material in the air of work areas; measurements of quantities of radionuclides in the body by whole-body counting; evaluation of quantities of radio-

5.2 Air monitoring equipment. *(a)* Continuous heavy-duty air sampler, rotary vane oil-less vacuum pump (Gast Manufacturing Co. model 0522). Weighs 27 lb., pulls 4 cfm at atmospheric pressure, 3.25 cfm under 5″ Hg, 0.95 cfm under 20″ Hg. Pump shown with holder for charcoal filter and rotameter for indicating air flow. *(b)* Light-duty diaphragm oil-less sampling pump (Cole-Palmer Instrument Co. Air Cadet). Pulls 0.52 cfm at atmospheric pressure, 0.4 cfm under 5″ Hg. (*c*) Hi-Volume Air Sampler, turbin-type blower uses 4″ diameter filters. Samples at 18 cfm with Whatman #41 filters (good efficiency for diameters down to 0.01 μm). Rotameter shows flow rates of 0–70 cfm. Similar units produced by Atomic Products Corp. (illustrated) and Staplex Co. *(d)* Personal air sampler (Mine Safety Appliances Co. Monitaire sampler). Rechargeable battery-operated diaphragm pump. *(e)* Charcoal filter cartridge, TEDA-impregnated for adsorption of radioactive iodine (Scott model 605018–03). *(f)* Filter holder for 50 mm diameter filters. *(g)* Gas drying tube. When filled with charcoal, as shown, can be used for sampling or removing radioiodine. *(h)* Flowmeter for field use. Available with or without metering valve. Standard ranges 0.1–200 scfh (Dwyer Manufacturing Co. Series 500). *(i)* Laboratory flowmeters (for calibration). Ranges of 0.0013–68.2 l/min (Manostat, 36–541 series, manufactured by Fisher and Porter).

nuclides excreted from the body, usually through urinalysis; or a combination of these measurements. The inhalation route for an individual worker is best monitored at the breathing zone. This is done with a personal air sampler, consisting of a filter and a small, battery-driven pump that draws air through it. The filter can be fastened to the lapel of the lab coat. Personal air samplers are widely used to monitor labeling operations with radioiodine that are accompanied by the release of free iodine. A glass filter impregnated with charcoal is used to collect the iodine. Examples of personal air-monitoring equipment are shown in Figure 5.2.

Although personal samplers have the advantage of monitoring the actual breathing zone of the worker, they have much lower sampling rates, and therefore much lower sensitivities, than environmental monitors. Personal samplers should be investigated thoroughly to see if they have adequate sensitivity for a particular operation.

The worker himself is usually the best monitor for radioiodine. Over 20 percent of the inhaled iodine finds its way to the thyroid, which provides a source close to the surface of the neck. The iodine can be detected efficiently with a NaI scintillator placed against the neck over the thyroid. A crystal 2 mm thick has good efficiency for the low-energy gamma photons emitted by iodine-125 along with a reduced response to the higher-energy photons in the background.

Gamma-emitting radionuclides in the lungs or more generally throughout the body can be monitored with whole-body counters. The greatest sensitivity is obtained with the low-background room type, but sufficient sensitivity can often be obtained with a shadow-shield type at a much lower cost (Palmer and Roesch, 1965; Orvis, 1970; Masse and Bolton, 1970).

Whole-body counters are ineffective for beta emitters (the most sensitive ones may detect bremsstrahlung from the higher-energy beta particles). Urine analysis is done instead, though the evaluation of body burden from measurements on urine can be quite complicated and uncertain (ICRP, 1988a, Publication 54). An exception is the bioassay for tritiated water, since the concentration is the same for water in the body as it is for water in urine. Bioassays for radioiodine and tritium are within the capabilities of research workers who use them. Analyses for most other radionuclides are more difficult (Harley, 1972) and it may be advisable to use commercial services to arrange for them.

NRC requirements for bioassays are not specifically stated in 10CFR20 but are incorporated into the NRC licenses. They generally follow the recommendations of the regulatory guides (NRC, 1993, Regulatory Guide 8.9).

In the event a bioassay measurement indicates a significant level of activity in the worker, it is necessary to determine if the intake by the worker exceeded the annual limit on intake or, equivalently, the annual limit for the committed dose. The kinds of data needed to determine compliance with the regulatory limits based on various bioassay measurements are described as retention and excretion functions. The retention functions relate the measurement of the activity in a systemic organ or the whole body to a single intake resulting in that activity, as a function of the time between the intake and the measurement. The excretion function relates the activity in a 24-hour urine sample, or 24-hour feces sample, to the intake as a function of the time of the sample after the intake. The NRC has published data to convert measurements with a whole-body counter, thyroid counter, lung counter, or measurements on urine samples to an estimate of intake by ingestion and inhalation (NRC, 1987a, Report NUREG/CR-4884). Table 5.2 gives values of the fraction in a 24-hour urine sample of an initial ingestion of phosphorous-32 as a function of days after intake, as taken from Report NUREG/CR-4884. It should be noted, however, that the data presented here, based on general modeling, cannot be expected to apply to an individual case with a high degree of accuracy. They present an effective way of determining compliance with regulatory standards, but an accurate determination of the dose to an individual resulting from a particular incident would entail a detailed assessment with regard to all the factors contributing to the exposure.

Table 5.2   Fraction of initial intake of phosphorus-32 in 24-hour urine samples.

Days after intake	Fraction	Days after intake	Fraction
1	0.112	7	0.00890
2	0.0504	10	0.00553
3	0.0273	20	0.00202
4	0.0183	30	0.000861
5	0.0137		

Source: NRC, 1987b, Report NUREG/CR-4884.

Example 5.2   A urine sample submitted by a postdoctoral research student to the Radiation Safety Office was analyzed for ^{32}P. It was found to be significantly contaminated, and a 24-hour urine sample was immediate prescribed. The activity of the sample was 0.52 μCi (0.0192 MBq). A detailed investigation concluded that the phosphorous had been ingested in a single episode, 30 days prior to the measurement. What was the intake of ^{32}P?

Reference to Report NUREG/CR-4884 gave the fraction in the 24-hour urine sample at 30 days as 0.000861 of the intake. The intake was 0.52 μCi/0.000861 = 604 μCi. The NRC annual limit on intake for phosphorous-32 is 600 μCi. Because of the seriousness with which the NRC regards intakes above the ALI, the incident required an intensive investigation to find as accurate a value as possible for the intake.

7   NRC AND ICRP VALUES FOR ANNUAL LIMITS ON INTAKE AND AIRBORNE RADIOACTIVITY CONCENTRATION LIMITS

The NRC regulates internal occupational exposure from inhalation of a radionuclide through setting a maximum activity that may be inhaled by a worker in a year (the annual limit on intake, or ALI) and the airborne concentration of activity that will result in that intake for standard working conditions (the derived air concentration, or DAC; see Part Two, sections 17 and 18). Values for the ALI and the corresponding DAC are given in 10CFR20.

The derivation of the occupational limits in 10CFR20 follows the approach described by the International Commission on Radiological Protection in 1979 (ICRP, 1979) and is based on a 50 mSv annual limit for the effective dose, provided the maximum organ dose does not exceed 500 mSv. The ICRP reduced the dose limits in 1990 to 100 mSv in 5 years—in essence, 20 mSv per year (ICRP, 1991b). While they retained the same risk-based dose limitation approach and computation scheme, they also changed the tissue weighting factors on the basis of new biological information and changes in the models used for the fate of the radionuclides in the body. As a result, the revised ICRP values are significantly lower than those given by NRC in 10CFR20, which remained unchanged. The 1990 ICRP values for the ALIs are given along with those of the NRC (based on the ICRP 1979 recommendations) in Table 5.3.

The NRC air concentration limits, as derived from the ALIs, are given in Table 5.4. An area where the air concentration is above the DAC or where the product of the time the worker is in the area and the air concentration exceeds 12 times the DAC is designated an airborne radioactivity area with special monitoring and control requirements. Where exposure is to more than one radionuclide, the fraction of the limit contributed by each radionuclide is determined. The sum of the fractions must be less than 1.

The NRC derives the occupational DAC by assuming an annual exposure time of 2,000 working hours (40 hours/week × 50 weeks) and an air inhalation rate of 20 liters per minute. This inhalation rate is a generic value applicable to working conditions of "light work."

The derived air concentration for members of the public is reduced from occupational limits first, by reducing the annual dose from 50 mSv to 1 mSv (a factor of 50), then by reducing another factor of 2 to take into account the exposure of children and a factor of 3 to adjust for the differences in exposure time (40 hours/week versus 168 hours/week) and inhalation rate between workers and members of the public. This gives a total reduction of 300 in the value of the DAC.

Example 3 A worker who performed syntheses with ^{125}I was given a thyroid scan, which showed an uptake of 2,960 Bq in his thyroid. His previous scan had been given 30 days earlier. Determine if his exposure was in compliance with the limits, assuming that the uptake was by inhalation.

In the absence of other data, it may be assumed that his uptake occurred immediately following the previous scan. For an effective half-life of 32 days, the elapsed time was 30/32 = 0.94 half-lives (fractional decay = 0.52), so the maximum uptake could have been 2,960/0.52 = 5,698 Bq. Assume the uptake in the thyroid is 30 percent of the activity getting into the body and, in the absence of other data, that all the

Table 5.3   Annual limits on intake.

Radionuclide	Occupational (NRC, 10CFR20)		Occupational (ICRP, 1991b)	
	Ingestion (mCi)	Inhalation (mCi)	Ingestion (MBq)	Inhalation (MBq)
HTO water	80	80	1,000	1,000
$^{14}CO_2$ gas		200		3,000
^{14}C compounds	2	2	40	40
^{22}Na	0.4	0.6	7	10
^{24}Na	4	5	50	60
^{32}P	0.6	0.9	8	10
^{35}S (sulfides, sulfates)	6	2	70	30
^{36}Cl	2	0.2	20	3
^{42}K	5	5	50	50
^{45}Ca	2	0.8	20	10
^{51}Cr	40	20	400	200
^{57}Co	4	0.7	60	8
^{55}Fe	9	2	100	30
^{59}Fe	0.8	0.3	10	5
^{60}Co	0.2	0.03	0.19	3
^{67}Ga	7	10	80	100
^{75}Se	0.5	0.6	10	9
^{82}Br	3	4	40	50
^{86}Rb	0.5	0.8	8	10
^{90}Sr	0.03	0.004	0.6	0.06
^{99m}Tc	80	200	1,000	2,000
^{109}Cd	0.3	0.04	9	1
^{111}In	4	6	50	90
^{123}I	3	6	200	90
^{125}I	0.04	0.06	1	2
^{131}I	0.03	0.05	0.8	1
^{137}Cs	0.1	0.2	1	2
^{201}Tl	20	20	300	400
Microspheres				
^{46}Sc	0.9	0.2	10	3
^{85}Sr	3	2	40	10
^{95}Nb	2	1	30	10
^{103}Ru	2	0.6	20	8
^{113}Sn	2	0.5	20	7
^{114m}In	0.3	0.06	3	1
^{141}Ce	2	0.6	20	8
^{153}Gd	5	0.1	50	5
Alpha emitters				
^{210}Po	0.003	0.0006	0.09	0.01
^{226}Ra	0.002	0.0006	0.09	0.009
^{228}Th	0.006	0.00001	0.3	0.0002
^{232}Th	0.0007	0.000001	0.00.05	0.02

Table 5.3 (continued)

Radionuclide	Occupational (NRC, 10CFR20)		Occupational (ICRP, 1991b)	
	Ingestion (mCi)	Inhalation (mCi)	Ingestion (MBq)	Inhalation (MBq)
^{238}U	0.001	0.00004	0.8	0.01
^{239}Pu	0.0008	0.000006	0.04	0.0003

Note: The values shown here are the annual intakes that would result in a committed *effective* dose of 50 mSv, unless this entailed a committed equivalent dose greater than 500 mSv to a tissue. In this case, the annual intake which results in the maximum equivalent dose allowed for the tissue of 500 mSv is given. See 10CFR20 for more details. To compare ICRP and NRC values in megabecquerels, multiply NRC values by 37.

radioactivity in the inspired air is deposited in the body. This gives an intake of 18,981 Bq. To compare with the limits in 10CFR20, we convert to microcuries to get 0.513 μCi. The derived air concentration is $3 \times 10^{-8} \, \mu$Ci/cc and the corresponding ALI is 60 μCi. The exposure is well within occupational limits, but the investigator should still examine the methods used to determine if the exposure could have been lowered with additional reasonable precautions.

The NRC regulations require monitoring of the occupational intake and assessment of the committed effective dose if the worker is likely to receive, in 1 year, an intake in excess of 10 percent of the occupational ALI. The intake in the working environment can be assessed in several ways—through measurement of airborne concentrations by air sampling and multiplication by the volume of air breathed; through determination of the activities of radionuclides in the body by whole-body counting and calculating the intake that resulted in that activity; or by measuring the radioactivity of urine or feces samples and calculating the intake from generic equations relating excretion to body content and intake. Monitoring is required for minors and declared pregnant women who are likely to receive, in 1 year, a committed effective dose in excess of 0.5 mSv. Records of surveys must be maintained in a clear and readily identifiable form suitable for summary review and evaluation Intakes greater than the annual limit must be reported to the NRC.

A workplace must be designated as an airborne radioactivity area if the concentration of airborne radioactive materials exceeds the DAC or if the intake of an individual present in the area without respiratory protective equipment could exceed, during the hours the individual is present in a week, 0.6 percent of the annual limit on intake (ALI). The area must be

Table 5.4   Concentration limits for radionuclides (in pCi/cc) promulgated by the Nuclear Regulatory Commission.

Radionuclide	In air, occupational	In air, environmental	In water, environmental
HTO water	20	0.1	1,000
$^{14}CO_2$ gas	90	0.3	
^{14}C compounds	1	0.003	30
^{22}Na	0.3	0.0009	6
^{24}Na	2	0.007	50
^{32}P	0.2	0.0005	9
^{35}S	0.9	0.003	100
^{36}Cl	0.1	0.0003	20
^{42}K	2	0.007*	60
^{45}Ca	0.4	0.001	20
^{51}Cr	8	0.03	500
^{57}Co	0.3	0.0009	60
^{59}Fe	0.1	0.0005	10
^{60}Co	0.01	0.00005	3
^{67}Ga	4	0.01	100
^{75}Se	0.3	0.0008	7
^{82}Br	2	0.005	40
^{85}Kr	100	0.7	
^{86}Rb	0.3	0.001	7
^{90}Sr	0.002	0.000006	0.5
^{109}Cd	0.01	0.0002	6
^{111}In	3	0.009	60
^{123}I	3	0.02	100
^{125}I	0.03	0.0003	2
^{131}I	0.02	0.0002	1
^{133}Xe	100	0.5	
^{201}Tl	9	0.03	200
Microspheres			
^{46}Sc	0.1	0.0003	10
^{85}Sr	0.6	0.002	40
^{95}Nb	0.5	0.002	30
^{103}Ru	0.3	0.0009	30
^{113}Sn	0.2	0.0008	30
^{114}In	0.03	0.00009	5
^{141}Ce	0.2	0.0008	30
^{153}Gd	0.06	0.0003	60
Alpha emitters			
^{222}Rn	0.03	0.0001	
^{226}Ra	3×10^{-4}	9×10^{-7}	0.06
^{228}Th	4×10^{-6}	2×10^{-8}	0.2
^{232}Th	5×10^{-7}	4×10^{-9}	0.03
^{238}U	2×10^{-5}	6×10^{-8}	0.3
^{239}Pu	3×10^{-6}	2×10^{-8}	0.02

Table 5.4 (continued)

Sources: NRC, 1991. The environmental limits are based on a committed effective dose limit of 1 mSv (0.1 rem), but they are actually determined for half that dose to take into account age differences in sensitivity in the general population.

Notes: Lowest values listed in the tables are given here and may be for soluble or insoluble forms. Sources should be consulted for details.

Permissible releases to sewers are 10 times the limits for environmental releases and may be averaged over a month.

Multiply by 0.037 to convert pCi/cc to Bq/cc.

posted, and controls must be imposed to restrict exposure. Control measures include control of access, limitation of exposure times, or use of respiratory protection equipment. Use of individual respiratory equipment requires institution of a special respiratory protection program.

It is convenient to express exposure limits in terms of the product of the DAC and the number of hours exposed for comparison with the actual exposure. Thus the annual limit for the generic working period of 2,000 hours at the DAC is 2,000 DAC-hours. Multiplying by 0.6 percent, or 0.006, gives the weekly limit for designating an airborne radioactivity area as (0.006)(2000) or 12 DAC-hours. If the product of the time an individual is present in an area times the air concentration, expressed as a fraction of the DAC, exceeds 12 DAC-hours, then the area is designated as an airborne radioactivity area.

Example 5.4 A nuclear medicine laboratory plans to use 370 MBq of ^{133}Xe per patient and will perform a maximum of 10 studies per week. What ventilation rate is required to ensure compliance with the regulations?

The maximum activity used per week is 3,700 MBq (100 mCi). Assume 25 percent of the activity used leaks to the room during the week and let the leakage rate of activity per hour equal L. Assume the concentration is not to exceed the value C, the room has a volume V, and the fraction of the air turned over per hour is r, that is, the number of air changes per hour is r. The leakage of radioactivity into the room must be compensated by ventilation to keep the concentration at the acceptable level, that is, $L = C \times V \times r$.

The occupational DAC for ^{133}Xe, in the traditional units employed by the NRC, is $10^{-4}\,\mu\text{Ci/ml}$, so

$$\frac{25{,}000\,\mu\text{Ci}}{40\,\text{hr}} = \frac{10^{-4}\,\mu\text{Ci}}{\text{ml}} \times V \times r$$

The volume of air replaced per hour $(rV) = \dfrac{25,000}{40 \times 10^{-4}}$ ml/hr. Ventilation rate $= 6.25 \times 10^6$ ml/hr $= 3.68$ cubic feet per minute (cfm).

It is desirable to limit the weekly intake to 0.6 percent of the ALI (or 12 DAC-hours per week) to avoid the airborne-radioactivity classification. This requires a ventilation rate of 40/12 × 3.68 or 12.27 cfm. If the room has a volume of 1,000 ft^3, $r = 12.27$ ft^3/min ÷ 1,000 ft^3 = 0.01227/min, and the required number of air changes is 60 × 0.01227 or 0.736/hr.

Example 5.5   An environmental monitor located at the boundary of a facility handling large amounts of ^{125}I indicated an average concentration reported as 2×10^{-10} μCi/cc (7.4 Bq/m^3). Estimate the dose to the thyroid of a member of the public resulting from exposure to this level for 1 year.

The occupational ALI is 60 μCi(2.22 MBq) and is rounded off from the value that gives an equivalent dose of 500 mSv/year to the thyroid (the lower limit when tissue weighting factors result in a higher dose). The effective dose is only 15 mSv, a fraction of the occupational limit of 50 mSv to the whole body. The reference volume of air breathed per day is 20 m^3. The annual intake by a member of the public staying continuously at the boundary would be 20 m^3/day × 10^6 cc/m^3 × 365 days/year × 2 × 10^{-10} μCi/cc = 1.46 μCi. The thyroid dose would be (1.46/60) × 500 mSv = 12.2 mSv. The effective dose is obtained by multiplying by the tissue weighting factor (NRC value is 0.03) to give 0.37 mSv. This is below the effective dose limit of 1 mSv. The measured level is also below the environmental limit for ^{125}I of 3 × 10^{-10} μCi/cc.

The NRC concentration limits in regulations promulgated in 10CFR20 prior to the revision in 1991 were set so that the activity accumulated in body organs after a working lifetime (50 years) of exposure never reached the activity limit that could be maintained continuously without exceeding the dose limits. These original limits were derived by specifying maximum allowable annual doses to single organs (150 mSv occupational, except 50 mSv to gonads, blood-forming organs, and total body) and were based on ICRP recommendations in 1959, 1962, and 1966. The progression to the risk-based system in 1991 based on producing a committed effective dose of 50 mSv and a maximum organ dose of 500 mSv generally resulted in an increase of the allowable occupational organ doses. The air concentration limits used by the NRC prior to 1991 and based on the 150 mSv organ dose were 10 times lower for ^{133}Xe and 6 times lower for ^{125}I.

7.1 Air Monitoring for Environmental Radioactivity

The air must be monitored if significant levels of airborne contaminants can occur in working areas. Air samplers are designed to remove contaminants quantitatively from the air by collection on a filter, absorbent, or solvent, and the sampling medium is then assayed for radioactivity (NCRP, 1978b; IAEA, 1971b). The air can be monitored continuously by making the radiation detector an integral part of the sampler. The sampler should be placed in the work area at breathing zone level.

In addition to the sampling medium, an air monitoring system requires a pump to draw the air, a meter for determining rate of flow or quantity of flow, and whatever controls are desired for adjusting flow. Flow-limiting devices can be used to fix the rate of air flow through the sample. Alternatively, gas meters in the line will give the total air volume sampled independent of flow rate (ACGIH, 1983). Air monitoring instrumentation is shown in Figure 5.2.

Oilless pumps are less messy than ones that must be oiled. Because of their portability, smaller pumps are useful if they must be used at several locations, but the resultant loss in sensitivity because of lower sampling rates must be checked. Flow rates can be monitored by inserting a rotameter between the sampler and the pump. Some types of rotameters can also be used to regulate the flow rate.

The flow rate will decrease if the sampling medium becomes clogged; eventually it may be reduced to such a low level that the pump will overheat and become damaged or ruined. This can be prevented by placing a relief valve in the line between the sampling medium and pump that opens to let in additional air if the resistance of the sampler gets too high.

There are times while sampling air containing particles above a particular size that the air velocity through the sampler must be adjusted to equal the velocity in the air stream. This condition is known as isokinetic sampling. If it is not met, there can be reduced intake of particulates and errors in determining particle concentration or size distribution (Mercer, 1973; Silverman et al., 1971).

8 POSTING OF AREAS

The following types of signs are required in areas where significant levels of radiation or radioactivity are present:

(a) "CAUTION, RADIATION AREA"—This sign is used in areas accessible to personnel in which a major portion of the body could receive in any 1 hr a dose of 0.05 mSv at 30 cm from the object containing the source or from any surface from which the radiation emerges.

(b) "CAUTION, RADIOACTIVE MATERIAL"—This sign is required in areas or rooms in which radioactive material is used or stored in an amount exceeding quantities listed in Table 5.5.

(c) "CAUTION, RADIOACTIVE MATERIAL" (label)—A durable, clearly visible label is required on any container in which is transported, stored, or used a quantity of any material greater than the quantity specified in Table 5.5 (licensed material). The labels must state the quantities and kinds of radioactive materials in the containers and the date of measurement of the quantities.

(d) "CAUTION, AIRBORNE RADIOACTIVITY AREA"—This sign is required if airborne radioactivity exists at any time in concentrations in excess of the derived air concentrations for 40 hours' occupational exposure (Table 5.4), or if during the number of hours in any week during which an individual is in the area he could receive an intake of 12 DAC-hours (0.6 percent of the ALI); that is, if the actual concentration $\times$ the hours spent is greater than DAC $\times$ 12 hours.

(e) "CAUTION, HIGH RADIATION AREA"—This sign is required if the radiation dose is in excess of 1 mSv in any 1 hr at 30 cm from the object containing the source or from any source from which the radiation emerges. These areas also require audible or visible alarm signals.

The signs must bear the three-bladed radioactive caution symbol (magenta or purple on yellow background).

Warning signs are essential, since individuals might otherwise be unaware of the presence of the radiation field. On the other hand, signs should not be used when they are not needed.

Nuclear Regulatory Commission licensees are also required to inform workers by posted notices of the availability of copies of 10CFR19 and 10CFR20 and of the regulations, the license and amendments, the operating procedures applicable to licensed activities, and notices of violations involving radiological working conditions (10CFR19.1011). The NRC also requires posting of a special form (NRC-3) notifying employees of the regulations and inspections.

9 LABORATORY FACILITIES

Laboratory facilities for handling unsealed radioactive materials must provide adequate containment and allow for ease of cleanup in the event of contamination incidents. All surfaces, especially the floors and walls around sinks, should be smooth and nonporous. Generally, a well-designed and well-maintained chemistry laboratory suffices. Glassware, tongs, and other equipment used to handle unsealed radioactive material should be segregated and given a distinct marking to prevent their use with nonradioactive materials. For handling exceptionally high levels of highly toxic radionuclides or strong gamma emitters, special glove boxes and radiation shields may have to be installed. If special facilities are required, the reader should consult available handbooks and guides for high-level operations (NCRP, 1964, Report 30; ICRP, 1977a, Publication 25; IAEA, 1973, Safety Series no. 38).

Hoods are necessary for controlling possible airborne contamination arising from work with radioactive materials. The airflow into the hood must be adequate, and the hood must be designed so the flowlines are all directed into the hood. Airflow into the hood should be between 100 and 125 linear feet per minute when the hood sash is at its normal open position during use (a recommended opening is 14 inches, to give eye protection as well as effective ventilation). Flows above 125 feet per minute may lead to turbulence and some release of hood air to the laboratory. If appreciable levels of activity are used, the hood should have its own exhaust system and not be connected into other hoods, as this could be a mechanism for the transmission of airborne contamination to other laboratories through improper baffling. The exhaust system should have provision for installing filters, if needed. The working surface should be able to support lead shielding. Controls for air, water, and so on should be located outside the hood. Even when hoods are used it is often worthwhile to collect and filter radioactive airborne particulates and vapors from the operation with local suction devices located near the source, since this can help minimize contamination of the hood and diminish the work required later for decontamination.

Even the best hoods do not completely isolate the area inside the hood from the laboratory, so there is a limit to the maximum amount of activity that can be handled. If the worker is very careful, he should be able to process solutions containing up to 1 mSv of the less hazardous beta emitters in the hood without serious contamination to himself or the surroundings. However, if he must perform complex wet operations with risk of serious spills, or dry and dusty operations, he may need to use a completely isolating system such as a glove box or a hot cell (if massive shielding is needed).

An arrangement that gives protection somewhere between that provided by a glove box and a hood is a small enclosure with ports for inserting the hands and a local exhaust. The exhaust from the enclosure is cleaned before it is discharged, preferably inside a hood. This method allows the use of smaller filters and charcoal adsorption beds for cleanup at

Table 5.5   Minimum quantities (in μCi) of some radioactive materials requiring warning signs.

Radionuclide	(*a*) Sign in room[a]	(*b*) Label[b]
^{3}H	10,000	1,000
^{14}C	1,000	100
^{22}Na	100	10
^{24}Na	1,000	100
^{32}P	100	10
^{35}S	1,000	100
^{36}Cl	100	10
^{42}K	10,000	1,000
^{45}Ca	1,000	100
^{51}Cr	10,000	1,000
^{57}Co	1,000	100
^{55}Fe	1,000	100
^{59}Fe	100	10
^{60}Co	10	1
^{64}Cu	10,000	1,000
^{65}Zn	100	10
^{75}Se	1,000	100
^{82}Br	1,000	100
^{85}Kr	10,000	1,000
^{86}Rb	1,000	100
^{90}Sr	1	0.1
^{109}Cd	10	1
^{111}In	1,000	100
^{123}I	1,000	100
^{125}I	10	1
^{131}I	10	1
^{133}Xe	10,000	1,000
^{198}Au	1,000	100
^{201}Tl	10,000	1,000
Microspheres		
^{46}Sc	100	10
^{85}Sr	1,000	100
^{95}Nb	1,000	100
^{103}Ru	1,000	100
^{113}Sn	1,000	100
^{114}In	100	10
^{141}Ce	1,000	100
^{153}Gd	100	10
Unidentified, but not an α emitter	0.1	0.01
Unidentified α emitter	0.01	0.001

Table 5.5 (continued)

Radionuclide	(a) Sign in room[a]	(b) Label[b]
Alpha emitters		
^{222}Rn	10	1
^{226}Ra	1	0.1
Th (natural)	1,000	100
U (natural)	1,000	100
^{239}Pu	0.01	0.001

a. These quantities are 10 times the values in column *(b)*, except for natural uranium and thorium. Caution signs are not required to be posted at areas or rooms containing radioactive materials for periods of less than 8 hr provided (1) the materials are constantly attended during such periods by an individual who shall take the precautions necessary to prevent the exposure of any individual to radiation or radioactive materials in excess of the limits established by the regulations; (2) such area or room is subject to the authorized user's control.

b. These are also minimum quantities requiring specific licenses from NRC when an institution does not have a broad license (see 10CFR30). These values in 10CFR20 were obtained by taking 1/10 of the most restrictive occupational ALI, rounding to the nearest factor of 10, and arbitrarily constraining the values listed between 0.0001 and 1,000 μCi. Values of 100 μCi have been assigned for radionuclides having a half-life in excess of 10^9 years (except rhenium, 1,000 μCi) to take into account their low specific activity.

Multiply by 0.037 to convert μCi to MBq.

much lower cost. Additional protection is obtained if the unit in turn is placed in a hood.

Another control method is to recirculate the air from the filter back into the box, at a flow rate chosen so the air turns over every 2–3 minutes. This serves to clean the air in the box, thus reducing the discharge to the environment. It is a particularly useful method when close to 100 percent filtration is needed but not readily achievable with a once-through filtering system.

10 PROTECTIVE CLOTHING

Suitable gloves must be worn whenever hand contamination is likely. Extra care should be exercised to prevent contamination of skin areas where there is a break in the skin. In addition to gloves, other protective clothing, such as coveralls, laboratory coats, and shoe covers, should be worn wherever contamination of clothing with radioactive materials is possible. Protective clothing must not be taken out of the local areas in which their use is required unless they are monitored and determined to be free of contamination. Under no conditions should protective clothing be worn in eating places.

There are many kinds of disposable gloves for users who do not care to bother with decontamination. Plastic gloves are the most inexpensive but are clumsy to use and are suitable only for the very simplest operations. Disposable surgeon's gloves are recommended when good dexterity is needed. Sometimes two pairs of gloves are worn when handling extra-hazardous materials to prevent skin contamination in the event of a break in one of the gloves.

The potential of contamination is very high when vials of high-specific-activity radionuclides are handled. A tiny droplet from these solutions carries a lot of activity and is easily carried through the air. Gloves should always be worn when opening vials, since the covers and vials may become contaminated, even with cautious handling.

11   Trays and Handling Tools

Work that can result in contamination of table tops and work surfaces should be done in trays with a protective liner.

Tweezers, tongs, or other suitable devices should be used as needed to handle sources with significant surface dose rates. Maintaining a distance of even a few inches with tweezers or tongs can cut down the exposure rate by orders of magnitude relative to handling small sources directly with the fingers, because of the inverse square law. Syringe shields are available that provide effective protection when personnel inject large quantities of beta or low-energy gamma emitters.

12   Special Handling Precautions for Radioiodine

The control of radioiodine is a problem because of its volatility and very low permissible concentrations. The following handling procedures are recommended when volatile species of radioiodine are processed.

1. Always work in a well-ventilated fume hood. The hood should be equipped with an activated charcoal stack filter if releases approach allowable limits.
2. Two pairs of gloves should be used because radioiodine can diffuse through rubber and plastic. The inner pair must be free of contamination.
3. Do not handle contaminated vials or items directly. To ensure a secure grip on containers, use forceps fitted with rubber sleeves,

such as one-inch lengths of 1/8-inch O.D. latex surgical tubing. The sleeves are replaced easily when contaminated.

4. Do not leave vials containing radioiodine open any longer than necessary and cap tightly when not in use.

5. Always open vials in a hood because the pressure of the radioactive vapor builds up in the vial while it is in storage.

6. Double-bag all contaminated materials.

7. Decontaminate spills using a solution consisting of 0.1 M NaI, 0.1 M NaOH, and 0.1 M $Na_2S_2O_3$. This helps to stabilize the material and minimize evolution of volatile species. Complete the cleanup with a detergent.

8. Do not add acids to radioiodine solutions. The volatility of ^{125}I is enhanced significantly at low pH.

9. If the quantities handled require better control than that provided by a hood, place in the hood a transparent enclosure (for example, Lucite) fitted with a blower unit that recycles the air through an activated charcoal filter. The enclosure is equipped with sliding doors that provide convenient access and can be adjusted to the minimum opening required for performing operations in the enclosure. For work with smaller quantities of radioiodine, a once-through filter cycle may be adequate, in which air flows from the room into the enclosure and is exhausted through the charcoal filter into the hood and up the stack.

10. Venting of vials through a charcoal trap is recommended before opening if there are likely to be volatile species in the vial airspace. A simple vent is constructed by placing charcoal in the barrel of a hypodermic syringe between glass-wool plugs. The syringe is fitted with an 18-gauge hypodermic needle for penetrating septa and closures. The needle should be protected with a plastic shield when it is not in use.

11. Iodine-125 should be monitored with a thin sodium-iodide detector. This has an efficiency of over 20 percent at contact compared with less than 0.5 percent from a G-M counter. Scintillation monitors for ^{125}I are available from commercial companies.

Large quantities of ^{131}I are handled as NaI in hospitals for diagnosis and treatment of diseases of the thyroid. The iodine is not very volatile in this form, provided the solution is not acidic. Studies have found some releases to the air, however, both in handling the radioiodine and through exhalation by the patient (Krzesniak et al., 1979). The airborne iodine exists as elemental iodine, organic iodine, and iodine adsorbed on aerosols. Patients administered about 740 MBq (20 mCi) exhaled between 0.003 and 0.07 percent, and concentrations measured in the air were several times greater than concentration guides on two occasions, but they were less than 20 percent of concentration guides when averaged over the year.

12.1 Use of Potassium Iodide as a Thyroid-Blocking Agent

The uptake of radioactive iodine by the thyroid can be blocked or significantly reduced by the administration of stable iodine. The iodine is generally given in the form of potassium iodide (KI). Because KI itself is not completely safe, the FDA and WHO recommend that it not be taken unless the dose to the thyroid gland can exceed 250 mGy (25 rem), while the WHO would block the thyroid at a 50 mGy projected dose in children. The prescription calls for 130 mg KI for adults and children above 1 year and 65 mg for children below 1 year of age, administered immediately before or immediately after exposure, and continuing it daily for the duration of the exposure, and perhaps for several days longer. The recommended duration for ^{131}I is 10–14 days (Crocker, 1984).

While blocking of the thyroid before exposure is the decisive factor in minimizing radioiodine uptake, when the longer-lived ^{125}I is used, or when the iodine is bound to molecules with a long biological half-life, the blocking agent may need to be administered for a considerably longer period after the exposure to prevent a significant dose to the thyroid (Reginatto et al., 1991).

The effectiveness of the KI in controlling the exposure as a function of the time of administration and the dietary level of iodine was studied by modeling iodine metabolism (Zanzonico and Becker, 2000). It was reported that in euthyroid adults 50–100 mg KI administered up to 48 hr before exposure to iodine-131 can almost completely block thyroid uptake. KI administered 96 hr or more before or 16 h after iodine-131 exposure has no significant protective effect. Administration of KI 2 hr after exposure had an 80 percent protective effect, which dropped to 40 percent 8 hr after exposure, provided the diet was not deficient in iodine. The protective effect was less and decreased more rapidly in iodine-deficient diets.

The World Health Organization states that KI prophylaxis should not be used where ingestion is the main source of exposure and intake can be prevented by changing the source of the food. Only when inhalation is the main route of entry should KI be used to block uptake.

Adverse effects in the administration of iodine are of concern and responsible for limiting its use. In the administration of 70 mg of KI to 10.5 million children and adolescents in Poland following the Chernobyl reactor accident, adverse reactions occurred at a frequency of about 4.5 percent. The most common reaction was vomiting, which could have been a psychological effect, dermatologic effects in 1.1 percent, and abdominal pain in 0.36 percent. In another study, most individuals who received KI in amounts greater than 30 mg/day for 8 days had a significant fall in serum thyroid hormone levels. Three of the five subjects who received 100 mg a day for 8 days had elevated TSH levels and biochemical hypothyroidism. There was considerable individual variation in response (Zanzonico and Becker, 2000).

13 HYGIENE

Eating, smoking, storing food, and pipetting by mouth cannot be allowed in areas where work with radioactive materials is being conducted, or in rooms containing appreciable loose contamination, because of the potential for ingestion of radioactivity. Personnel working in areas containing unsealed sources of radioactivity must "wash up" before eating, smoking, or leaving work and must use an appropriate detection instrument to monitor hands, clothing, and so on, for possible contamination. Unnecessary exposure or transfer of activity from undetected contamination can be avoided by making a habit of "washing up" and "surveying."

14 TRIAL RUNS

For nonroutine or high-level operations, the user should conduct a trial run with inactive or low-activity material to test the adequacy of procedures and equipment.

15 DELIVERY OF RADIONUCLIDES

All packages of radionuclides must be carefully checked upon receipt for evidence of damage or leakage (see NRC Regulatory Guide 10.8; NRC, 1987a). A record of receipts of material must be maintained.

Packages labeled as containing radioactive material or showing evidence of potential contamination, such as packages that are crushed, wet, or damaged, must be monitored for contamination and radiation levels. The monitoring must be done no later than 3 hr after the package is received at the licensee's facility if it is received during the licensee's normal working hours, or 3 hr from the beginning of the next working day if it is received after hours. If removable beta-gamma radioactive contamination in excess of 22,000 dis/min per 100 cm^2 of package surface is found on the external surfaces of the package, or if the radiation levels exceed 2 mSv/hr (200 mrem/hr) at the surface of the package or 0.1 mSv/hr (10 mrem/hr) at 1 meter from the surface, the licensee must immediately notify the final delivering carrier and by telephone and telegraph, mailgram, or facsimile the appropriate NRC Inspection and Enforcement office.

Rules regarding the handling of shipments will vary depending on the local circumstances. Measures must be taken to ensure that the packages are always placed in designated, secure locations until they are opened and processed. Institutions have reported to the Nuclear Regulatory Commission that unsecured radioactive-materials packages delivered to research laboratories have been accidentally thrown out by housekeeping personnel as ordinary trash or have disappeared for unknown reasons. Packages ideally should be received at a central radiation facility, where the contents of the package are inspected, monitored, and logged by trained personnel and the material secured until picked up by the authorized user. If delivery is made to a general receiving area of the institution, the package should be logged in and then transferred with dispatch to the user, the radiation facility, or another secured, controlled, and protected area established for storage of radioactive materials. Highly visible signs should be posted in the receiving area giving specific instructions for handling packages. The receiving area must have a record of the name of the person receiving the package and the person to whom it was transferred or who placed it in a locked area. This procedure allows tracking down of any packages that might be misplaced after being received.

Packages must not be left on the floor or unsecured and unattended on a bench top when they are delivered to a laboratory. They must be placed in a designated secure location and the responsible person must be promptly notified if he is not present to receive it.

16   STORAGE AND CONTROL OF RADIONUCLIDES

The NRC requires that stored licensed materials must be secure from unauthorized removal or access. This means that storage areas must be locked and placed under the control of responsible individuals only. The radionuclides must be stored in suitable containers that are adequately shielded. It is usually practicable and desirable to shield stored materials so the radiation level at one foot from the surface of the shield is less than 0.05 mSv/hr (5 mrem/hr). In any event, the level should be less than 1 Sv (100 mrem) in 1 hr. Otherwise the area is a high-radiation area and must be equipped with visible or audible alarm signals. Sources must be properly labeled and area signs posted. The radiation-protection office must be kept informed of any transfer of a source to new storage or use areas.

Some radioactive materials must be stored under refrigeration, and those that also contain flammable solvents constitute an explosion hazard. Explosions have occurred in refrigerators, ignited by a spark in the controls or switches, and have resulted in extensive physical damage, starting of fires, and bodily injury. Flammable materials must be stored in explosion-proof refrigerators, that is, refrigerators with controls and other potential spark-producing components mounted on the outside.

Material not secured must be under the control and constant surveillance of the licensee.

17 STORAGE OF WASTES

Radioactive wastes may be stored only in restricted areas. Liquid waste should be stored in shatterproof containers, such as in polyethylene bottles. If circumstances make this impracticable, an outer container of shatterproof materials must be used.

Flammable wastes should be kept to a minimum in the laboratory. Waste containers must be metallic. A fire extinguisher must be located in the vicinity and a sign posted giving its location.

During storage there must be no possibility of a chemical reaction that might cause an explosion or the release of chemically toxic or radioactive gases. This is usually accomplished by the following precautions: *(a)* liquids must be neutralized (pH 6 to 8) prior to placement into the waste container; *(b)* containers of volatile compounds must be sealed to prevent the release of airborne activity; and *(c)* highly reactive materials (such as metallic sodium or potassium) must be reacted to completion before storage.

18 WASTE DISPOSAL

A limited quantity of wastes may be disposed of by release to the atmosphere, inland or tidal waters, sewerage, or by burial. The limits are established by federal and state agencies. Short-lived radionuclides are often stored and allowed to decay until they can be disposed of as nonradioactive wastes. The NRC requires a minimum decay time of 10 half-lives before release as ordinary waste for waste designated for decay storage. This period reduces the activity to less than 0.1 percent of the original value. Longer decay periods are necessary if significant levels of radioactivity are detected after this time. When special waste disposal problems occur, disposal through a commercial company licensed to handle radioactive material often constitutes the most satisfactory approach. Records must be kept of the disposal of all radioactive wastes as evidence that the regulations have been observed.

The management of the most hazardous radioactive wastes—such as plutonium and other long-lived alpha emitters, that are produced in large quantities in the production of nuclear power—is of particular concern to regulatory officials and the public. The approach here is to concentrate them, immobilize them by incorporation into a glass or other nonleachable medium, and then confine them to a repository that is isolated from water sources and has insignificant risk of being breached as a result of earthquakes or other geological disturbances.

18.1   Disposal of Gases to the Atmosphere

The Nuclear Regulatory Commission requires that the radioactivity concentrations in gases released through a stack to an unrestricted area must not exceed limits specific to each radionuclide. The simplest way to ensure compliance with the regulations is to limit the concentration at the stack discharge point to the maximum allowed if the effluent were breathed continuously by a person standing at the point of discharge. The concentrations may be averaged over a period not exceeding one year. If the discharge is within a restricted area, the limit may be applied at the boundary by using appropriate factors for dilution, dispersion, or decay between the point of discharge and the boundary. The user may petition the NRC to allow higher concentrations at the discharge point by demonstrating that it is not likely that any member of the public will be exposed to concentrations greater than those allowed by the regulations.

Some values for maximum permissible concentrations in air for unrestricted areas are given in Table 5.2. If the discharged gas contains combinations of radionuclides in known amounts, a limit may be derived for the combination by determining the ratio between the quantity present in the combination and the limit allowable when it is the sole constituent. The sum of the ratios determined in this manner for each of the constituents may not exceed unity.

Example 5.6   A department of nuclear medicine in a metropolitan hospital is conducting studies with xenon-133 and releasing the recovered xenon through a hood to a discharge point on the roof of the building. The face velocity of flow of gas through the hood, when the area of the opening is 3.5 ft^2, is 125 linear ft/min, as measured with a velometer. Assuming releases are controlled on a weekly basis, what is the maximum permissible weekly discharge rate? What are the restrictions when ^{14}C and ^{131}I are also discharged through the hood?

The flow rate of air through the hood is 125 × 3.5 or 437.5 ft^3/min. This gives $1.24 \times 10^7 \times 1{,}440 = 1.78 \times 10^{10}$ cm^3/day. From Table 5.4, the maximum permissible concentration of ^{133}Xe in air in unrestricted areas is 0.5 pCi/cc or 5×10^{-7} μCi/cc. The permissible discharge is thus $1.78 \times 10^{10} \times 5 \times 10^{-7} = 8.9 \times 10^3$ μCi/day or 8.9 mCi/day (329 MBq/day). The maximum weekly discharge for control purposes is 62.3 mCi (2305 MBq).

Similarly, one could dispose of 3.69 mCi (37 MBq) ^{14}C or 2.49 μCi (0.92 MBq) of ^{131}I per week, if either one were the only radionuclide discharged. If in a particular week it were necessary to discharge 2 mCi of ^{14}Co$_2$ and 3 μCi of ^{131}I, 2/3.69 + 3/24.9 or 0.662 of the permissible discharge would be used up. Thus, 0.337 of the permissible discharge would still remain, allowing release of 0.337 × 62.3 = 21.0 mCi (777 MBq) of ^{33}Xe, if this were the only other radionuclide to be released.

Example 5.7   A radiochemist accidentally released 1 curie (37,000 MBq) of tritiated water through a hood while performing a synthesis. The air face velocity was 100 ft/min with a 1-ft opening. The width of the opening was 4.5 ft. Were the radioactivity release limits exceeded? Assume no other radionuclides were released through the hood during the year.

The flow rate is 1.84×10^{10} cc/day. The maximum permissible concentration at the point of release to the atmosphere is 1×10^{-7} μCi/cc. The maximum permissible daily release is 1.84 mCi, and the annual limit is 0.67 Ci. An NRC licensee would have to notify the NRC in a written report within 30 days after learning of the incident since the accident resulted in a release in excess of the annual limit. Release of activity through the hood is prohibited for the remainder of the year as a result of the accident.

The NRC must be notified of incidents that result in doses or releases exceeding the limits. The time allowed for reporting the incident depends on the amount by which limits are exceeded; it ranges from immediate notification to twenty-four hour notification to submittal of a report within 30 days after learning of the incident. Details are in 10CFR20.

18.2   Disposal of Liquids to Unrestricted Areas

The NRC regulations limit the release of gaseous and liquid effluents at the boundary of unrestricted areas. The concentration in liquid effluents discharged to inland or tidal waters is limited to the maximum permitted in drinking water consumed by the public. The maximum concentration may be evaluated for the boundary of the restricted area and averaged over a month. As with discharges to the atmosphere, the NRC may accept higher limits if it is not likely that individuals would be exposed to levels in excess of applicable radiation-protection guides, but any action taken by the NRC is based on the condition that the user first take every reasonable measure to keep releases of radioactivity in effluents as low as practicable. Concentration limits in water in unrestricted areas are given in Table 5.6.

18.3   Disposal of Liquid Wastes to Sanitary Sewerage Systems

The regulatory limit for the concentration of the activity of a radionuclide in the sewerage discharged from an institution to a sanitary sewerage system is determined by dividing the most restrictive annual limit on intake by ingestion by 7.3×10^6 ml (ten times the annual water intake of "Reference Man"). Since the calculation of the ALI for ingestion is based on a committed dose equal to the occupational limit of 50 mSv, the use of 10 times the reference annual water intake results in a committed effective dose of 5 mSv if the sewerage released by the licensee were the only source of water ingested by a reference man during a year. The concentration value used with respect to determining compliance is the concentration averaged over a month. In any event, the total quantity of licensed and other radioactive material that the licensee releases into the sanitary sewerage system in a year must not exceed 185 GBq (5 curies) of hydrogen-3, 37 GBq (1 curie) of carbon-14, and 37 GBq (1 curie) of all other radioactive materials combined.

The material discharged into the sanitary sewerage must be readily soluble in water or readily dispersed if it is biological material.

Concentration limits for selected radionuclides are given in Table 5.6. If several radionuclides are being discharged, the determination of compliance with the limits is made as described in section 18.1. Excreta from individuals undergoing medical diagnosis or therapy with radioactive material are exempt from these limitations.

Although disposal through the sewerage system is permitted for the trace amounts of radioactivity remaining after counting experiments, it should never by used for disposal of highly concentrated solutions, such as master solutions used in radionuclide synthesis. These should be disposed

Table 5.6 Concentrations (monthly average) allowed by the Nuclear Regulatory Commission for release to sewerage.

Radionuclide	Concentration (pCi/cc)	Radionuclide	Concentration (pCi/cc)
HTO water	10,000	^{111}In	600
^{14}C	300	^{123}I	1,000
^{22}Na	60	^{125}I	20
^{24}Na	500	^{131}I	10
^{32}P	90	^{201}Tl	2,000
^{35}S	1,000		
^{36}Cl	200	*Microspheres*	
^{42}K	600	^{46}Sc	100
^{45}Ca	200	^{85}Sr	400
^{51}Cr	5,000	^{95}Nb	300
^{57}Co	600	^{103}Ru	300
^{59}Fe	100	^{113}Sn	300
^{60}Co	30	^{114m}In	50
^{67}Ga	1,000	^{141}Ce	300
^{75}Se	70	^{153}Gd	600
^{82}Br	400		
^{86}Rb	70	*Alpha emitters*	
^{90}Sr	5	^{226}Ra	0.6
^{109}Cd	60	^{228}Th	2
		^{232}Th	0.3
		^{238}U	3
		^{239}Pu	0.2

Sources: NRC, 1991. The limits were derived by dividing the most restrictive occupational stochastic ingestion ALI by the annual water intake by "Reference Man" (7.3×10^6 ml) and a factor of 10, such that the concentrations, if the sewage released by the licensee were the only source of water ingested by a reference man during a year, would result in a committed effective dose of 5 mSv (0.5 rem). The discharged radioactivity must also meet strict requirements with respect to its solubility in water.

of in their original containers through a commercial company as bulk waste. If they are short-lived, they can be stored for decay.

Complications arise when the material to be disposed of is also a chemical hazard. For example, flammable solvents that are not miscible with water should not be flushed down the drain. They should be poured into a solvent can (properly labeled) in a hood and disposed of ultimately by evaporation, by incineration, or through a commercial disposal company.

It is sometimes desirable to reduce the volume of significant quantities of liquid wastes or partially to clean up the liquid prior to disposal. The ultimate disposal arrangement is dependent on individual circumstances, and consultation with a specialist in radiation protection is advisable.

Example 5.8   Determine how much ^{125}I and ^{32}P can be dumped into the sewer if the water flow to the sewerage (according to the water bill) is 1.2×10^7 ft^3/yr.

The maximum concentrations for ^{125}I and ^{32}P are 20 pCi/cc (0.74 Bq/cc) and 90 pCi/cc (3.33 Bq/cc), respectively. The average monthly water flow is 2.83×10^{10} cc. The monthly limits are 566 mCi (20.9 GBq) for ^{125}I and 2,547 mCi (94.2 GBq) for ^{32}P, if the particular nuclide is the sole constituent of the waste. Otherwise, the analysis must be made in terms of the fraction of the maximum discharge limits, as discussed in section 18.1. It must be borne in mind that notwithstanding the limits calculated, the maximum gross activity that can be released into the sewer for the year is 1 curie (37 GBq).

18.4   Solid Wastes

Covered metal cans (the type equipped with foot-operated lids is convenient for small volumes) should be used to contain nonflammable solid wastes in low-level laboratories. The cans should be easily distinguishable from cans for ordinary trash to prevent accidental disposal of radioactive materials into the regular trash, and they should display a "radioactive materials" label. A plastic bag should be used as a liner. Hypodermic needles and other sharp objects should be placed in special containers; even a mere scratch to a person handling the bag can result in serious infection or disease. When the contents of the can are to be disposed of, the plastic bag is sealed and a tag stating the upper limits to the contents is attached for the information of the disposer. Materials contaminated with radioiodine should be enclosed in two bags before discarding. Animal carcasses are best stored in a freezer prior to final disposal. If sufficient storage or freezer space is available, the shorter-lived nuclides may be allowed to decay to insignificant levels. Otherwise, the wastes must be disposed of by burial, by incineration, or through a commercial company. When the wastes collected from a laboratory are to be shipped out in drums, and the volume generated is large enough, it is sometimes convenient to keep the shipping drum in the laboratory for use also as a waste receptacle. For all practical purposes current regulations do not allow disposal of radioactive solid waste through public sanitation departments.

18.5   Disposal on Site by Incineration and Other Methods

Disposal on site by incineration or burial or other methods having environmental impacts that are not covered in the regulations require special approval by the NRC. The only exceptions are limited amounts of ^{3}H and ^{14}C. The NRC regulations consider up to 1,850 Bq (0.05 microcurie) per gram of medium used for liquid scintillation counting or per gram of animal tissue, averaged over the weight of the entire animal, as nonradioactive, so these may be incinerated on site without NRC approval or disposed of commercially as nonradioactive, provided records are maintained. However, regulations requiring the installation of the best available air cleaning equipment to meet rigorous standards have made the operation of incinerators very costly and incineration that is not done on a commercial scale is not cost effective for most research institutions.

Incineration may be attractive when the waste material is a fuel itself. For example, the scintillation fluids prepared for liquid scintillation counting consist mainly of toluene or xylene, solvents with high heat content. The incineration of toluene consumes about 15 liters of air per milliliter. (The reaction $2CH_2 + 3O_2 \rightarrow 2CO_2 + 2H_2O$ requires 3.4 g O_2 to oxidize 1 g of CH_2, or the oxygen in about 12 l of air.) The combustion of 20 ml of scintillation fluid (volume in a typical vial) containing 20,000 dis/min of ^{3}H would result in a gaseous effluent with a concentration of 1.3 dis/ml. This is only three times the limit for continuous exposure of a member of the public. Levels orders of magnitude below the limits for environmental releases are attained readily by mixing the scintillation fluid with fuel used for incineration, heating, or power production. Alternatively, even if the toluene is burned directly, concentrations can be reduced to trivial levels by diluting the effluent with large amounts of air prior to discharge from the stack.

Burial of limited quantities of radioactive waste on site was once authorized by the NRC but is no longer permitted.

18.6   Government Regulation of the Disposal of Hazardous Wastes

The technologically advanced societies produce enormous quantities of hazardous waste materials as by-products of manufacturing activities. According to the Environmental Protection Agency (EPA, 1978; Crawford, 1987), about 247 million metric tons of hazardous waste subject to regulation by the Environmental Protection Agency under the Resource Conservation and Recovery Act are generated annually. These wastes must be properly handled, transported, treated, stored, and disposed of to safeguard public health and the environment. In addition, billions of tons of mining, agricultural, and other wastes and about 246 million tons of municipal waste are produced (Abelson, 1987). Thoughtless and irresponsible waste disposal practices have led to the contamination of groundwater supplies, the condemnation of wells and other sources of drinking water, and tragic illness for many persons living near waste disposal sites.

The Resource Conservation and Recovery Act (RCRA) passed by Congress in 1976 seeks to bring about development of comprehensive state and local solid-waste programs that include regulation of hazardous wastes from the point of generation through disposal. It includes institution of a manifest system to track these wastes from point of generation to point of disposal and organization of a permit system for waste treatment, storage, and disposal facilities. Standards have been prepared that cover record keeping, labeling of containers, the use of appropriate containers, the furnishing of information on waste composition, and the submission of reports to the EPA or authorized state agencies. Public participation is encouraged and provided for in the development of all programs, guidelines, and regulations under the act.

Research and development activities also produce hazardous wastes that, although in quantities nowhere near those produced in manufacturing, nevertheless constitute part of the total inventory and therefore are subject to the same controls. When the waste is both radioactive and hazardous, it is classified as "mixed" waste, and oftentimes there is no clear approach to disposing of this type of waste. Facilities designed to handle hazardous waste may not be licensed to receive radioactive wastes, and radioactive waste sites may not be allowed to accept or be interested in accepting hazardous wastes. Thus, users of radioactive materials that are also classified as hazardous have to devote considerable time, energy, and money to finding ways to dispose of these wastes. They cannot throw materials of any kind into the radioactive waste disposal barrels without compunction, seal the drum, and feel relieved of them, just by paying the waste disposal company a fee. The waste disposal company is merely a transporter; it usually does not have the authority to look into the barrels given to it and is not responsible for the safe packaging of the contents. The mark of the originator of the waste remains with it unto perpetuity. Furthermore, the shipper must look into the credentials of the waste disposal company to determine that the wastes will be handled properly. Waste disposal companies have gone bankrupt, causing the wastes to revert back to the shipper; others have disposed of wastes in ways that hardly conformed with the regulations (Raloff, 1979). For example, the nation's largest handler of solid and chemical wastes was accused of mixing toxic wastes with used motor oil that was then handed over without charge to contractors as surfacing material for roads. Other companies have engaged in "midnight dumping," dumping the wastes covertly by the side of the road when no one was looking. In these instances, the original shipper could well be responsible for paying the costs of correcting the situation. Until the political process works out practical means of disposing of mixed wastes, on-site storage may be the only option in some instances.

18.7   Volume Reduction in Waste Disposal

With disposal costs continually rising, and the continued availability of sufficient capacity at existing sites to handle all the waste generated uncertain, it is important that users actively work to reduce the volume of radioactive waste generated. Storage for decay is a practical management method for short-lived wastes. The NRC requires a minimum storage period of 10 half-lives, which is usually manageable for short-lived radionuclides like iodine-131 and phosphorus-32 but is more difficult for iodine-125 (requiring a minimum of 600 days in storage). It usually pays for institutions to set up central storage facilities if significant volumes of wastes contaminated by iodine-125 are produced. Such facilities require careful management to store the wastes efficiently and retrieve them after the required decay period. The wastes may be stored in fiber drums and then incinerated as nonradioactive trash if they do not contain hazardous chemicals. Drums must be monitored before they are released to ensure that the residual radiation is insignificant. Complete and accurate records must be kept.

Compaction on site leads to large reductions in volumes of solid waste that is disposed of by shipment to a commercial burial site. Some commercial companies utilize supercompactors to minimize the volume to be buried.

Users can reduce waste designated for disposal as radioactive by carefully monitoring all waste generated and designating as radioactive only those wastes that give positive readings. Glassware containing trace amounts of radioactivity can be rinsed and disposed of with nonradioactive laboratory waste after being monitored. An effective volume reduction program will require special processing by the institution for each of the different types of radionuclides, so wastes disposed through the institution should be segregated by the user and labeled according to the radionuclide and other information required. Solid waste should be packed in clear plastic bags to allow for inspection of the contents. Sharp objects should be packed in puncture-proof containers.

Some institutions have realized large savings in disposal of scintillation vials by acquiring vial crushers, separating the contents from the crushed vials, rinsing the fragments and disposing of them as nonradioactive while the contents of the vials are disposed as bulk liquids. The use of mini-vials produces substantial savings in both volumes of scintillation cocktail used and in the volume of vials for disposal.

These and other methods can reduce volumes to a small fraction of that disposed of without treatment. The effort to reduce volume is labor intensive and therefore not popular in a busy research laboratory. Both the regulatory process and the economics of waste disposal, however, mandate the establishment of radioactive waste management programs that take advantage of decay-in-storage, special packaging techniques, and increased training and awareness. The appropriate allocation of time and resources can effect large volume reductions in radioactive waste. The tracking of a program over the years through metrics, such as number on staff, volume of waste processed annually, and the amount of waste generated annually per radiation worker and per laboratory, can be very helpful in evaluating and increasing its effectiveness (Ring et al., 1993).

18.8   The Designation of *De Minimus* Concentrations of Radioactivity

The high cost of disposal of only slightly radioactive waste points out the waste of resources when controls designed for relatively small volumes of truly hazardous wastes are applied to the much larger volumes of materials with minimal radioactivity resulting from physical, biological, and medical research with radioactive tracers. Users who must dispose of materials with trace amounts of radioactivity desperately need standards to define levels that pose no significant radiation hazard to the public or the environment, standards which allow for disposal in accordance with the regulations for comparable nonradioactive substances. Such levels are known as *de minimus* levels (*de minimus* comes from the Latin maxim *de minimus non curat lex*, "the law does not concern itself with trifles").

So far, government regulations have not attempted to define *de minimus* levels, although the following numerical guides to meet the criterion "as low as is reasonably achievable" (ALARA) for radioactive material in light water nuclear power reactor effluents may be used to suggest appropriate values (10CFR50, Appendix I).

1. The calculated annual total quantity of all radioactive material above background to be released from each light-water-cooled nuclear power reactor to unrestricted areas will not result in an estimated annual dose or dose commitment from liquid effluents . . . in excess of 3 mrem to the total body or 10 mrem to any organ.

2. The calculated annual total quantity of all radioactive material above background to be released from each light-water-cooled

nuclear power reactor to unrestricted areas will not result in an estimated annual air dose from gaseous effluents at any location near ground level which could be occupied by individuals in unrestricted areas in excess of 20 mrem for gamma radiation or 20 mrem for beta radiation.

3. The calculated annual total quantity of all radioactive iodine and radioactive material in particulate form above the background to be released from each light-water-cooled nuclear power reactor in effluents to the atmosphere will not result in an estimated annual dose or dose commitment from such radioactive iodine and radioactive material in particulate form for any individual in an unrestricted area from all pathways of exposure in excess of 15 mrem to any organ.

Another approach is to set *de minimus* levels at some level comparable to variations in the natural environment. Recommended values have been in the range of 1 mrem per year to the total body or 3 mrem per year to individual organs (Rodger et al., 1978). Guidance on *de minimus* values may also be obtained from values of concentrations exempt from a license (10CFR20.14), quantities exempt from a license (10CFR20.18), and the regulations of the Department of Transportation (49CFR173.389, par. 5e). Here it is stated that materials in which the estimated specific activity is not greater than 0.002 microcuries (74 Bq) per gram of material, and in which the radioactivity is essentially uniformly distributed, are not considered to be radioactive materials.

18.9 Natural Radioactivity as a Reference in the Control of Environmental Releases

While the use of radioactive materials must be strictly controlled to prevent excessive releases to the environment that can affect the public health, it should be borne in mind that the environment is naturally very radioactive. One of the ways to assess the significance of the disposal of radioactive waste materials is by comparison with naturally occurring radioactivity. It includes radioactivity in the air (as radioactive gases or particles), in the ground, in rainwater, in groundwater, in building materials, in food, and in the human body. The levels vary appreciably in different locations. The naturally occurring radionuclides also differ greatly in their toxicities; some radionuclides rank among the most hazardous but others rank among the least hazardous.

Natural radioactivity in the environment originates from a variety of sources. The most significant are the radionuclides potassium-40, uranium-238, and thorium-232, which were produced when the universe was created some ten billion years ago and remain in significant quantities today because of their long half-lives (greater than a billion years). When they decay they are followed by many additional radioactive products with shorter half-lives, such as radium-226 (1,960 years), radon-222 (3.8 days), polonium-214 (10^{-5} sec), and polonium-210 (120 days). Except for potassium-40, the preceding radionuclides emit alpha radiation and are considered to be highly toxic.

All of the radionuclides listed except one are solids and are distributed throughout the ground, from which they are taken up by vegetation or dissolved in groundwater. One radioactive decay product, radon-222, is a noble gas. It originates from the decay of the radium in the ground, but it diffuses out of the ground and reaches significant concentrations in the atmosphere, particularly when the air is still. It also leaks into buildings, where the levels reached depend on the concentration in the ground, on cracks and other openings in the building, on the building's ventilation, and on pressure differentials between the building and the soil. Heat conservation measures in buildings in cold climates result in minimizing air exchange with the environment and serve to increase radon levels. Many studies have been and are being conducted of radon levels in buildings throughout the world, and unacceptably high levels of radon in many homes, schools, and commercial buildings have been found. Indoor radon pollution is now recognized as a major public health problem requiring remedial action. Most of the dose from radon is not caused by the decay of radon but of the subsequent decay products. The decay of each radon atom is followed by six successive decays, producing radionuclides which emit alpha, beta, and gamma radiation. The decay products form radioactive aerosols in the air, which are breathed in and retained in the lungs and which are also responsible for contamination of the ground, food, and water.

Radionuclides are also generated continuously from the action of cosmic radiation on elements in the atmosphere. The most significant are carbon-14 and hydrogen-3 (tritium). Both emit very low energy beta radiation and are among the least hazardous of radioactive materials.

The cosmic radiation and the gamma radiation emitted by radioactive materials in the ground are responsible for large differences in external radiation doses in different places (NCRP, 1987a, Report 94). For example, at 1.6 km (1 mile) altitude, the cosmic ray annual dose of 0.45 mSv is 0.17 mSv/yr greater than the dose at sea level. Neutrons contribute an additional 0.30 mSv at 1.6 km and 0.06 mSv at sea level. Values of annual doses in the United States (including both terrestrial and cosmic radiation but not neutrons) range in various locations from 0.32 mSv to 1.97 mSv, a total difference of 1.65 mSv. Residents of the city of Denver receive a whole-body dose of 1.25 mSv/yr, compared with 0.65 mSv/yr to inhabitants of the Atlantic and Gulf coastal states and 0.80 mSv/yr to the majority of the U.S. population. There are also large differences in radioactivity in the air, primarily due to the naturally occurring radioactive gas radon-222. Concentrations of ^{222}Rn in outdoor air range from 0.74 to 37 pCi/m^3 (0.027 to 1.4 Bq/m^3). The corresponding average dose rates to the lungs (from the radon decay products) range from 0.2 to 10 mSv/yr. Much higher levels are found indoors and lung doses of many tens of millisieverts per year can be imparted from continuous exposure to radon in some homes. Variations in radium-226 content in the diet produce variations in the dose to bone of about 0.10 mSv/yr around an average annual dose of about 1 mSv/yr.

Despite the large differences in radiation levels, very few people give any thought to natural radioactivity in selecting a place to work or live. There is no evidence that these variations are significant in affecting the incidence of cancer or other diseases. In fact, one can select areas throughout the country where the cancer incidence goes down as the natural radiation level increases (Frigerio and Stowe, 1976). Yet, the maximum whole-body doses resulting from the Three Mile Island nuclear power plant accident (which caused so much concern) were not much different from variations in levels in various parts of the country, and these maximum doses were imparted to only a few individuals.

Because of the natural abundance of radioactive materials, the disposal of sufficiently small quantities of radioactive materials via the ground and the air would not produce any significant change in the existing levels. Typical levels of radioactivity in the ground and in the air are given in Table 5.7. These are quite significant, and it must be noted that the radionuclides are not contained but are accessible to groundwater, to food crops, and to the atmosphere. Discharges that contribute only a small fraction to the activity already present in the environment should have no noticeable effect on the public health. Of course, the existence of natural levels of radioactivity does not give a license to pollute indiscriminately. The release of low levels of pollutants should be weighed against the benefit to society of the activities that produced the pollutants. In any event, the releases should be reviewed for compliance with the ALARA principle, which requires that the discharge of pollutants to the environment be kept as low as reasonably achievable and not merely in compliance with air-pollution regulations.

Table 5.7 Activity in the environment and in people of naturally occurring long-lived radionuclides.

Radionuclide	Half-life (years)	Global inventory (millions of curies)	Activity in soil to depth of 2 meters		Concentration in		Activity in body (pCi)
			1 acre (mCi)	1 km² (mCi)	Air (pCi/m³)	Water (pCi/m³)	
Alpha emitters							
Uranium-238	4.5 billion		10	2,520	0.00012		26
Thorium-232	14 billion		10	2,520	0.00003		
Radium-226	1,600		10	2,520	0.00012	1,000–10,000	120
Radon-222	3.82 days	25 (atmosphere)			70	(well water)	
Polonium-210	138 days	20	13	3,240	0.0033	100	200
Beta emitters							
Potassium-40	1.3 billion		175	43,200			130,000
Carbon- 14	5,730	300					87,000
Hydrogen-3 (tritium)	12.3	34 (natural) 1,700 (fallout)		0.038	6,000–24,000		
Lead-210	21	20	13	3,240	0.014	100	

Sources: UNSCEAR, 1977; NCRP, 1975. For a comprehension treatment, see Eisenhud and Gesell, 1997.

Notes: 1 acre = 4,047 m²; 1 km² = 247 acres. Assume soil density of 1,800 kg/m³. Mean value of radon emanation rate from soil is 0.42 pCi/m²-s, range 6×10^{-3} to 1.4 pCi/m²-s.

Multiply mCi by 37 to convert to MBq; multiply pCi by 0.037 to convert to Bq; multiply Ci by 37 to convert to GBq.

19 USE OF RADIOACTIVE MATERIALS IN ANIMALS

Injection of radioactive materials into animals should be performed in trays lined with absorbent material. Cages housing animals injected with radioisotopes should be labeled as to radionuclide, quantity of material injected per animal, date of injection, and user. Metabolic-type or filter cages should be used if contamination is a problem. These cages should be segregated from those housing other animals. Animal excreta may be disposed of via the sewer if the concentration is in accordance with limits applicable to liquid waste and the excreta are not mixed with sawdust or wood shavings; otherwise, the excreta may be placed in plastic bags and disposed of as solid wastes.

Adequate ventilation must be provided in instances where animals are kept after an injection with radioactive materials that may become volatilized and dispersed into the room at significant levels. Animal handlers must be indoctrinated by the responsible investigator as to the dose levels, time limitations in the area, and the handling requirements of the animals and excreta.

20 TRANSPORTATION OF RADIONUCLIDES

20.1 Transportation within the Institution

Within institutional grounds, all radionuclides must be transported in nonshatterable containers or carrying cases with the cover fastened securely so it will not fall off if the case is dropped. Shielding of containers should follow federal transportation regulations, which limit dose rates to less than 2 mSv/hr (200 mrem/hr) in contact with the container and 0.1 mSv/hr (10 mrem/hr) at 3 ft from the surface of the container (dose rates should be reduced as much below these limits as practicable in accordance with the ALARA principle). There should not be any removable radioactive contamination on the surface of the container, but in the event there is contamination it should be below 2,200 dis/min per 100 cm^2 for beta or gamma contamination and below 220 dis/min per 100 cm^2 for alpha contamination (limits for non-fixed radioactive contamination of the U.S. Department of Transportation).

A route should be chosen to encounter minimal pedestrian traffic. The cart used for transportation should be completely leak proof. Otherwise, should any leakage occur inside the cart, highly radioactive contamination could be dripped throughout the route of the cart with most distressing results, mandating herculean efforts for cleanup.

20.2 Mailing through the U.S. Postal Service

Government regulations pertaining to the packaging and shipment of radioactive materials are quite complicated. The Postal Service accepts the DOT definition of material that is not considered as radioactive, material with a specific activity less than 74 Bq/g (0.002 μCi/g). If the specific activity of material to be shipped is greater than 74 Bq/g (0.002 μCi/g), the shipper has the problem of inquiring into the existence and content of applicable regulations.

The U.S. Postal Service does not allow the mailing of any radioactive materials by air, or any package that bears any of the Department of Transportation's "Radioactive" labels (white-I, yellow-II, or yellow-III) by domestic surface transportation or international mail, but it does allow the mailing of "small quantities" of radioactive materials and certain radioactive manufactured articles that are exempt from specific packaging, marking, and labeling regulations prescribed for higher levels of radioactivity by the Department of Transportation. The regulations may be reviewed on the postal services website, *www.usps.gov*. The package limits are one-tenth the limits for packages designated as "limited quantity" by the DOT and are given for selected radionuclides in Table 5.8. Note that two categories

Table 5.8 Maximum mailable quantities of some radionuclides in solid form. These values are one-tenth the upper limits for the designation of "limited quantities" assigned to radionuclides by the Department of Transportation. Values for liquids are one-tenth those listed here.

Radionuclide	Radiochemicals —normal form (mCi)	Sealed sources —special form (mCi)
^{3}H (gas, luminous paint, adsorbed on solid)	100	100
^{3}H (water)	100	100
^{3}H (other forms)	2	2
^{14}C	6	100
^{24}Na	0.5	0.5
^{32}P	3	3
^{35}S	6	100
^{36}Cl	1	30
^{42}K	1	1
^{45}Ca	2.5	100
^{51}Cr	60	60
^{55}Fe	100	100
^{59}Fe	1	1
^{60}Co	0.7	0.7
^{64}CU	2.5	8
^{65}Zn	3	3
^{67}Ga	10	10
^{75}Se	4	4
^{82}Br	0.6	0.6
^{85}Kr (uncompressed)	100	100
^{86}Rb	3	3
^{90}Sr	0.04	1
^{109}Cd	7	100
^{111}In	2.5	3
^{125}I	7	100
^{131}I	1	4
^{133}Xe (uncompressed)	100	100
^{198}Au	2	4
^{201}T1	20	20
Microspheres		
^{46}Sc	0.8	0.8
^{85}Sr	3	3
^{95}Nb	2	2
^{103}Ru	2.5	3
^{113}Sn	6	6
^{114}In	2	3
^{141}Ce	2.5	30
^{153}Gd	10	20

Table 5.8   (continued)

Radionuclide	Radiochemicals —normal form (mCi)	Sealed sources —special form (mCi)
Alpha emitters		
^{222}Rn	0.2	1
^{226}Ra	0.005	1
^{228}Th	0.0008	0.6
^{232}Th	Unlimited	Unlimited
^{238}U	Unlimited	Unlimited
^{239}Pu	0.0002	0.2

Note: Normal form is defined as material that could be dispersed from the package, contaminate the environment, and present an inhalation and ingestion problem. Typically this class includes liquids, powders, and solids in glass, metal, wood, or cardboard containers. *Special form is* defined as material that is encapsulated and is not likely to be dispersed, contaminate the environment, and present an inhalation and ingestion problem. The hazard is only from direct radiation from the source.

Up to 10,000 times the quantities listed here can be shipped by common or contract carriers as type-A packages.

Multiply mCi by 37 to convert to MBq.

are identified, "special form" for sources that are encapsulated and meet stringent test requirements (49CFR173.469) and "normal form" for all other items. These exemptions allow the mailing of up to 37 MBq (1 millicurie)[10] of the less hazardous beta-gamma emitters in common use, provided the following conditions are met:

1. Strong, tight packages are used that will not leak or release material under typical conditions of transportation. If the contents of the package are liquid, enough absorbent material must be included in the package to hold twice the volume of liquid in case of spillage.
2. Maximum dose rate on surface is less than 0.5 mrem/hr.
3. There is no significant removable surface contamination (that is, less than 2,200 dis/min/100 cm^2 beta-gamma; 220 dis/min/100 cm^2 alpha).
4. The outside of the inner container bears the marking "Radioac-

10. Some exceptions are 0.04 (mCi) for ^{90}Sr; 2 Ci of tritium per article as a gas, luminous paint, or absorbed on solid material; or 7 mCi of ^{125}I. The regulations should be checked for the limits set for a specific radionuclide. For a brief summary of mailing regulations, request the U.S. Postal Service pamphlet *Radioactive Matter.*

tive Material—No Label Required." The identity or nature of the contents must be stated plainly on the outside of the parcel. The full name and address of both the sender and addressee must be included on the package.

20.3   Shipment of "Limited Quantities"

The packaging and transportation in interstate or foreign commerce of radioactive materials not shipped through the postal service are governed by regulations issued by the U.S. Department of Transportation (DOT),[11] website *www.dot.gov*. The NRC has identified in its regulations (10CFR71.5) those sections in the DOT regulations of most interest to users of radionuclides. They include packaging, marking and labeling, placarding, accident reporting, and shipping papers, and the most recent regulations should be checked before making any shipments. If shipments can be limited to what DOT defines as limited quantities, regulations are much simpler, as these quantities are exempt from specific packaging, marking, and labeling requirements. The containers must be strong, tight packages that will not leak under conditions normally encountered in transportation. The radiation level may not exceed 0.5 mrem/hr (0.005 mSv/hr) at any point on the surface and the removable contamination on the external surface may not exceed 2,200 disintegrations per minute (dpm)/100 cm^2 beta-gamma and 220 dpm/100 cm^2 alpha averaged over the surface wiped. The outside of the inner packaging must bear the marking "Radioactive" and a notice must be included in the package that includes the name of the consignor or consignee and the following statement: "This package conforms to the conditions and limitations specified in 49CFR173.421 for excepted radioactive material, limited quantity, n.o.s., UN2910" ("n.o.s." stands for "not otherwise specified"). There are other exceptions for instruments and articles. Maximum quantities that can be shipped as limited quantities depend on the radionuclide shipped and whether the material is in "special" or "normal" form (see section 20.2, above). Limits for solids are ten times those for liquids, and the limit for most of the beta-gamma radionuclides used in tracer research is greater than 10 mCi as solids and 1 mCi (37 MBq) as liquids. Specific limits for shipment as liquids include 2 mCi (74 MBq) for hydrogen-3 in organic form, 6 mCi (222 MBq) for carbon-14, 7 mCi (259 MBq) for iodine-125, and 1 mCi (37 MBq) for iodine-131. Limits for selected other radio-

---

11. The regulations of the Department of Transportation incorporate recommendations of various government agencies, including the Nuclear Regulatory Commission, Federal Aviation Agency, Coastguard, and Post Office. See DOT, 1983; and the Code of Federal Regulations, Title 49 (Transportation).

nuclides are 10 times the values give in Table 5.7. It is not possible to present the detailed regulations here; a copy of the regulations should be obtained if a shipment must be made (see 49CFR173.421–443).

20.4   Shipment of "Low-Specific-Activity" Materials

If the amount of activity to be shipped is greater than a limited quantity, some requirements of the regulations are still exempted if the material can be classified as "low-specific-activity" (LSA).[12] The simplest requirements apply if the shipment is sent "exclusive use" or "sole use." This means that the shipment comes from a single source and all initial, intermediate, and final loading and unloading are carried out in accordance with the direction of the shipper or the receiver. Any loading or unloading must be performed by personnel having radiological training and resources appropriate for safe handling of the shipment. Specific instructions for maintenance of exclusive-use shipment controls must be issued in writing and included with the information that accompanies the shipment (49CFR173.403).

LSA materials shipped as exclusive use are excepted from specific packaging, marking, and labeling requirements. The materials must be packaged in strong, tight packages so that there will be no leakage of radioactive material under conditions normally incident to transportation. The exterior of each package must be stenciled or otherwise marked "Radioactive—LSA". There must not be any significant removable surface contamination and external radiation must meet limits applicable to radioactive packages.

When LSA materials are part of another shipment, they must be contained in packaging that meets the DOT specifications for type-A packages, with just a few exemptions (49CFR173.425).

Objects of nonradioactive material that have surface radioactive contamination below $1\ \mu Ci/cm^2$ (37,000 Bq/cm^2), averaged over 1 square meter, for almost all radionuclides can be shipped unpackaged, provided the shipment is exclusive use and the objects are suitably wrapped or enclosed (49CFR173.425).

20.5   Shipment of Type-A Packages

Most shipments in quantities or concentrations above the "exempt" level will fall into the type-A category. Typical packaging includes fiberboard boxes, wooden boxes, and steel drums strong enough to prevent loss

12. Specific activity less than 0.3 mCi/g for the less hazardous beta-gamma emitters. Details are given in 49CFR173.392.

or dispersal of the radioactive contents and to maintain the incorporated radiation shielding properties if the package is subjected to defined normal conditions of transport. The maximum quantities that can be shipped as type-A packages are 10,000 times the values in Table 5.7. Containers certified to meet type-A requirements are available from commercial suppliers. Type-B packaging is for high-level sources and is designed to withstand certain serious accident damage test conditions.

If radioactive material is transported in a cargo-carrying vehicle that is not exclusively for the use of the radionuclides, the dose rate cannot exceed 2 mSv/hr (200 mrem/hr) at the surface of the package and 0.1 mSv/hr (10 mrem/hr) at 1 meter. If the vehicle is for the radionuclides only and the shipment is loaded and unloaded by personnel properly trained in radiation protection, the dose rate can be 10 mSv/hr (1,000 mrem/hr) at 1 meter from the surface of the package, 2 mSv/hr (200 mrem/hr) at any point on the external surface of the vehicle, 0.1 mSv/hr (10 mrem/hr) at 2 m from the external surface of the vehicle, and 0.02 mSv/hr (2 mrem/hr) in any normally occupied position in the vehicle. Special written instructions must be provided to the driver.

The following labels must be placed on packages containing radioactive materials unless the contents are exempt as "limited quantities." Packages carrying these labels are not mailable: *(a)* a radioactive white-I label, if the dose rate at any point on the external surface of the package is less than 0.005 mSv/hr (0.5 mrem/hr) and the contents are above a "limited quantity"; *(b)* a radioactive yellow-II label if the dose rate is greater than 0.005 mSv/hr (0.5 mrem/hr) but less than 0.5 mSv/hr (50 mrem/hr) on the surface and less than 0.01 mSv/hr (1 mrem/hr) at 1 meter; and *(c)* a radioactive yellow-III label if the dose rate on the surface is greater than 0.5 mSv/hr (50 mrem/hr) or greater than 1 mrem/hr at 1 m. Each package in an exclusive-use LSA shipment must be marked "Radioactive—LSA." There are labeling exemptions for instruments and manufactured articles containing activity below prescribed limits (49CFR173.422) and for articles containing natural uranium or thorium (49CFR173.424).

The yellow labels have an entry for the "transport index." This is the maximum radiation level in millirem per hour (rounded up to the first decimal place) at one meter from the external surface of the package. The number of packages bearing radioactive yellow-II or radioactive yellow-III labels stored in any one storage area must be limited so that the sum of the transport indexes in any individual group of packages does not exceed 50. Groups of these packages must be stored so as to maintain a spacing of at least 6 meters from other groups of packages containing radioactive materials.

Packages shipped by passenger-carrying aircraft cannot have a transport index greater than 3.0.

A vehicle has to be provided with *radioactive*[13] signs if it is carrying packages with yellow-III labels or is carrying LSA packages as an exclusive-use shipment. Users sending radioactive material by taxi should ensure that the taxi will not carry passengers and the package is stored only in the trunk (49CFR177.870). Users should also determine their responsibilities as shipper, including provision of shipping papers and shipper's certification. Users who have a license to transport radioactive materials in their own cars should be aware that their insurance policy may contain an exclusion clause with regard to accidents involving radioactive materials.

In the event of a spill, DOT regulations state (173.443) that vehicles may not be placed in service until the radiation dose rate at any accessible surface is less than 0.5 mrem/hr and removable contamination levels are less than 2,200 dpm/100 cm^2 beta-gamma and 220 dpm/100 cm^2 alpha.

20.6 Shipping Papers and Shipper's Certification

All radioactive-materials shipments must be accompanied by shipping papers describing the radioactive material in a format specified by the Department of Transportation. For "limited quantities," the information must include the name of the consignor or consignee and the statement, "This package conforms to the conditions and limitations specified in 49CFR173.421 for excepted radioactive material, limited quantity, n.o.s., UN2910." Similar wording applies to several other types of excepted articles. Incidents of decontamination (such as of vehicles or packages) associated with the shipment must be reported.

Shipping papers for activities greater than "limited quantities" must include the proper shipping name and identification number in sequence; the name of each radionuclide; physical and chemical form; activity in terms of curies, millicuries, or microcuries; the category of the label, for example, radioactive yellow-II; and the transport index. Abbreviations are not allowed unless specifically authorized or required. The following certification must also be printed on the shipping paper: "This is to certify that the above-named materials are properly classified, described, packaged, marked and labeled, and are in proper condition for transportation according to the applicable regulations of the Department of Transportation."

13. For current regulations as to the exact wording and dimensions of the sign, see Code of Federal Regulations, Title 49.

21   CONTAMINATION CONTROL

Loose contamination should not be tolerated on exposed surfaces, such as bench tops and floors, and should be removed as soon as possible. Work areas should be monitored for contamination before and after work with radioactive materials. Library books, periodicals, or reports must not be used in areas where there is a reasonable possibility of their becoming contaminated with radioactive materials. Contaminated equipment must be labeled, wrapped, and stored in a manner that constitutes no hazard to personnel, and there must be no possibility of spread of contamination.

All spills of radioactive material must be cleaned up promptly (IAEA, 1979). A survey must be made after cleanup to verify that the radioactive material has been removed. Cleaning tools must not be removed or used elsewhere without thorough decontamination. (Instructions for handling spills and other accidents are given in Appendix A to this part.)

The hazard from a contaminated surface is difficult to evaluate. One mechanism of intake of the contamination by humans is through dispersion of the contamination into the air and subsequent inhalation. Some of the contamination may be transferred from the surface to the hands and then from the hands to the mouth to be swallowed.

How contaminated can we allow a surface to be without worrying about a hazard to individuals? In most cases the question becomes academic, as other considerations force removing the contamination to as great a degree as possible. At least in research laboratories, where contamination on surfaces can spread to counting equipment and complicate low-level measurements, there is strong motivation to keep contamination levels as low as possible, well below levels that could cause harm to individuals. Where low-level counting is not a factor, it is still accepted practice to keep surfaces as clean as practicable. Where work is done with "hot particles," that is, with particles of such high specific activity that single particles small enough to be inhaled could produce appreciable local doses, contamination control has to be very stringent.

21.1   Monitoring for Contamination

The most widely used monitor for beta-gamma contamination is the Geiger-Mueller (G-M) counter. The pancake G-M tube is the most commonly used detector. It is in the form of a short cylinder, about 5 cm in diameter by 1 cm high, with a window thickness equivalent to 0.03 mm unit-density material. The area of its window, four times the area of the 1 inch end-window G-M tube (which preceded it), and its relatively low radiation background, about 70 counts per minute (cpm), make it very effective in monitoring laboratory surfaces for beta contamination. A doubling of the background counting rate might be considered a positive indication of contamination. Monitor performance checks are discussed in section 4.4.

The best monitor for alpha contamination is one employing a gas flow proportional counter with a very thin window for the detector. The bias is adjusted so only the pulses due to alpha particles are counted. Scintillation detectors are easier to use, but care must be taken to prevent light leaks in the detector covering, as they produce spurious counts, and the background is higher.

Monitoring for contamination is done by slowly moving the detector over the suspected surfaces (Clayton, 1970). It is very useful to have an aural signal, such as from earphones or a loudspeaker, since small increases of radiation above the background are detected most easily by listening to the clicks. It is also easier to pay attention to the surface being monitored if the meter does not have to be watched. Measurements of beta-gamma contamination with a G-M counter are taken with and without interposing a shield that stops beta particles. The difference between the readings gives the contribution from beta radiation.

For monitoring loose contamination, an operation known as a wipe test is performed. A piece of filter paper is wiped over an area of approximately 100 cm^2 and then counted with a shielded end-window G-M detector. It has also been found convenient to use liquid scintillation counting by inserting the filter paper into a liquid scintillation vial. This method is attractive because liquid scintillation counting systems are equipped with automatic sample changers and printouts and are very efficient for processing a large number of samples.

21.2 Decontamination of Equipment and Buildings—Limits for Uncontrolled Release

The removal of radioactive contamination from surfaces is a battle against chemical and physical binding forces; the weapons include chemical and physical methods of decontamination. The literature on decontamination and cleaning is voluminous (Ayres, 1970; Lanza, Gautsch, and Weisgerber, 1979; Nelson and Divine, 1981; Osterhout, 1980). Radiological health handbooks (BRH, 1970; Slaback, Birky, and Shleien, 1997) list detailed cleaning procedures. Technicians report that cleaning agents normally used in the laboratory and even good household cleaners work quite well for most routine problems. Bleach should not be used to decontaminate radioactive iodine as it acts to release the iodine to the air.

The effort required for decontamination depends very strongly on how clean property with potential residual activity must be before it can be released from regulatory control. Considerable thought has been given to acceptable limits (Fish, 1967; Healy, 1971; Clayton, 1970; Shapiro, 1980; HPS, 1988, 1999). The preferred approach in setting a standard is to base it on the limitation of dose in accordance with regulatory requirements. However, the potential for personnel exposure from contaminated facilities is very site specific and does not lend itself to generalized modeling or rigorous technical analysis. Accordingly, one approach to regulatory control is to set performance standards, to choose the lowest limits that by consensus are achievable, using state of the art practices. This process usually incorporates enough conservatism to assure risks well below those normally encountered in daily life.

Performance standards promulgated by the Nuclear Regulatory Commission for decontamination of facilities are given in NRC Regulatory Guides 1.86 (NRC, 1974) and 8.23 (NRC, 1981b). The Health Physics Society issued a performance standard in 1988 (HPS, 1988), "Surface Radioactivity Guides for Materials, Equipment and Facilities to be Released for Uncontrolled Use." Because the regulatory organizations looked upon exposure assessment as an important factor in supporting the setting of standards, the Society followed the performance standard with the development of a dose-based standard. A dose limit of 10 microsieverts (1 mrem) per year that could be incurred as the result of release of materials or equipment with potential residual radioactivity was adopted. This was consistent with the recommendations of the International Atomic Energy Agency and was selected for consistency with international commerce. The standard differed from previous work in providing only a single limit for surface contamination instead of separate limits for fixed and removable contamination, since the scenario analyses used assumed all the material to be in removable form. However, it noted that measurements of removable surface contamination might be appropriate and included as part of survey programs. The scope of the standard was also expanded to include both surface and volume contamination. A critical group of potentially exposed persons likely to have the closest contact with the released material was identified. Limits for any potential residual activity were then derived, based on several exposure pathway assessments, that would ensure that an average member of the group could not receive an annual dose greater than $10 \, \mu Sv$. The standard developed by the committee went through the many steps required by the American National Standards Institute (ANSI) and was approved and issued in August 1999 as an official ANSI standard, ANSI/HPS N13.12-1999, "Surface and Volume Radioactivity Standards for Clearance." The standard was the culmination of an effort that had begun 35 years earlier and had produced during its development the 1988 Health Physics Society performance standard.

While release limits were calculated for each of the radionuclides considered, they were not applied separately. Instead, each radionuclide was placed into one of four broad groups for screening prior to release, with activity limits differing by factors of 10. The use of a small number of groups was also the format followed in the development of the previous performance standards. It had the advantage of ease of application and reflected the broad range of limits that were derived for the different exposure scenarios. The group with the highest limits contained radionuclides which, on the basis of dose, could have had much higher limits, but they were not necessary or operationally justifiable. The limits promulgated in the ANSI/HPS standard are given in Table 5.9, along with limits developed previously as performance standards.

22 Personnel Contamination and Decontamination

When hands, body surfaces, clothing, or shoes become contaminated in the absence of injuries, steps should be taken as soon as possible to remove loose contamination (BRH, 1970). If injuries occur, medical care has priority and must include measures to prevent contamination on the body from spreading or from getting into wounds. Washing with a mild soap or a good detergent and water is generally the best initial approach. This is followed by harsher methods when necessary, such as mild abrasive soap, a paste made up of 50 percent cornmeal (abrasive) and 50 percent Tide, rubbing briefly and rinsing, and white vinegar (works on phosphorous-32 nucleotides). Scrubbing of skin should stop when it gets red so it is not pierced. Researchers can suggest concoctions that remove contaminants by exchange. Specific instructions for personnel decontamination in an emergency are given in Appendix A to this part.

When monitoring of hands indicates that the tips of the fingers are contaminated, clipping the fingernails may remove most of the residual activity after washing. When other measures still leave residual contamination on the hands, it may be worthwhile to wear rubber gloves for a day or so. The induced sweating has been reported as very effective in certain instances.

23 Leak Tests of Sealed Sources

Sealed radioactive sources must be checked for leakage when received and on a regular schedule thereafter. The source is wiped or "smeared" with a filter paper or other absorbent material, which is then counted for

Table 5.9 Surface radioactivity guides.

Radioactive material	Dose-based standard ANSI/HPS N13.12, 1999[a] (dpm/100 cm² or dpm/100 g)	Performance standards, removable/total HPS standard, 1988 (dpm/100 cm²)	NRC regulatory guide 8.23 (dpm/100 cm²)
Radium, thorium, transuranics; Po-210, Pb-210, and the total activity of the decay chains	600		
All alpha emitters except natural or depleted uranium and natural thorium; Pb-210, Ra-228[a]		20/300	20/100[e]
Uranium; Na-22, Mn-54, Co-58, Co-60, Zn-65, Sr-90, Nb-94, Ru-106, Ag-110m, Sb-124, Cs-134, Eu-152, Eu-154, Ir-192.	6,000		
Sr-90; I-125,126,129,131[b]		200/5,000	200/1,000[f]
Na-24, Cl-36, Fe-59, Cd-109, I-131, I-129, Ce-144, Au-198, Pu-241	60,000		
All beta and gamma emitters not otherwise specified except pure beta emitters with $E_{max} \leq 150$ keV[c]		1,000/5,000	1,000/5,000
H-3, C-14, P-32, S-35, Ca-45, Cr-51, Fe-55, Ni-63, Sr-89, Tc-99, In-111, I-125, Pm-147	600,000		
Natural or depleted uranium, natural thorium, and their associated α-emitting decay products[d]		200/1,000	1,000/5,000[g]

Notes: The standards should be consulted if their recommendations are to be adopted in a specific application. All values of disintegrations per minute per 100 cm² (dpm/100 cm²)are to one significant figure.

ANSI/HPS N13 does not provide separate total and removable values for contamination. All contamination is considered removable in pathway exposure assessment. The screening levels shown are used for either surface activity concentrations (in units of dpm/100 cm²) or volume activity concentrations (in units of dpm/100 g). For decay chains, the screening levels represent the total activity (i.e., the activity of the parent plus the activity of all decay products) present. Multiple surface measurements are averaged over a surface area not to exceed 1 m². For items with a surface area less than 1 m², an average 100 cm² over the entire surface area shall be derived for each item. Multiple volumetric measurements are averaged over a total volume not to exceed 1 m³ or a mass of 1 metric ton. For items with mass less than 1 metric ton, an average over the entire mass shall be derived for each item. No single measurement made to calculate an average surface activity shall exceed 10 times the surface screening level.

The performance standards allow averaging over one square meter provided the maximum surface activity in any area of 100 cm² is less than three times the guide values.

a. Pb-210 is included because of the presence of an alpha emitter, Po-210, in its decay chain, and Ra-228 is included because of the presence of another alpha emitter, Th-228, in its decay chain.

b. This category lists the radionuclides that are considered to present the greatest hazards as surface radioactivity among those undergoing beta or electron capture decay.

c. The pure beta emitters with maximum energy less than 150 keV are excluded because detection by direct methods is not practical and they must be treated on a case-by-case basis.

d. Unat and Thnat include gross alpha desintegration rates of natural uranium, depleted uranium, uranium enriched to less than 10 percent U-235, Th-232, and their decay products.

e. NRC Guide 8.23 includes in this category Ac-227, I-125, and I-129.

f. NRC Guide 8.23 places I-125 and I-129 in the more restrictive alpha-emitters category. It includes here, in addition to those listed, Th-nat, Th-232, Ra-223, Ra-224, U-232.

g. NRC Guide 8.23 places Th-nat into the more restrictive Sr-90 category.

radioactivity. If the surface dose rate from radiation with significant ranges in air is excessive, for example, greater than 1 rem/min, the wiping of the source must be done with long-handled tools or other adequate means of protection. For the smaller sources, a medical swab may be satisfactory. Often the swab is moistened with ethanol to improve the transfer of any contaminant.[14] A common limit used by regulatory agencies for removable contamination is 0.005 μCi.

Leak tests must be performed at intervals generally not exceeding 6 months. Alpha and beta sources are particularly vulnerable to developing leaks in the covering, which must be thin enough to allow penetration of the particles.

24 Notification of Authorities in the Event of Radiation Incidents

Notification of radiation-protection authorities is required in the event of accidents involving possible body contamination or ingestion of radioactivity by personnel, overexposure to radiation, losses of sources, or significant contamination incidents. Conditions requiring notification of the NRC by its licensees are presented in 10CFR20. Users must report an accident to the radiation-protection office at their institution, which in turn will notify the appropriate government agencies.

25 Termination of Work with Radionuclides

The radiation-protection office must be notified when work with radionuclides is to be terminated at a laboratory. The laboratory must be surveyed thoroughly and decontaminated, if necessary, before it may revert to unrestricted use. The radioactive material in storage must be disposed of or transferred to another authorized location.

One occasionally comes across areas in laboratories or even homes that were contaminated years earlier and were left contaminated without notification to proper authorities. One of the more notorious episodes involved a residence in Pennsylvania that had been severely contaminated with radium by a radiologist before being turned over to an unsuspecting family. Employees of the Pennsylvania Department of Health learned of the possibly contaminated house through hearsay. When they came to sur-

14. A special leak test is used for radium. It is based on the detection of the noble gas radon, a decay product of radium (see Wood, 1968).

vey the house, they found that their meters were reading off scale even before they entered the driveway. With the controls that exist today, episodes of this type are very improbable.

APPENDIX A: EMERGENCY INSTRUCTIONS IN THE EVENT OF RELEASE OF RADIOACTIVITY AND CONTAMINATION OF PERSONNEL

The following instructions cover only the radiation aspects of accidents. If injuries occur, the procedures must be coordinated with appropriate first aid measures and priorities assigned to provide necessary medical care.

A.1   Objectives of Remedial Action

In the event of an accident involving the release of significant quantities of radioactive material, the objectives of all remedial action are to:

(a) Minimize the amount of radioactive material entering the body by ingestion, by inhalation, or through any wounds.

(b) Prevent the spread of contamination from the area of the accident.

(c) Remove the radioactive contamination on personnel.

(d) Start area decontamination procedures under qualified supervision. Inexperienced personnel should not attempt unsupervised decontamination.

A.2   Procedures for Dealing with Minor Spills and Contamination

Most accidents will involve only minor quantities of radioactivity (that is, at the microcurie level).

(a) Put on gloves to prevent contamination of hands. (Wash hands first if they are contaminated as a result of the accident.)

(b) Drop absorbent paper or cloth on spill to limit spread of contamination.

(c) Mark off contaminated area. Do not allow anyone to leave contaminated area without being monitored.

(d) Notify the radiation-protection office of the accident.

(e) Start decontamination procedures as soon as possible. Normal cleaning agents should be adequate. Keep cleaning supplies to the minimum needed to do the job and place them into sealed bags after use. Proceed from the outermost edges of the contaminated area inward, reducing systematically the area that is decontaminated. (This principle may not apply in decontamination of highly radioactive areas, which would require supervision by a radiation-protection specialist.)

(f) Put all contaminated objects into containers to prevent spread of contamination.

(g) Assign a person equipped with a survey meter to follow the work and to watch for the accidental spread of contamination.

A.3 Personnel Decontamination

If personnel contamination is suspected, first identify contaminated areas with survey meter. Do not use decontamination methods that will spread localized material or increase penetration of the contaminant into the body (such as by abrasion of the skin). Decontamination of wounds should be accomplished under the supervision of a physician.

Irrigate any wounds profusely with tepid water, and clean with a swab. Follow with soap or detergent and water (and gentle scrubbing with a soft brush, if needed). Avoid the use of highly alkaline soaps (may result in fixation of contaminant) or organic solvents (may increase skin penetration by contaminant).

Use the following procedures on intact skin:

(a) Wet hands and apply detergent.

(b) Work up a good lather, keep lather wet.

(c) Work lather into contaminated area by rubbing gently for at least 3 minutes. Apply water frequently.

(d) Rinse thoroughly with lukewarm water (limiting water to contaminated areas).

(e) Repeat above procedures several times, gently scrubbing residual contaminated areas with a soft brush, if necessary.

(f) If radiation level is still excessive, initiate more-powerful decontamination procedures after consultation with the radiation-protection office.

For additional details, see Saenger, 1963, and BRH, 1970.

A.4 Major Releases of Airborne Radioactivity as a Result of Explosions, Leakage of High-Level Sealed Gaseous and Powdered Sources

Since it is not possible to present recommendations that apply to all types of accidents, readers are referred to specialized texts (Lanzl, Pingel, and Rust, 1965; NCRP, 1980a, Report 65). Personnel working with high-level sources must receive training from radiation-protection specialists and proceed in accordance with previously formulated accident plans and emergency measures based on hazard analysis of possible types of accidents, potential airborne radioactivity levels, and dose rates.

(a) If possible, cut off the release of radioactive materials from the source to the environment but avoid breathing in high concentrations of radioactive material. Close windows.

(b) Evacuate room and close doors. Remove contaminated shoes and laboratory coats at laboratory door to avoid tracking radioactive material around.

(c) Report incident to radiation-protection office.

(d) Shut off all ventilation, heating, and air-conditioning equipment that can transport contaminated air from the laboratory to other parts of the building.

(e) Shut off hoods if they are connected to other hoods in building or if they are not equipped with exhaust filters.

(f) Seal doors with tape if airborne material is involved and if there is no net flow of air into room (that is, as a result of exhaust through hoods).

(g) Lock or guard the doors and post appropriate signs warning against entry.

(h) Assemble in nearby room with other personnel suspected of being contaminated. Wash off possibly exposed areas of the skin, if there is a delay in performing a survey. Do not leave the control area until you have been thoroughly surveyed for contamination. (Personnel decontamination measures should be instituted promptly if significant contamination is found.)

(i) Major decontamination jobs should be attempted only by personnel experienced in radiation protection.

APPENDIX B: THE REGULATORY PROCESS

The control of radiation is exercised at national, state, and local levels. The regulatory process starts with legislation that provides for a designated authority to develop and enforce regulations (Marks, 1959). The passage of this legislation can be a long, drawn-out process because many interests are involved—the worker, the citizen (as consumer and as guardian of the environment), the industrialist, the politician, and so on. The records of legislative hearings often make fascinating reading and provide valuable reference material in the field of radiation protection.

B.1   Radiation Control at the Federal Level

The main federal agencies now concerned with radiation control are the Nuclear Regulatory Commission (NRC); Department of Health, Education, and Welfare, through the Food and Drug Administration (FDA) and its National Center for Devices and Radiological Health (NCDRH); Environmental Protection Agency (EPA), through the Office of Radiation Programs; Department of Transportation (DOT); Department of Labor through the Occupational Safety and Health Administration (OSHA); and Department of Interior, through the Mining Enforcement and Safety Administration (MESA). Areas of jurisdiction sometimes overlap, and conflicts are generally resolved by agreements between the parties involved on division or delegation of authority. At times, judicial resolution of conflicting interests has been required.

The rule-making procedure by the regulatory agencies is designed to allow input from concerned parties to identify problem areas. Provisions are made for receipt of comments and for public hearings. The NRC also has a mechanism for the initiation of rule-making procedures by members of the public through "Petitions for Rule Making" and "Requests for a Hearing." The rules are accompanied by a rationale that describes the public hearings, analyses, and inputs by interested persons prior to adoption. The details of the regulations are much too extensive and undergo changes too frequently for inclusion here. Concerned persons should request the latest regulations from local, state, and federal agencies. Following are the most important acts of the Congress setting up federal control agencies, selected material in the acts giving purposes and methods of administration, and references to the regulations in the Federal Register.

Atomic Energy Act of 1946, as Amended (Including Energy Reorganization Act of 1974)

The U.S. Congress determined that the processing and utilization of *source material* (natural uranium and thorium), *special nuclear material* (plutonium, uranium-233, or uranium enriched in the isotopes 233 or 235), and by*product material* (radioisotopes produced in the operation of a nuclear reactor) must be regulated in the national interest, in order to provide for the common defense and security, and to protect the health and safety of the public. It therefore provided in this act for a program for government control of the possession, use, and production of atomic energy. The control is exercised by the Nuclear Regulatory Commission, whose members are appointed by the President. The act specifies that no person may manufacture, produce, transfer, acquire, own, process, import, or export the radioactive material identified above except to the extent authorized by the commission.

The act does not give the commission authority to regulate such naturally occurring radioactive materials as radium and radon, accelerator-produced radioisotopes (such as cobalt-57), or machine-produced radiation (for example, radiation resulting from the operation of an accelerator or an x-ray machine). Thus the NRC does not control exposure to uranium ore while it is in the ground but assumes responsibility as soon as it leaves the mine. Occupational radiation exposure of uranium miners (and other miners such as phosphate and coal miners) exposed to naturally occurring radioactive substances is regulated by the Mine Safety and Health Administration of the Department of Labor. The MSHA also regulates occupational radiation exposure of workers in the uranium milling industry.

The NRC licenses users, writes standards in the form of regulations, inspects licensees to determine if they are complying with the conditions of their license, and enforces the regulations. The regulations are issued as Title 10 (Atomic Energy) of the Code of Federal Regulations. The most important sections for users of radionuclides are: Standards for Protection Against Radiation (Part 20); rules of General Applicability to Licensing of Byproduct Material (Part 30); General Domestic Licenses for Byproduct Material (Part 31); and Human Uses of Byproduct Material (Part 35).

The NRC is also empowered to enter into agreements with state governments that want to take on the licensing and regulatory functions over the use of radioisotopes and certain other nuclear materials within the state. A condition of the agreement is that both the NRC and the state put forth their best efforts to maintain continuing regulatory compatibility.

Radiation Control for Health and Safety Act of 1968 (Public Law 90-602 Amendment to the Public Health Service Act)

The purpose of this act is to protect the public health and safety from the dangers of exposure to radiation from electronic products. The Secretary of the Department of Health, Education, and Welfare was directed to establish a program that included the development and administration of radiation safety performance standards to control the emission of radiation from electronic products and promoted research by public and private organizations into effects and control of such radiation emissions. "Electronic product radiation," as defined by the act, includes both ionizing and nonionizing electromagnetic and particulate radiation, as well as sonic, infrasonic, and ultrasonic waves.

The Center for Devices and Radiological Health, an agency of the Food and Drug Administration, conducts the regulatory program. Its authority is limited to regulating the manufacture and repair of equipment and is exercised through its promulgation of performance standards. Standards pertaining to ionizing radiation have been issued for diagnostic x-ray systems and their major components, television receivers, gas discharge tubes, and cabinet x-ray systems, including x-ray baggage systems. It performs surveys on the exposure of the population to medical radiation. It conducts an active educational program on the proper use of x rays for medical purposes and has completed teaching aids for x-ray technicians, medical students, and residents in radiology. It has published recommendations for quality assurance programs at medical radiological facilities to minimize patient exposure. It has also issued standards pertaining to nonionizing radiation, such as microwaves and laser beams.

Pure Food, Drug and Cosmetic Act of 1938 and Amendments (1962, 1976)

It took five years of often intense controversy to pass this act. The original legislation limited the Food and Drug Administration to regulating the safety of drugs offered for interstate commerce through the control of product labeling. Later legislative amendments in 1962 extended its authority to include control over the manufacture and efficacy of drugs, including radioactive drugs, and (in 1976) control over the manufacture and distribution of medical devices. The latter include cobalt-60 irradiators for radiation therapy and gamma scanners for use in nuclear medicine. The FDA does not have the authority to control the use of radioactive (or other) drugs on patients by physicians once they are approved for routine use. The FDA requires the manufacturer to carry out investigational programs, including clinical trials, to establish the safety and efficacy of new drugs. The FDA does regulate the investigational use of radioactive drugs on human subjects before they are approved for routine use.

Occupational Safety and Health Act of 1970 (Public Law 91-596)

The purpose of this act was "to assure so far as possible every working man and woman in the Nation safe and healthful working conditions and to preserve our human resources." Under the jurisdiction of the Secretary of Labor, it provided for the establishment of the Occupational Safety and Health Administration with power to set and enforce mandatory occupational safety and health standards. However, the act specifically excludes the OSHA from jurisdiction in areas covered by other legislation, including the Atomic Energy Act. This limits the OSHA, as far as occupational radiation exposure is concerned, to radiation from x-ray machines, accelerators, and other electronic products; natural radioactive material (such as radium); and accelerator-produced radioactive material. A National Institute for Occupational Safety and Health (NIOSH) was established in the Department of Health, Education, and Welfare to conduct research needed for the promulgation of safety and health standards and to support training and employee education programs promoting the policies presented in the act. An Occupational Safety and Health Review Commission was also established to carry out adjudicatory functions under the act. Regulations of the Occupational Safety and Health Administration pertaining to the control of radiation sources are given in the Federal Register under Title 29, specifically Part 1910.96.

The Environmental Protection Agency (EPA)—Product of a Presidential Reorganization Plan (1970)

Unlike the previous agencies, which were established by acts of Congress, the EPA was the result of a presidential reorganization plan that transferred to one agency responsibilities in the area of environmental protection that had been divided previously among several other agencies. The EPA assumed the functions of the Federal Radiation Council, which had been authorized to provide guidance to other federal agencies in the formation of radiation standards, both occupational and environmental (the FRC was abolished); assumed the responsibility from the Atomic Energy Commission for setting both levels of radioactivity and exposure in the general environment, an authority created under the Atomic Energy Act and applying only to material covered under that act; and took from the Public Health Service the responsibility to collate, analyze, and interpret data on environmental radiation levels. The EPA also has authority under various statutes to regulate the discharge in navigable waters of radioactive materials, like radium, not covered by the Atomic Energy Act; to establish national drinking-water standards and to protect supplies when states fail to do so; to regulate the recovery and disposal of radioactive wastes not covered by the Atomic Energy Act; and to control airborne emissions of all radioactive materials. It has instituted a limited program of monitoring general environmental radiation levels with the help of the states through a monitoring network called the Environmental Radiation Ambient Monitoring system, and conducts special field studies where an environmental problem has been identified. It maintains a special interest in population exposure from natural radioactivity, an area that is outside the jurisdiction of other agencies. It publishes an annual report, "Radiation Protection Activities," which summarizes the work and findings of the federal agencies.

National Environment Policy Act of 1969 (Public Law 91-190)

This act served as a declaration of national policy "to create and maintain conditions under which man and nature can exist in productive harmony, and to fulfill the social, economic, and other requirements of present and future generations of Americans." It required that a detailed statement of environmental impact be included in every recommendation or report on proposals for legislation and other major federal actions significantly affecting the quality of the human environment. The statement was also required to include any adverse environmental effects that cannot be avoided should the proposal be implemented; alternatives to the proposed action; the relationship between local short-term uses of man's environment and the maintenance and enhancement of long-term productivity; and any irreversible and irretrievable commitments of resources that would be involved in the proposed action should it be implemented. The act also created in the executive office of the President a three-member Council on Environmental Quality (CEQ), whose functions were to assist the President in the preparation of an annual Environmental Quality Report and to perform a variety of other duties, such as information gathering, program review, and policy, research, and documentation studies.

Thousands of "environmental impact statements," some thousands of pages long and costing millions of dollars, have been produced under the law. Hundreds of lawsuits have been brought, alleging violations of the act by the federal agencies responsible for building, financing, or permitting various kinds of projects, including nuclear power plants, oil pipelines, dams, and highways. A report issued in 1976 by the CEQ on the first six years' experience with NEPA concluded that it was working well. It has been observed that judges have had to become the true enforcers of NEPA "because of the lack of commitment on the part of those in power, and that perhaps only if controls on economic growth are instituted to curb consumption and resource depletion, will NEPA become truly effective" (Carter, 1976).

Low-Level Waste Policy Act of 1980

This act and the Low-Level Radioactive Waste Policy Amendments Act of 1985 made each state responsible for the disposal of waste generated within its borders. States were authorized to form compacts for the establishment and operation of regional facilities for the disposal of low-level radioactive waste generated within the compact states. While various deadlines for action were included in the act, by the year 2000 there were still no low-level waste burial sites developed under the provision of the act. Meanwhile, the cost of disposal and the uncertainties in the ability of the political process to produce additional burial sites for the waste caused users of radioactive materials to make strong efforts to minimize the production of waste and to find alternative means of disposal. As a result, production of wastes became low enough that existing commercial sites seem to have enough capacity to process them.

Nuclear Waste Policy Act of 1982

This act and the Nuclear Waste Policy Amendments Act of 1987 represented another attempt to solve the disposal of high-level, long-lived radioactive wastes by legislative mandates and deadlines. Here the challenge is to develop a geologic repository for the storage of high-level waste and spent nuclear fuel to ensure the safety for thousands of years. Uncertainties in accounting for the acceptable performance of a repository under potential challenges of earthquakes, water contact, and other geologic processes over an enormous time span have hindered the establishment of a site that passes muster technically and is politically acceptable.

B.2   Radiation Control at the State Level

The basic responsibility for the protection of workers and the public against occupational and environmental hazards lies with the individual states, although the federal government has intervened through the various acts of Congress previously described and may preempt control authority previously exercised by the states. Thus, there is a provision in the Radiation Control for Health and Safety Act that no state or political subdivision of a state can establish or continue in effect any standard that is applicable to the same aspect of performance of an electronic product for which there is a federal standard, unless the state regulation is identical to the federal standard. It also appears that states are preempted from saying anything about radiation hazards from materials covered under the Atomic Energy Act (*Northern States Power Co. v. Minnesota,* 1971). Situations where state and federal authorities both maintain strong interests can result in considerable confusion. In the end, the limits of the powers of the various levels of government in regulation must be resolved by the judicial branch of government.

Communication between states in matters of radiation control is maintained through the Conference of Radiation Control Program Directors, which is supported by NRC, EPA, and FDA. This group holds annual conferences to discuss current problems. The proceedings of these conferences, which cover policy, technical, and regulatory matters, are published by the Center for Devices and Radiological Health.

B.3   Inspection and Enforcement

Inspections are an essential part of the regulatory process. The frequency of inspections of a facility depends on the significance of the radiation hazard. The U.S. Nuclear Regulatory Commission's inspection program for materials licensees (which does not include nuclear reactors) gives first priority to radiopharmaceutical companies and manufacturers of radiographic sources, scheduling inspections at least once a year. Radiographers, large teaching hospitals, and universities using large amounts of activity also receive high priority and are inspected on approximately an annual basis. Inspections of industrial radiation sources by state enforcement agencies appear to be less frequent. Industrial facilities containing x-ray machines have been inspected every 7 years on the average, and those containing radium sources every 4 years (Cohen, Kinsman, and Maccabee, 1976).

The nature of enforcement varies depending on the powers granted in the enabling legislation and the severity of violations. The OSHA has powers to impose severe financial penalties; the NRC can revoke licenses as well as impose fines. Minor violations may draw only a citation from the inspection agency, pointing out the infraction and requesting that the regulatory agency be informed when the condition is corrected. One of the most effective enforcement methods is the publication of inspection results and disciplinary actions as part of the public record, thereby creating the possibility of considerable embarrassment to the licensee.

The conduct of inspections varies for different agencies and even for different inspectors from the same agency. Therefore it is not possible to describe a standard inspection in detail, but it generally proceeds along the following lines:

The inspector reviews the organization of the radiation-protection office to get an overall impression of the scope of the program. He notes the attitude prevailing toward compliance with regulations and safety measures. This guides him in the actual conduct of the inspection, which has, as major components, review of records, questioning of all levels of personnel, observations of work areas, and independent measurements. He determines whether citations and unresolved items remaining from previous inspections were corrected. He then considers other items, such as training of workers, safety procedures that are being followed, adequacy of personnel monitoring and bioassays, waste disposal, records and reports, and environmental monitoring. He examines the license to insure that special license conditions are being fulfilled. Many inspectors talk directly with workers to determine their habits and the degree of compliance with the regulations. The involvement of management through its awareness of the program and the extent of quality assurance measures are also reviewed. Following the inspection there is a closing session with management at a high level to explain the findings and to obtain commitments from management to correct faults. These findings, and the commitments from management, are also documented in a letter sent to the licensee.

Noncompliance items are grouped into the categories of violations, infractions, and deficiencies. A violation is very serious. The criterion for a violation is a large exposure or potential for a large exposure. An infraction is a condition that could lead to exposure if allowed to continue; this category includes failures to make surveys, conduct training sessions, and so on. A deficiency is a violation of regulations that could not lead to overexposure, such as an error in record keeping. The letter requests a response in 30 days in which the licensee tells what has been done to correct the conditions, what the organization will do to prevent recurrences in the future, and when it will be in full compliance with the regulations.

If the inspection indicates serious control problems, NRC management will phone or meet with licensees. If the outcome is unsatisfactory, there may be a letter from headquarters in Washington; this can be followed by a civil penalty. Civil penalties are used if (1) doses have occurred or could occur that are large enough to cause injury (that is, no training to workers) and (2) the licensee shows complete recalcitrance—there is no communication with the commission or effort made to correct conditions. Fines have been extremely effective in these situations, not only because of the financial loss but also because of the accompanying publicity. All correspondence and actions of the Nuclear Regulatory Commission are a matter of public record. In extreme cases, the license is modified or revoked completely.

There are typical and recurring citations of violations or poor practices: lack of personal involvement by the management; excessive reliance on outside consultants; inadequate surveys; inadequate records; lack of security of sources; inadequate leak testing; improper procedures for handling packages. Problems also arise in institutions where control is exercised through departments rather than a central authority, while the license is issued to the institution. Special license conditions may be forgotten, particularly when a license is transferred from one person to another who does not check special conditions. There is also greater risk of use of radionuclides by unauthorized personnel. Central to satisfactory compliance with the regulations and good practice is an adequate quality assurance program and periodic audit by management of radiation-protection operations.

APPENDIX C: CONTROL OF AIRBORNE RELEASES TO THE ENVIRONMENT

Airborne contamination from laboratory operations with radioactive materials is in the form of vapors, gases, or particles. The operations must be performed in a hood if significant exposure of workers is possible. The contaminated air can be disposed of by dilution and discharge through a stack; it is further diluted by spreading out as it flows away from the stack. Alternatively, the air can be cleaned prior to discharge with the use of filters (for particles) and absorbers (for vapors and gases). These must then be disposed of as radioactive waste.

C.1   Dilution in the Atmosphere

The transport of radioactive emissions through the atmosphere subsequent to discharge from a stack is governed primarily by the properties of the prevailing air currents. Many pathways can be taken by the discharged material. One need only look at the plumes from smoke stacks over a period of time and under various weather conditions to see how many possibilities there are. The smoke leaving a stack can proceed straight up or travel horizontally; it can meander at constant height above the ground, fan out, loop down—the possibilities seem endless.

The objective of the analysis of stack discharges is to determine the concentration of the radioactivity as a function of time and distance from the stack and the resulting doses to exposed personnel. The equations, while based on diffusional processes, are largely empirical and describe the concentrations vertically and laterally to the effluent flow in terms of the normal error curve.

Consider the discharge into a stable atmosphere in Figure 5.3. A quantity Q MBq is released over a period of τ seconds and mixes with the air stream, which is flowing by at an average velocity of u m/sec. In τ seconds, a column of air $u\tau$ meters long will have flowed by and therefore acquire a contamination loading of $Q/u\tau$ MBq/m. The activity along a differential element dx *is* $Q/u\tau$dx. The activity will spread out laterally and horizontally as it is carried along by the air stream. The degree of spread is determined in the horizontal direction primarily by shifting wind directions, and in the vertical direction by the temperature gradient. We wish to fit the spreading to a form given by the normal curve of error. This is $\Phi(y) = (1/\sqrt{2\pi}\sigma)e^{-y^2/2\sigma^2}$, where σ is the standard deviation. The area under this curve,

$$\int_{-\infty}^{\infty} \Phi(y)\mathrm{d}y = 1$$

Thus we can let y represent a horizontal coordinate and spread the activity out laterally to give:

$$\frac{Q}{\mu\tau\sqrt{2\pi}\sigma_y}e^{-y^2/2\sigma_y^2}\,\mathrm{d}y\mathrm{d}x$$

5.3 Transport of activity discharged from stack.

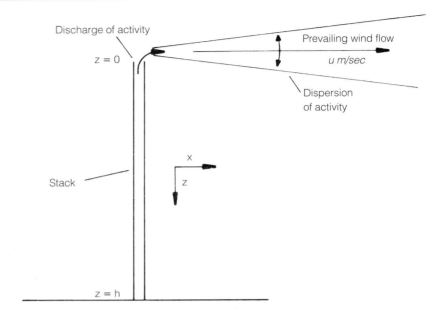

or we can spread it out both vertically (z direction) and laterally:

$$\frac{Q}{\mu\tau\sqrt{2\pi}\sigma_y \ \sqrt{2\pi}\sigma_z} \cdot e^{-y^2/2\sigma_y^2} \cdot e^{-z^2/2\sigma_z^2} \ dxdydz$$

Since the area under the curve is still unity we have not changed the total activity to be expected at any distance x downwind from the stack. If this has the concentration ψ,

$$\psi_{x,y,z} = \frac{Q}{2\pi\mu\tau\sigma_y\sigma_z} e^{-y^2/2\sigma_y^2} e^{-z^2/2\sigma_z^2}$$

The concentration is doubled when the plume reaches the ground ($z = h$) because of reflection from the surface.

The wind fluctuates constantly about its mean direction. An empirical expression has been derived for σ_y as a function of the extent of the variation of wind direction. If the angle of fluctuation from the mean direction is given by θ, and σ_θ is its standard deviation (in radians), then,

$$\sigma_y^2 = \frac{Ax}{u} - \frac{A^2}{2(\sigma_\theta u)^2}(1 - e^{-2(\sigma_\theta u)^2 x/Au})$$

A is related to scale of turbulence. An empirical expression for A is $A = 13.0 + 232\sigma_\theta u$. For $u = 5$ m/sec, $\sigma_\theta u = 0.05$ for instantaneous release and 0.2 for intermediate (1 hr) release. In the vertical direction:

$$\sigma^2 = a(1 - e^{-k^2 x^2 / u^2}) + bx/u$$

where a, k^2, b are functions of degree of stability; $a = 97$ m^2, $k^2 = 2.5 \times 10^{-4}$ sec^{-2}, $b = 0.33$ m^2/sec for moderately stable conditions.

The effective height of the chimney is somewhat greater than the actual height if the gas is discharged at a significant velocity and/or high temperature. One expression for the effective height is

$$H = h + d\left(\frac{\eta}{\mu}\right)^{1.4} (1 + \frac{\Delta T}{T})$$

where

h = actual chimney height, meters
d = chimney outlet diameter, meters
η = exit velocity of gas, meters/second
μ = windspeed, meters/second
ΔT = difference between ambient and effluent gas temperatures
T = absolute temperature of effluent gas

For additional data applicable to a wide range of meteorological conditions, see Slade, 1968; NCRP, 1996.

Much of the analysis presented here is for dispersion over long distances and over smooth terrain. Local dispersion, say in a congested area surrounded by buildings of various sizes, is much more complex and not amenable to a simplified treatment. While there is a tendency to apply the equations presented here to such conditions, the results are suspect. Some attention has been given to this problem but no simple calculational methodology has been found (Smith, 1975, 1978). Accordingly, releases in congested areas should be monitored carefully and kept as low as possible.

C.2 Filtration of Particles

When air laden with particles is incident on a fiber, the air will flow around it but the particles, because of their inertia, will tend to continue in a straight line and impact on the fiber. Filters are made by packing the fibers in a mat, and the filtering action (for small particles) is the result of the action of the individual fibers rather than a sieving action. After impact, the particles are held to the fibers by physical short-range forces (van der Waals forces). As particles get smaller, their inertia decreases, and they are more prone to follow the stream lines. At still smaller sizes, they again stray from the stream lines as the result of collisions with the individual air molecules (Brownian motion). This increases the probability of contact with the fibers and removal from the stream. Thus, a graph of the efficiency of a fiber filter versus decreasing particle size shows a decrease at first as the inertia decreases, but a gradual turnaround and increase as diffusion due to Brownian motion becomes important. The particle size for minimum filtration efficiency depends on the type of filter material and air velocity and is in the range of 0.1 to 0.3 μm (Dennis, 1976, p. 62). Equations for the collection efficiency of isolated fibers, which must take into account diffusion, inertia, and gravity effects, are extremely complicated (Iinoya and Orr, 1977).

Paper filters used for laboratory analysis have a fiber diameter of about 20 μm and a void fraction of 0.70. The thickness is usually between 0.2 and 0.3 mm. Their collection efficiency is less than 90 percent for 0.3 μm particles. High-efficiency particulate air (HEPA) filters are used for very toxic radioactive dusts and bacterial particulates. They are composed of fine glass fibers (0.3 to 0.5 μm in diameter) mixed with coarser fibers for support. They show at least 99.97 weight percent collection efficiency for particles having diameters of 0.3 μm.

Filters resist the flow of air and the pressure drop across a filter under anticipated air velocities is critical in evaluating its suitability for a particular application. Even if a filter is superbly effective in preventing the passage of particles, it is of no use if it also presents such a high resistance that only a trickle flow of air is possible. The pressure drop will increase as the filter becomes clogged with dust. Eventually the flow rate will be affected and will decrease to a point where the filter is not usable. Thus the pressure drop must be followed to determine when the filter must be replaced or cleaned.

The particles that initially contact a fiber do not necessarily adhere to it and are not necessarily permanently bound to it if they do adhere initially. They can be knocked off by other particles and turbulent air currents. An interesting reentrainment mechanism occurs in the filtration of particles that carry alpha radioactivity. When an alpha particle is emitted, the resulting recoil can be sufficient to knock a small particle off a fiber (Ryan et al., 1975; Mercer, 1976). Reentrainment is generally not very important, but the possibility must be considered for very hazardous substances such as plutonium, where extremely high filter efficiency is required (McDowell et al., 1977).

There are times in sampling air containing particles above a particular size that the air speed through the sampler should be adjusted to be equal to the speed in the airstream. This condition is known as isokinetic sampling and its need should be determined in specific cases (ACGIH, 1983).

C.3 Adsorption of Gases and Vapors on Charcoal

Pollutant gases and vapors can be removed by adsorbents. The adsorbent usually used is activated charcoal (Smisek and Cerny, 1970). Beds containing ten tons of charcoal or more are not uncommon as part of the air-cleaning systems for radioactive noble gases and iodine in nuclear power plants. Smaller adsorption systems, containing charcoal in panels, are used in hood exhausts or, on a larger scale, in the building exhausts of research institutions and hospitals. Canisters containing 10 to 100 g of charcoal are used in sampling in contaminated atmospheres, and cartridges of charcoal are often used in respirators.

The contaminating molecules reach the interior surfaces of the charcoal grains by diffusion from the airstream. This surface is extremely irregular, consisting of innumerable tiny cavities and tortuous channels, and presents an enormous area to the diffusing pollutant. For example, a high-quality activated charcoal made from coconut shell can have a surface area as high as 1,500 m^2/g. Chemically inert molecules, such as krypton and xenon, that reach the surface are held briefly by physical forces (van der Waals forces), escape, struggle from point to point, become lost in the passageways in the charcoal, but eventually make their way down the bed. The charcoal bed serves to hold up the flow of these contaminants, a delaying action that is effective in the control of radioactive noble gases with half-lives short compared to the holdup time of the bed. This method works well in the nuclear power industry in the control, through radioactive decay, of discharges of short-lived nuclides such as ^{85m}Kr ($T^{\frac{1}{2}} = 4.4$ hr), ^{133}Xe (5.3 d) and ^{131m}Xe (12 d).

The mean holdup volume of a charcoal bed for noble gases can be determined by a simple test. A pulse of the radioactive gas (for example, ^{85}Kr) is injected into a stream of air flowing at a constant rate into the bed. The effluent is passed through a gas flow ionization chamber or other radiation detector that gives a reading proportional to the concentration of the radioactivity in the effluent gas, C_v, and the reading is monitored continuously. Following the injection of the pulse of radioactive gas, some time will elapse until the monitor detects radioactivity in the effluent. The reading will rise to a maximum and then fall back to background when all the radioactive gas has passed through the bed. A gas meter is also connected in the line and the total volume flow V recorded periodically. Since the flow rate is constant, the volume is known as a continuous function of time (and radiation reading). The mean holdup volume is given by the equation $\overline{V} = \int_0^\infty VC_v \, dV / \int_v^\infty C_v \, dV$.

The mean holdup volume is proportional to the mass of charcoal, and the ratio can be considered to be an adsorption coefficient, k. If a radioactive gas with half-life T passes through a charcoal bed of mass m in a carrier gas with volume flow rate f, the mean residence time of the charcoal is k/f and the ratio of the effluent and input concentrations is given by the expression $C/C_0 = e^{-0.693 km/Tf}$. Adsorption coefficients are affected by temperature, pressure, relative humidity, and other adsorbed materials. Thus it is important to determine them under actual operating conditions and to test the bed periodically after it is in use. Rough estimates of required bed size may be made by using the relationships: 1 gram charcoal per 40 cc throughput of air containing krypton and per 520 cc air containing xenon (at room temperature).

Additional insight into bed performance can be obtained from breakthrough curves. These are measurements of effluent concentration versus time (or volume flow) on a sample of the adsorber during continuous passage of the contaminated gas. The curve is S-shaped and is closely fit by the cumulative distribution curve of the normal distribution (Grubner and Burgess, 1979). The form of the curve follows from the random nature of the penetration of individual contaminant molecules through the bed, which causes some to get through more quickly and others less quickly than the average velocity for the flow. The volume flow at which the effluent concentration is 50 percent of the influent concentration has particular significance. The product of this volume and the input concentration, less the amount of contaminant in the pores of the adsorber, gives the amount of contaminant that is actually held by the bed. This result follows from the symmetry of the breakthrough curve. The data can be used to design a bed to give a desired degree of performance, or to estimate how long a bed can be used before breakthrough occurs, provided the bed is used under the same conditions as during the test. The presence of certain contaminants, such as organic vapors, or the occurrences of unusual temperature or humidity conditions in the field can serve to reduce significantly the actual performance and service time.

Breakthrough curves are useful when the forces of adsorption are primarily physical. When the adsorption mechanisms are largely chemical in nature (for example, adsorption of iodine on charcoal), the action of the bed is quite different. The contaminants are adsorbed at sites on the surface by a mechanism that is essentially irreversible. At the low mass concentrations characteristic of radioactive contaminants, the concentration of the contaminant decreases exponentially with distance down the bed.

The basic determining factor for removal is the time t in contact with the bed. If V = bulk volume of the sorption material and f = volumetric flow rate of the gas, then $t = V/f$ at constant flow velocity.

The ratio between the influent and effluent concentrations, C_0/C, known as the decontamination factor, is given by the expression $C_0/C = e^{Kt}$, where K is referred to as the K-factor or performance index and depends on many factors that cannot be accounted for in a given bed. It must be determined experimentally for each batch of material used. However, fresh impregnated activated carbon should have a K-factor of at least 5 in a test with radioactive methyl iodide at 90–100 percent relative humidity, 30°C, and anticipated face velocity (usually 20–50 cm/sec).

The sites at which chemical adsorption occurs become unavailable for further adsorption. As already stated, radioactive contaminants are not likely to affect a significant fraction of the sites, but if appreciable quantities of the nonradioactive form are in the air, they will gradually immobilize all the sites. As the available sites become exhausted, the contaminant will finally break through and be transported through the bed at essentially the speed of the transporting air.

Certain trace impurities in the air that cannot be anticipated or adequately accounted for can have significant effects on the performance of a bed. Rapid decreases of removal efficiencies can result from adsorption of solvents and oil vapors (as from newly painted surfaces or recently cured plastic materials). Beds subjected to unknown and possibly deleterious atmospheres should be tested periodically. One possible test procedure is to place a sample of the charcoal in a test bed that is followed by one or more backup beds to trap the effluents being removed. Air is passed through at a typical face velocity (about 20 cm/sec). The labeled contaminant is then injected into the input air. Several injections are preferable at 15 minute intervals. Following the injections air is passed through the bed for an additional hour. The activity of the contaminant in the test bed and in the backup beds is then measured and the retention computed. The test should be carried out at relative humidities of 70 and 95 percent.

Radioactive contaminants are carried in the airstream in different forms (particulate, vapor) and compounds. It is often useful to determine the various forms and their penetration through the bed. This can be done by constructing the backup bed of several components, each having a selective absorption for one of the contaminants.

C.4 Adsorbers for Radioiodine

Radioactive iodine is released in laboratory operations either as elemental iodine (I_2) or as part of some organic molecule. Elemental iodine is removed very effectively from an airstream by activated coconut charcoal (8/16 mesh) even in humid air. It is held quite adequately to the charcoal by physical (van der Waals) attractive forces. Organic iodine is not bound as well on charcoal, which loses effectiveness at high humidity. The charcoal must be impregnated with a chemical with which the organic iodine will combine. Two compounds have been widely used as impregnants, each providing a different mechanism of binding—potassium iodide (KI) and triethylenediamine (TEDA). The KI works through an isotopic exchange mechanism; the organically bound iodine exchanges with the iodine on the KI. If the compound is methyl iodide, the reaction is $CH_3{}^{131}I$ (air) + $K^{127}I$ (charcoal) → $CH_3{}^{127}I$ (air) + $K^{131}I$ (charcoal). The methyl-iodide molecule itself is not trapped by the exchange mechanism. On the other hand, when TEDA is used, the amine group reacts with the methyl iodide converting the radioiodine molecule to an ionic form (R_3N + $CH_3{}^{131}I$ → R_3N + $CH_3{}^{131}I^-$). The quaternary iodide produced is a stable nonvolatile compound that is adsorbed strongly on the charcoal. TEDA-impregnated charcoals are effective even with airstreams of high moisture content. The highly polar water molecules are strongly attracted to each other in competition with the nonpolar carbon surface, which adsorbs selectively the larger, less polar organic iodides. Impregnated activated carbon is tested at 95 percent relative humidity, 80°C, a face velocity of the gas of 20 cm/sec in a bed 50 mm deep (Burchsted et al., 1976). The bed is challenged with a concentration of approximately 2 mg CH_3I/m^3 for 2 hr. Beds should be designed to provide a minimum contact time of 0.25 sec. A good quality impregnated charcoal should give greater than 99 percent methyl-iodide retention at 80°C, 95 percent relative humidity, and 1 atmosphere pressure. Even higher performance is obtained in removing elemental iodine.

PART SIX

Ionizing Radiation and Public Health

1 FORMULATION OF STANDARDS FOR RADIATION PROTECTION

Up to this point, we have been concerned with the technical aspects of radiation protection, involving studies of the properties of radiation; methods of defining, calculating, and measuring radiation exposure; and practical measures for radiation control. Our objective has been to prepare readers to prevent excessive exposure to themselves and others when working with radionuclides and radiation machines, and, as guides for excessive exposure, we have used maximum levels specified in government regulations. Exposure of healthy individuals below these levels is legally permissible; the levels represent the current basis for the protection of the public. It is important that every user of radiation understand the evolution of present radiation control standards and the degree of protection they offer. It is also important to appreciate the type of information that must be continuously developed in order to appraise and revise standards of control when necessary. There is a need not only to provide adequate control but also to deal with the complementary problem of avoiding excessive controls or restrictions that will prevent the development of technology or medical care for the maximum benefit of the public.

In order to better understand the present radiation standards, let us chronicle some of the important findings and studies that led up to them. X rays were discovered in 1895 and radioactivity in 1896. By 1920, many of the early radiologists and technicians had developed skin cancer and others had died of anemia and probably leukemia. The concept of a "tolerance dose" was developed to protect radiation workers. One approach to establishing a tolerance dose was to consider the exposure that produced reddening (erythema) of the skin. This reddening followed the exposure within a period of about a week. It was suggested that exposure in any monthly period be limited to a small fraction of the "threshold erythema" exposure, less than 1 percent. In terms of present values of erythema doses expressed in roentgens, this provided a monthly limit of 5 R, or a daily limit of 0.2 R. The reasoning behind the establishment of these limits was the belief that the body could repair any damage that might occur at these levels. The production of genetic effects and permanent residual injury from radiation exposure was not known at that time.

In 1934 the International X-ray and Radium Protection Commission, established by the Second International Congress of Radiology in 1928, also recommended a "tolerance dose" of 0.2 R/day. This recommendation was based on observations of the health of fluoroscopists, x-ray therapy technicians, x-ray therapy patients, and radium therapists and technicians, and it was concluded that deleterious effects had not been shown in individuals exposed at these levels. The same value was recommended in the United States by the Advisory Committee on X-ray and Radium Protection. In 1936 the advisory committee reduced the "tolerance dose" to 0.1 R/day at the suggestion of Dr. G. Failla.

There were no definite data to indicate that the previous level should be reduced by a factor of 2, but the genetic hazards of radiation were becoming apparent and there was a general feeling the limit should be reduced. Dr. Failla suggested the 0.1 R/day level because he had found no effect on blood cell counts taken over periods of three and four and a half years from two technicians who had been working with a radium source. Estimating that their exposure had averaged around 0.1 R/day, he noted, "This could hardly be considered satisfactory evidence [for a safe level] but it was better than anything available at the time" (Failla, 1960, p. 203).

In 1949 the U.S. National Committee on Radiation Protection (NCRP) recommended that the permissible dose be reduced to 0.3 rem per week. The International Commission on Radiological Protection (ICRP) adopted the same recommendation a year later. (Both these committees had evolved from the X-ray and Radium Protection committees referred to earlier.)[1]

---

1. The NCRP was granted a Congressional charter in 1964 and has since operated as an independent organization financed by contributions from government, scientific societies, and manufacturing associations. The ICRP operates under rules approved by the International Congress of Radiology. The members are selected by the ICRP from nominations submitted to it by the National Delegations to the International Congress of Radiology and by the ICRP itself. The selections are subject to approval by the International Executive Committee of the Congress. The ICRP receives financial support from the World Health Organization, the International Atomic Energy Agency, The United Nations Environment Program, the International Society of Radiology, and other international and national sources.

Again, it was stated that this reduction was supported not by concrete evidence that the original levels had been harmful, but by a desire to be on the safe side. Higher-energy sources were coming into use, and the radiations penetrated the body more readily and irradiated it more uniformly than the radiations from the lower-energy x-ray machines for which the earlier limits had been devised. The increased internal body dose relative to surface dose imparted by the higher-energy radiations could be compensated for in part by lowering the limits.

By 1956 the prospects for the large-scale use of nuclear power had grown significantly, and the potential for irradiation of a large number of workers was growing. The committees on radiation protection were uneasy about allowing the irradiation of large numbers of individuals without further reduction of dose levels, and accordingly they reduced the limits for whole-body irradiation by approximately a factor of three by limiting the accumulated whole-body dose to $5(N-18)$ rem, where N was the age of the worker.

Limits also had to be set for the possible incorporation of radioactive materials in the body. The concept of "critical organ" was used—that is, the organ receiving the greatest exposure as a result of ingestion of a particular radionuclide. It was decided that irradiation of the critical organ from radionuclides in the body should not be allowed to exceed the dose that had been set for exposure of single organs from external radiation, 0.3 rem per week.[2]

As yet there were no official recommendations concerning the exposure of individuals nonoccupationally exposed to radiation and therefore not receiving special monitoring or other protection. In 1956 the National Council on Radiation Protection and Measurements (the new name for the NCRP, with the same initials) and the ICRP both recommended that, for individuals nonoccupationally exposed, the levels be set at one-tenth the occupational limits, that is, at 500 mrem/yr for whole-body exposure.

The need to establish limits of exposure for the childbearing segment of the population in order to restrict genetic damage was also recognized, and by 1959 specific recommendations were formulated by the ICRP. For an entire population, a limit equal to that which had been recommended by the Genetic Committee of the National Academy of Sciences (NAS) of 10 R to the reproduction cells was adopted. The 10 R limit applied from conception to age 30 and was in addition to the dose from the natural background radiation. The value had been chosen by the NAS committee as an

2. In the special case of bone, the limit of 0.1 μCi for occupational exposure to radium-226, established in 1941, has stood as a separate standard. This limit, which is supported by extensive detailed data on humans with radium body burdens, serves as a reference point for evaluating the hazards of other radionuclides that localize in bone.

acceptably small fraction of the dose (30–80 R) that was believed to double the number of harmful mutations that occurred spontaneously in the population over the same period.[3]

For practical reasons, the "10 R in 30 yr" recommendation developed into an allocation of 5 R for medical practice and 5 R for radiation exposure associated with all other man-made sources. This amounted to specifying separate average annual limits of 170 mR to the population from medical radiation and from all other man-made sources.

At about the same time, the Federal Radiation Council, which had been formed in the United States to provide a federal policy on exposure of human beings to ionizing radiation, reviewed existing standards. It reaffirmed the standard for the exposure of individual members of the public, which it expressed as an annual Radiation Protection Guide (RPG) of 0.5 rem. It also introduced an operational technique which specified that the RPG would be considered met if the average per capita dose of a suitable sample of the population did not exceed one-third the individual guide of 0.5 rem, or 0.17 rem. Thus, through two different avenues, the figure of 170 mrem/yr was associated in 1960 with a maximum level of exposure for large numbers of individuals to nonmedical man-made sources.

Accompanying the specifications of upper limits was the special recommendation that radiation exposure should be kept as far below the limits as practicable and, in particular, that medical exposure be kept as low as consistent with the necessary requirements of modern medical practice.

The NCRP issued a new report on basic radiation-protection criteria in 1971 (NCRP, 1971a). The report essentially reaffirmed the limits recommended a decade earlier, except for minor modifications and additions. The limits in the new report applied primarily to the dose incurred in any one year. Both occupational exposure and exposure of the public were covered. The report also provided guidance for regulatory agencies concerned with the preparation of regulations for the control of exposure. In presenting dose limits for workers, the public, emergency cases, and the families of patients receiving radioactivity, the NCRP cautioned that the application of the limits was conditioned substantially by the qualifications and comments provided in its report.

The ICRP revised its basic recommendations in 1977 as a result of new information that had emerged in the previous decade.[4] The basic limit of 5 rem in one year for occupational exposure to uniform whole-

---

3. NAS/NRC, 1956. For additional discussion on "doubling dose," see Part Two, section 15. NCRP (1993a) cites values between 1 and 4.4Sv as the doubling dose.

4. ICRP, 1977b, Publication 26. The previous basic recommendations were published in 1966 (ICRP, Publication 9) followed by amendments in 1969 and 1971.

body radiation was retained, now expressed in SI units as 50 mSv. The total dose could be accumulated in a single occupational exposure (except that women diagnosed as pregnant should work only under conditions in which it was unlikely that the annual exposure could exceed three-tenths of the equivalent dose limit, that is, 15 mSv). The commission did not make a separate definitive recommendation for women of reproductive age, but expressed its belief that appropriate protection during the essential period of organogenesis would be provided by limitation of the accumulation of the dose equivalent (as determined by the maximum value in a 30 cm sphere) to an approximately regular rate of 50 mSv/yr. This would make it "unlikely that any embryo could receive more than 5 mSv during the first 2 months of pregnancy." The commission dropped all previous recommendations on exposures of parts of the body or single organs, including the concept of critical organ. Instead, it recommended that an accounting be made of the doses in all the irradiated tissues and the associated risk (of inducing fatal cancer) be determined. Partial-body exposures were to be considered within allowable limits if the associated risk was less than that incurred from exposure of the whole body to the allowable limit.

To account for the differences in the risk of induction of cancer to the organs in the body from exposure to radiation, individual organ doses were assigned tissue weighting factors. These were multiplying factors that converted each organ dose to a whole-body dose that produced an equivalent excess risk of fatal cancer and hereditary disease. The contributions of the weighted doses from all irradiated parts of the body were summed to give the "effective dose," and the effective dose was governed by the same 50 mSv limit set for uniform radiation of the whole body. The weighting factors were based on the best estimates of the risk of production of cancer from single-organ exposure, in comparison with the risk of malignancy from whole-body exposure. The values assigned were: gonads, 0.25; breast, 0.15; red bone marrow and lung, 0.12 each; thyroid and bone surfaces, 0.03 each; and remainder, 0.30 (assignable as 0.06 to each of five remaining organs or tissues receiving the highest doses).

If only a single organ were exposed, this procedure could lead to levels of exposure that, while not producing a risk of cancer greater than that produced by the acceptable level for whole-body exposure, could result in unacceptable damage to the organ (such as cataract of the lens of the eye; nonmalignant damage to the skin, blood vessels, or other tissue; abnormal blood counts; or impairment of fertility). To prevent this, the commission set an upper limit of dose for any individual organ of 500 mSv, except the thyroid where the limit was set at 300 mSv.

An important application of the single dose limit is in determining limits for internal emitters. These are expressed in terms of annual limits on intake (ALI), levels that will prevent organ doses that, when multiplied by the appropriate weighting factors and summed, will exceed the whole-body limit. The formula considers external and internal exposures together in evaluating conformance with the limits and was designed to remove the ambiguity in applying the previous separate limits for internal and external exposure when both external and internal exposure occurred.

The commission allowed additional exposures (from external radiation and intakes) in exceptional cases, provided the total dose did not exceed twice the relevant annual limit in a single event and five times the limit in a lifetime (equivalent to the 25 rem emergency exposure in previous recommendations). Such exposures could be permitted only infrequently and only for a few workers.

Recognizing the difficulty in specifying an equivalent dose when external radiation produced nonuniform internal dose distributions, the commission specified that its limits would be met for external exposure if the maximum value of the dose equivalent in a 30 cm sphere (called the dose equivalent index) were less than 50 mSv.

The commission emphasized that its limits were intended to ensure adequate protection even for the most highly exposed individuals, and that these typically constituted only a small fraction of the working force. It advised that the planning of exposures close to the annual limits for extended periods of a considerable proportion of workers in any particular occupation would be acceptable only if a careful cost-benefit analysis had shown that the higher resultant risk would be justified.

Individual members of the public were not to receive whole-body doses more than 5 mSv in any year, but average doses over many years should not average more than 1 mSv/yr. (In 1985, the ICRP affirmed 1 mSv/yr as its principal dose limit for members of the public; ICRP, 1985b.) Higher limits were allowed for partial-body irradiation in accordance with the weighting factors discussed previously. In any case, an overriding annual dose equivalent limit of 50 mSv for single organs applied.

Dose limits were not set for populations; here each man-made contribution had to be justified by its benefits. The commission felt that its system of dose limitation and other factors were likely to insure that the average dose equivalent to the population would be much less than the limits for individuals, that is, less than 10 percent (0.50 mSv/yr).

When the NCRP issued new recommendations in 1987, it adopted, in principle, the risk-based, effective dose equivalent system used by the ICRP but modified and updated this approach in several respects. It recognized that recent information indicated that risk estimates utilized in setting the standards might be low "by an undetermined amount, perhaps a factor of two or more" (NCRP, 1987b), but, because of the incomplete nature of the new data and analyses, and the loose coupling between risks and limits, it was not ready to make any changes in the ICRP limits.

Major reviews by national and international advisory bodies of the effects of radiation on human populations were published by the United Nations Scientific Committee on the Effects of Atomic Radiation (UNSCEAR, 1988) and the U.S. National Research Council of the National Academy of Sciences (NAS/NRC, 1990) following the issuance of the 1987 recommendations of the NCRP. These reviews concluded that the epidemiological studies of the Japanese survivors of the atomic bombs provided by far the most complete data source for external low-LET radiation and that risk estimates derived from them were broadly supported by the results of the other studies reviewed. The emerging data coupled with adjustments to the doses imparted to the exposed population showed the continuing appearance of excess cancers many years after the dropping of the atomic bombs in 1945. Furthermore, these cancers were appearing at a rate consistent with the multiplicative projection model—that is, the rate of excess cancers was proportional to the rate of cancers in the population. As anticipated in the 1987 NCRP report, the latest findings resulted in an increase in the estimates of the projected risk of exposure to radiation.

The new biological information and trends in the setting of safety standards compelled the ICRP to issue a completely new set of recommendations in 1990 (ICRP, 1991a). In addition to lowering the limits, its aim was also to improve the presentation of the recommendations and the nomenclature (see Part Two, sec. 12.1). Dose equivalent was changed to equivalent dose (accompanied by slightly different definitions), effective dose equivalent to effective dose, quality factor to radiation weighting factor (also defined differently but equally applicable), and weighting factor to tissue weighting factor.

The occupational limit on effective dose was reduced to 20 mSv per year, although it could be averaged over 5 years with a maximum of 50 mSv in any one year. The ICRP set separate annual limits of 150 mSv for the lens of the eye and 500 mSv for the skin, averaged over any 1 cm^2 area. The limit for a woman who had declared pregnancy was an equivalent dose of 2 mSv to the surface of the woman's abdomen for the remainder of the pregnancy and a limit on intakes of radionuclides to about 1/20 of the ALI. The limit for public exposure was expressed as an effective dose of 1 mSv in a year, with the provision that in special circumstances, a higher value of effective dose could be allowed in a single year, provided that the average over 5 years did not exceed 1 mSv per year. The annual limit for the lens of the eye was 15 mSv and for the skin, 50 mSv, averaged over any 1 cm^2, regardless of the area exposed.

Following the publication of the reports by the United Nations and the

National Research Council, the National Council on Radiation Protection and Measurements addressed the problem of making specific recommendations on converting their risk estimates for high-dose and high-dose-rate exposure to low doses and dose rates applicable to radiation protection, as well as the problem of translating the risks from a Japanese population to a United States population. A scientific committee of the National Research Council, Scientific Committee 1–2, *The Assessment of Risk for Radiation Protection Purposes,* reviewed the UNSCEAR and NRC reports, and the results of their review were published in 1993 as Report 115, *Risk Estimates for Radiation Protection.* This report served as the basis for the revision of its 1987 recommendations, published that same year in Report 116, *Limitation of Exposure to Ionizing Radiation.* The basic framework of this report and the approach to dose limitation were based on the earlier report, but much greater consideration was given to discussing the degree of protection achieved under the recommended limits. The NCRP incorporated in general the recommendations and concepts in the 1991 ICRP report, deviating in a few cases where it felt that greater flexibility could be obtained at similar or lesser risk (for example, the occupational dose limits) or where increased protection was considered to be warranted (a monthly exposure limit for the embryo or fetus).

The philosophy of NCRP in setting occupational limits was that the risk to an individual of a fatal cancer from exposure to radiation should be no greater than that of fatal accidents in safe industries, a risk taken to be 1/10,000 per year, and should be kept as much below that risk as reasonably achievable (ALARA philosophy). It was recognized that many arbitrary choices and uncertainties were inherent in this approach. For example, it based the risk assessment not on the maximum dose that could be received by a worker, which was the promulgated limit, but on the associated average dose to workers, which was only a fraction of the established limit. Thus the risk assessment was based on an annual average dose, chosen as 10 mSv, rather than the 50 mSv annual limit. The NCRP concluded that the average annual dose should result in a contribution to the lifetime risk from each year's exposure of between 1 in 50,000 and 1 in 5,000. The dose limit was based on uniform exposure of the whole body, so doses that did not meet this condition had to be converted to a dose with the equivalent risk, namely the effective dose. The effective dose was the sum of the effective dose from external irradiation and the committed effective dose from internal exposures. One complication introduced by this approach was that the fatal accident rates in the various industries considered were decreasing with time at the rate of nearly 3 percent per year.

The NCRP concluded that those few individuals who were exposed close to the allowable limits over their working life would accumulate an annual risk of fatal cancer no greater than the annual risk of accidental death of a worker at the top end of the safe worker range (between 1/10,000 and 1/1000).

Special exposure limits were set to protect the embryo or fetus, which showed especial susceptibility to both mental retardation and cancer. Occupational limits were to be in effect only once the pregnancy was known. Women not known to be pregnant were governed by the same limits as male workers.

Limits recommended for the general public were reduced by a factor of ten from the occupational limits. Exposures to individuals under 18 years of age were acceptable for educational or training purposes but were limited to less than 1 mSv per year and accompanied by guidance on control of exposure. Medical exposures were excluded because they were assumed to result in personal benefit to the exposed individual. Regardless of the limits, the overriding considerations were *justification* and *ALARA*. Where significant exposures came from a single source or set of sources under one control, the recommendations specified that the exposure of a single member of the public did not exceed 25 percent of the annual effective dose limit unless the source operator could ensure that the total annual exposure from man-made sources did not exceed 100 mrem.

The NCRP also defined an annual negligible individual dose (NID) of 0.01 mSv effective dose below which it was not considered with relationship to the limits (and presumably any potential effects).

Highlights of the NCRP and ICRP recommendations are given in Table 6.1.

Regulations of the Nuclear Regulatory Commission in the United States are generally based on the recommendations of ICRP and NCRP, although they lag them by a few years. The Nuclear Regulatory Commission published revised standards for protection against ionizing radiation in 1991 (NRC, 1991), putting into practice the 1977 recommendations of the ICRP and subsequent ICRP publications. While the new concepts of dose limitation provided a more logical approach to dose limits, they allowed considerably higher single-organ exposures than allowable under the previous NRC limits.

Limits set by regulatory agencies for the exposures of the public from the radiation produced by nuclear power plants are much lower than those given in the standards. They are set in accordance with the principle of keeping radiation exposures *as low as reasonably achievable* (ALARA). The Nuclear Regulatory Commission specifies as design objectives for planned releases from a single commercial nuclear power plant annual whole-body doses of 5 mrem from airborne effluents and 3 mrem from liquid effluents. Annual organ doses caused by a single reactor are limited to 15 mrem from

Table 6.1   Dose-limiting recommendations of NCRP (1993) and ICRP (1991).

Exposure	NCRP (1993)	ICRP (1991)
Occupational exposure		
Whole body, external and internal, effective dose	50 mSv annual *and* 10 mSv × age (y) cum.	50 mSv annual *and* 100 mSv in 5 y cum.
Lens of eye, equivalent dose	150 mSv annual	150 mSv annual
Skin, hands and feet, equivalent dose	500 mSv annual	500 mSv annual
Commited dose (internal exposure)	20 mSv, annual[a]	20 mSv annual
Exposure to public		
Whole body, external and internal, effective dose	1 mSv annual *and* 5 mSv annual for infrequent exposures	1 mSv annual *and* 5 mSv in 5 y, if higher annual limit is needed
Lens of eye, equivalent dose	50 mSv	15 mSv
Skin, hands and feet, equivalent dose	50 mSv	50 mSv
Embryo or fetus		
Equivalent dose, once pregnancy is known	0.5 mSv in a month	2 mSv to abdomen
Intake of radionuclides		Limit to about 1/20 of an ALI
Negligible individual dose, effective dose	0.01 mSv annual per source or practice	

Source: NCRP, 1993b.
Note: See NCRP and ICRP reports for exposure guidance in emergencies and special situations.
a. NCRP specifies the annual committed dose as a reference level for design purposes rather than as a limit.

airborne effluents and 10 mrem from liquid effluents. The Environmental Protection Agency has a whole-body dose limit for planned releases from all nuclear power operations of 25 mrem for both the whole body and individual organs except the thyroid, which has a limit of 75 mrem.

1.1   Standards for Protection of the Public against Radioactive Contamination

Any material or property used in operations involving harmful chemicals or radioactivity is susceptible to contamination and must be monitored and remedied in accordance with standards for protection of the public prior to release to the environment. Standards are generally of two types: performance standards and standards based on dose limits to potentially exposed individuals. The dose standards are currently preferred for regulating exposure to radioactively contaminated environments. In the absence of effective methods of evaluation of dose, performance standards are used. These specify state-of-the-art technology to clean up contamination situations to the lowest levels practicable. The question then becomes, "How clean is clean?" Because of the different analyses that are invariably employed by experts with respect to difficult problems, regulatory standards usually are based on a consensus among participants in the standard-setting process. However, political considerations and perceptions of the public often modify the recommendations of professional bodies. The outcome can have profound implications regarding the resources required for environmental controls and the actual reduction in risk achieved with those resources.

It can take a long time to produce a consensus standard. Agreement requires literature reviews, meetings, ballots, voting changes, and trial periods. The development of a consensus standard for the release of property with surface radioactivity is discussed in Part Five, section 21.1. The standard provides screening levels for the potential release but emphasizes that the screening levels are designed only to comply with the primary dose criterion and are not designed to be used to authorize any decommissioning or cleanup projects. These are more amenable to cleanup and decommissioning requirements that are specific to each site and thus normally handled by regulatory agencies on a case by case basis. Screening levels for surface and volume radioactivity to comply with a dose limit of 0.01 mSv to a critical group of potentially exposed persons, consisting of individuals likely to have the closest contact with the released material, are given in Table 5.9 of Part Five.

1.2 Standards for the Cleanup of Sites Contaminated with Radioactivity

Standards developed by EPA for the cleanup of radiation-contaminated sites are designed to limit radiation exposure to levels that incur only a very small risk of cancer, a lifetime risk of cancer of less than 1 in 10,000. EPA cites the effective dose equivalent that it considers to be associated with this risk as 15 mrem per year (Luftig and Weinstock, 1997). In addition to a dose limit from all pathways combined, EPA has a 4 mrem per year EDE above background for the drinking water pathway in implementation of the Safe Drinking Water Act of 1974. The dose standard cited by NRC for decommissioning of contaminated sites is somewhat higher, an EDE of 25 mrem per year. Because of the large uncertainties inherent in estimates of risk versus dose, the standards may be considered to be comparable. Provisions usually exist for relaxation of standards on a case by case basis, if cost-benefit analyses show that imposed limits cannot be reasonably achieved and effective controls are incorporated to assure the public health. However, any decisions and actions taken by governmental agencies on acceptable cleanup levels also come under the watchful eye of the public, and are often significantly influenced by public and media concerns.

Thus, permissible levels of contamination are strongly dependent on dose models (and public acceptance of the results of the modeling). The most significant exposure pathways in the dose models are external exposure to penetrating radiation, inhalation, and ingestion.

The calculation of external exposure rates is straightforward, as it uses the same analytical methods used in radiation shielding once the source term is characterized. Determining doses from inhalation is somewhat more complicated, requiring reasonable and quantitative mechanisms for resuspending into the air contamination on the ground and other sources, but the dosimetry pertaining to inhalation of radioactive aerosols is well established. Ingestion is much more complex. Modeling groundwater contamination involves pathways that depend on detailed knowledge of the hydrogeologic features of the region, knowledge which is not available in most situations.

1.3   Protective Actions for Exposures of the Public from Long-Term and Unattributable Sources

The development of radiation standards and the enforcement actions of the regulatory agencies are primarily directed toward protecting people from excessive radiation exposure as the result of work and practices by other identifiable people who work with radiation sources. The target of protective action and enforcement measures is clear, the permissible exposure limits well defined. But what are the limits, what are the interventional procedures, when the habitat of a population is an unusually high radiation background, whether from natural sources, or from contamination arising out of previous work of indeterminate origin, or from a major accident? What action is taken when the air in people's homes contains high levels of alpha-emitting radionuclides from radon gas seeping out of the ground, well above normal exposure limits, but the exposure does not appear to affect the quality of their lives? What resources should be expended in reducing the dose or, where this is not an option, at what point is serious attention given to evacuation of the exposed individuals?

These questions presented a most challenging problem to the ICRP and resulted in the provision of guidance on the application of their system of radiological protection to prolonged exposure situations affecting members of the public (ICRP, 1999). The ICRP stated that its quantitative recommendations must be interpreted with extreme caution and within the context of the information given in the report. Accordingly, limits presented here should be taken as simply indicative of the contents of the report and should not be applied to a specific situation.

A generic reference level of ~10 mSv/yr was given for exposures from high natural background radiation or from radioactive residues that were "a legacy from the distant past," levels that had persisted in the environment for many years and could not readily be reduced. Below this level, intervention was not likely to be justifiable. Intervention was almost always justifiable for exposures above 100 mSv/yr. The additional annual limit attributable to all relevant practices was given as 1 mSv, with a target dose of less than approximately 0.3 mSv. Where exposures could occur from more than one source and it was not possible to provide a reliable dose estimate, it would be prudent to limit exposure from a long-lived source to the order of 0.1 mSv. An exemption limit for a source accompanying the exercise of an occupation was given as 0.01 mSv.

The report also discussed interventional measures for radon in existing buildings. An action level for intervention was recommended at 3–10 mSv/yr. This corresponded to values for the radon concentrations between 200 and 600 Bq/m^3 and an annual occupancy of 7,000 hours, the calculations including an equilibrium factor between radon and its decay products of 0.4 (ICRP, 1993b). For new buildings in a radon prone area, the commission noted that proven protective actions against radon in indoor air were readily available and that construction codes and building guides should be devised that consistently achieved low concentrations of radon in the completed buildings.

2 MEDICAL FINDINGS ON HUMANS EXPOSED TO RADIATION

The nature of the interaction of radiation with human tissue is such that any level of exposure may pose a risk of initiating a serious disease such as leukemia or other cancer or of producing detrimental genetic effects that show up in future generations. On the other hand, some radiation exposure of human beings is an unavoidable consequence of the benefits of radiation used in medicine and technology. Thus, in establishing standards of radiation exposure, it is necessary to weigh both benefits and risks of specific radiation uses. The benefit-risk decision may be made for exposure of a specific individual, as in connection with medical treatment or occupational exposure; or for society as a whole, as in connection with the development of nuclear power.

Benefit-risk decisions, by their nature, must be based on judgment. A very important element in any decision is the knowledge of the actual extent of the risk and harm to individuals or society resulting from particular radiation uses. Thus it is extremely important that individuals in a position to be exposed—or in a position to expose others—understand clearly what is known and what is not known about the effects of low-level exposure.

2.1   Sources of Human Exposure Data

Data on the effects of low-level radiation on humans come from the following sources:

(a) Survivors of the atomic bombing of Hiroshima and Nagasaki. This source provides the largest sample of individuals exposed to date. There have been difficulties in determining the actual doses to individuals and in defining control populations for comparisons. However, the data have contributed toward determining the values for the risk of leukemia and other cancers from exposure to a single significant dose of radiation.

(b) Children exposed prenatally as a result of abdominal x-ray examination of the mother during pregnancy. These irradiations occurred at the most radiation-sensitive time during the lifespan of the individual. Thus the results yield an upper limit to the risk.

(c) Children treated for enlarged thymus glands by irradiation of the thymus. The treatments resulted in high-level irradiation of a relatively small region of the body.

(d) Adults who underwent x-ray treatment to the spine for ankylosing spondylitis. These exposures resulted in substantial irradiation of much of the bone marrow. Smaller but significant doses were also given to other organs.

(e) Adults who received radioactive iodine for treatment of thyroid conditions. When compared with data on the irradiation of the thyroid gland in children, the results provide valuable data on the sensitivity of the thyroid to radiation as a function of age.

(f) Individuals with body burdens of radium. Epidemiologic studies of this group provided the basis for setting the limits of exposure to the bone by radium and by other bone-seeking radionuclides.

(g) Uranium miners exposed to high levels of radioactive gases and radioactive particles. These occupational exposures are providing data on the development of lung cancer from irradiation of regions of the lung.

The findings from these and other significant studies are given in Table 6.2, which should be studied carefully. The data in the table do not include studies of detrimental effects in offspring of irradiated parents (genetic effects). Such studies are extremely difficult to conduct and interpret, and those that have been made have been inconclusive. The genetic risk associated with radiation exposure in humans is discussed separately in section 2.3.

Table 6.2 Findings of epidemiologic studies of cancer in irradiated populations.

Subject of study and citation	Follow-up time (yr)	Dose (R or rad)	No. of persons[a]	Form of cancer	No. of cases Observed in exposed group[b]	No. of cases Expected if not exposed[c]
1. Hiroshima and Nagasaki A-bomb survivors (bomb exploded 1945). Ichimaru and Ishimaru, 1975.	26	0–1	61,263	Leukemia	36	36 (controls)
		1–99	39,093	Leukemia	43	23
		> 100	6,046	Leukemia	61	3.6
2. Hiroshima and Nagasaki A-bomb survivors (women), in age groups shown at time of bombing. All exposed to more than 100 rad and observed 1950–1969. McGregor et al., 1977.	5–24	Age: 0–9 yr	429	Breast	0	0.1
		10–19 yr	1,048	Breast	10	1.4
		20–34 yr	887	Breast	13	3.8
		35–49 yr	685	Breast	7	3.5
		50+ yr	309	Breast	4	0.9
		Total	3,358	Breast	34	9.7
3. Hiroshima and Nagasaki A-bomb survivors exposed prenatally. Jablon and Kato, 1970.	10	>64,500 person-rad[d]	1,292	All cancers	1	0.75
4. Japanese A-bomb survivors exposed within 1400 m of detonation (died 1950–1962). Angevine and Jablon, 1964.	17	—	1,215 autopsies	All cancers except leukemia	61	56.8
5. Japanese A-bomb survivors (died 1950–1990). Preston et al, 1997	45	0(<0.5)	36,459	Solid cancer	3013	3013
		0.5–10	32,849		2795	2761
		10–20	5467		504	475
		20–50	6308		632	557
		50–100	3202		336	258
		100–200	1608		215	145
		>200	679		83	34
		Total	86,572		7578	7244
6. Children exposed prenatally due to abdominal x ray to mother (exposed 1945–1956 and died before end of 1958). Court Brown et al., 1960.	2–12	—	39,166	Leukemia	9	10.5
7. Children exposed prenatally (born in 1947–1954 and died before end of 1960). MacMahon, 1962.	4–13	1–2	77,000[e]	All cancers	85	60

Table 6.2 (continued)

Subject of study and citation	Follow-up time (yr)	Dose (R or rad)	No. of persons[a]	Form of cancer	No. of cases — Observed in exposed group[b]	No. of cases — Expected if not exposed[c]
8. Infants who received irradiation of chest before age 6 mo. in treatment for enlarged thymus (treated 1926–1957; follow-up to 1971). Hemplemann et al., 1975.	13–45	119 av. to thyroid; air dose 225	2,872 (69,402 person-yr at risk)[f]	Thyroid	24	0.29
				Thyroid (benign)	52	3.42
				Leukemia or lymphoma	8	3.97
				Other	14	8.1
				Other (benign)	69	34
Subgroup: high doses, large ports, apparently more susceptible, genetically.		399 av. to thyroid; air dose 461	261 (8,088 person-yr at risk)	Thyroid	13	0.04
				Thyroid (benign)	20	0.4
				Leukemia	2	0.49
				Other	5	1.84
				Other (benign)	16	2.24
9. Infants irradiated routinely with x rays to anterior mediastinum through small (4 × 4 cm) port, 7 days after birth, as "apparently harmless and perhaps beneficial procedure" (x ray 1938–46, follow-up 1956–58). Conti et al., 1960.	10–20	75–450 (most 150)	1,401, including 244 with enlarged thymus	Thyroid carcinoma	0	0.03
				Leukemia	0	0.95
10. Children treated with x rays to head, neck, or chest for various benign conditions, mainly "enlarged" thymus and adenitis, treated before age 16 and followed till age 23. Saenger et al., 1960.	>11 (83%)	<50 (4%) 50–200 (36%) 200–600 (33%)	1,644	Thyroid	11	0
				Thyroid (benign)	7	0
11. Children treated before age 16 with x rays for enlarged thymus, pertussis, and head and neck diseases and died before age 23 (treated 1930–1956; follow-up 1940–1956). Murray et al., 1959.	Up to 23	Not given	3,872	Leukemia	7	1.4
12. Children irradiated for ringworm of the scalp. Up to 3 treatments if relapses occurred. Follow-up to 1973. Modan et al., 1974.	12–23	350–400 (140 to brain; 6.5 to thyroid)	10,902	Brain	8	1
				Brain (benign)	8	1
				Thyroid	12	2
				Scalp	1	0
				Leukemia	7	5
				Lymphoma	8	5
				Breast	2	0

Table 6.2 (continued)

Subject of study and citation	Follow-up time (yr)	Dose (R or rad)	No. of persons[a]	Form of cancer	No. of cases Observed in exposed group[b]	No. of cases Expected if not exposed[c]
13. Patients (most between ages 10–40) treated with x rays for benign lesions in neck, mainly for tuberculous adenitis (treated between 1920–50). Hanford et al., 1962.	10–40	100–2,000	295	Thyroid	8	0.1
14. Patients ages 20–70, treated with x rays to thyroid for benign disorders. DeLawter and Winship, 1963.	10–35	1,500–2,000	222	Thyroid	0	
15. Hyperthyroid patients, ages 20–60, treated with x rays (treated 1946–53; follow-up 1959–61). Sheline et al., 1962.	5–15	Not given	182	Thyroid (probable)	1	
				Multiple benign thyroid nodules	7	
16. Patients single treatment with x rays to spine for ankylosing spondylitis (treated 1935-54, died to 1983) Darby, Doll et al. 1987; Weiss, Darby et al. 1994(doses)	5–28	438 555 254 454 438 59	14,106 (183,749 person-yr)	Leukemia Aplastic anemia Esophagus Lung Bone Lymphomas Breast Cent. Nerv. Sys.	36 7 27 224 4 21 26 23	11.9 0.96 12.7 185 1.36 10.9 16.1 14.2
17. Female tuberculosis patients who received many fluoroscopic examinations. Boice and Monson, 1977.	21–45	400+ 300–399 200–299 100–199 1–99 0	62 65 177 251 469 717	Breast Breast Breast Breast Breast Breast	4 3 12 12 10 15	1.1 1.5 4.8 5.7 9.6 14.1
Classified by age at first exposure.		Age: <15 yr 15–19 yr 20–24 yr 25–29 yr 30–34 yr 35+	99 242 263 200 105 138	Breast Breast Breast Breast Breast Breast	2 13 9 9 4 4	0.9 3.4 5.4 5.5 3.3 4.8
18. Women treated with x rays for acute postpartum mastitis. Shore et al., 1977.	19–34	50–1,065 (air); 377 average to irradiated breast	571	Breast Breast (benign)	37 29	11.3 15
19. Patients treated with x rays for cancer of the cervix. Hutchison, 1968.	4–8 (31%) <4 (69%)	300–1,500 av. to bone marrow	27,793 (57,121 person-yr)	Leukemia Lymphatic malignancy	4 6	5.1 6.3

Table 6.2 (continued)

Subject of study and citation	Follow-up time (yr)	Dose (R or rad)	No. of persons[a]	Form of cancer	No. of cases Observed in exposed group[b]	No. of cases Expected if not exposed[c]
20. American radiologists (died 1948–1964). Lewis, 1970.	Through 1964		530 deaths	Leukemia	13	3.91
				Multiple myeloma	5	1.01
				Aplastic anemia	5	0.23
21. American radiologists. Warren and Lombard, 1966.	Through 1960		5,982	Leukemia		
				1940–44	4	0.5
				1945–49	7	0.86
				1950–54	6	1.26
				1955–60	7	2.05
22. American radiologists. Seltser and Sartwell, 1965.	1935–58		3,521 (48,895 person-yr)	Ages 35–49		
				Leukemia	2	1.9
				Other cancer	9	7.3
				Total deaths	79	61.5
				Ages 50–64		
				Leukemia	8	1.1
				Other cancer	54	32
				Total deaths	339	271.5
				Ages 65–79		
				Leukemia	9	4.7
				Other cancer	72	48
				Total deaths	438	295
23. Hyperthyroid patients treated with ^{131}I (treated 1946–64; follow-up through June 1967). Saenger et al., 1968.	3–21	7–15 rads to bone marrow (9 mCi ^{131}I av.)	18,370 (119,000 person-yr at risk)	Leukemia	17	11.9
Comparison group treated by surgery and not given ^{131}I. Saenger et al., 1968.			10,731 (114,000 person-yr at risk)	Leukemia	16	11.4
24. Population of Marshall Islands accidentally exposed to radioactive fallout from test of 17-megaton thermonuclear device, 1945. Conard, 1976.	22	200 (av)	243	Thyroid	7	0
25. Children in Utah and Nevada exposed to fallout in the 1950s. Rallison et al., 1974.	12–15	46 av.; >100, max.	1,378	Thyroid	0	1.05 (0.64)[g]
				Thyroid (benign)	6	4.2 (3.9)
				Adolescent goiter	22	12.6 (21.2)
				Hyperthyroidism and misc.	5	3.1 (2.6)

Table 6.2 (continued)

Subject of study and citation	Follow-up time (yr)	Dose (R or rad)	No. of persons[a]	Form of cancer	No. of cases Observed in exposed group[b]	Expected if not exposed[c]
26. Radium dial painters and others who ingested radium. Evans, 1974; Evans et al., 1972.	40–50	1–100	381	Bone	0	
		100–1,000	122	Head	0	
		1,000–5,000	42	Bone	9	
				Head	2	
		5,000–50,000	25	Bone	3	
				Head	5	
27. Radium dial painters (began dial painting before 1930). Evans, 1967; Rowland et al., 1978.	50	^{226}Ra equivalent entering blood (μCi)				
		100 to >2,500	115	Bone	38	
		0.5–99	439	Bone	0	
		<0.5	205	Bone	0	
		25 to > 1,000	134	Head	17	
		0.5–24.9	388	Head	0	
		<0.5	227	Head	0	
28. Uranium miners (white) exposed to radon gas and decay products; started mining before 1964. Mortality follow-up to September, 1974. Archer et al., 1976.	>10	WLM[h]	3,366 P-YR[f]	(a) smokers		
		> 1,800	5,907	Respiratory	60	
		360–1,799	16,331	Respiratory	55	
		1–359	14,031	Respiratory	25	
				(b) Nonsmokers		
		> 1800	1,437	Respiratory	2	
		360–1,799	3,488	Respiratory	3	
		1–359	4,918	Respiratory	1	
29. Women with scoliosis who received multiple x rays. Hoffman et al., 1989.	26 (av)	0–9	466	Breast	4	3.12
		10–19	298	Breast	3	1.66
		>20	187	Breast	4	1.17
Total population		12.8 (av)	95	Breast	11	6.06

Note: See UNSCEAR, 2000, Vol. II, Annex I for a comprehensive tabulation of epidemiological studies.

a. This refers to the number of individuals at risk unless otherwise specified.

b. The information needed to determine the number of cases is obtained by either following a designated study population or by working back from a review of all death certificates in a defined geographical area.

c. The numbers in this column are based on available statistical data for unexposed populations.

d. The term *person-rad* pertains to the dose imparted to a population and is equal to the sum of the doses incurred by the individuals in the population.

e. This figure is based on a systematic sampling of the population rather than a review of all the records.

f. Person-years at risk is the sum of the number of years in which the disease could develop in each member of the group.

g. The first number is based on 1,313 children who moved into the area after ^{131}I fallout decayed. The number in parentheses is based on 2,140 children in southeastern Arizona, an area remote from the fallout.

h. Working Level Months (see Part Three, section 5.6).

Most of the medical findings in Table 6.2 consist of comparisons of the "observed" morbidity or mortality rates of malignancies in the irradiated groups and "expected" rates in comparable unirradiated groups. Values for the expected rates precise enough to allow comparison with the rates observed at low exposure levels can be obtained only by studying the medical histories of very large groups of individuals. The control groups should be identical in composition to the exposed group in all respects except that they did not receive the type of exposure being evaluated. The difficulties in defining and obtaining incidence data for suitable control groups are enormous, and the conclusions must often be viewed with less than total confidence.[5]

Studies in which a population is identified in advance for determination of the incidence of a specific disease are called prospective studies. Also called cohort studies, these last for many years because of the long delay in effects that result from radiation exposure. They rank highest for quality and have the least susceptibility to bias. When it is not practical to establish and follow such a population, a retrospective study, also called a case-control study, may be undertaken. In such a study, the individuals with the disease in question are designated as a group and their histories are obtained. Histories are also obtained for a control group as comparable as possible to the former group in all respects except that they do not have the disease under study. A comparison is made of the frequencies with which the factor suspected of affecting the incidence of the disease occurred in both groups, and if the frequency is significantly higher in the group with the disease, it is considered a candidate for a causative role. In making a test for a specific factor, it is necessary to correct for the effect of other characteristics that differ in the two groups. As an example, in studies made on the association of prenatal x rays to the fetus and the subsequent appearance in childhood of leukemia or other cancer, some differences that had to be considered (between the x-rayed and non-x-rayed groups) included birth order, sex ratio, and maternal age.

Retrospective studies are more susceptible to bias and are more difficult to interpret than prospective studies and the results usually do not merit the same degree of confidence. However, the information already exists to be mined. They require fewer resources and smaller populations and sometimes represent the only feasible approach. The results of two retrospective studies concerned with testing the association between excess risks of childhood cancer and prenatal x-ray exposure are presented in the next section.

5. For discussion of the methodology and problems of obtaining population statistics, see Grove and Hetzel, 1968.

The least reliable studies, but quite popular, are ecologic or geographical studies in which the incidence of disease is compared in areas with different average radiation levels. While epidemiologists are highly critical of any conclusions drawn from these studies, they can provide perspective in assessing the significance of exposure to selected radiation sources and in providing leads to the presence or absence of effects for further investigation.

In examining the human exposure data presented in Table 6.2, note the numbers of subjects involved, the actual number of cases of disease, and the follow-up times.[6] Bear in mind the factors that serve to weaken conclusions drawn from the results—effects on patients who are ill to start with; the generally better than average health of workers in occupational studies; more intensive screening in exposed groups; interview bias; recall bias; and the contributory effects of cigarette smoking or other harmful exposures. Regardless of their limitations, these data form the main resource for evaluating the effects of subjecting major segments of the population to radiation.

2.2 Epidemiological Studies of Leukemia and Other Cancers

2.2.1 Japanese Survivors of the Atomic Bomb

The general consensus from reviewing human exposure data (Table 6.2, items 1, 15) is that the risk of induction of leukemia in a period of 13 years following an x- or γ-ray dose of 100 rads to a substantial part of the blood-forming organs averages out to 1 or 2 in 10,000 in each year following the exposure (ICRP, 1966b, p. 4). The experience of the Japanese atomic bomb survivors indicates that the risk peaks 6–7 years after the exposure and diminishes substantially after a period of perhaps 15 years (Fig. 6.1), but it does not become negligible even 30 years later (Okada et al., 1975).

The incidence of leukemia was much greater at Hiroshima than at Nagasaki in the lower dose range of 10–99 rad. It was originally believed that the nature of the radiation exposures in the two cities was quite different, that at Hiroshima, 20 percent of the dose (kerma) was contributed by neutrons, whereas at Nagasaki less than 0.3 percent was contributed by neutrons. This was interpreted to mean that the leukemias at low doses were induced primarily by the neutrons and the data appeared to be a valuable resource for evaluating the relative biological effectiveness of neutrons (Rossi and Mays, 1978). In the mid-1970s, however, researchers working on computer simulations realized the neutron component had been sub-

6. Leukemia incidence should be followed for about 15 years after exposure; cancer incidence should be followed for at least 30 years.

6.1 Incidence of leukemia in atomic bomb survivors and controls (Okada et al., 1975).

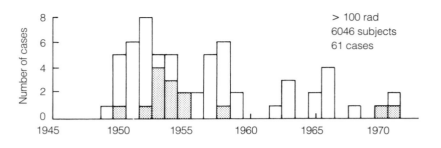

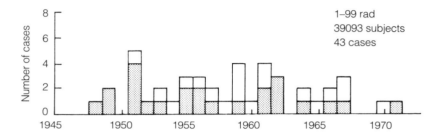

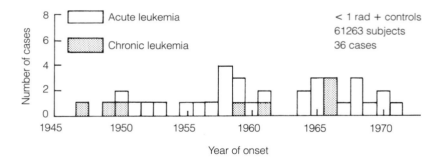

stantially overestimated. This finding stimulated a joint U.S.-Japanese reassessment of the atomic bomb radiation dosimetry. The findings were published in 1987 (RERF, 1987). It was concluded that the attenuation effects of high humidity and the bomb shielding material on the neutrons had not been taken into account adequately, and that the neutron dose at Hiroshima was 10 percent of the previous estimate. The gamma dose was 2 to 3.5 times higher, depending on the distance from the hypocenter. The changes were smaller at Nagasaki and had little effect on the evaluation of dose. The estimate of the dose to the survivors was further reduced when the enhanced shielding effect of clusters of houses, which had been neglected previously, was taken into account. Refinements were also made in organ dosimetry. These revisions served to eliminate the Hiroshima data as a reference for the RBE of neutrons and pointed to significant increases in the estimates of risk of exposure to gamma radiation—a two-fold increase for leukemia and a 50 percent increase in risk for other cancers (Roberts, 1987).

The induction of solid cancers was also followed in the Japanese survivors. Statistics on mortality, starting with 1950, are given in Table 6.2, items 4 and 5. Item 4 data are for a follow-up period of 17 years, item 5 data for a follow-up period of 45 years, indicating the need for watching an exposed population for the duration of their lifetimes. Surveillance for incidence was not begun until 1958. The collection of incidence statistics is not as encompassing as mortality statistics, which are based on a family registration system covering the whole population. Incidence data provided a much greater number of solid-cancer cases than the mortality data, despite limitations in the collection of the data (Mabuchi et al., 1997). Cancer of the female breast was the major site for incidence, with 95 cases out of 507. The mortality series gave 30 cases of breast cancer out of 306 excess deaths attributable to radiation, the fourth highest proportion (lung cancer, with 69 deaths out of 306, is in first place).

2.2.2 Adult Patients Given Radioiodine Therapy for Hyperthyroidism

The incidence of leukemia was studied in adult patients who had been given radioiodine as a treatment for hyperthyroidism (Table 6.2, item 23). The treatment resulted in a dose to the bone marrow of 7–15 rad. The appearance of leukemia in the patients was about 50 percent greater than that expected in a similar untreated group. However, the leukemia incidence was also greater by about the same amount in a group of hyperthyroid patients who had been treated surgically and not given radiation. The investigators concluded that the increased incidence of leukemia was due to some factor associated with hyperthyroidism rather than to radiation exposure. Regardless of the true reason for the increase, the study indicated dramatically the possibilities of error when incidence in a population treated for an abnormal condition is compared with healthy controls.

In a major epidemiological study of diagnostic iodine-131 involving a mean dose to the thyroid gland of 1.1 Gy (100 rad), 67 thyroid cancers developed in a population of 34,000, representing a standard incidence ratio (SIR) of 1.35, that is, the incidence of thyroid cancer was 35 percent greater than expected (Hall et al., 1996). The mean age at first exam was over 40 years and less than 1 percent of the patients were under age 10 at the time of the first exam. Most of the elevated risk occurred between 5 and 10 years after exposure and appeared to be confined to patients who had been referred for the examination because of suspected thyroid tumors. No excess risk or dose response was found among patients referred for other reasons (Ron, 1997).

While there is tremendous interest in the effects of iodine-131 dose to the thyroid in children, data are extremely limited. No studies have enabled using the data on the carcinogenic effectiveness of external radiation to provide a reasonable assessment of the effects of deposition of iodine-131 in the thyroid (Ron, 1997). One looks forward to the epidemiology of the effects of the Chernobyl reactor accident to cast some light in this area.

2.2.3   Irradiation of the Thyroid in Children

While thyroid cancer rarely develops following adult exposure to radiation, it can readily be induced by relatively small doses of external radiation during childhood. At 1 Gy, the relative risk is about six times higher when exposure occurs before age 15 years than after age 15 (Ron et al., 1995).

Statistics of one study of the results of acute exposure of the thyroid of infants to therapeutic doses of x rays are given in Table 6.2, item 8. An increased incidence of both leukemia and thyroid cancer was observed. In addition, later studies found an increased risk of breast cancer in infants irradiated at 6 months of age for enlarged thymus, the cancers appearing after 40 years. The observed numbers of thyroid cancers in the irradiated group were almost 100 times greater than expected and over 300 times greater than expected in a selected high-risk subgroup. A minimum latent period of 5 years was observed in the appearance of thyroid cancer, and 10 years for benign lesions. It appears that even after 30 years, an excess number of cases was still appearing. The investigators concluded that if there were a threshold dose for the induction of thyroid nodules, it was below 20 rad (Hempelmann, 1968). The risk of tumor induction remained elevated beyond the age at which the leukemia risk appeared to disappear. The risk for the induction of thyroid cancer was 2.5 cases per million per year per roentgen. At this level, the probability of producing cancer of the thyroid 30 yr after a dose of 300 rad would be $2.5 \times 10^{-6} \times 300 \times 30$, or about 1 in 40. The prevalence of all spontaneous thyroid nodules per 100 persons at a given age is about 0.08 times the age; about 12 percent are malignant (Maxon et al., 1977, p. 972).

The testing of nuclear weapons in the atmosphere resulted in the release of enormous quantities of ^{131}I (see section 4.5.1). Studies were made of possible effects on the thyroid of children living in the region of the heaviest fallout around the test site during the testing period in the 1950s (Table 6.2, item 25). Estimates of thyroid doses were 46 rad for all children in Utah between 1952 and 1955 and over 100 rad for children residing in southwestern Utah. No significant differences were found between the children in Utah and Nevada exposed to the highest fallout level and control groups living in Utah and Arizona at the time of the study. However, the follow-up time was only about ten years and data for longer follow-up time are needed. One interesting outcome of the study was that only 6 of the 201 children with thyroid disease knew of their disease prior to the examination. According to the authors, this indicated the need for more attention to the thyroid in routine medical checkups of children.

The BEIR III committee (NAS/NRC, 1980) expressed the cancer risk to the thyroid from external photon radiation applicable to all ages as four carcinomas per rad per million persons per year at risk. Benign adenomas are also induced by radiation, with an absolute risk of 12 adenomas per 10^6 PY per rad. The BEIR V committee (NAS/NRC, 1990) preferred a relative risk model with a relative risk for all ages of 8.3 at a dose of 1 Gy. It could not provide a value of lifetime risk based on an excess-absolute-risk model because of the variability of risk estimates in different populations. While the thyroid is one of the organs more susceptible to radiation-induced cancer, there may be some comfort in the knowledge that radiation does not appear to produce the highly malignant anaplastic carcinoma that is responsible for most deaths from thyroid cancer, but rather carcinomas of the papillary and follicular types, which are fatal in only a small percentage of the cases. The risks associated with the use of radioactive iodine for treatment of thyroid conditions appear to be much lower than those attributed to external irradiation on an organ-dose basis, perhaps by a factor of 70 for induction of thyroid cancer (Maxon et al., 1977).

2.2.4   Exposure of the Fetus in Diagnostic X-Ray Examinations

There is evidence that diagnostic x-ray examinations resulting in exposure of the fetus increase the chances of leukemia or other cancer during childhood. The only studies providing numbers of cancer cases on a scale large enough to demonstrate the possible existence of a risk from doses characteristic of diagnostic x rays were retrospective in nature. The records of children who died of cancer were reviewed to see if they received x-ray exposures in excess of those expected for a similar sample that was free of cancer. The results of two major retrospective studies by Stewart and Kneale (1970) and by MacMahon (1962) are given in Table 6.3.

MacMahon concluded that mortality from leukemia and other cancers was about 40 percent higher among children exposed in a diagnostic x-ray

Table 6.3 Occurrence of prenatal x rays in children dying of cancer.

Reference	No. of cancer deaths		No. x-rayed in uterus	No. expected as determined from cancer-free controls
Stewart and Kneale, 1970	7,649		1,141	774
		By number of films:		
		1	274	218
		2	201	151
		3	103	58
		4	60	28
		>4	65	29
		No record	438	290
MacMahon, 1962[a]	556		85	63

a. This study is also entered in Table 6.2 (item 7), as the information provided in the study enabled an estimate of the population at risk.

study in utero than among children not so exposed. It was estimated that prenatal x rays produced an excess of 3.03 cancer deaths (including leukemia) before the age of 14 for every 10,000 live births (the expected mortality derived from the unexposed control group was 3.97 from leukemia and 3.31 from other cancers).

Stewart and Kneale, after reviewing over 7,600 cancer deaths and an equal number of cancer-free controls, concluded that the risk of childhood cancer depended on the number of x-ray films (and on the fetal dose). The risk was expressed as 300–800 extra deaths before the age of 10 among one million children who received a dose of one rad shortly before birth.

A large prospective study (Table 6.2, item 6) failed to demonstrate the appearance of leukemia or cancer in children after prenatal diagnostic x rays. However, although over 39,000 cases of fetal x rays were followed, the sample size was not large enough to rule out with a high degree of confidence as much as a 50 percent increase in leukemia.

It has been reported that postnatal exposure also increases the risk of leukemia (Graham et al., 1966). The relative risk from postnatal x rays was said to vary from an increase of 50 percent to 400 percent, with the greatest risk in children exposed to both medical and dental x rays. The number of cases involved in the study, however, was small. Only 19 leukemia cases were in the combined dental plus medical x-ray group receiving the highest exposures, whereas 31 cases were found in a large control group. The conclusions were based on a retrospective study. The parents of children who had died of cancer were questioned concerning the exposure of their children during infancy, and their replies were compared with those of the parents of healthy controls. The two groups of children were matched as closely as possible. One major weakness of the study was reliance on the memory of the parents to determine the frequency of x rays. The results of several independent studies suggest the need for continuing investigations into the effects of mass medical and dental x rays on the population, particularly in children.

2.2.5   Exposure of the Female Breast

The relative importance of cancer of the breast in women as a consequence of exposure to radiation is increasing as data become available for longer follow-up times. Among the Japanese survivors, the total number of breast cancers attributed to radiation has already exceeded the number of leukemias (Table 6.2, item 2; McGregor et al., 1977). During the period 1950–69, women exposed to 100 or more rads had 3.3 times the breast cancer incidence of women exposed to less than 10 rad. The most sensitive age group with regard to radiation-induced breast cancer was 10–19 years. It was the only age group that showed strong evidence of increased risk of breast cancer in the intermediate dose range (10–99 rad). These figures are for a population that shows a very low natural incidence compared with other countries; the rate is about one-fifth that of U.S. white women. Thus it is quite possible that the radiation risk to American women could be significantly greater than that found among the Japanese, if the risk is enhanced by environmental factors or life style.

Data on the sensitivity of the breast to radiation-induced cancer are also provided from surveys of women with tuberculosis who received frequent fluoroscopic examinations for artificial pneumothorax (Table 6.2, item 17) and postpartum mastitis patients treated with radiotherapy (Table 6.2, item 18).

The study of tuberculosis patients, who were fluoroscopically examined an average of 102 times with average doses to the breast of 150 rad, indicated that the greatest absolute excess breast cancer risk occurred among exposed women who were first treated between the ages of 15 and 19 years. Among those women 30 years of age and older at the time of first exposure, no elevated breast cancer risk was detected; however, the failure to observe an excess did not exclude a risk increased by as much as 50 percent. In the mastitis patients, who received an average dose of 377 rad to the irradiated breast, it was concluded that the overall relative risk of breast cancer was 2.2 for years 10–34 post-irradiation and 3.6 for years 20–34. Women over age 30 years at radiation treatment had as great an excess risk of breast cancer as did younger women, an apparent contradiction to the findings for the tuberculosis patients. If this were a real difference, the authors suggested that the reason may have been that the breasts of the mastitis patients were actively lactating at the time of radiation, whereas they were quiescent in the other studies. A linear or near linear relationship for cancer versus dose was observed. In contrast, no significant excess of lung cancer was found despite average doses of 1 Gy (Howe, 1995; Davis et al., 1989.)

2.2.6   Exposure of the Spine in Treatment of Ankylosing Spondylitis

The late incidence of cancer has been followed in patients administered high doses of x rays to the spine as a treatment for ankylosing spondylitis (Table 6.2, item 16). In the first decade following irradiation of the patients, interest focused on the incidence of leukemia, which was about double that of all other cancers combined. After a decade, the incidence of excess cancers attributed to the irradiation rose significantly at some sites, while the leukemia incidence dropped (Table 6.4). The ratio of observed to expected cancer deaths reached a maximum of 1.7 between 10 and 12.4 years after irradiation and then declined (Darby, Doll, et al., 1987). When the lower doses to the sites away from the spine are taken into account, the data suggest that, for a uniform whole-body dose, the number of fatal malignancies (other than leukemia) to appear eventually may reach 10 times the number of leukemias occurring in the first 15 years (Tamplin and Gofman, 1970).

Table 6.4.   Change in rate of induced malignant disease with duration of time since exposure in persons irradiated for the treatment of ankylosing spondylitis (from data in table VI of Court Brown and Doll, 1965).

	Cases per 10,000 person-years at risk	
Years after irradiation	Leukemia + aplastic anemia	Cancers at heavily irradiated sites
0–2	2.5	3.0
3–5	6.0	0.7
6–8	5.2	3.6
9–11	3.6	13
12–14	4.0	17
15–27	0.4	20
Total of expected cases in 10,000 persons in 27 yr, calculated from rates given	67	369

Source: ICRP, 1969. See Weiss et al. for later statistics.

2.2.7 Breast Cancer in Women with Scoliosis X-Rayed Many Times in Childhood

Scoliosis, a condition involving abnormal lateral curvature of the spine, is generally recognized in childhood and followed through childhood and adolescence with diagnostic x rays. The x rays impart a significant dose to the breast at a time it is developing, and as a result the breast is at an enhanced risk of cancer. The incidence of cancer was studied in a population of 1,030 women with scoliosis who had received many x-ray examinations (Hoffman et al., 1989). The results are presented in Table 6.2, item 29. The subjects (on average) were diagnosed with scoliosis at 12.3 years of age, had 41.5 radiographs of the spine over 8.7 years, received a dose of 13 rad to the breast, and were followed for a period of 26 years. An excess risk of breast cancer was found in the patients in the study. The highest risk was among those followed for more than 30 years (standardized incidence ratio 2.4). The risk also increased with the number of x rays and with the estimated radiation dose to the breast. The authors noted that although the exposures in current practice appear to be much lower than what they were 25–30 years ago, the magnitude of the potential radiation risk is still a concern, and it is recommended that exposures should be reduced whenever possible, without sacrificing needed diagnostic information. The authors concluded that radiation exposure from scoliosis radiographs should be minimized whenever possible, especially among young girls who may be at especially sensitive stages of development.

2.2.8 Second Cancers Following Radiotherapy for Cancer

The results of studies of second cancers following radiotherapy for cancer provide a source of dose-response data for different tissues at high doses, since the radiation fields are well defined. A cohort of patients treated with surgery alone (no radiation) can serve as a control group when available. The studies also provide statistics of cancers at lower doses for regions that are outside the beam.

The risk of leukemia following partial-body radiotherapy is about double the background rate over a wide dose range (Curtis, 1997). The incidence of the second cancers is well below the rate to be expected on the basis of the atomic bomb survivors. It is difficult to account for the differences. For example, local therapy doses are considerably higher than the whole-body doses from the atomic bombs; on the other hand, therapy doses are protracted whereas most of the dose from the bomb was delivered essentially instantaneously.

Most investigators have found no association between radiotherapy for breast cancer and the appearance of a cancer in the second breast. One reason may be that the risk of breast cancer from radiation decreases with age and is especially low for women over the age of forty. Risk of lung cancer appears to be increased by a factor of two in survivors of breast cancer. Smoking history is a confounding factor and has not been fully accounted for, although smoking is likely to increase the risk substantially.

An increased understanding of the risks of inducing second cancers in radiotherapy for cancer will come with longer follow-up time, particularly as survival of cancer patients continues to improve.

2.3   Risk of Cancer from Exposure to Radiation

It is difficult to develop assessments of the risk of cancer throughout the lifetime of the individual as a function of dose. The analysis must consider a minimal latent period, the rate of appearance of cancer with time following the latent period, and the period of time over which the cancers will appear. There may be differences in susceptibility as a function of age at exposure and of sex. The doses of particular interest are the regulatory limits established for occupational and environmental exposures and doses imparted in diagnostic radiology, whereas significant epidemiological data exist only for much higher levels. At a dose of 2 Gy, there is approximately a doubling of the risk, a finding that can be stated with some confidence. At 1 Gy, the hypothesis that the risk is proportional to the dose deals with increases of the order of 40 or 50 percent, which epidemiology also can deal with. But at 0.1 Gy (10 rem), twice the value of the annual occupational dose limit, linear interpolation gives an excess risk of only 5 percent and it is extremely difficult to obtain confirmatory data from epidemiological studies. Dose-response relationships must be developed to infer an effect (Boice, 1997).

Assumptions as to the relationship between dose and effect greatly influence the risk estimates for low doses. The linear dose-effect model is believed to be conservative and quite suitable for purposes of radiation protection. However, the linear-quadratic model (for low-LET radiation) is generally felt to be more consistent with both knowledge and theory. It takes the form $I(D) = (\alpha_0 + \alpha_1 D + \alpha_2 D^2)\exp(-\beta_1 D - \beta_2 D^2)$, where $I(D)$ is the cancer incidence in the irradiated population at radiation dose D. The modifying exponential function represents the competing effect of cell killing at high doses. Values derived from this model for the risk at low doses are less than those derived from the linear model and may be used to define the lower limits of risk from low-dose, low-LET radiation (Fabrikant, 1980; NAS/NRC, 1980).

The projections of lifetime risk of cancer can vary considerably depending on whether an *absolute* or *relative* risk model is used. The "absolute" or "additive" projection assumes that the exposure produces an excess risk of cancer incidence per year that remains constant following a latent period. The risk may end at a specified age or continue throughout the lifetime of the person. The relative model considers an excess risk that may increase gradually throughout the life of the individual and is proportional to the spontaneous risk, which increases with age for nearly all cancers. This model has gained support over the additive model as more complete epidemiological data has been acquired with time. It doubles the risk estimate as determined from the additive model, resulting in a greater projected total cancer occurrence.

The Advisory Committee on the Biological Effects of Ionizing Radiations of the National Academy of Sciences, in their BEIR V report (NAS/NRC, 1990), derived separate models for leukemia, respiratory cancer, and digestive cancer for the risk of radiation-induced cancer as a function of dose. The model for leukemia was a relative risk model with terms for dose, dose squared, age at exposure, time after exposure, and interaction effects. A minimum latency of 2 years was assumed. For cancers other than leukemia, all cases less than 10 years after exposure were excluded on the assumption of a 10-year minimum latency period. The relative risk of cancer of the respiratory tract was considered to decrease with time after exposure. The important modifying factors for breast cancer were age at exposure and time after exposure. The risk factor for digestive cancer was increased seven-fold for those exposed when they were less than 30 years old. Not enough data were available to develop models specific to the other sites of cancer, and the excess risk was obtained by using a linear dose-response model in extrapolating from the available high-dose data to estimates at low doses, incorporating a negative linear effect by age at exposure at ages greater than 10.

Estimates by the BEIR Committee of excess cancer mortality at different sites as a function of age at exposure are given in Table 6.5.[7] The leukemia estimates contain an implicit dose-rate reduction factor. No dose-rate reduction factor was applied in extrapolating the risk estimates for solid cancers. The risks estimated in BEIR V were substantially higher

---

7. Some investigators have conducted studies which they claim indicate that the risks of low-level radiation are considerably greater than normally assumed (Bross, Ball and Falen, 1979; Bross and Natarajan, 1972, 1977; Mancuso, 1978; Mancuso, Stewart, and Kneale, 1977, 1978; Najarian and Colton, 1978). However, the validity of their results has been questioned (Anderson, 1978; Boice and Land, 1979; Reissland, 1978; Sanders, 1978) and more work is needed before results in this most important area can be reported with confidence.

Table 6.5 Excess cancer mortality per million persons per mSv by age at exposure.

Age at exposure	Males					
	Total	Leukemia	Nonleukemia	Respiratory	Digestive	Other
5	127	11.1	116.5	1.7	36.1	78.7
25	92.1	3.6	88.5	12.4	38.9	37.2
45	60	10.8	49.2	35.3	2.2	11.7
65	48.1	19.1	29.0	27.2	1.1	0.7
85	11.0	9.6	1.4	1.7	—	—
Average[a]	77.0	11.0	66.0	19.0	17.0	30.0

Age at exposure	Females						
	Total	Leuk.	Nonleuk.	Respiratory	Digestive	Breast	Other
5	153.2	7.5	145.7	4.8	65.5	12.9	62.5
25	117.8	2.9	114.9	12.5	67.9	5.2	29.3
45	54.1	7.3	46.8	27.7	7.1	2.0	10.0
65	38.6	14.6	24.0	17.2	5.2	—	16
85	9.0	7.3	1.7	1.5	0.4	—	—
Average	81.0	8.0	73.0	15.0	29.0	7.0	22.0

Source: NAS/NRC, 1990.

a. Averages are weighted for the age distribution in a stationary population having U.S. mortality rates. The BEIR V Committee presented the data as risks per 100,000 per 100 mSv (10 rem) and did not incorporate a dose-rate reduction factor for solid tumors in extrapolating from high doses. Risk data are converted to risk per million per mSv in this table to use a standard method of expressing the results of different risk assessments for ready comparison, though the appropriate extrapolation is much less certain at the lower dose.

than those presented in the earlier BEIR III report (NAS/NRC, 1980), which were cited in the previous edition of *Radiation Protection.* The leukemia cancer risk is about 4 times higher and the nonleukemia risk about 3 times higher (comparing relative risk models). These differences are due, in part, to the use of a linear dose-response model for cancers other than leukemia rather than a linear-quadratic model with an implicit dose-rate effectiveness factor (DREF) of 2.5 used by BEIR III. Other reasons were new dosimetry for the Japanese Life Span Study (LSS) data, additional years of follow-up, and changes in the structure of the fitted models. The BEIR Committee concluded that on the basis of the available evidence, the population-weighted excess risk of death from cancer following an acute dose equivalent to all body organs of 100 mSv (presented as 100 mGy of low-LET radiation in their table) is 8 in 1,000 (7 from solid cancers, and 1 from leukemia) and that accumulation of the same dose over weeks or months was expected to reduce the lifetime risk appreciably, possibly by a factor of 2 or more. There was no attempt to express the risk for lower doses, say 1 mSv, at which dose a linear extrapolation would give an estimate of 80 excess cases in 1,000,000 per mSv (no DREF).

Another source of risk data is the International Commission on Radiological Protection. The data were contained in a report published to encourage medical professionals to become aware of and utilize basic principles for radiation protection of the patient in diagnostic radiology (ICRP, 1993c). They gave nominal risks of excess fatal cancers (averages for a population of equal numbers of men and women) in individual exposed organs and, by extension, total induced cancers attributable to a whole-body dose or effective dose per mSv. They incorporated a DREF of 2 for the dose from diagnostic x rays. The data are presented in Table 6.6. It should be emphasized that since all these estimates of cancer risk are averages over a large population of exposed subjects, the risk to any individual can dif-

Table 6.6 Nominal risks for doses from diagnostic x rays.

Effect	Excess fatal cancers per million persons per mSv	Total cancers per million persons per mSv
Cancers		
Leukemia (active bone marrow)	5	5.05
Bone surface	0.5	0.7
Breast (females only)	4	8
Lung	8.5	8.9
Thyroid	0.8	8
Colon	8.5	15.3
Esophagus	3	3.15
Skin	0.2	100
Stomach	11	12.1
Liver	1.5	1.58
Bladder	3	6
Ovary (females only)	2	2.8
Other (combined remaining tissues and organs)	5	9
Total for whole-body irradiation, average for male and female	50 (1 in 20,000 per milligray)	
Baseline cancer mortality	0.15 (1 in 6.7) to 0.25 (1 in 4)	

Note: The nominal risks are average values for a population comprised of equal numbers of males and females of all ages (except for the breast and ovary, which are for females only). The extrapolation from high doses includes a DREF of 2. The ICRP expressed the risks in terms of mGy, which are equal to the risks in terms of mSv for x rays, as presented here. The risk of cancer in infants and children is very likely 2 to 3 times higher than the value given.

Source: Table modified from ICRP, 1993c; also in ICRP, 1991c.

fer significantly from the population risk, depending on genetic and other factors.

No review of epidemiological studies is complete without citing studies that do not show significant effects to low levels of radiation (Boice, 1997). Studies of workers at nuclear installations have shown no or very small excesses in cancer. Comparison of elderly Chinese women living in areas with high and low levels of natural background radiation did not find increases in cancer or thyroid nodular disease in subjects with chronic lifetime exposures in the high-background areas, although there was a significant difference in chromosome translocations.

2.3.1.   *Modeling Specific Procedures—Risk from X Rays in Adolescent Idiopathic Scoliosis*

Adolescent idiopathic scoliosis (AIS) is the most prevalent orthopedic disorder among adolescents. Management of this condition involves the administration of multiple full-spinal radiographs over many years (see sec. 2.2.7). A detailed study of organ doses incurred by 2,181 patients was performed at a large pediatric hospital (Levy et al., 1994) and the lifetime risk of excess cancer inferred on the basis of risk values in BEIR V (NAS/NRC, 1990). On average, 12 radiographs were taken per subject over a 3-year follow-up period. The thyroid gland and the female breast received the highest mean cumulative doses (about 300 mGy). Using a life table procedure, it was estimated that about 10 excess cancers were caused out of a total of 399 projected, and about four excess deaths occurred out of 247 projected over the lifetime of the 1,847 women in the cohort. This was equivalent to an excess lifetime risk of about 1 to 2 percent among women.

2.4   Effects on the Developing Embryo

The potential of giving birth to an abnormal child as a result of exposure of the developing fetus to radiation is a major concern of pregnant women as well as the regulatory agencies charged with implementing radiation control programs. It would be expected that the epidemiological studies on the atomic bomb survivors would have shown an increase in congenital malformations but this was not observed. However, 30 instances of severe mental retardation were identified in the high-dose exposure groups, with the highest incidence occurring in developing embryos between the eighth and fifteenth weeks of gestation at the time of the bombing. Some of the abnormalities were probably not caused by the radiation exposure. It was concluded that the developing embryo was at greatest risk from exposure to ionizing radiation 8 to 15 weeks after conception.

This is the time when the neuroblasts of the central nervous system proliferate at the greatest rate and most neuroblast migration from the proliferative zones to the cerebral cortex takes place (Little, 1993). The risk of severe mental retardation was assessed to be about 4 percent per 100 mSv (10 rem). It dropped to 1 percent per 100 mSv (10 rem) at 16–25 weeks. The data, considered along with findings in other animal and human studies, supported the existence of an effective threshold at around 100 mSv (10 rem) outside the 8–15 weeks stage of most rapid development of the brain. Lesser degrees of mental impairment were also observed as manifested by declines in intelligence test scores and school performance. One nonscientific but cogent argument for controlling radiation exposure to the fetus to levels that are as low as practicable is the high rate of genetic defects that occur spontaneously, approximately 4 to 8 percent of all live births, and the even higher statistic that 25 percent of conceptions normally result in prenatal death. While it may be difficult to present a convincing argument that low levels of radiation exposure were responsible for an abnormal outcome of a pregnancy, a history of unnecessary radiation exposure to the fetus can provoke feelings of *mea culpa* and other consequences that often stem from tragic events.

2.5 Genetic Risks

The hereditary effects of radiation are the result of mutations in the human genome, the code that determines the development, day-to-day functioning, and reproduction of every human being. Mutations can be induced in any of the 46 chromosomes (23 pairs) that make up the genome, and their complement of 3–7 billion base pairs of DNA making up the 50–100 thousand genes that are expressed through the functions they perform. This multitude of elements that are vulnerable to disruption is contained in each of the myriad cells in the body and are at risk of mutation by physical, chemical, or biological agents.

It is much more difficult to develop risk data for genetic effects than for somatic effects. Only when the mutation occurs to a dominant gene is the abnormality readily apparent in the affected individual. If a recessive gene is affected the result is generally much more subtle, the harm produced may be more difficult to identify, and its appearance may be delayed for one or more generations. Because of the lack of pertinent human data, the assessment of risk is based on indirect evidence, most of which comes from experimental studies of mice. Thus, the estimates are tenuous—they should be regarded only as indicators of radiation effects rather than as reliable predictors of the genetic consequences of exposure. Fortunately, the data from long-term studies on children and grandchildren of atomic bomb survivors indicate that the risk of hereditary effects from acute exposure to radiation is not as high as earlier estimates and its effect on the public health is considered to be much less serious than the somatic effects, and particularly the carcinogenic effects, of radiation.

The population carries a heavy burden of inherited disorders and traits that seriously handicap the afflicted persons at some time during their lives. Estimates of spontaneously occurring genetic or partially genetic diseases vary. One study found that genetic or partially genetic diseases with serious health consequences occurred in 5.3 percent of the population before the age of 25. An additional 2.6 percent had serious congenital anomalies that were nongenetic in origin, giving a total of 8 percent of all births with significant genetic diseases (Baird et al., 1988). If the population had received an average genetic dose of 10 mSv per generation, the additional risk of significant genetic effects would have been (upper limit) 0.5 percent in the first generation and 2 percent after many generations (NAS/NRC, 1990). According to the ICRP (1993c, 1991a), the risk of serious hereditary ill-health within the first two generations following exposure of either parent is given as 1 percent for 10 mSv. The risk of damage to later generations is an additional 1 percent.

Animal experiments indicate that the risk of transmitting mutations to offspring for a given dose to the parent prior to conception is considerably reduced in females who are irradiated at low dose rates or who do not conceive until several months after exposure at high dose rates (Russell, 1967). A delay in conception of about two months following irradiation of the male is also likely to reduce the risk of abnormalities in the offspring.

2.6   Basic Mechanisms in the Genesis of Cancer by Ionizing Radiation

The assessment of the risks of exposure to radiation at levels much lower than can be determined directly from epidemiological studies is a difficult exercise. The current approach, as discussed previously, is one of sophisticated curve-fitting, extrapolating the data on risk versus exposure either as a simple linear relationship or as a quadratic or linear-quadratic equation, or even possibly more complex equations. Inevitably, any evaluation stirs up controversy in view of the major economic, psychological, regulatory, and even public health consequences that can stem from any course that is taken. In time, the development and application of basic science may have the best chance of proceeding from the known to the unknown, as has been demonstrated so effectively in the past. Such scientific research is directed to the effects of radiation at the molecular and cellular levels.

2.6.1 Molecular Mechanisms

The removal of an electron from a molecule by ionizing radiation is an oxidizing process and results in raising the molecule to a high energy level, at which it has the capability of inducing strong chemical changes. The radiation may ionize and damage a DNA molecule directly, but since tissue consists of about 80 percent water, most of the ionizations occur in water molecules and lead within less than a microsecond to the production of highly reactive H* and OH* free radicals. These, in turn, can produce major damage in DNA. In any event, the end result may be a mutation that can lead to the formation of a cancerous cell.

What determines whether a mutation will result in the development of a cancer? The steps leading from the initial damaging event to the growth of a malignant tumor are many and complicated and still not completely understood (Little, 2000). However, DNA damage can often be repaired, if the body has the appropriate enzymes and other defense mechanisms. Thus, the potential for radiation exposure to initiate cancer in a specific individual depends on the countermeasures that can be mustered by the body's defense system.

It has been found in laboratory studies that the frequency of malignant transformations can be markedly increased by certain chemicals, known as tumor promotors, after exposure to a carcinogen. On the other hand, the production of malignant transformations is also almost entirely suppressed by several classes of chemicals, classified as chemopreventive agents. These discoveries have promising implications in cancer prevention through the identification of chemicals that can repair radiation damage before the steps toward the development of cancer can begin or the avoidance of chemicals that promote the development of cancer once the radiation has produced the initial lesions. Both observations also indicate that the progression from the initiating molecular event in a cell to a clinically recognizable malignant tumor is a complex process that accounts for the decades that often ensue between exposure to radiation and the diagnosis of cancer in an individual.

2.6.2 Production of Chromosome Aberrations

A most sensitive way to detect signs of radiation exposure in a person is to look for chromosome aberrations in the nuclei of lymphocytes (one of the types of white blood cells). Researchers have found chromosome damage in individuals many years after exposure to radiation. Although the clinical significance of an increased number of aberrations is not understood, their presence demonstrates the long-term persistence of the effects of radiation. Chromosome damage has been reported five years or more after x-ray therapy for ankylosing spondylitis (Buckton et al., 1962) and 17–18 years after the exposure of the Japanese to the atom bomb (Doida et al., 1965). Aberrations may be seen after only a few roentgens of exposure.[8] Most chromosomal damage results in the death of the cell or in the cells following division. However, the interchange of different genes between the two strands of a DNA molecule (reciprocal translocations) and small deletions may be transmitted over many generations. Certain chromosomal translocations have resulted in the activation of oncogenes with unfortunate consequences to the carrier. Normally present but passive in DNA, the oncogenes, when activated, drive the uncontrolled multiplication of their host cells and the growth of a cancer.

3   RISKS TO HEALTH FROM EXPOSURE TO ALPHA RADIATION

The utilization of the results of epidemiological studies and the development of standards of protection are different for low-LET radiation than for high-LET radiation. Most of the epidemiology for low-LET radiation derives from studies of the effects of irradiation of the body from external sources, such as exposure to gamma radiation from the atomic bomb or from x rays in radiation therapy. The epidemiology of high-LET alpha radiation comes primarily from the effects of exposure to two radioactive nuclides: to radon-222 by inhalation (induction of lung cancer, Table 6.2, item 28) and to radium-226 by ingestion (production of bone cancer, items 26 and 27). As a result, protection standards for low-LET radiation are based on the assessment of *absorbed dose* to organs in the body, while standards for alpha emitters are based on the measurement of *activities* of radon-222 and/or its decay products in air and radium-226 in bone. Dosimetry analysis applied to these measurements is used to extend the epidemiology of high-LET radiation to exposure conditions and populations other than those under which they were obtained.

3.1   Evolution of Protection Standards for Radon Gas and Its Decay Products

When the uranium mining industry underwent a great expansion in the 1950s in the United States, the regulatory agencies were faced with some grim statistics. Very high rates of lung cancer had been found in the late nineteenth and early twentieth centuries in miners working in the ra-

---

8. A detailed review is given in UNSCEAR, 2000.

dium mines of Schneeberg, Germany, and in the neighboring region of Joachimsthal, Czechoslovakia. The cancers were responsible for 50 percent of the deaths of the miners and for cutting their lives short at an early age. Exposure to radiation from radon gas was believed to be the main cause of the disease. A standard to provide a safe working level had been proposed in 1940 (Evans and Goodman, 1940) for application in industrial plants, laboratories, or offices. It was based on investigations of the radon concentration in the European mines, which averaged, in current units, 107,300 Bq/m^3 (2,900 pCi/l). On the basis that 37,000 Bq/m^3 (1,000 pCi/l) presented some risk in causing lung cancer, it was proposed that a concentration that was lower by a factor of 100, 370 Bq/m^3 (10 pCi/l), would provide adequate safety. With uranium mines in the United States exceeding the "safe" concentration by orders of magnitude, the experience of the European miners was a portent of the grim fate that lay in store for American uranium miners.

In spite of the high rate of lung cancer in miners exposed to radon gas, the potential for causing cancer by inhalation of radon and the radioactive products resulting from the decay of radon atoms in the lung was not supported either by laboratory experiments with animals exposed to radon gas or by calculations of dose resulting from the decay of radon gas in the lungs (Stannard, 1988). This remained a puzzle until William F. Bale, Professor of Radiation Biology and Biophysics at the University of Rochester, and John Harley, a research scientist at the Health and Safety Laboratory (HASL) of the Atomic Energy Commission (AEC), separately recognized the potential hazard of inhalation of the short-lived decay products that accompanied radon in air. Dr. Bale was on temporary assignment with the Division of Biology and Medicine of the AEC and John Harley was investigating the properties of the decay products in a chamber at HASL as part of a Ph.D. thesis at Rensselaer Polytechnic Institute. In contrast to the radon, which was a noble gas and thus not absorbed to any significant degree or retained in the lungs, the decay products constituted a radioactive aerosol that was deposited in the lungs and built up to significant activities, resulting in a dose much higher than that attributed to radon and its decay products in the lung (Bale, 1951; Harley, 1952, 1953). The dominant role of the dose produced by the radon decay products in air was confirmed by extensive studies on the deposition in animals and supporting theoretical analyses by a graduate student working on a Ph.D. thesis under Dr. Bale (Shapiro, 1954, 1956a, 1956b). These findings required a whole new approach in setting and evaluating standards.

By 1955, occupational limits focused on the concentrations of the short-lived radon decay products produced by the decay of 3,700 Bq/m^3 (100 pCi/l) of radon in air. In the technical language of radioactivity, this was the concentration of short-lived decay products in equilibrium with

3,700 Bq/m^3 (100 pCi/l) of radon in air, decay products that emitted 1.3 × 10^5 MeV of alpha energy in complete decay. This concentration of decay products, and of other combinations of decay products radiologically equivalent to them (Part Three, sec. 5.6, radon) was given the name working level (Holaday et al., 1957) and exposure at this level for 170 hours in a month (assuming a 40 hour work week) designated as a working level month (WLM). Thus two quantities and their units came to characterize exposure to air containing radon. One quantity referred to the concentration of the radon itself in terms of becquerels per cubic meter (or picocuries per liter). The other referred to the radioactivity of the aerosol established in air by the radon short-lived decay products, in terms of working levels. The magnitude of the working levels produced by a given concentration of radon in air varied according to the fraction of the radon decay products that was airborne, which was normally around 40 to 50 percent of the maximum (equilibrium) activity because of the proclivity of the decay atoms to deposit on surfaces, reducing their concentration in the air. A radon concentration of 3,700 Bq/m^3 (100 pCi/l) could theoretically be accompanied by a maximum of 1 working level of decay products, if none were removed from the air. In practice, estimates of the decay product concentration typically ranged between 0.4 and 0.5 WL in confined environments, such as in room air and in mines.

By 1964, rapid increases in lung cancers in uranium miners were reported and epidemiological studies intensified. The lung cancer incidence in miners was six times that of nonminers (Stannard, 1988). The occupational limit in mines was set tentatively by the Federal Radiation Council at 1 WL (12 WLM per year). In 1970, the Secretary of Labor issued a decree that the limit in mines be reduced to 0.3 WL (4 WLM per year). This is the current standard in the NRC regulations (10CFR20) and is also comparable to the standard set by the International Commission on Radiation Protection.

Example 6.1   A miner worked in a mine where the concentration of decay products during the month was 14 WL for 12 hours, 2 WL for 80 hours, and 1 WL for 30 hours. What was his cumulative exposure in working level months for that month?

The miner accumulated (14 WL × 12 hr) + (2 WL × 80 hr) + (1 WL × 30 hr) or 358 working level hours during the month. A working level month is 170 hr at 1 WL or 170 working level hours. Thus the miner accumulated 358/170 = 2.11 WLM. This is half his yearly allotment of 4 WLM. The levels in the mine would have to be severely reduced if he were to work there for the rest of the year, or he could wear a respirator to filter out the decay products.

Exposure to radon is not limited to uranium miners. The public is also exposed to significant concentrations in homes, concentrations that can reach and exceed those found in mines in some homes. Concern over exposures to elevated levels of radon in homes first focused on homes in Colorado built on fill containing uranium mill tailings (Cole, 1993). The tailings, left over after the uranium had been extracted from the ore, had enhanced concentrations of radium, the parent of radon, and hence were strong radon sources, producing high radon levels in the homes. The U.S. Congress passed the Uranium Mill Tailings Radiation Control Act in 1978, requiring the EPA to establish radiation standards for homes built on fill containing tailings. In 1983, the EPA set 148 Bq/m^3 (4 pCi/l) as an objective, with a maximum limit of 222 Bq/m^3 (6 pCi/l). It called this standard an "optimized cost-benefit" alternative. Fill derived from residues of phosphate mines in Florida was also identified as a significant source of radon in homes because of its radium content. Then, in December 1984, a worker at a nuclear plant under construction in Pennsylvania tripped every alarm through which he passed. His radioactive contamination was traced, not to radioactivity in the plant, but from radioactivity in his home. This prompted measurements of radon levels in his home, levels which were found to be as high as 100,000 Bq/m^3 (2,700 pCi/l). The family was moved to a motel, where they remained for 6 months until the radon levels were reduced to acceptable levels. This episode initiated a monitoring program in homes in the surrounding area. Forty percent were found to be above acceptable levels. The monitoring program then was extended to cover the whole country. Elevated radon levels were widespread, above 148 Bq/m^3 (4 pCi/l) in perhaps one out of fifteen homes, according to EPA. Radiation dose to the lungs from indoor radon was declared a major public health problem, possibly responsible for thousands of fatal cases of lung cancer each year. The Environmental Protection Agency pursued with vigor a program focused on research, public education, and mitigation of excessive levels. In 1988, Congress passed the Indoor Radon Abatement Act, which declared as the long-term national goal that all buildings in the United States "should be as free of radon as the ambient air outside of buildings."

The goal of an essentially radon-free environment is technologically impractical to achieve, even if the nation were willing to expend the enormous financial resources needed to attain it. EPA recognized the need to set a practical limit for control when it set 148 Bq/m^3 (4 pCi/l) as a radon level above which steps for mitigation should be taken. A standard based on the practical considerations was perhaps a unique approach to standard setting. Its implementation required a defensible determination of the risk. Whatever the outcome, some would consider the risk assessment too low. Others would question the assessment as being exaggerated. Others would criticize EPA as manufacturing a problem which did not exist (Cole, 1993).

3.2 Risk of Lung Cancer from Extended Exposure to Radon and Its Short-Lived Decay Products

Radon has been and continues to be the subject of intensive worldwide research to assess the risk of exposure and to develop suitable protection standards and control measures. The major sources of data for assessing the risk of lung cancer from extended exposure to air containing radon gas come from studies of the health of workers in mines in the United States and in Europe. They were mostly uranium mines, but also included mines for other minerals, including iron, tin, and fluorspar. Radon decay product levels in mines in the United States ranged generally between 10 and 100 Working Levels before 1960, with an average cumulative exposure of the miners of 1,180 Working Level Months (WLM).

The risk of dying from lung cancer varies with age at exposure, period of exposure, smoking history, and the success rate of medical treatment (which is currently very low). Expressions of risk can range from simple estimates based on total working level months of exposure to complex equations with parameters for exposure history, exposure rate, attained age, and elapsed length of time since exposure.

A simple statistic is the risk of fatal lung cancer from lifetime inhalation of radon in air. A committee of the National Research Council adopted a value for the risk of lung cancer of 1.6 in 10,000 to a mixed population of smokers and nonsmokers exposed to radon continuously at a concentration of 1 Bq/m^3 (NAS/NRC, 1999b). At the EPA limit of 148 Bq/m^3 (4 pCi/l) for instituting measures for mitigation, the risk is then 237 in 10,000 or a little over 2 percent. In a guide prepared for the public, the EPA stated the risk at 148 Bq/m^3 as 2 in 1,000 to people who never smoked and about 2.9 percent (or 29 in 1,000) to smokers (EPA, 1992a).

3.2.1   Risk Assessment of the BEIR Committees

Two comprehensive reports on the health risks of exposure to radon were issued by committees on the biological effects of ionizing radiation (BEIR) of the National Research Council, BEIR IV (NAS/NRC, 1988) and BEIR VI (NAS/NRC, 1999a). The committees drew extensively on findings from molecular, cellular, and animal studies in developing a risk assessment for the general population. Review of cellular and molecular evidence supported the selection of a linear nonthreshold relation between lung-cancer risk and radon exposure. Although a linear-nonthreshold model was selected, however, the committee recognized that a threshold could exist and not be identifiable from the available epidemiologic data. The BEIR IV committee used the data for 4 cohorts of miners; the BEIR VI committee was able to combine additional data from a total of 11 cohorts, involving 68,000 miners and 2,700 deaths from lung cancer. The statistical methods were similar in both studies.

The BEIR VI report fitted the epidemiological data with an equation expressing the age-specific excess relative risk (ERR), the fractional increase in lung cancer risk resulting from exposure to the radon decay products. The equation had parameters to account for the length of time that passed since the exposure, the concentration, and the attained age:

$$\mathrm{ERR} = \beta \, (\theta_{5-14}\omega_{5-14} + \theta_{15-24}\omega_{15-24} + \theta_{25+}\omega_{25+})\phi_{age}Y_z$$

where $\beta = 0.0768$; ω_{5-14} is the cumulative exposure in WLM between 5 and 14 years of age; $\theta_{5-14} = 1.00$; $\theta_{15-24} = 0.78$; $\theta_{25+} = 0.51$; $\phi_{<55} = 1.00$; $\phi_{55-64} = 0.57$; $\phi_{64-74} = 0.29$; $\phi_{75+} = 0.09$; Y_z, is here an exposure concentration effect-modification factor and z is equal to the concentration in working levels; $Y_{<0.5} = 1.00$; $Y_{0.5-1.0} = 0.49$; $Y_{1.0-3.0} = 0.37$; $Y_{3.0-5.0} = 0.32$; $Y_{5.0-15.0} = 0.17$; $Y_{15+} = 0.11$.

The committee derived risk estimates for two different models, one which included the concentration (exposure rate) as a parameter and one which included the duration of exposure as a parameter. The equation for the exposure-age-concentration model is given here.

The accumulated exposure at any particular age is grouped into exposures accumulated in three age intervals: 5–14 yr, 15–24 yr, and 25+. The exposure in the 5 years preceding the attained age is excluded as not biologically relevant to cancer risk. The rate of exposure is accounted for through the parameter Y_z, which acts to increase the effect of a given exposure with decreasing exposure rate, as indexed either by the duration of exposure or the average concentration at which exposure was received. The ERR also declines with increasing age, as described by the parameter ϕ_{age}. The new model is similar in form to the BEIR IV model but has the additional term for exposure rate and more-detailed categories for the time-since-exposure windows and for attained age.

The radiation dose pattern to the lungs and the resultant dose to the critical cells in the lungs can differ in the mine and home environments for identical exposures in working level months. Differences that significantly affect the dose are incorporated into the analysis through the use of a K factor that characterizes the comparative doses to lung cells in homes and mines for the same exposure. Some assessments concluded that the dose in the home environment was about 30 percent less than the dose in the mine environment for the same working level concentration. However, based on a model that incorporated new information, the value of K was calculated to be about 1, and no correction is made for differences in the exposure environment in the BEIR VI report.

The fact that the lung cancer rate in smokers is so much larger than in nonsmokers, of the order of 10 times as large, means that the effect of exposure to radon will also be proportionately larger. The actual assessment, based on the meager epidemiological data, is that there is a synergistic effect for the two exposures. The BEIR VI committee preferred a submultiplicative relation, that is, the number of cancers occurring is less than expected if they were the product of the risks from exposure to radon and smoking individually, but more than if the joint effect were the sum of the individual risks. However, the committee also determined the effect of a full multiplicative relation as done by the BEIR IV committee.

BEIR VI applied the same multiplier for exposure to radon to the background cancer rate for women as well as for men, although the miners, who contributed the epidemiological data, were all men. Since current cancer rates were much lower for women than men, the results of the exposure to radon were also relatively less.

Since the calculations give the relative increase to the lung cancer incidence in the absence of radon, the radon data needs to be accompanied by epidemiological data on the occurrence of lung cancer in the population, the background rates. Lung cancer rates for smokers and never-smokers are given in Table 6.7.

The age-specific data determined from the epidemiology were used to compute the lifetime relative risk resulting from exposure to indoor radon beyond the risk from exposure to outdoor-background concentrations. Multiplying the epidemiologic data in Table 6.7 by the lifetime relative risk data gives a measure of the effects of elevated levels of radon exposure on lung cancer incidence. The results are given in Table 6.8.

There is some question on the applicability of the risk of lung cancer in the mines to the risk from a comparable dose in homes. For example, arsenic, a known pulmonary carcinogen, is present in varying amounts in uranium mines and it may have an impact on the risk of lung cancer from

Table 6.7   Lung cancer rates per 100,000 in male never-smokers and smokers.

Age	Smokers	Never-smokers
45	40	1.9
50	100	5.8
60	340	12
70	920	31
75	1,130	33

Source: NAS/NRC, 1999a, Table C-5, Figure C-5.

Table 6.8 Estimated lifetime relative risk (LRR) of lung cancer for lifetime indoor exposure to radon.

Exposure (Bq/m³)	WL	WLM/yr	Male		Female	
			Ever-smoker	Never-smoker	Ever-smoker	Never-smoker
25	0.003	0.10	1.081	1.194	1.089	1.206
50	0.005	0.19	1.161	1.388	1.177	1.411
100	0.011	0.39	1.318	1.775	1.352	1.821
150	0.016	0.58	1.471	2.159	1.525	2.229
200	0.022	0.78	1.619	2.542	1.694	2.637
400	0.043	1.56	2.174	4.057	2.349	4.255
800	0.086	3.12	3.120	7.008	3.549	7.440

Source: NAS/NRC, 1999a, Table ES-1.

Note: Data are for exposure-age-concentration model and are higher than values obtained for exposure-age-duration model. Calculations based on 70% home occupancy and 40% equilibrium between radon and its decay products. To convert Bq/m³ to pCi/l, divide by 37.

exposure to radon. Differences in smoking habits between the miner population and the rest of the public is also a source of error. In fact, in some animal studies, in which the animals were exposed to both cigarette smoke and radon, the cigarette smoke was found to have a protective effect. One explanation is that the cigarette smoke caused a thickening of the mucous lining the airways in the lung, producing increased shielding of the alpha particles.

Many epidemiological studies have been and are being performed on the risks of cancer from exposure to indoor radon, but the results so far are largely inconclusive, primarily because of the lack of information on the exposures received by the study populations. A case-control study of lung cancer patients who had lived in their homes for at least 20 years prior to diagnosis of the disease gave excess odds of 0.5 relative to an exposure of 148 Bq/m³ (4 pCi/liter) during this period. However, the 95 percent confidence limits (0.11, 3.34) were quite large (Field et al., 2000). Thus, the risks determined for occupational exposure are the main resource for evaluating risks of exposure for radon in homes and the need for corrective action.

3.2.2   *Government Action against Radon's Threat to Public Health*

The number of lung cancer deaths per year attributable to indoor radon are estimated to be 15,400 by one model and 21,800 by another model, according to BEIR VI (NAS/NRC, 1999a), although consideration of the uncertainties in the calculations indicate the deaths may be as high as 33,000 or as low as 3,000. Most of the fatal lung cancers occur in smokers;

the committee's best estimate of the annual radon-related lung cancer deaths among people who never smoked is between two and three thousand.

Congress recognized exposure to radon as a significant public health problem through passage of the Indoor Radon Abatement Act in 1988. The act instructed the EPA to keep the public informed about radon through a *Citizen's Guide,* which covers health risks, testing, and methods of reducing indoor concentrations. In addition, the EPA was to assist states with radon programs, assess the extent of radon in the nation's schools, designate regional radon training centers, and propose construction features in new buildings in high-radon areas that would minimize the cost of mitigating radon levels if they were found to be excessive.

The result of the legislation and subsequent programs has been to make available nationally extensive resources for identifying and controlling exposure of the population to radon. EPA publications provide guidance and technical information, including recommendations to home buyers and sellers (EPA, 1992a, 1992b, 2000). Most radon problems are easily remedied at low or moderate cost. The most common method, referred to as soil depressurization, is to run a pipe from underneath the basement floor up through the house to the roof. A blower in the attic draws air from the soil beneath the house, reducing the pressure and minimizing the entry of the air under the basement with its high radon concentration into the house. Other methods, which are not as effective but may work, include ventilating the basement with heat recovery ventilators (to reduce heating costs in winter); closing large openings, like holes for sump pumps; and sealing floor and wall cracks. Details can be found in EPA publications (EPA, 1991) as well as in standards and codes prepared by professional or trade associations (ASTM, 2001).

States committed to undertaking vigorous action on radon have received financial assistance from EPA through State Indoor Radon Grants, enabling them significantly to increase the resources they can commit to protecting the public against radon. When the city of Worcester, Massachusetts, screened all their school buildings in 1995, they found radon concentrations of nearly 74,000 Bq/m^3 (2,000 pCi/l) in some classrooms located in an underground level in one building. Initial efforts by local contractors to seal the cracks and ventilate the area had a minimal effect on the levels. The Radiation Control Office of the Massachusetts Department of Public Health, which had an active radon control program, was then contacted. It took considerable analysis and the installation of 22 suction points exhausted through 6 stacks, but the result was that a survey of the building at the conclusion of the project found no levels above 74 Bq/m^3 (2 pCi/l; Bell and Anthes, 1998).

3.3 Exposure of Bone to Alpha Radiation

The bone data (table 6.2, items 25 and 26) show apparent evidence of a threshold, that is, there were no excess cases of bone cancer found below the long term retention of 0.1 μg Ra-226 fixed in bone and an average skeletal dose of 2000 rad of alpha radiation (Stannard, 1988). Minimum doses resulting in findings of head carcinomas were a factor of two lower.

4 IMPLICATIONS FOR HUMANS FROM RESULTS OF ANIMAL EXPERIMENTS

It is difficult to set radiation-protection standards for humans on the basis of animal experiments, but such studies do provide guidance in the interpretation and extension of data on human exposure. Animal experiments provide valuable answers to such questions as: Is there a threshold dose below which no effects appear? Is susceptibility to cancer induction dependent on age, sex, and hereditary factors? What are the general effects of dose rate on the risk of cancer and other radiation-induced diseases? Are there chemical agents that can reduce injury from radiation or that enhance or promote injury subsequent to radiation exposure (such as cigarette smoke)? What is the relative effectiveness of the different types of radiation, such as gamma rays, neutrons, and alpha particles, on tumor induction and other life-shortening effects? What is the effectiveness of partial-body irradiation versus whole-body irradiation? What is the toxicity from ingestion or inhalation of a particular radionuclide? At what period in the life span are genetic effects most pronounced?

In an authoritative review of animal exposure data prepared by the National Committee on Radiation Protection (NAS/NRC, 1961b), the following conclusions were presented:

(a) A rather high degree of correlation exists between results from animal experiments and those from man.

(b) An unusually high susceptibility seems to exist in some experimental animal species or strains for certain diseases, for example, ovarian tumors and lymphatic leukemia in mice and mammary tumors in rats.

(c) Most animal experiments, usually performed on relatively homogeneous populations, have demonstrated that there are dose levels below which no detectable increase in incidence of certain neoplasms can be found; the dose-effect relationship is not linear. On the other hand, a few experiments with relatively homogeneous populations of animals have shown that for some tumors in certain species or strains of animals the in-

cidence is increased at such low dose levels that there may not be a practical threshold for the production of an increased incidence of tumors.

In a study using 576 mice per exposure group at five levels of single whole-body exposures between 50 and 475 R, it was concluded that, in plots of life span versus dose, "the fit to a straight line was very good, and the intersection with the vertical axis showed that there was no apparent threshold for life shortening within the experimental error . . . The mice lost 5.66 ± 0.2 weeks (5%) of their lifespan/100 R" (Lindop and Rotblat, 1961). It was also concluded that life shortening by irradiation was due, not to the induction of a specific disease, but to the advancement in time of all causes of death.

In another study of effects at low doses (Bond et al., 1960), it was found that at short exposures as low as 25 R, 5 out of 47 female rats developed breast cancers within 11 months. This was an incidence of 12 percent compared to an incidence of 1 percent in the controls. The dose-effect relationship appeared linear over a range of 25–400 R. However, the data did not allow conclusions concerning the presence or absence of a threshold below 25 R. The authors cautioned that results of rat experiments could not be extrapolated to humans inasmuch as other experiments showed that there were hormonal influences on the induction of tumors that differed in the two species. The animals used in these experiments had a high natural susceptibility to breast cancer. In one series of studies of virgin female Sprague-Dawley rats allowed to live out their lives,[9] between 51 and 80 percent developed breast tumors, of which 12 percent were malignant (that is, a natural cancer incidence between 5 and 10 percent) (Davis et al., 1956). Accordingly, the results cannot be applied to species, or individuals in species, with a higher resistance to cancer induction. On the other hand, they point to the possibility of a much higher susceptibility to cancer production by radiation in individuals who have a genetic tendency to develop cancer.

Another organ having high radiosensitivity is the mouse ovary. In the RF mouse, a significant increase in the incidence of ovarian tumors was produced by a dose as low as 32 R (Upton et al., 1954). At the lowest exposure rate thus far systematically investigated (0.11 R/day), mice exposed daily throughout their lives to radium gamma rays exhibited a slightly increased incidence of certain types of cancer (Lorenz et al., 1955). Because of statistical limitations, however, the data do not enable confident extrapolation to lower dose levels.

9. The average lifespan of Sprague-Dawley rats given in one series of studies is 739–792 days, with a range between 205 and 1,105 days.

One of the most critical questions that must be considered in decisions on the large-scale development of nuclear power is the toxicity of plutonium. While it is universally agreed that plutonium is extremely toxic, there are large differences of opinion on how great a risk to the health of the public and the safety of the environment it in fact represents (Edsall, 1976). While one cannot find very much data on the hazards of plutonium to humans, health and safety standards are necessary and must be soundly based if a reasonable judgment on the acceptability of nuclear power from a health and safety point of view can be made. In the absence of adequate human data, considerable reliance must be placed on animal data for guidance in setting standards. Studies have been conducted with many different species of animals (ICRP, 1980), particularly the beagle. All dogs exposed to 3–20 nCi/g lung developed malignant tumors after periods of 6–13 years. It was concluded that a dose of more than 1 nCi ^{239}Pu/g could cause premature death from a lung tumor in a beagle. Nine of thirteen beagles each injected with 48 nCi of ^{239}Pu developed bone sarcomas over a period that averaged 8.5 years. The experiments indicate that the 40 nCi limit for plutonium may be several-fold less safe than the 100 nCi limit for radium (Bair and Thompson, 1974). It has been estimated that a maintained skeletal burden of 0.04 μCi would lead to possibly 2 cases of leukemia and 13 cases of bone sarcoma per 1,000 persons over 50 years (Spiers and Vaughan, 1976, p. 534). Estimates based on rat data indicate that 1 in 8 persons might develop lung cancer from a lung burden of 16 nCi (Bair and Thomas, 1975). Thus there have been suggestions to reduce the standard for plutonium (0.04 μCi), ranging from a factor of 240 (Morgan, 1975) to 9 (Mays, 1975).

5 SOURCES PRODUCING POPULATION EXPOSURE

The standards of radiation protection allow for the imparting of low radiation doses to the public as a result of the use of radiation in technology, although a slight increase in the incidence of cancer, leukemia, or birth defects may result. The justification is that the benefit from the use of radiation outweighs the risk of injury. The decision as to the level of exposure at which benefit outweighs risk is a matter of judgment, and by its nature will always be controversial. Certainly if experts disagree on this question it must be extremely difficult for the layman to make an objective judgment. In fact, the permissible levels for population exposure will be set by the most persuasive elements of our society.

In the absence of conclusive data on the effects of very low level radiation exposure of large numbers of people, the basic yardstick for measuring the significance of population exposure from man-made sources is the exposure already being incurred by the population from natural sources. As a result, considerable effort has been expended by scientists on the evaluation of this exposure.

5.1   Natural Sources of External Radiation

It is convenient to divide the exposure of the population into exposures from external and internal sources, although the effects of the radiation on the body depend only on the dose imparted, regardless of origin. The external background radiation comes from the interactions in the atmosphere of cosmic rays from outer space and from gamma photons emitted by radioactive minerals in the ground (Fig. 6.2).

5.1.1   Radiation from Cosmic-Ray Interactions in the Atmosphere

About 2×10^{18} primary cosmic-ray particles (mainly protons) of energy greater than one billion electron volts are incident on the atmosphere every second. They interact with atoms in the atmosphere and produce a large variety of secondary particles. At sea level, essentially all the original particles have disappeared. The cosmic-ray dose is produced by the secondary particles. Figure 6.2 shows the progeny particles produced by a single energetic cosmic-ray particle. The penetration of a single proton of relatively low energy into the atmosphere may result in the ultimate appearance of only a single particle (a muon, usually) at ground level. A very energetic proton may produce a shower containing hundreds of millions of particles, including muons, electrons, photons, and some neutrons.

5.1.2   Radiation Emitted by Radioactive Matter in the Ground

Radionuclides that were part of the original composition of the earth, and additional radionuclides formed as a result of their decay, emit gamma radiation, which contributes a large share of the environmental radiation dose. The extent of the exposure at any particular location depends on the amount and distribution of radioactive material. The distribution of radioactive elements in the ground has been studied extensively. The major sources are provided by potassium, of which the typical concentration is a few grams per hundred grams of ground material, and thorium and uranium, the typical concentration of which is a few grams per million grams. Data on concentration, expressed as fractional weight and activity per gram for selected locations, are presented in Table 6.9.

6.2 Penetration of radiation to surface of earth.

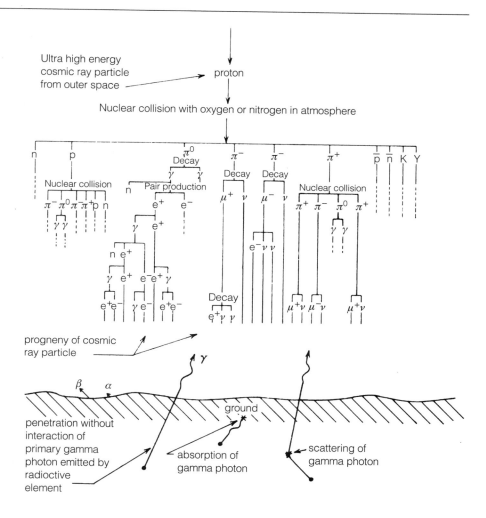

While the values in Table 6.9 are typical, there are areas with unusually high levels. For example, uranium concentrations in phosphate rock in the United States reach levels of 400 g per million grams (267 pCi/g; EPA, 1977). The commercial phosphate fertilizers derived from these rocks also contain a high uranium content, and runoff from fertilized areas has resulted in the increase of the uranium content of North American rivers.

Most of the gamma photons emitted by radioactivity in the ground are absorbed in the ground. The environmental exposure is produced primarily by photons originating near the surface. Estimates of the doses from the

Table 6.9 Concentration of radioactive materials in the ground.

| Type of rock | Concentration (fraction by weight) | | | | Activity (pCi/g) | |
	Potassium ($\times 10^1$)	Uranium ($\times 10^6$)	Thorium ($\times 10^6$)	Potassium	Uranium	Thorium
Igneous	2.6[a]	4	12	21.6	1.33	1.31
Sandstone	1.1	1.2	6	9.1	0.40	0.65
Shale	2.7	1.2	10	22.5	0.40	1.09
Limestone	0.27	1.3	1.3	2.25	0.43	0.14
Granite	3.5–5	9–12	36–44	>29	>3.0	>3.9

Sources: For igneous, sandstone, shale, and limestone, UNSCEAR, 1958, p. 52; for granite, Adams and Lowder, 1964.

a. Fraction is 0.26, shown here as 10 times its value.

major contributors, for uniform concentrations in the ground, expressed in picocuries per gram, are given by the following formulas:

	Dose (in mrad/yr at 1 m above ground)
For ^{238}U + decay products	$17.8 \times$ pCi/g
For ^{232}Th + decay products	$25.5 \times$ pCi/g
For ^{40}K	$1.6 \times$ pCi/g

The gamma photons emitted in the decay of the natural atmospheric radioactivity also contribute to the external dose. The external dose from radon and its decay products is given approximately by the expression: yearly dose (in mrad) = 14 × concentration (in pCi/l). At a concentration of 0.1 pCi/l, this adds 1.4 mrad/yr to the dose. The contribution from thoron and its decay products is generally much less.

5.1.3   Doses from External Sources

Representative doses from external natural sources are:

- From cosmic radiation with low linear energy transfer (photons, muons, electrons): 0.28 mGy/yr
- From cosmic-ray neutrons (high linear energy transfer): 0.0035 mGy/yr
- From gamma photons originating in the ground: 0.5 mGy/yr

The low-LET cosmic-ray dose rises to 0.53 mGy/yr in Denver (1,600 m) and to 26.3 mGy/yr at 12,000 m (40,000 ft). The neutron dose rises to 0.031 mGy/yr (0.31 mSv/yr) in Denver and 1.93 mGy/yr at 12,000 m.

Measurements of radiation levels of natural origin in several cities in the United States are given in Table 6.10.

5.1.4 Cosmic Radiation Exposure of Airline Crews and Passengers

The higher one rises above the earth, the less shielding to cosmic radiation is provided by the atmosphere and the higher the radiation dose, a steady component of galactic origin and a variable component from the sun. The doses are lower at the equator, where the earth's geomagnetic field is nearly parallel to the earth's surface and the charged particles that approach below a minimum momentum are deflected back into space. The dose rate at the poles, where the magnetic field is nearly vertical and the number deflected much less, is between 2.5 and 5 times the dose rate at the equator (Goldhagen, 2000). As aviation technology enables flights at higher and higher altitudes, the personnel dose increases per flight to the point where the International Civil Aviation Organization has recommended that all airplanes intended to be operated above 15,000 m (49,000 ft) be equipped to measure dose rate and cumulative dose from ionizing and neutron radiation on each flight and crew members. The dosimetry is quite difficult and uncertain because of the complex nature of the radiation field. From a nominal dose rate of 0.038 μSv/hr at sea level, the total effective dose rate increases to 0.5 μSv/hr at an altitude of 5 km, to 4 μSv/hr at 10 km and to 11 μSv/hr at 15 km. The ICRP (1991a) recommends that air crew members be classified as radiation workers, and frequent flyers might well approach annual doses that would classify them as radiation workers by regulatory standards.

Table 6.10 Radiation exposure levels in different cities.

| | Exposure rate (μR/hr) | | | |
	Terrestrial sources (potassium, uranium, thorium)	Cosmic rays	Total (μR/hr)	Total (mrad/yr)
Location				
Denver	7.1–15.2	5.9–6.3	13–21.5	114–188
New York City	6.3	3.6	9.9	87
Conway, NH	10.9	3.6	14.5	127
Burlington, VT	5.2	3.6	8.8	77

Sources: For Denver and New York, Beck et al., 1966; for Conway and Burlington, Lowder and Condon, 1965.

Note: Natural gamma dose rates at Pelham, NY, fluctuated between 7 and 8.2 μR/hr between 5/3/63 and 7/29/65 (Beck et al., 1966). The variability is attributed primarily to variations in the moisture content in the ground. In the past, fallout added several μR/hr to the external exposure levels.

5.2   Natural Sources of Radioactivity within the Body

Section 19 of Part Two presents information on radioactivity in the body. The radioactivity results primarily from ingestion of radioactive nuclides that occur naturally in food and drinking water, with some contribution from inhalation of radioactivity in the air. The most important radionuclides that are ingested are potassium-40, radium-226, and the decay products of radium-226. Carbon-14 and tritium are also of interest, although they make only minor contributions to the absorbed dose. The major contributors to dose from inhalation are the radioactive noble gases, radon (radon-222) and thoron (radon-220), and their decay products.

Potassium is relatively abundant in nature, and some data on the potassium content in the ground has already been presented in Table 6.9. The element contains 0.0119 percent by weight of the radioactive isotope potassium-40. The body content of potassium is maintained at a fairly constant level, about 140 g in a person weighing 70 kg. This amount of potassium has a potassium-40 content of approximately 3,700 Bq (0.1 μCi).

Some results of measurements of the concentrations of other radionuclides in food and water are presented in Table 6.11. Note the high variability in the levels in different sources.

Carbon and hydrogen in the biosphere contain radioactive ^{14}C and tritium (^{3}H) that result from the interaction of cosmic-ray neutrons with the nitrogen in the atmosphere. Because production of these radionuclides has been proceeding at a constant rate throughout a period much longer than their half-lives, the world inventory is essentially constant, and the atoms are decaying at rates equal to the rates at which they are being produced. Libby (1955) calculated that 1.3×10^{19} atoms of ^{14}C are produced each second, yielding a world inventory at equilibrium of 13,500 million GBq (365 million Ci). The rate of tritium production is about 10 percent that of ^{14}C, and the world inventory of tritium is about 1,110 million GBq (30 million Ci).

Table 6.11   Concentrations of radionuclides in food and water.

Substance	Concentration (pCi/kg)			
	^{226}Ra	^{210}Po	^{90}Sr	^{137}Cs
Cereal grains (N. America)	1.5–4	1–4	15–60	50
Nuts	0.5–2000		3–120	
Beef	0.3–0.9	3–300	1	100–5000
Cow's milk	0.15–0.25	1	3–15	20–200
Water (Great Britain, some locations)	3.3			

Source: Mayneord and Hill, 1969, Table VII.

The ^{14}C produced in the atmosphere is rapidly oxidized to radioactive carbon dioxide and ultimately appears in the carbon compounds in the sea and in all living matter. The measured value of specific radioactivity is constant at 16 disintegrations per minute (dis/min) per gram of carbon and agrees well with calculations based on dispersion of the ^{14}C in a carbon reservoir of 42×10^{18} g. The tritium is maintained at an equilibrium concentration of 1 tritium atom for 2×10^{17} hydrogen atoms, or about 0.036 dis/min per milliliter of water.

From a radiation-protection point of view, of more interest than the concentration of the activity in water and various foods is the daily intake. The intake varies widely with locality. For example, values for daily intake of radium-226 in Great Britain vary by a factor of 500, with an upper limit of 0.22 Bq (5.9 pCi). The actual significance of daily intake depends on the absorption of the materials by the gastrointestinal tract; ^{40}K, ^{137}C, and ^{131}I are almost completely absorbed. Significant absorption also occurs for ^{210}Pb and ^{210}Po. The absorption of ^{14}C and tritium depends on the metabolism of the molecules in which they are incorporated. The rest of the radionuclides ingested by man are only poorly absorbed.

Data on environmental levels of the long-lived naturally occurring radionuclides are given in Table 6.12. All the alpha emitters are part of the radioactive decay series that originates from primordial uranium and thorium and that includes many members (see sections 5.5 and 5.6 in Part Three). Most of these radionuclides are taken up in bone and retained for long periods of time, producing alpha radiation doses to the bone marrow, bone-forming cells, and cells lining the bone surfaces. Daily intakes and absorbed doses to these tissues are given in the table. The absorbed doses of the alpha emitters must be multiplied by 20 to obtain the equivalent dose. The radionuclide lead-210 listed in the table is a low-energy beta emitter in the radium decay series rather than an alpha emitter, but its dosimetric significance results primarily because it decays into the alpha emitter polonium-210. Since ^{210}Pb and ^{210}Po follow the decay of the noble gas radon-222 in the ^{238}U–^{226}Ra decay series, the initial distribution in the environment is determined by the distribution of radon gas in the ground and in the atmosphere following its production from the decay of ^{226}Ra (UNSCEAR, 1977; Jaworowski, 1969). The airborne radioactive gases are discussed in the following section. The environmental contamination levels of the naturally occurring long-lived radionuclides and their decay products are a useful reference for assessing the significance of disposal of manmade radioactive wastes. Note that the internal whole-body dose from naturally occurring radionuclides adds about 0.2 mSv/yr or 25 percent to the external dose, and that localized regions in bone receive even higher doses.

The data in Table 6.12 can be used to provide estimates of the total

Table 6.12 Activity in the environment and in people of naturally occurring long-lived radionuclides.

Radionuclide	Alpha emitters (except ^{210}Pb)						Beta emitters			
	^{238}U	^{232}Th	^{226}Ra	^{228}Ra	^{210}Pb	^{210}Po	^{40}K	^{14}C	^{3}H	^{87}Rb
Half-life (yr)	4.5×10^9	1.4×10^{10}	1,600	5.8	21	0.38	1.26×10^9	5,730	12.3	4.8×10^{10}
Activity in soil (pCi/kg)	700	700	700	700[o]	900[a]	900[a]	12,000	6,100/kg carbon	146/kg hydrogen	
Activity in air (10^{-5} pCi/m^3)	7[b]	7[b]	7[b]	7[b]	1,400[c]	330[c]				
Activity in water (pCi/1)	<0.03				3[d]	0.5[d]			16 (natural) 84 (fallout)	
Daily intake										
Inhalation (pCi)	0.0014	0.001	0.001		0.3[c]	0.07[c]				
Ingestion (pCi)	0.4[f]	0.1[g]	1[h]	1	3	3				
Activity in tissues										
Soft tissue (pCi/kg)	0.2	0.05	0.13	0.1	6	6	1,600			220
Bone (pCi/kg[j])	4	0.5	8[i]	2.4[k]	80[k]	64[k]	400[k]			640
Whole body (pCi)	26		70	21	600	200	130,000	87,000	700 (natural) 27,000 (fallout)	29,000
Annual absorbed dose										
Soft tissue (mrad)	0.02	0.004[l]	0.03	0.06		0.6	17	0.6	0.001 (natural) 0.06 (fallout)	0.4
Red marrow (mrad)	0.02	0.05	0.09	0.18		0.7	27	2.2	0.001	0.4
Bone lining cells (mrad)	0.12[m]	0.8	0.7[n]	1.08[n]	3	3	15	2.0	0.001	0.9

Sources: UNSCEAR, 1977; NCRP, 1975; Spiers, 1968; Holtzman, 1977.

a. Average activity in top layer, 20 cm thick. ^{210}Pb is a beta emitter, a sixth-generation decay product of ^{226}Ra and the precursor of ^{210}Po. Contributions in pCi/kg are 200 from dry deposition, 100 from wet deposition, and 600 from radium in ground under the assumption that 90 percent of radon is retained in soil. The corresponding area concentration is 68 mCi/km^2 from dry deposition, 37 mCi/km^2 from wet deposition. For calculational purposes, a soil density of 1.8 g/cm^3 and porosity of 60 percent may be used. The total radium activity in a layer of rock 1 m thick in continental U.S. (excluding Alaska), assuming specific gravity of 2.7 = 700×10^{-12} Ci/kg $\times 2700$ kg/m^3 $\times 1$ m $\times 8 \times 10^{12}$ m^2 = 1.5×10^7 Ci.

b. Based on resuspension of dust particles as main natural source in atmosphere, dust loading of 100 μg/m^3 in surface air of populated areas, and dust specific activity of 700 pCi/kg.

c. Results primarily from decay of ^{222}Rn in air.

d. In rain water; less than 0.1 pCi/l ^{210}Po in drinking water.

e. Cigarettes contain 0.6 pCi ^{210}Pb, 0.4 pCi ^{210}Po. Estimated intakes (10% ^{210}Pb and 20% ^{210}Po in cigarettes) for 20 cigarettes/day are 1.2 pCi ^{210}Pb and 1.6 pCi ^{210}Po.

f. Intake primarily from food in diet; activity in tap water usually less than 0.03 pCi/l, which can be neglected, but concentrations as high as 5,000 pCi/l have been measured in wells.

g. Makes negligible contribution to body content because of very low absorption of thorium through gastrointestinal tract.

h. Populated areas with high concentrations of thorium and uranium in their soil show higher daily intakes: 3 pCi ^{226}Ra and 160 pCi ^{228}Ra along coast of Kerala in India; 10–40 pCi ^{226}Ra, 60–240 pCi ^{228}Ra in 196 individuals in the Araxa-Tapira region in Brazil.

i. UNSCEAR (1977) expresses activity concentrations in bone as per unit mass of *dry* bone, given as 5 kg, which comprises 4 kg of compact bone and 1 kg of cancellous bone; the total bone surface in an adult man is 10 m^2 with 5 m^2 in compact bone and 5 m^2 in cancellous bone.

j. ICRP model of the metabolism of alkaline earths in adult man gives an average bone-to-diet quotient of 6 pCi/kg per pCi/day intake.

k. There is 5 pCi/kg ^{210}Pb, 4 pCi/kg ^{210}Po, 3,600 pCi/kg ^{40}K in red bone marrow. Lower value of ^{228}Ra compared to ^{226}Ra results from shorter half-life combined with biological half-life of 10 yr.

l. Activity concentrations of ^{230}Th (in ^{238}U decay series) and ^{232}Th are about equal, and thus levels of ^{230}Th are similar in man and make similar contributions to dose.

m. ^{234}U in equilibrium with ^{238}U adds approximately equal dose.

n. Based on average retention factor in skeleton and soft tissues of 0.33 for ^{222}Rn and of 1.0 for ^{220}Rn and a uniform concentration of radium and its short-lived decay products over the total mass of mineral bone.

o. ^{228}Ra is a beta emitter; alpha particles are emitted by its decay products.

quantities of natural radioactive material at different depths in soil. For example, consider the continental United States (excluding Alaska), with an area of 8×10^{12} m^2, an average rock density of 2.7×10^6 g/m^3, and a concentration of uranium-238 (specific activity 1.23×10^{-5} GBq/g) of 2.7×10^{-6} g per gram of rock. A layer of rock 1 m thick in the continental U.S. contains 58.3×10^{12} g uranium with a total activity of 718×10^6 GBq (19.4×10^6 Ci). A layer of rock 600 m thick (the projected depth for burial of high-level radioactive wastes) contains 44×10^{10} GBq (1.2×10^{10} Ci). Equal quantities of radioactivity are contributed by the 13 radioactive decay products of ^{238}U, including ^{234}U, ^{234}Th, ^{226}Ra, and ^{210}Po. Similar contributions are made by the thorium series. While we have focused attention on the long-lived alpha emitters as the constituents in high-level waste of most concern, their activity is matched by the beta-emitting ^{40}K in the ground. Its activity is about 20 times that of uranium, and therefore is comparable to the whole uranium series. Of course, the alpha energy imparted by the radiation from the uranium and thorium series is much greater than the beta energy from potassium.

5.2.1   Atmospheric Aerosols

There is constant leakage from the ground of two alpha-emitting noble gases, radon-222 (radon) and radon-220 (thoron). These gases are produced in the uranium and thorium decay series. They are responsible, through their decay products, for the most significant exposures of the U.S. population (NCRP, 1984a; NAS/NRC, 1988). The amount that emanates from the ground at any given place and time depends not only on the local concentrations of the parent radionuclides but also on the porosity of the ground, moisture content, ground cover, snow cover, temperature, pressure, and other meteorological conditions.

It is estimated that the total radon emanation rate from land areas is 1,850 GBq/sec (50 Ci/sec), which produces an equilibrium activity in the atmosphere of 925 million GBq (25 million Ci). Measured emanation rates from soil are as high as 51.8 mBq/m^2-sec (1.4 pCi/m^2-sec) with a mean value of 15.5 mBq/m^2-sec (0.42 pCi/m^2-sec) (Wilkening et al., 1972). The concentration of radon at ground level depends strongly on meteorological conditions, so it can vary greatly, typically by a factor of 4, both daily and seasonally. Average statewide ambient radon levels measured in the U.S. range from 5.3 Bq/m^3 in New Mexico to 21.7 Bq/m^3 in South Dakota. The average value nationally is about 15 Bq/m^3 (NAS/NRC, 1999b). The total thoron inventory in the atmosphere is much less because of its much shorter half-life (55 sec versus 3.8 days for radon). The decay products of both radon and thoron are also radioactive and undergo a series of additional decays by alpha and beta-gamma emission. The decay products, atoms of polonium, lead, and bismuth, are breathed in and deposited in the lungs either as free atoms, part of a cluster of atoms, or attached to particles in the air. They are responsible for most of the dose to the lungs resulting from breathing in an atmosphere containing radon and thoron (see Part Three, sec. 5.6, for a detailed treatment of the dosimetry).

Radon concentrations will build up to appreciable levels indoors (Moeller and Underhill, 1976). Most of it comes from soil that is beneath the basement or foundation. Where radon exists in significant concentrations in groundwater, well water used in homes serves as a source of radon, which is released to the air during activities such as showering, washing clothes and flushing toilets and results in a dose to the lungs from inhalation. The assessment of exposure from inhalation of radon released to the air is through the concept of the *transfer coefficient*, defined as the average fraction of the initial average radon concentration in water that is contributed to the indoor airborne radon concentration. A recommended value for the transfer coefficient is 0.0001 (NAS/NRC, 1999b).

Example 6.2   The indoor concentration of radon in a house is found to be 37 Bq/m^3 in the absence of any contribution from the use of well water. What concentration in well water would raise the exposure assessment to 148 Bq/m^3?

The additional airborne concentration would be $148 - 37$ Bq/m^3 = 111 Bq/m^3. The water concentration to produce this level is calculated by dividing 111 by the transfer coefficient, and would be equal to $10,000 \times 111$ Bq/m^3 = 1,110,000 Bq/m^3.

The committed effective dose from drinking water containing radon depends on the age of the subject, with calculations ranging from 4.0×10^{-8} Sv/Bq for an infant to 3.5×10^{-9} Sv/Bq for an adult. The associated age- and gender-averaged cancer death risk from lifetime ingestion of radon dissolved in drinking water at a concentration of 1 Bq/m^3 is 2×10^{-9} (NAS/NRC, 1999b).

Measurements reported for indoor levels vary over several orders of magnitude. A ten-state survey by the Environmental Protection Agency found that 21 percent of the 11,600 homes sampled during the winter—or more than one home in every five—had levels exceeding EPA's action level of 148 Bq/m^3 (4 pCi/l) of air. One percent of the homes had levels exceeding 740 Bq/m^3 (20 pCi/l) and a few had levels exceeding 5,550 Bq/m^3 (150 pCi/l; *Science News,* August 15, 1987). A representative value for the average ^{222}Rn concentration indoors is about 48 Bq/m^3 (1.3 pCi/l), giving an annual absorbed dose (assuming a decay product equilibrium factor of 0.5) averaged over the whole lung of about 0.3 mGy (30 mrad) and an annual equivalent dose of 6 mSv (600 mrem). The absorbed dose to the basal cells of the bronchial epithelium is 1.6 mGy (160 mrad) and the equivalent dose is 32 mSv (3,200 mrem) (UNSCEAR, 1977, p. 79). These dose rates produce the highest organ doses (in mrem) from natural sources. While attention has centered on exposure to radon and its decay products, the exposure from thoron can also be significant. Much less thoron leaks out of the ground than radon because of its shorter half-life (55 sec versus 3.8 days), but the thoron decay product chain has a longer half-life ($\sim$10 hr) than the radon decay product chain ($\sim$1 hr). As a result, there is a much higher buildup of radioactivity from the thoron than from the radon decay products in the lung for the same activity concentration and the potential for a significant dose.

Small concentrations of other radionuclides are also found in air. These originate primarily from airborne soil. Some reported levels (NCRP, 1975, p. 77) in mBq/m^3 air are: ^{238}U, 4.44 $\times$ 10^{-3}; ^{230}Th, 1.67 $\times$ 10^{-3}; ^{232}Th, 1.11 $\times$ 10^{-3}; and ^{228}Th, 1.11$\times$ 10^{-3}. Radium levels are comparable to the uranium levels. A significant portion is probably contributed by coal-burning power plants (Moore and Poet, 1976). Lead-210 and ^{210}Po, the long-lived decay products of ^{222}Rn, also contribute to airborne radioactivity, reaching levels of 0.37 mBq/m^3 and 0.037 mBq/m^3, respectively. The ^{210}Po levels result from releases from soil of about 185 $\times$ 10^{-6} mBq/m^2 on calm, clear days to 2,590 $\times$ 10^{-6} mBq/m^2 when the air is dusty (Moore et al., 1976).

5.3   Population Exposure from Medical and Dental X Rays

The most important contributions to exposure of the population from man-made sources of radiation are from the medical and dental professions, which far exceed contributions to date by the nuclear power industry and the military. We shall discuss the exposure of the population from medical radiation machines in this section and, in section 5.4, examine internal exposure resulting from administration of radioactive materials in medical diagnosis.

Statistics from a variety of sources indicate the extent of the radiation exposure and its rate of increase in the United States (NCRP, 1989a) and throughout the world (UNSCEAR, 1993). These include the number of x-ray machines, annual sales of x-ray film, numbers of hospital x-ray examinations, numbers of x-ray visits, and estimated number of diagnostic x-ray procedures. Similar statistics apply to nuclear medicine procedures. A good

Table 6.13 Resources utilized in the administration of x rays in the United States.

Resource	1970–1972	1980–1982	After 1982
Number of medical diagnostic x-ray machines	110,000	127,000	
Number per 1,000 population	0.53	0.55	
Number of dental diagnostic x-ray machines	126,000	204,000	
Number per 1,000 population	0.60	0.89	
Estimated annual medical x-ray film sales (in millions)	584	845	845 (1986)
Sheets per capita	2.79	3.64	3.50 (1986)
Dental x-ray films (in millions)	300	380	
Number of x-ray examinations (in millions)			
Chest (radiographic)	48.6	64.0	
Chest (photofluorographic)	10.4		
Upper GI (barium meal)	6.7	7.6	
Kidney, ureters, bladder (KUB)	4.0	3.4	
Pyelograms	3.7	4.2	
Biliary	3.4	7.9	
Barium enema	3.4	4.9	
Lumbosacral spine	8.6	12.9	
Head CT		2.7	
Body CT		0.6	
Total	136	180	
Rate per 1,000 population	670	790	
Dental x-ray examinations (in millions)	67	101	
Rate per 1,000 population	330	440	

Source: NCRP, 1989a. Three-year interval chosen to cover different years in which survey was made. Where use was surveyed in more than one year interval, latest year was recorded.

correlation appears to exist between the number of x-ray examinations per unit of population and the number of physicians per unit of population.

A sampling of the statistics over time for x-ray examinations in the United States is given in Table 6.13.

The rate of radiographic examinations increased from 670 to 790 per 1,000 population from 1970 to 1980 and the dental rate from 330 to 440. The quantity of medical x-ray film sold in the United States annually is another index of the exposure of the population to medical x rays. The number of sheets increased from 263 million in 1963 to 845 million in 1986, although it did not change much between 1980 to 1986. Similarly, the number of sheets per capita increased from 1.38 in 1963 to 3.5 in 1986 (NCRP, 1989a).

Introduction of computed tomography provided a new powerful x-ray diagnostic tool with accompanying increases to population dose. Total scans as determined by questionnaire increased from 2,337,000 in 1981 to 4,303,000 in 1983 (Evans and Mettler, 1985). Mammography of the female breast increased fivefold between 1970 and 1980 from 246,000 mammograms (2.4 per 1,000 female population) in 1970 to 1,260,000 (11 per 1,000) female population) in 1980 although only a small fraction of the female population was examined by this procedure. Prescription of other diagnostic procedures using x rays or radionuclides also increased markedly over this period. These included cardiac, biliary and pelvic imaging procedures (NCRP, 1989a).

The collective effective dose to the population from a given procedure depends on both the dose imparted by the procedure and the number of patients examined. The tracking of collective effective dose is of considerable public health interest (NCRP, 1989a). The data are used to compare sources of radiation exposure, look for trends in population dose, identify problem areas, and initiate programs to minimize population dose where efficacious. The highest percent contributions to the collective dose come from upper gastrointestinal and barium enema examinations—21.5 and 21 percent, respectively, in U.S. hospitals in 1980 (NCRP, 1989a); and 29 and 7 percent, respectively, in a later Japanese study (Maruyama et al., 1992). Other examinations that make significant contributions to the population dose from medical diagnostic radiation include IVP (intravenous pyelogram), biliary, pelvis, lumbar spine, chest, KUB (kidneys, ureters, bladder), and computed tomography. Details of exposures in various x-ray procedures are given in Part Two, sec. 23, and Part Three, sec. 9. The mean annual effective dose to the U.S. (1980) population from diagnostic x-ray examinations is estimated as 0.4 mSv. The mean effective dose (1982) from diagnostic nuclear medicine procedures is estimated to be 0.14 mSv (NCRP, 1989a).

We noted in the discussion on x-ray protection (sec. 25, Part Two) that exposures vary considerably depending on the techniques and equipment used. Under the best conditions, without the use of an image intensifier fluoroscopy exposes skin to about 5 R/min, but exposure is one-third as much *with* image intensifiers. Chest x rays made with fast, full-size x-ray film in direct contact with an intensifying screen produce no more than 10 mR at the skin, as compared with exposures up to 100 times as high with other commonly used techniques. Current exposures, however, are only a small fraction of the typical exposures received by a patient in the first decades following the introduction of x rays. Some rather startling facts are given in a paper by Braestrup (1969). "Within the first few years of Roentgen's discovery, the application of x-rays in diagnosis required doses of the order of 1000 times that required today. Radiographs of heavy parts of the body took exposures 30–60 minutes long. Maximum allowable exposures were set by the production of skin erythemas (300–400 rad). Thus the skin served as a personal monitor. The Wappler fluoroscope, manufactured around 1930–1935, produced 125–150 R/min at the panel. Skin reactions were produced and in some cases, permanent injury. To minimize hazard, a 100 R per examination limit was set in the New York City hospitals."

5.4 Population Exposure (Internal) from Radiopharmaceuticals

With the growth of nuclear medicine, radioactive drugs have become the largest man-made source of internal exposure of the population. New applications occur continually for diagnosing abnormal conditions, evaluating organ function, and imaging organs in the body. Some data on the frequency of examinations using radiopharmaceuticals are given in Table 6.14.

The major reason for the phenomenal success of nuclear medicine techniques in diagnosis was the introduction in the mid-1960s of technetium-99m, a short-lived (6 hr) radionuclide that can be tagged on to many diagnostic agents. Because of its short half-life, ^{99m}Tc produces much lower doses than the much longer lived agents that it has replaced. Many medical institutions are largely abandoning the use of ^{131}I as a diagnostic tool, replacing it with shorter-lived ^{99m}Tc pertechnetate or ^{123}I for scanning. Examples of the doses imparted in nuclear medicine are given in Part Three, section 8.

5.5 Environmental Radiation Levels from Fallout from Past Weapons Tests

Next to medical x rays, the only other significant source of external radiation to the population from man-made sources has been fallout from nuclear bombs exploded in the atmosphere. The fallout emits gamma photons of various energies, with an average of about 0.9 MeV. The radiation

Table 6.14 Number of diagnostic examinations with radionuclides (in millions).

Examination	Year			
	1972	1975	1978	1981
Brain	1.250	2.120	1.546	1.038
Liver	0.455	0.676	1.302	1.445
Bone	0.081	0.220	1.160	1.613
Respiratory	0.332	0.597	1.053	1.095
Thyroid (scans and uptakes)	0.356	0.627	0.699	0.664
Urinary	0.108	0.154	0.205	0.402
Cardiovascular	0.025	0.049	0.160	0.708
Other	0.441	0.316	0.281	.234
Rounded total	3.1	4.8	6.4	7.2

Source: Mettler et al., 1985, in NCRP, 1989a.

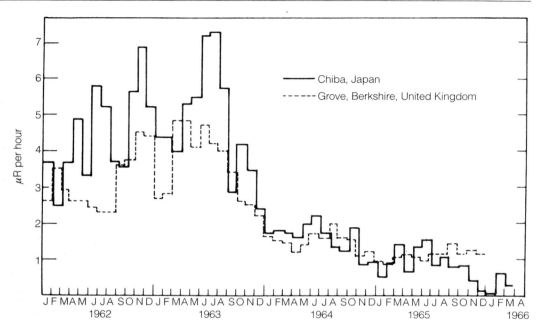

6.3 External gamma exposure rates due to fallout (UNSCEAR, 1966, p. 68).

levels produced by fallout on the ground depend on a variety of factors, including terrain, pattern of the fallout, and the particular radioactive nuclides that dominate the fallout at that particular time. The main mechanism for the deposition of fallout is through rainfall and snowfall, and thus the radiation levels from fallout in any region are strongly dependent on the precipitation. Other factors, such as latitude, wind direction, and distance from the original detonation, are also important (Glasstone, 1962).

In 1963, at Argonne, Illinois, fallout increased natural background levels by as much as 30 percent (UNSCEAR, 1966). In subsequent years, the dose rate decreased because of the decay of the shorter-lived contributors, ^{95}Zr, ^{144}Ce, and ^{106}Ru. Thereafter, the main contributor was ^{137}Cs. Examples of the seasonal variation of gamma fallout levels in Japan and the United Kingdom following the peak testing period in 1962 are shown in Figure 6.3.

The total external dose imparted by gamma radiation resulting from complete decay of fallout will amount to about 2.47 mGy (247 mrad), of which 1.33 mGy (133 mrad) will be due to ^{137}Cs and will be delivered over many decades. The other 1.14 mGy (114 mrad) is from the shorter-lived nuclides and has already been delivered essentially in full. The actual dose to the gonads and blood-forming organs is estimated at 20 percent of the environmental level because of shielding by buildings and screening by body tissue.

While the average population dose from external gamma radiation as a result of bomb testing has been quite low, there have been fallout patterns considerably above the average in some local areas. For example, a very unusual deposit was observed in Troy, New York, in April 1954 following a violent electrical storm that occurred after a 43 kiloton[10] bomb test (Simon) in Nevada, 36 hr earlier and 2,300 miles to the west (Clark, 1954). Representative gamma levels were of the order of 0.4 mR/hr, 1.1 days after onset, with levels as high as 6 mR/hr found on a pavement near a drain. It was estimated that the accumulated gamma exposure in 10 weeks was 55 mR.

Close to the Nevada test site, in towns in Nevada and Utah, fairly high dose rates occurred as a result of atmospheric tests of nuclear fission bombs in the 1 to 61 kiloton range in 1952 and 1953 (Tamplin and Fisher, 1966). Dose rates 24 hr after the detonation were: 30 mrem/hr in Hurricane, Utah; 26 mrem/hr in St. George, Utah, and 40 mrem/hr at Bunkerville, Nevada. Other measurements off the test site ranged as high as 115 mrem/hr (24 miles west of Mesquite, Nevada). It should be noted that the half-life of fission products varies approximately as the time since the detonation, so the measurements given are for radioactive nuclides of fairly short half-lives, that is, half-lives of 24 hr one day after the explosion, rising to 7 days after 1 week. A 40 mrem/hr dose rate at 20 hr implies a dose of about 1,580 mrem in the next 7 days, increasing to a maximum of 4,800 mrem when essentially all the atoms have decayed.

The fallout levels discussed above were not high enough to produce demonstrable injury to the small populations exposed. However, in at least one test, at Bikini Atoll on March 1, 1954, fallout produced severe radiation injuries (Eisenbud, 1963). The high radiation exposures resulted from fallout from a megaton bomb.[11] The highest exposures were incurred by 64 natives of Rongelap Atoll, 105 miles east of Bikini, and 23 fishermen on the Japanese fishing vessel Fukuru Maru about 80 miles east of Bikini. The exposed individuals had no knowledge of the hazards of the fallout and therefore took no protective measures. They incurred external whole-body

---

10. A 1-kiloton nuclear test is equivalent in energy released to the detonation of 1,000 tons of TNT.

11. A yield of 1 megaton is equivalent in explosive power to the detonation of 1 million tons of TNT.

doses between 200 and 500 rem plus much higher local beta exposures. Skin lesions and epilation resulted.

The fallout was originally detected on the Island of Rongerik, 160 miles east of Bikini, where 28 American servicemen were operating a weather station. Evacuation procedures for the Americans on Rongerik were put into effect some 30 hr after the detonation, and 23 hr after the fallout was detected. They received whole-body doses estimated at 78 rem.

The experience at Bikini demonstrated the potential severity of fallout from megaton explosions produced with thermonuclear weapons. It provided a dramatic portent of the holocaust that would result if the major nuclear powers were to unleash their nuclear arsenals in the madness of war.

5.6 Potential External Exposure to the Population from Large-Scale Use of Nuclear Power

The development of a nuclear power industry to supply a fraction of the power needs of the world will result in the production of enormous and unprecedented quantities of radioactive wastes. It will be necessary to control and confine these wastes to prevent excessive exposure of the world's population. Although there are strong differences of opinion as to society's capacity to confine the radioactivity resulting from such power production, it is not within the scope of this book to examine the practicability of controlling those wastes that might accidentally escape to the environment. However, there are gaseous wastes (from the fission process) that may be released deliberately and routinely to the environment. The most significant is the radioactive rare gas, krypton-85, which is not metabolized when inhaled from the air and accordingly is treated for purposes of dose calculations as an external source of radioactivity, distributed throughout the atmosphere. The calculation of the dose at a location in an infinite volume of ^{85}Kr was described in Part Three, section 5.4. Estimates of ^{85}Kr releases from various facilities are given in Table 6.15.

5.7 Population Exposure (Internal) from Environmental Pollutants

The atmosphere provides a convenient receptacle for the discharge of waste materials. It carries offensive wastes away from the polluter's environment and, if the wastes are particulate, spreads them thinly before dumping them back on the ground. In the case of gaseous wastes, the enormous capacity of the reservoir provided by the atmosphere dilutes the discharges to such low levels that the effects of individual discharges are

Table 6.15 Krypton-85 release from various nuclear power facilities and processes.

Facility /process	Amount released
Production in fission reactors per MW(e) or 3 MW(t)[a]	480 Ci/yr
Release from a water-cooled nuclear power plant—1000 MW(e)[b]	200 Ci/hr
Release from a single fuel reprocessing plant[c]	6300 Ci/day
Production in nuclear explosions using fusion, 10 Kt fission/Mt fusion[a]	240 Ci/Mt

a. See note to Table 6.17 for data used in calculations. MW(e), megawatts of electrical energy; MW(t), megawatts (thermal); Mt, megatons; kt, kilotons.
b. Upper limit estimated from discharge rates (Goldman, 1968, p. 778; Kahn et al., 1970, table 4.1).
c. Davies, 1968.

seldom significant. Usually, it takes a continual accumulation of wastes, with the absence of any accompanying removal process, to achieve levels that cause concern.

The initial dilution of waste introduced into the atmosphere near the surface of the earth takes place in the lower part of the atmosphere (the troposphere—an unstable, turbulent region 10–15 kilometers in depth). It is capped by a stagnant region known as the tropopause. Measurements on radioactive fallout indicate it takes about 20 to 40 days for half the particulate matter in the troposphere to settle out.

The region above the tropopause is called the stratosphere. Under normal discharge conditions, it does not receive wastes. It takes something like a megaton hydrogen bomb explosion to propel the contaminants into the stratosphere. Because of the relative isolation between stratosphere and troposphere, the activity deposited in the stratosphere is returned to the lower levels at a low rate. The time for half the fallout from air-burst bombs to leave the stratosphere is about eight months. Data on volumes available for diluting wastes injected into the earth's atmosphere are presented in Table 6.16.

5.7.1 Fallout from Bomb Tests

A considerable amount of radioactivity has been injected into the atmosphere as a result of nuclear bomb tests. The radionuclides in the resultant fallout that present the greatest ingestion hazard are ^{131}I, ^{90}Sr, ^{89}Sr, and ^{137}Cs. Other radionuclides of interest are ^{14}C, ^{3}H, ^{239}Pu, and ^{238}Pu. Data on the quantities of these radionuclides produced in various nuclear operations are given in Table 6.17. Power reactor data are included for later reference.

A summary of the bomb tests conducted in the atmosphere is given in

Table 6.16   The earth as a global dump.

Radius of earth	6.37×10^8 cm
Surface area of earth	5.11×10^{18} cm^2
Area of earth between 80° N and 50° S	4.48×10^{18} cm^2
0–80°N	2.51×10^{18} cm^2
0–50°S	1.96×10^{18} cm^2
Height of troposphere	1.28×10^6 cm
Volume of troposphere	6.5×10^{24} cm^3
Volume of gas in troposphere at 76 cm, 20°C	2.9×10^{24} cm^3

Table 6.17   Magnitudes of sources (megacuries) associated with various nuclear operations.

Radionuclide (half-life)	^{131}I (8 days)	^{137}Cs (30 yr)	^{89}Sr (51 days)	^{90}Sr (28 yr)	^{14}C (5700 yr)	^{3}H (12.3 yr)	^{85}Kr (10.8 yr)
1 megaton fission explosion in atmosphere	113	0.169	29.6	0.181	0.020[a]	0.0007[b]	0.024
1 megaton fusion explosion in atmosphere					0.020[a]	10–50[b]	
1000 megawatt (thermal) reactor							
After 1 mo operation	22.5	0.095	13.6	0.129		0.00039	0.014
After 1 yr operation	25.1	1.13	40.2	1.24		0.0046	0.169
After 5 yr operation	25.1	5.90	40.2	5.60		0.021	0.720
Equilibrium level	25.1	49.5	40.2	49.5		0.084	2.51

Note: Calculations for the yield of fission products are based on 1.45×10^{26} fissions /megaton fission energy (Glasstone, 1962), where 1 megaton is the energy equivalent of the explosion of one million tons of TNT or the complete fissioning of 56,000 g of ^{235}U. Other megaton equivalents are 10^{15} calories, 1.15×10^9 kilowatt-hours, and 1.8×10^{12} BTU. The fission yields for the radionuclides presented in the table in atoms per hundred fissions are ^{90}Sr, 5.9; ^{89}Sr, 4.8; ^{137}Cs 5.9; ^{85}Kr, 0.3; ^{3}H, 0.01; ^{131}I, 1.9 (Etherington, 1958). The large thermonuclear explosions produced during weapons testing provided the energy approximately equally from fission and fusion. Calculations for the 1,000 megawatt reactor are based on the conversion factor, 1 watt (thermal) = 3.1×10^{10} fissions/sec.

a. USAEC, 1959.

b. Jacobs, 1968.

Table 6.18. The explosive power from fission alone was over 200 megatons, a small percentage of the postulated megatonnage for a hypothetical full-scale nuclear engagement between the major nuclear powers.

The behavior of fallout has been studied closely, and extensive data are available on its deposition, incorporation into the food chain, and subsequent ingestion and metabolism by man. The radionuclide that has been investigated in greatest detail is ^{90}Sr because of its abundance, long half-life, and long-term incorporation in bone (see Fig. 6.3). A total of 15 megacuries fell to the earth prior to January 1970. This represents almost

Table 6.18 Summary of nuclear weapons tests conducted in the atmosphere.

Inclusive years	Country	No. of tests	Megaton range	Fission yield (megatons)	Total yield (megatons)
1945–1951	U.S.	24	0	0.8	0.8
	U.S.S.R.	3	0		
1952–1954	U.S.	27	3		
	U.K.	3	0	38	60
	U.S.S.R.	3	0		
1955–1956	U.S.	22	1		
	U.K.	6	3	13	28
	U.S.S.R.	11	3		
1957–1958	U.S.	78	2		
	U.K.	12	7	40	85
	U.S.S.R.	38	14		
1959–1960	Moratorium on air-burst bomb tests by U.S.A. and U.S.S.R.				
1961	U.S.S.R.	31	?	25	120
1962	U.S.	36	?	16	217
	U.S.S.R.	37	?	60	
1964–1965	China	2	0	.06	.06
1966–1968	China	6	2	4.1	6.6
	France	13	2	3	4.7
1969	China	1	1	1.5	3
1970–1971	China	2	1	1.5	3
	France	8	3	1	2.5
1972–1973	China	3	1	1.5	2.5
	France	7			
1974	China	1	1	0.6	0.6
	France	8	1		1.5
1976	China	3	1	2.2	4.2
1977–1980	China	5	1	0.25	1.25

Sources: Telegadas, 1959; Hardy, 1970; FRC, 1963; Telegadas, 1977; Carter and Moghissi, 1977; UNSCEAR, 2000.

Note: The largest nuclear test had a total yield (fission plus fusion) of 58 megatons, and was conducted by the U.S.S.R. in 1961. In 1962 the USSR again conducted two high-yield tests, about 30 Mt each. The largest test conducted by the U.S. was 15 Mt in 1954.

See UNSCEAR, 2000, for detailed statistics on nuclear weapons tests and environmental contamination from weapons tests, nuclear power production, and nuclear accidents.

all the fallout that will occur unless major testing is resumed. Estimates of the fallout deposited per unit area on the surface of the earth are made by averaging over the area between 80°N and the equator for deposition in the Northern Hemisphere, and between 50°S and the equator for deposition in the Southern Hemisphere, since this includes over 97 percent of the deposition (UNSCEAR, 1966, p. 6). The fraction of the fallout that deposits in each hemisphere depends on the site of the tests, the nature of the explosion, and the time; so far, about 76 percent of the fallout has been deposited in the Northern Hemisphere; the Northern and Southern hemispheres received 12.1 and 3.9 megacuries, respectively, through 1975.

The amount of fallout in different areas within a hemisphere varied considerably. For example, Figure 6.4 shows the annual deposition of ^{90}Sr in the Northern Hemisphere as a whole and in New York City. Taking the year in which the greatest deposition occurred, 1963, we note that the deposition in the Northern Hemisphere was 2.6 megacuries and the deposition in New York City, 24 mCi/km^2. Averaged over 0–80°N, where almost all the fallout occurred (an area of 2.51×10^8 km^2), the average deposition in the Northern Hemisphere was 10.3 mCi/km^2, or less than half the deposition per square kilometer in New York City.

Values of the doses received by the world population as a result of ex-

6.4 Worldwide dissemination of ^{90}Sr from weapons tests. *(a)* From HASL, 1970, Report 223; *(b)* from Volchok, 1967; *(c)* from UNSCEAR, 1969, 1977; and *(d)* from FRC, 1963, Report 4; Carter and Moghissi, 1977.

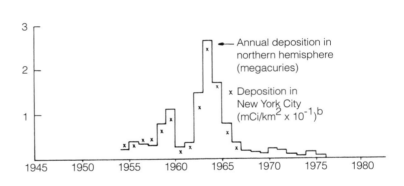

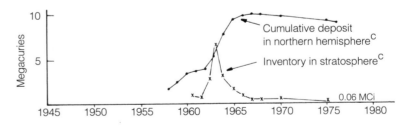

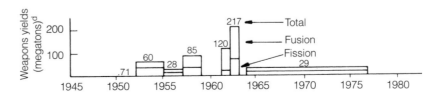

ploding nuclear bombs in the atmosphere are presented in Table 6.19. Both external and internal sources are included for comparison. The doses are expressed in terms of dose commitments, defined here as the dose delivered during complete decay of the radioactivity (except for ^{14}C, which is calculated to year 2000).

From Table 6.19 it can be seen that the dose commitment from fallout is low. However, large local deviations from the average deposition levels have been found. For example, in the Minot-Mandan region of North Dakota, ^{90}Sr levels in milk increased steadily from 33 pCi/g calcium in August 1957 to 105 in the spring of 1963 (Pfeiffer, 1965). This is four times the levels shown in Figure 6.4. The higher levels, however, are still less than those believed to be cause for concern by regulatory agencies in the United States (that is, the Federal Council on Radiation Protection specified 200 pCi/g calcium as the point at which intake should be monitored, the lower limit of their so-called Range III).

More serious deposition incidents occurred in the 1950s in the United States from fallout injected into the lower atmosphere following tests of kiloton weapons in Nevada. The fallout was high in many cities far removed from the tests. For example, data from the gummed film network operated by the USAEC gave fallout levels as high as 80 μCi/m^2 in Albany, New York, 75 in Salt Lake City, 65 in Roswell, New Mexico, and 25 in Boston (Tamplin and Fisher, 1966).

One of the highest fallout incidents occurred in Troy, New York, in

Table 6.19 Dose commitments to population in north temperate zone from bomb tests.

| | | Dose commitment (mrad) | | |
| | | | Internal | |
Source	External	Gonads	Cells lining bone surface	Bone marrow
^{137}Cs	62	27	27	27
^{90}Sr[a]			120	84
^{89}Sr				0.4
^{14}C[b]		7	29	32
Short-lived fission products	48			

Source: UNSCEAR, 1977, p. 153.

a. The evaluation of the dose to bone from an intake of ^{90}Sr is a complex procedure (see Spiers, 1968). For long-term ingestion, a concentration of 1 picocurie of ^{90}Sr per gram calcium in the skeleton will give a dose of 1.4 mrad/yr to bone-lining cells and 0.7 mrad/yr to bone marrow.

b. Dose commitment calculated to year 2000.

1954, as discussed earlier in connection with external radiation exposure. Levels as high as 13×10^6 dis/min per square foot of ground area were observed, and it was estimated that about 50 Ci of fission products were introduced into the Tomhannock Reservoir alone the first day. Activity of tap water (not from the Tomhannock supply) was 2.62 pCi/ml at 1 day after arrival. However, the activity was short-lived, as is characteristic of fresh fallout, and was down to 0.034 pCi/ml 16 days later.

Fallout that arrives within days of a nuclear detonation has a high content of radioactive iodine, which becomes concentrated in the thyroid after ingestion and produces a high local dose. It has been estimated that a contamination of forage by ^{131}I of 1 μCi/m^2 will lead to a dose of 30 rad to the thyroid of an infant fed milk of cows that grazed in the contaminated pasture. Particularly high levels of iodine were deposited in Nevada and adjacent states after many of the bomb tests conducted at the Nevada test site. According to Tamplin and Fisher (1966), thyroid doses above 100 rads were probably received by children who drank milk from cows that had grazed in the areas exposed to these levels of fallout.

Serious contamination occurred as an aftermath to the 1954 Bikini Atoll thermonuclear bomb test discussed previously (in sec. 4.3.2). The individuals living on Rongelap Island were in close contact with the fallout for many days before they were removed from the contaminated environment. As a result, they accumulated body burdens of radionuclides: 6 μCi of ^{131}I, 3 μCi of ^{140}Ba, and 2 μCi of ^{89}Sr. The iodine produced substantial thyroid doses: about 10–15 rad, which was small, however, in comparison with the dose from the external radiation. It may be noted that the ^{131}I activity was approximately that normally administered in thyroid function tests, which are very common in medical practice. The accumulated body burdens were considered surprisingly low in view of the extremely heavy contamination that existed, and the amounts of radionuclides deposited in tissues did not contribute appreciably to the overall effects observed. After three and a half years, the Rongelap inhabitants were allowed to return to their island. Because of the residual contamination, they continued to accumulate ^{90}Sr, although the levels are well below current recommended limits.

Strontium-90. The uptake of ^{90}Sr in living matter has been studied intensively. In foods and in the body ^{90}Sr levels are almost always given in terms of strontium activity per unit mass of calcium (picocuries of strontium per gram calcium), since the strontium and calcium follow similar pathways. However, the body discriminates against strontium, and the concentration of strontium relative to calcium in the body is about 25 percent of that in food.

Levels in the bones of children in the 1–4 yr age group as high as 11.8

pCi ^{90}Sr/g Ca were reported from Norway in 1965. Measurements made in New York City in 1965 showed a peak of 7 pCi/g in the bones of children between 1 and 2 years of age. The levels decreased to 4 pCi/g in 1967 and 1.6 in 1975 in children 4 years old. Levels in the 5–19 yr age group were about 3 pCi/g in the 1956–1968 period and down to 1.4 in 1975 (UNSCEAR, 1969, 1977). Levels for the ^{90}Sr-Ca ratio in milk measured in New York City peaked at 26 pCi/g in 1963 but decreased to 9 pCi/g by 1968 and to 4 pCi/g in 1977 (Hardy and Rivera, 1965; Bennett and Klusek, 1978). In Norway and Ireland levels several times higher were found.

From consideration of the uptake data, it has been possible to make some general correlations between amount of ^{90}Sr deposited and subsequent appearance of this nuclide in milk. One relationship (UNSCEAR, 1966) is

$$\text{Yearly average } ^{90}\text{Sr/Ca ratio in milk supplies, pCi/g}$$
$$= 0.3 \times (\text{total accumulated } ^{90}\text{Sr deposit in soil, mCi/km}^2)$$
$$+ 0.8 \times (\text{yearly fallout rate of } ^{90}\text{Sr in given year, mCi/km}^2.)$$

The daily intake of ^{90}Sr by residents of New York City and San Francisco for the period 1960–1978 is shown in Figure 6.5. The yearly intake, calculated from data on the ^{90}Sr content of the foods in a representative diet, was 350 g Ca out of a total mass of 637,000 g. Dairy products (200 kg/yr) provided 58 percent of the calcium; vegetables, 9 percent; fruit, 3 percent; grains, 20 percent; and meat, fish, and eggs, 10 percent (Bennett and Klusek, 1978). The maximum intake was 35 pCi/g Ca in 1963.

Carbon-14. The distribution of carbon-14 has been followed in some detail. The radiocarbon is produced in significant quantities in atmospheric thermonuclear explosions as a result of absorption of neutrons by the nitrogen in the air. Since it occurs mainly as carbon dioxide (a gas), it

6.5 Daily intake of strontium-90.

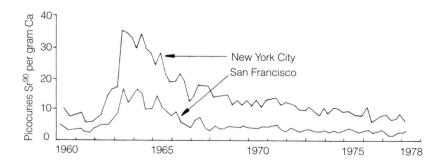

does not settle to the ground as do the particles that comprise fallout; eventually it is incorporated into living tissue through the pathways involved in the carbon cycle. Dilution of ^{14}C is limited initially to the atmosphere, surface waters of the ocean, and living matter. Eventually, after many years, the ^{14}C becomes further diluted by transfer to the much larger mass of carbon in the deeper part of the ocean. The magnitudes of the carbon reservoirs and of ^{14}C inventories produced by both natural processes and bomb tests are given in Table 6.20.

The transfer to different compartments is reflected in measurements of atmospheric ^{14}C generated from bomb tests. The peak introduction of ^{14}C approximately doubled the natural atmospheric activity, but the level is gradually decreasing due to absorption in the oceans. By the year 2040 the activity of ^{14}C will be only 3 percent above the normal atmospheric level—if there are no further air-burst bomb tests.

Iodine-131. Iodine-131 is a fission product that becomes concentrated in the thyroid after ingestion and produces a high local dose. The fission products also contain other radioisotopes of iodine, but these are much less important. The highest levels of radioiodine in people occurred during the early years of weapons testing in the United States in the 1950s, in parts of Utah and Nevada in the path of fallout from the Nevada test site. However, there was no serious program to measure environmental levels and

Table 6.20 Global carbon inventories.

Region	Natural content prior to bomb tests	
	Elemental carbon (10^{18}/g)	^{14}C (10^{27} atoms)
Atmosphere	0.61	40
Biosphere	0.31	19
Humus	1.0	55
Surface water of ocean (above thermocline)	0.92	55
Remainder of ocean	38	2,000
Total	41.4	2,170
	Excess ^{14}C as a result of bomb tests (percent above normal)	
	1965	1980 (est.)
Troposphere	69	28
Ocean (surface)	13	17

Source: UNSCEAR, 1977, p. 119.

a. Natural production rate of ^{14}C is 3.7×10^{26} atoms/yr; 2.3 atoms/cm²-sec; 0.038 MCi/yr.

doses. The short-lived ^{131}I disappeared during the three-year moratorium beginning in 1958, but when testing resumed with a vengeance in the fall of 1961 detailed studies of the transport and uptake of ^{131}I were made (Bustad, 1963). These included the monitoring of radioiodine in milk nationwide through the Pasteurized Milk Network (PMN) of the U.S. Public Health Service. The highest levels were reported for Palmer, Alaska, with a peak monthly average (in September 1962) of 852 pCi/l and Salt Lake City with a peak monthly average (in July 1962) of 524 pCi/l. The maximum concentration found in the milk in Palmer was 2,530 pCi/l (Dahl et al., 1963).

Other milk samples contained, in pCi/l, 2,000 (Salt Lake City, Utah), 1,240 (Spokane, Washington), 700 (Dallas, Texas), and 660 (Wichita, Kansas). These amounts were associated with levels of gross beta activity in precipitation that ranged from 100,000 to 200,000 pCi/m^2 and concentrations in the air that ranged from 10 to 800 pCi/m^3. The air concentrations were averaged over a 24 hr period (Machta, 1963).

About 23 billion curies of ^{131}I were injected into the atmosphere from the 200 megatons of fission explosive power produced in weapons tests. Most of the dose to humans from the resultant fallout occurred from the contamination of forage and ingestion through the cow-milk chain. It has been estimated that a contamination of forage by ^{131}I of 1 μCi/m^2 will lead to a dose of 30 rad to the thyroid of an infant fed milk of cows that grazed on this contaminated pasture.

In 1962, the concentration of radioiodine in milk in the United States averaged over the year was 32 picocuries (pCi) per liter, resulting in an annual dose of 200 mrad to the thyroid glands of infants 6–18 months old (consumption 1 l/day, uptake 30 percent, thyroid mass 2 g, effective half-life 7.6 days; Eisenbud, 1968; FRC, 1961). This annual dose was probably typical over a decade of weapons testing. While most of the dose resulted from ingestion of milk, there was also a contribution from inhalation. Levels in air were 3.8 pCi/m^3 in October 1961. Since a 1-year-old inhales 1 m^3 air per day, this gives an annual dose of 24 mrad. The average daily concentration of 32 pCi/l results in an integrated annual milk concentration of 32 × 365 = 11,680 pCi/l, or 11.7 nCi-day/l. Integrated annual milk concentrations resulting from later periods of weapons testing were, in nCi-day/l: 3.7 (Houston, 1967); 0.9 (Nashville, 1972); and 27 (Buenos Aires, 1966). When other data were not available, the integrated concentration was taken as 10 times the highest observed concentration in nCi/l. To convert these figures to dose to infants, UNSCEAR (1977) assumed a daily milk consumption of 0.7 l and the conversion factor, nCi-day/l × 11.5 = mrad to thyroid. Thus the associated infant doses were 42.6, 10.4, and 311 mrad, respectively, for the year.

Little is known about doses to the fetus from ^{131}I, and the data available show a wide variation in the relative dose to the fetus compared with the dose in children and adults. Thus, the concentration in pCi/g was measured to be 9 times greater in a 12-week fetus than in the thyroids of children, the ratio dropping sharply with fetal age for other measurements (Eisenbud, 1968). On the other hand, measurements made on a pregnant woman who died suddenly indicated a fetal concentration that was 30 percent greater than in the mother (Beierwaltes et al., 1963).

Because of the failure to monitor for radioiodine in the early 1950s, the dose to the thyroids of infants during that period must be inferred from available data on gross beta activity in air (Pendleton et al., 1963). The 1962 data indicated that a beta activity of 3,400 pCi/m^3 in air was associated with an intake in infants of 58,000 pCi ^{131}I, with a resultant dose of 1 rad to the thyroid. The beta activity was determined from the maximum concentration averaged over 24 hr and corrected to 1 day after the detonation, assuming the activity decreased as $1/\sqrt{t}$ (where t was the period between the time of measurement and 1 day after detonation). On this basis, a 24-hr average air concentration of 287,000 pCi/m^3 measured in St. George, Utah, in 1953 was assumed to have resulted in an average infant thyroid dose of 84 rad. The highest dose evaluated for 1962 was 14 rad (800,000 pCi intake).

5.7.2   Release of Plutonium to the Environment

Over five thousand kilograms (320 kCi) of plutonium have been injected into the stratosphere and subsequently deposited worldwide as a result of testing programs related to the development of nuclear weapons. In addition, one kilogram (17 kCi) of plutonium-238, which was used as fuel for a power pack, vaporized into the atmosphere when a United States Snap 17A satellite burned up. The environmental consequences of these releases may be summarized as follows (Wrenn, 1974):

- Total explosive power in weapons tests, 1945–1973:
     Equivalent to more than 200 million tons of TNT (200 Mt)
- Plutonium released (as insoluble particles of oxide):
     5,480 kg, 440 kCi (58% ^{239}Pu, 39% ^{240}Pu , 3% ^{238}Pu)
- Plutonium deposited near the testing sites:
     ^{239}Pu, 1,039 kg, 64 kCi
     ^{240}Pu, 180 kg, 43 kCi
     ^{238}Pu, 0.19 kg, 3.3 kCi
- Plutonium deposited worldwide from stratosphere (residence half-time about 1 yr):
     ^{239}Pu, 3,117 kg, 191 kCi

^{240}Pu, 567 kg, 129 kCi

^{238}Pu, 0.57 kg, 9.9 kCi

- Deposited in the soils of conterminous United States:
  10–15 kCi
- Still suspended in atmosphere as of 1975:
  Less than 1 kCi
- Additional contribution from burnout of SNAP power supply:
  ^{238}Pu, 0.98 kg, 17 kCi
- Additional alpha activity from americium-241 (25 percent of plutonium activity):
  ^{241}Am, 110 kCi

The plutonium alpha activity deposited from fallout may be compared with the activity from the actinides occurring naturally in U.S. soil. Since soils contain typically 1 pCi/g of both uranium and thorium, it is estimated that the top 2 cm contain 1.6 million curies of uranium and thorium and 4.4 million curies of all alpha emitters (including the decay products). The concentration and amount of activity from plutonium-238, -239, and -240 in surface soil (top 2 cm) is about 1 percent of the natural background activity (Wrenn, 1974).

The worldwide release of plutonium resulted in the following local levels (in New York) about a decade after the peak activity (UNSCEAR, 1977). These levels were subsequently changing at only a slow rate.

- Cumulative deposition density, 1974: 2.68 mCi/km^2
- Average deposition density rate, 1972–1974: 0.017 mCi/km^2-yr
- Air activity, 1972: 0.031 fCi/m^3 (note that peak activity, in 1963, was 1.7 fCi/m^3)
- Cumulative intake by inhalation, 1954–1975: 43 pCi
- Average annual dietary intake, 1972–1974: 1.6 pCi
- Maximum activity in body, 1974: 2.4 pCi (in 1964, it was 4 pCi)
- Average plutonium contents of body organs, 1972–1973:

Lung, 1 kg	0.27 pCi
Liver, 1.7 kg	1.16 pCi
Lymph nodes, 0.015 kg	0.17 pCi
Bone, 5 kg	1.55 pCi

- Cumulative organ doses to year 2000 from inhalation:

Lungs	1.6 mrad
Liver	1.7 mrad
Bone-lining cells	1.5 mrad

- Bone dose from ingestion: The dose rate from plutonium fixed in bone is 0.098 mrad/yr-pCi. Assume 1.6 pCi/yr ingested in diet (steady state, resulting from cumulative deposit of 2.65 mCi/km^2),

0.003 percent (or 3×10^{-5}) absorbed through GI tract, and 45 percent transported to bone. Assume ingestion for 70 years, or an average residence time of 35 years. Total dose in 70 yr is $35 \times 1.6 \times 3 \times 10^{-5} \times 0.45 \times 0.098 = 7.4 \times 10^{-5}$ mrad.

The accumulation of ^{239}Pu is uneven; the areas with greater rainfall generally have higher fallout. Typical cumulative deposits (by 1971), in mCi/km^2, are 2.3 for the eastern half of the country and 0.8 for the coastal part of California. The vertical distribution of plutonium in soil has been measured at several sites. About 80 percent was found to be deposited in the top 5 cm in a sandy loam sample from New England, and measurable quantities were found down to 20 cm (^{137}Cs exhibits a similar behavior; ^{90}Sr is retained somewhat less in the top soil and can be found down to 30 cm).

5.7.3 *Atmospheric Pollutants from a Nuclear Power Industry*

With the discontinuance of large-scale atmospheric testing of nuclear weapons, the major man-made contaminants introduced into the environment in the absence of nuclear warfare will result from the use of nuclear power reactors to produce electricity or from the possible employment of nuclear explosives for excavation or other industrial activities. Contamination of the atmosphere on a large scale occurs from these sources primarily as a result of the release of radioactive gases such as ^{85}Kr, ^{14}C, and tritium. Data on the production of these atmospheric contaminants are presented in Table 6.17. Exposure from ^{85}Kr was discussed in section 4.3.3, since it is primarily a source of external exposure.

Interest in atmospheric sources of internal exposure has been centered on tritium and ^{14}C. Tritium is produced as a waste by-product in both nuclear power reactors and thermonuclear explosions. In nuclear fission reactors, the tritium nucleus is one of the products of fission, with a yield of 1 atom in 10,000 fissions. It is also produced in large amounts in reactions of the thermal neutrons in the reactor with lithium-6 and boron, which may occur either as impurities or as neutron absorbers for reactor control. In thermonuclear detonations, tritium is released primarily as a residual component of a deuterium explosive. Tritium release to the atmosphere is not likely to be a limiting factor in considerations involving large-scale peacetime use of thermonuclear explosives.

Tritium released to the air is diluted within a few years in the earth's circulating water, which has a volume of 2.74×10^{22} cm^3. However, because meteorological processes tend to favor initial retention of the wastes in the latitudes in which they are released, most of the initial dispersion has been limited to the Northern Hemisphere, with perhaps half the tritium depositing on 10 percent of the earth's surface.

About 1,700 megacuries of tritium were released to the atmosphere as a result of the testing of nuclear bombs, adding to the 60–125 megacuries already present (mainly in the waters of the earth) from natural sources. An interesting comparison of the world tritium inventories projected for a nuclear power economy, and inventories of tritium produced naturally and as a result of past bomb tests, is given in Figure 6.6. In a nuclear-power economy based on fission reactors, tritium may pose some local contamination problems, but its effect on worldwide population dose will be negligible.

Carbon-14 production is of significance only as a result of the detonation of thermonuclear explosives in the atmosphere, as it is produced from reactions of the many neutrons created in these explosions with the nitrogen in the air.

6 POPULATION EXPOSURE FROM RADIATION ACCIDENTS

In spite of the great attention to safety in the design, construction, and operation of nuclear devices and facilities, accidents have occurred. We shall review several that have resulted in significant environmental releases, including chemical explosions involving nuclear weapons, fires at a plutonium processing plant, and the overheating of the core of a nuclear power reactor. The details of these accidents and the responses to them offer an invaluable lesson to all concerned with the large-scale use of nuclear energy and radiation, whether from the viewpoint of accident prevention, emergency planning, or political and social responsibility.

6.1 Windscale, England—The First Major Nuclear Reactor Accident Causes Significant Environmental Contamination

Windscale (now Sellafield), on the coast in northwest England, was the site of two large air-cooled nuclear reactors for the production of plutonium from natural uranium for use in nuclear weapons or as a fuel for nuclear power plants. The reactors were able to use uranium as it occurs in nature, with its very small concentration of uranium-235 (0.71 percent), as the fuel by using very pure graphite, a crystalline form of carbon, as the moderator. The function of the moderator was to slow down the fast neutrons produced in the fission of the uranium atoms so they could be efficiently absorbed by the uranium-235 and cause additional fissions. The

6.6 Projected activity of tritium from nuclear power production (Peterson et al., 1969).

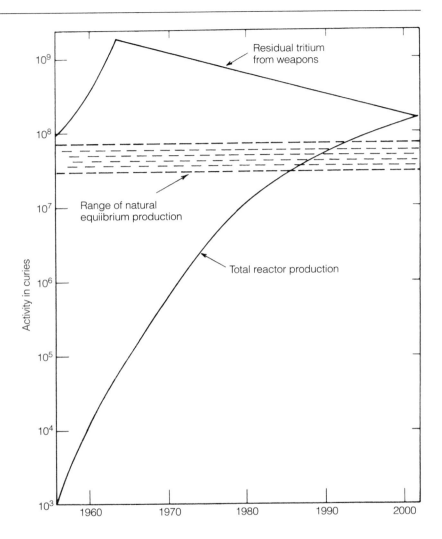

neutrons slowed down upon colliding with the carbon atoms in the crystal lattice. Part of the energy imparted to the carbon atoms as a result of the neutron collisions displaced them from their normal positions in the lattice, and as a consequence a large amount of energy was stored in the lattice. If all this energy were released at one time, it would raise the temperature of the graphite high enough to ignite it. To prevent combustion, the graphite was annealed periodically—that is, the reactor was deliberately caused to operate above its normal operating temperature to slowly cause the carbon atoms to revert to their normal position and release the stored energy at a low rate.

In October 1957, during a scheduled annealing procedure, the release of energy in a portion of the graphite caused it to rise in temperature above design levels, and the excessive heat was in a portion of the core that was not detected by the core instrumentation (Arnold, 1992). The temperature of the graphite continued to rise, overheating the uranium fuel. High levels of radioactivity were released through the 400 ft high stack discharging the air that cooled the reactor to the environment. Because the stack effluents were filtered, the release of fission products in particulate form was greatly curtailed, and the only radionuclide that escaped that resulted in significant dose to the public was iodine-131.

The accident was discovered initially by a reading that increased to ten times the normal rate on an air sampler about a half-mile from the reactor stack (Eisenbud and Gesell, 1997). Increases in readings of other environmental monitors confirmed that a release of radioactivity to the atmosphere was occurring. It was possible to visualize the core through a loading hole in the face of the reactor, and uranium cartridges were observed to be glowing at a red heat in about 150 fuel channels. Fuel elements adjacent to the affected region of the core were removed, limiting the extent of the fire, but efforts to extinguish the slowly burning core were ineffective. A day later, the core was cooled by flooding it with water, resulting in a total loss of the reactor.

Estimated releases of radioactivity to the environment were: ^{131}I, 599 TBq (16,200 Ci); ^{137}Cs, 45.9 TBq (1,240 Ci); ^{89}Sr, 5.07 TBq (137 Ci); and ^{90}Sr, 0.22 TBq (6 Ci). Measures were quickly taken to monitor activity levels in milk from cows in the vicinity, and milk containing more than 0.1 μCi/l was discarded. Children and adults living downwind from Windscale were scanned for iodine uptake. The highest child's dose to the thyroid was estimated to be 16 rad (160 mGy) and the highest adult dose was estimated to be about 60 percent as much.

Radioactivity discharged to the atmosphere was detected by monitors located throughout Europe, and although the readings were well below levels of health concern, the increase in activity was indicative of the severity of the accident.

6.2 Palomares, Spain—Atomic Bombs Drop from the Sky, Igniting and Contaminating a Countryside

A portentous incident occurred over Palomares, Spain, in January 1966, when a U.S. Air Force bomber carrying four atomic bombs collided with a refueling plane. The bombs fell to the ground. No nuclear explosion occurred, but the high-explosive components of two weapons detonated upon impact, one in low mountains and the other on land used for agriculture, setting the weapons on fire. Clouds of plutonium were released into the air and dispersed by a 35 mph wind, contaminating the countryside with highly radioactive plutonium particles. An area one-half mile long and one-sixteenth of a mile wide was contaminated with 50 to 500 $\mu g/m^2$. Low levels of plutonium were detectable to a distance of approximately two miles.

The United States and the Spanish governments agreed that soil contaminated with more than 32 $\mu Ci/m^2$ (521 $\mu g/m^2$) would have the top 10 cm removed for disposal in the United States; removed soil would be replaced where needed with soil free of plutonium from the incident; and crops in fields with contamination levels above 5 $\mu g/m^2$ would be removed and destroyed.

The incident caused a tremendous stir. To quote from Flora Lewis's book, *One of Our H-Bombs Is Missing* (1967, chapter 7): "By two weeks after the accident there was a solidly based community of 64 tents, 747 people, a motor pool, a kitchen with 22 cooks who baked 100 loaves of bread a day, a PX, and a nightly open-air movie [p. 95] . . . In the end, 604 acres (nearly a square mile) were treated either by soil removal or by plowing, with topsoil carried away wherever the reading showed anything above 60,000 counts per square meter . . . The decontamination job took eight weeks . . . 4879 blue metal 55 gallon drums were filled with contaminated soil, loaded on barges that came up on the beach, and transferred at sea to the USNS Boyce" for transfer to the United States and burial.

About 90 percent of the activity appeared to have been in the top 15 cm of soil. Surface activities following decontamination were as high as 5,000 dis/min/g, with 120 dis/min/g reported as an average. Levels in plants (tomatoes, maize, beans, alfalfa) ranged between 0.02 and 6 dis/min/g (wet), in fruit about 0.003 dis/min/g. Measurements on control plants in Spain ranged from 0.02 to 0.2, and on fruit between 0.003 and 0.0009. In the United States, values were reported between 0.0007 and 0.02. Urine samples were taken from the 100 villagers most likely affected and "insignificant levels" of the order of 0.1–0.2 dis/min per 24 hr sample were obtained from 30 samples. Chest counts were also taken to determine the presence of activity in the lungs (at the level of detection) and no positive counts were found.

Six years after the accident, there appeared to be little change in the community, according to the U.S. Atomic Energy Commission. Farming habits had changed, but mostly as a consequence of other factors, such as drought, flash flooding, and economics. Follow-up studies indicated there had been little change in exposed persons, and none was expected.

6.3 Thule, Greenland—A Bomber Crashes and Its Nuclear Weapons Ignite

In January 1968, a U.S. Air Force plane carrying four nuclear weapons crashed on ice in the Arctic, near Thule, Greenland, while attempting an emergency landing necessitated by an on-board fire (EPA, 1974). The weapons were unarmed and no nuclear explosion occurred, but the high-explosive components of all four weapons detonated on impact, igniting the fuel and producing an intense fire. All this occurred while the debris produced by the crash was propelled at a high forward velocity. The fire continued to burn for at least 220 min, producing a cloud that reached a height of approximately 2,400 ft and a length of about 2,200 ft. The burning plutonium was converted largely to extremely insoluble oxides and dispersed as fine insoluble particles, from fractions of a micron to several microns in diameter. Particles of plutonium oxide were impinged into all bomb and plane surfaces struck by the high-explosive shock wave, entrained and carried forward in the splashing fuel, blown into the crushed ice at the impact point, and carried aloft in the smoke plume along with the combustion products of the burning fuel.

The impact and momentum of debris produced a long patch of black discolored ice extending away from the aircraft's impact point, 100 m wide by 700 m long. About 99 percent of the plutonium (between 2,500 and 3,700 g) was on the surface in the blackened area. The contamination level was about 0.9 mg/m^2 at the edge, extending up to 380 mg/m^2 averaged over the most contaminated portion. The calculated mass median diameter (Silverman, Billings, and First, 1971) of the particles bearing the plutonium was about 4 microns, about 4 to 5 times larger than the plutonium particles themselves. Road graders windrowed the black material and mechanized loaders placed it in large wooden boxes for removal from the contaminated area. Eventually sixty-seven 25,000-gallon fuel containers were filled with this material and four additional such containers were required to store contaminated equipment and gear. This material was shipped to the United States for final disposal. Low-level surface contamination was measured on land masses in the near vicinity of the crash site, but the risk to inhabitants or to their ecology was believed to be insignificant. Investigators concluded that only a small percentage of the total plutonium involved in the accident escaped as an airborne aerosol for distribution away from the local area of the accident.

6.4   Rocky Flats, Colorado—A Case History in Environmental Plutonium Contamination from an Industrial Plant

The Rocky Flats Nuclear Weapons Plant, which has processed large quantities of plutonium, is located 15 miles northwest of Denver, Colorado, on federally owned land two miles square. Two creeks on the boundaries drain into public water supplies. The terrain is typically prairie-arid and sparsely vegetated except where it is irrigated. Windstorms occur frequently during the fall and winter months with gusts over 45 m/sec (100 mph) recorded. Accidental releases occurred from four separate accidents since the plant began operation in 1953: two major fires in 1957 and 1969, an accidental release of plutonium to the air in 1974, and leakage (about 500 gallons) of cutting oil contaminated with plutonium from corroded barrels that had been stored outdoors since 1958. Leakage of the barrels was first detected in 1964, and it was decided to transfer the material to new drums. A small building was constructed for the operation and the last drum was removed four years later in 1968. Subsequently, the storage area was monitored and alpha activity levels were found from 2×10^5 to 3×10^7 dis/min/g, with penetration of the activity from 1 to 8 inches. Fill was applied the following year to help contain the activity, and the actual area on which barrels had been stored, a 395 by 370 ft rectangle, was covered with an asphalt pad completed in November 1969. Additional fill was added around the pad in 1970 when soil samples containing from tens to hundreds of dis/min/g were obtained. Soil stabilization studies were started for the entire area, and a revegetation program was begun (Hammond, 1971).

The 1969 fire started with the spontaneous ignition of plutonium metal in a glove box and resulted in the burning of several kilograms of plutonium. Large amounts of smoke were seen to leave the stack and spread to surrounding areas. The community was greatly concerned about the possibilities of environmental contamination. A Rocky Flats subcommittee of the Colorado Committee for Environmental Information expressed disbelief in the contentions of plant management that no significant amount of plutonium had been released during the fire. Subsequently the plant collected some 50 soil samples in August 1969, but postponed analyzing them or even developing an analytical method for them until they had completed other environmental samples. In the meantime Ed Martell and Stewart Poet of the National Center for Atmospheric Research in Boulder collected soil and water samples in the area and analyzed them in their laboratory. Soil samples from 15 locations mostly east of the plant ranged from 0.04 dis/min/g (background) to 13.5 dis/min/g of plutonium, for samples from the top centimeter, and seven water samples ranged from 0.003 to 0.4 dis/min/l.

The following description and commentary is taken from an article in *Ramparts* magazine (May 1970): "The contamination of Denver ranged from 10 to 200 times higher than the plutonium fallout deposited by all atomic bomb testing. And it was nearly 1000 times higher than the amount plant spokesmen said was being emitted . . ." The article quoted Dr. Arthur Tamplin as follows:

> A study by Dr. Edward Martell, a nuclear chemist with the National Center for Atmospheric Research in Boulder showed about one trillion pure plutonium oxide particles have escaped from Rocky Flats. These are very hot particles. You may only have to inhale 300 of them to double your risk of lung cancer. Inhaled plutonium oxide produces very intense alpha radiation dose to lung tissues, thousands of times higher than the intensity for radioactive fallout particles and millions of times more intense than the dose from natural alpha radioactivity. An inhaled plutonium oxide particle stays in your lungs for an average of two years, emitting radiation that can destroy lung tissue. If the plutonium from the May 11 fire is being redistributed as Martell suggests, then it could increase the lung cancer rate for Denver by as much as 10 percent.

The article does not mention that although the radiation is intense, the mass of tissue affected is very small, and the risk of cancer is much less than that implied by the commentator. However, the hazard of contamination by intensely radioactive particles presents a condition very different from other sources of radioactivity discussed previously, and because the implications are not well understood, contamination of this sort remains a highly controversial subject.

The Health and Safety Laboratory of the Atomic Energy Commission (now the Environmental Measurements Laboratory of the Department of Energy) conducted an independent study of plutonium contamination in the area in February 1970 (Krey and Hardy, 1970). The air concentrations decreased with half-times of approximately 1–2 years, reflecting (according to the authors) the decreasing availability of plutonium, probably as the result of penetration into the soil and/or changes in the particle size. The data were obtained from soil samples collected to a depth of 20 cm, which was considered sufficiently deep to account for total deposition of plutonium. Levels as high as 2,000 mCi/km^2 were found offsite near the plant boundary. Later it was found (Krey et al., 1977) that over 90 percent of the activity was in the first 10 cm.

Other samples were taken from shallower depths, on the basis that the plutonium contamination was more available for resuspension and inhalation. Samples were studied from the top 10 cm (Krey, 1976), top 5 cm (Krey et al., 1976a), and the top centimeter (Poet and Martell, 1972). Surface soil particles were collected by vacuuming (Krey et al., 1976b), by brushing the superficial soil from within the top 0.5 cm (Johnson et al., 1976), and by using sticky paper to collect the very top layer (Krey et al., 1977). Attempts were also made to identify the respirable fraction in the sampler. Sampling directly for the respirable fraction in soil was also attempted. Selective sampling for only a fraction of the activity may be more representative of the potential airborne hazard, but also gives results that are more variable and more difficult to interpret and generalize.

Sampling results were given in terms of activity per unit mass of soil (dis/min/g) averaged over various soil depths and as deposition per unit area (mCi/km^2). The value in dis/min/g depends on the mass over which the activity is averaged. Obviously, averaging over a 20 cm depth will give a lower value than over a 10 cm depth and will not be as valid if most of the activity is in the first 10 cm. It is useful to obtain the activity per gram of soil that is resuspended into the air because this provides better data for determining the amount that is actually inhaled. The Colorado State Health Department proposed in 1973 an interim standard for land for residential development of 2 dis/min/g in soil taken at a depth of 0–0.5 cm, which favored the part that tended to become airborne. Because the hazard of soil contamination is such a complex problem, intensive and continuous air monitoring may turn out to be the best public health measure for control and evaluation (Volchok et al., 1977). In any event, the studies conducted at Rocky Flats represent a most valuable resource for evaluating the hazards of soil contamination.

6.5   Gabon, Africa—Site of Nature's Own Nuclear Reactor

The half-life of uranium-235 is 7.1×10^8 yr. The half-life of uranium-238 is 4.51×10^9 yr. Natural uranium is 0.71 percent ^{235}U. Two billion years ago the concentration of ^{235}U in natural uranium was 3.7 percent, similar to the enrichment of ^{235}U in the fuel of a light water nuclear power reactor.

One might expect, then, that under the right circumstances a natural uranium ore-body at some time in the past could actually have undergone sustained nuclear fission reactions with the release of energy and the production of fission products. Evidence is strong that just such a phenomenon occurred two billion years ago at a place now called Oklo in the southeastern part of the Gabon republic, near the Equator, on the coast of West

Africa (Cowan, 1976). The power produced was similar to that from a six-reactor complex. Studies on the abundance of plutonium in the ground indicate the generation of some fifteen thousand megawatt years of fission energy, producing 6 tons of fission products and 2.5 tons of plutonium. Apparently there was very little migration of the plutonium. Mother Nature was not only a good power engineer, but also a good environmental engineer. Those who are saddled with the problem of the safe disposal of nuclear wastes surely cannot escape the feeling that if Mother Nature can do it, they should be able to do it as well.

6.6 Three Mile Island, Pennsylvania—A Nation Confronts the Awesome Presence of the Atom

Three Mile Island is the site of two nuclear power plants. The plants are nearly identical and each has the capacity to generate about 900 megawatts of electricity. The island is situated in the Susquehanna River 12 miles southeast of Harrisburg, Pennsylvania, which has a population of 68,000 (1970). The site is surrounded by farmland within a radius of ten miles.

The fuel in water-cooled nuclear power reactors is uranium enriched in ^{235}U from its concentration in natural uranium of 0.71 percent to between 2 and 5 percent. It generates heat by nuclear fission of the ^{235}U, which releases about 200 MeV per atom fissioned.[12] The heat is used to produce steam, which in turn drives a turbine to produce electricity. The Three Mile Island plants are known as pressurized water reactors (Fig. 6.7). The uranium atom splits up in more than 40 different ways, yielding over 80 primary fission products. These are highly radioactive, and on the average go through three subsequent decay stages before a stable species is formed. Thus there are over 200 radioactive species present among the fission products after a short time. While the reactor is operating, the decay of the fission products produces a significant fraction of the operating power, about 8 percent. This decay heat decreases after shutdown, rapidly at first, and more slowly as time passes. At 10 seconds, the decay power is about 5 percent of the reactor operating power, and this decreases to 1 percent after

---

12. The energy is divided among the kinetic energy of the two fission fragments, 166 MeV; 2.5 neutrons (average) released per fission, 5 MeV; prompt gamma rays, 7 MeV; beta and gamma energy released at a later time in the decay of the radioactive fission products, 12 MeV; and energy carried out of the reactor by the neutrinos accompanying beta decay, 10 MeV. The loss of the energy carried by the neutrinos is largely made up by the instantaneous and decay energy resulting from the capture of neutrons in the structure of the reactor. Thus it is commonly assumed that at equilibrium 200 MeV of heat are produced per fission. This converts to 3.1×10^{10} fissions/sec per watt of energy. The fission of one pound of uranium or plutonium provides as much energy as 8,000 tons of TNT, 1,000 tons of high-quality coal, or 6,000 barrels of oil.

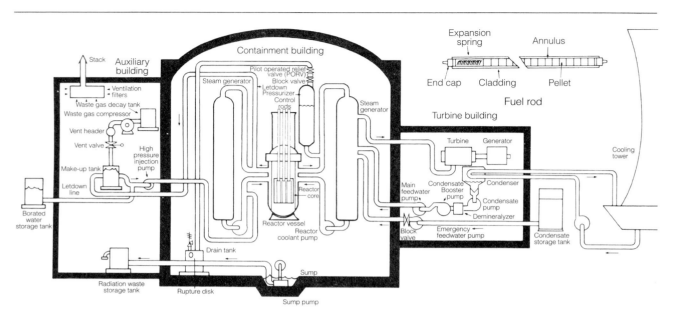

6.7 Elements of the pressurized water nuclear power plant at TMI. The heat is produced by fission in the fuel rods, which are loaded with uranium oxide pellets. The fuel rods are assembled into square arrays, which are combined to form the core of the reactor. The water is heated as it flows through the core and transfers its heat to the secondary system in the steam generator, where steam is produced to drive the turbine. The reactor has two outlet nozzles, each leading to a steam generator. The outlet of each generator on the primary side is connected with two coolant pumps, each of which is connected to an inlet nozzle at the reactor vessel. (Other designs have four loops, each with its own steam generator and coolant pump). The function of the pressurizer, which is connected to the "hot leg" of the primary coolant circuit, is to maintain the pressure of the primary coolant near the design value. Too high a pressure could result in rupture of the piping; too low a pressure, to boiling and the formation of steam in the reactor. The pressurizer volume is occupied partly by water and partly by steam; it has heaters for boiling water and sprayers for condensing steam, as needed to regulate the pressure. Emergency cooling systems, including a passive accumulator system and a high-pressure injector system, are incorporated for supplying coolant to the core through the "cold leg" in the event that the primary system fails.

To operate a plant of this capacity—1,000 Mw(e)—requires an initial loading of fuel derived from 452 tons of U_3O_8. Annual refueling at 7 percent capacity requires 200 tons of uranium ore without recycling (125 tons with recycling) and 6,400 tons during the 30-year lifetime of the plant (4,080 tons with recycling of plutonium). Fresh fuel has 3.2% ^{235}U, spent (design) has 0.9% ^{235}U, 0.6% $^{239,241}Pu$ (produced from ^{238}U at the ratio of 0.6 atoms of Pu per fission of either ^{235}U or ^{239}Pu).

Typical dimensions and specifications:

(a) Fuel pellet. UO_2, enriched to 3.2 percent in ^{235}U; cylindrical, 0.37″ diameter × 0.75″ long. Total of about 9 million in reactor.

(b) Fuel rod. Tube of zirconium-aluminum alloy, 0.0265″ thick × 0.43″ outer diameter; filled with pellets; active length 144″. Fuel assembly consists of 208 fuel rods.

(c) Core. 12′ high by 11.4′ in diameter; contains 177 fuel assemblies. Total fuel mass is 98 metric tons (average enrichment 2.6 percent). Maximum design fuel central temperature is 4,400°F and the cladding surface temperature at design power is 654°F.

(d) Pressure vessel. Typically 14′ OD, 40′ high, carbon steel walls, 8″ or more thick.

(e) Primary coolant system. Water enters core at 2,200 psia, 554°F leaves at 603°F. The total flow is 131.3×10^6 pounds per hour (124.2×10^6 pounds per hour effective for heat transfer).

1 hr. By one day, it is down to approximately 0.5 percent, the actual amount depending on the time the reactor was operating.[13]

The fission products represent an enormous amount of radioactivity. The activity of an individual radionuclide at a given time depends on its fission yield, half-life, and the operating history of the reactor. The inventories of fission products likely to produce most of the exposure of the local population in the event of an accident are given in Table 6.21 for the thermal power level of 3,000 MW required to produce 900 MW of electricity. The fuel must be encased in special cladding material to prevent the escape of the fission products to the coolant and ultimately to the environment. A large fraction of the radioactivity is contributed by radionuclides that are noble gases and build up to high pressures in the fuel during operation. Even small pinholes in the cladding will result in the escape of large amounts of these gases, since they diffuse so easily. Because it is volatile, radioactive iodine is another fission product that tends to escape from the fuel when the cladding is breached. Fission products that are normally nonvolatile solids leak out much less readily.

Because of the decay heat generated by the fission products, a nuclear reactor must be cooled even after it is shut down to prevent overheating and damage to the cladding and fuel. The cooling must continue until the radioactivity has died down to a point where it cannot produce temperatures high enough to damage the core. A catastrophic accident is possible if a reactor operating at full power loses the coolant; even if it is shut down, enough decay heat is generated to damage the cladding or even produce a meltdown, resulting in the release of the fission products.

Should a major accident occur, the final barrier to the escape of the radioactivity from the plant is the containment structure. This houses the reactor, steam generators, reactor coolant pumps, and pressurizer. The containment shell at Three Mile Island was designed to limit leakage of the radioactivity from the building to 0.2 percent per day at a maximum design pressure of 60 psi gauge. This leakage rate was used by the Nuclear Regulatory Commission for evaluating the suitability of the site in relation to the consequences of a major accident. The evaluation considered the factors of population density and land use in the region around the site.

It is routine in such evaluations to consider three particular geographical units: the exclusion area, which is the immediate area around the plant within the complete control of the reactor licensee; the low population

---

13. The rate of emission of beta and gamma energy following fission is approximately equal to $(2.8 \times 10^{-6})t^{-1.2}$ MeV/sec-fission. The decay heat power P at a time t days after startup of a reactor that was operating at a power level P_o for T_o days is readily derived from this expression to give $P = P_o \times 6.1 \times 10^{-3} [(t - T_o)^{-0.2} - t^{-0.2}]$. Here $t - T_o$ is the time in days after shutdown, that is, the cooling period.

Table 6.21 Inventories of fission products important to accident considerations after one year of reactor operation at 3,000 megawatts (thermal).

Fission product	Half-life	Inventory (MCi) At shutdown	Inventory (MCi) 1 day after shutdown	Comments
Xenon		316	186	Rare gases, large quan-
-131 m	12 days	0.9	0.9	tities released, mainly
-133 m	2.3 days	3	2.1	external gamma hazard
-133	5.27 days	162	141	
-135 m	15.6 min	48	0	
-135	9.2 hr	102	42	
Krypton		147	1	
-83 m	114 min	9		
-85	10.27 yr	0.3	0.3	
-85 m	4.4 hr	24	0.6	
-87	78 min	45	0	
-88	2.8 hr	69	0.3	
Bromine-83	2.3 hr	9	0	High volatility, external gamma hazard
Iodine		708	160	High volatility, ingestion
-129	17×10^6 yr	3×10^{-6}	3×10^{-6}	hazard (thyroid)
-131	8.1 days	75	69	
-132	2.3 hr	114	0	
-133	21 hr	165	78	
-134	52 min	189	0	
-135	6.7 hr	165	13	
Cesium-137	26.6 yr	3.8	3.8	Moderately volatile, ingestion (whole body)
Tellurium	25 min–105 days	551	107	Moderately volatile,
-132	77 hr	112	90	decays to ^{123}I
Ruthenium		82	82	High volatility under
-103	41 days	77	77	strongly oxidizing
-106	1 yr	4.6	4.6	conditions
Strontium		274	150	Relatively low volatility,
-89	54 days	117	117	hazard to bone and
-90	28 yr	3.6	3.6	lung
-91	9.7 hr	153	29	
Barium-140	12.8 days	159	144	

Source: Parker and Barton, 1973.

Note: Radionuclides with half-lives shorter than 25 min not included. Activity at shutdown calculated from equation $A = 0.0084 YM (1 - e^{0.693/T^h})$, where A is the activity in megacuries, Y is the percent fission yield (atoms/fission $\times$ 100), M is the continuous power level in megawatts, T^h is the half-life, and t is the length of time the reactor was operating.

zone, which contains a population small enough so that it can be evacuated quickly in the event of a serious accident; and the nearest population center. The exclusion area must be large enough so a person standing at the boundary would not receive more than 25 rem whole body or 300 rem to the thyroid from iodine exposure during the two-hour period immediately following the release of the fission products. The outer boundary of the low population zone must be at a sufficient distance so an individual located at any point on its outer boundary who is exposed to the radioactive cloud resulting from a postulated fission-product release (during the entire period of its passage) would not receive a total radiation dose to the whole body in excess of 25 rem or a total radiation dose to the thyroid in excess of 300 rem from inhalation of radioactive iodine.[14] The nearest boundary of a population center must be at least one and one-third times the distance from the reactor to the outer boundary of the low population zone.

Thus the examination of site suitability includes an exercise in dose calculations. A fission-product release must be assumed that is the largest that would result from any credible accident. Typically it is assumed that 100 percent of the noble gases, 50 percent of the halogens, and 1 percent of the solids in the fission-product inventory are released inside the containment building. One-half the iodines are assumed to plate rapidly on surfaces within the reactor building. One percent per day of the reactor building's contents is assumed to leak to the outside atmosphere. The calculation of the atmospheric dispersion from the reactor building to occupied areas is based on the meteorological characteristics of the area (DiNunno et al., 1962). The Three Mile Island exclusion distance was 2,000 feet and the low population zone (pop. 2,380) extended 2 miles from the plant (ONRR, 1976). The population center, Harrisburg, Pennsylvania, was at a distance well beyond the minimum required.

On March 29, 1979, at 4 A.M., Unit 2 on Three Mile Island underwent an accident involving a major loss of coolant (Kemeny, 1979; Rogovin, 1980). The feed pumps that sent condensed steam back to the steam generators were automatically shut down (tripped) for reasons not yet clearly understood. This caused the turbine to trip. With the loss of water flow, the steam generators ceased to remove heat from the reactor coolant. As a result, the coolant heated up, the pressure rose quickly, and the reactor shut down. At the same time, auxiliary pumps were supposed to turn on automatically to send water into the steam generators so they could con-

14. The NRC emphasizes that these dose limits are not intended to constitute acceptable limits for emergency doses to the public, but only to serve as reference values for the evaluation of sites with respect to potential reactor accidents of exceedingly low probability of occurrence and low risk of public exposure to radiation.

tinue cooling the water flowing through the reactor, but nothing happened; the water was blocked by two valves that had been left closed after a recent maintenance check, in violation of operating procedures. As the pressure rose in the primary system above 2,255 psi, the pilot-operated relief valve (PORV) opened and discharged water from the primary coolant loop into a quench tank to relieve the pressure. The valve was supposed to close automatically when the pressure dropped to 2,205 psi and, indeed, electric power to the solenoid that activated the valve did shut off at this point. However, the valve stuck open and water continued to pour out of the system, unknown to the operators. They were assuming that the valve was shut because the only signal at the control panel monitoring its condition, a pilot light indicating whether electric power was being supplied to the solenoid, was indicating that the power was off. The true condition of the valve was not recognized for almost two and one-half hours, during which water continued to pour out of the primary system, ultimately overflowing the quench tank and flowing onto the floor of the containment vessel. When the pressure dropped to 1,600 psi, the high-pressure injection system (HPI) automatically pumped water into the system to make up for the water being lost; however, in response to a variety of control-panel indicators, the operators wrongly decided to override this automatic emergency action and sharply reduced the HPI flow, flow that was not resumed until three hours later. About one hour into the incident, the primary coolant pumps were vibrating badly, and they were turned off manually (the vibrations were due to a lack of water in the primary system, a condition not recognized by the plant operators). At this point, water should have continued to flow by natural convection but did not, because of air in the system. Without passage of coolant water through the core, it overheated, the fuel cladding failed, and fuel melting began. A large portion of the upper part of the core was apparently uncovered at this point for an unknown period of time. After the circulation pumps were restarted, a gas bubble 1,000 cubic feet in volume was detected. It was suspected to contain hydrogen and this produced a new worry about a possible hydrogen explosion. There was also concern that the presence of the bubble could force water from the core and expose it again. However, the bubble was eventually eliminated. The reactor was cooled with the primary pumps and one steam generator until April 27, and after that the core was left to cool by convection.

The fraction of iodine-131 released from the reactor building was not 0.25, which is used in the Safety Analysis Report in accordance with assumptions required by the Nuclear Regulatory Commission, but about 0.000000003. The coolant on the floor of the containment vessel contained large amounts of radioactive noble gases and some radioactive iodine. A valve should have shut, isolating all this water within the containment. Instead, the water was erroneously pumped for a short time to an auxiliary building. This building was shielded and equipped with air filters but radioactive gases were released to the environment, exposing persons in the area. This release plus others in the following month added up to a total activity calculated to be 2.5 million curies, almost all of which consisted of noble gases (Rogovin, 1980, vol. II, part 2). Fortunately, the maximum dose to any person offsite from these releases was estimated to be less than 100 mrem. The only radionuclide released in significant quantities was xenon-133. Also released were some xenon-135 and about 15 Ci of iodine-131.

The events of Three Mile Island constituted a drama of epic proportions played before the people of the world through television and newspaper accounts. For perhaps the first time, the public became acutely aware of the almost supernatural power locked up in a nuclear reactor and the possibility of loss of control of that power. Every hint of trouble, every concern for danger, every feeling of anxiety by those operating the plant or responsible for the safety of the public was relayed almost instantaneously to the world. The President of the United States and his wife visited the plant, donned protective clothing, and entered the control room to demonstrate that the plant was under control. The Governor of Pennsylvania kept his constituents abreast of emergency plans and finally suggested evacuation of pregnant women and children near the plant. The Director of the Office of Nuclear Reactor Regulation of the Nuclear Regulatory Commission reported to the public continually from the site on the problems and progress in coping with the accident. Some reporters sensationalized the dangers of radiation exposure even though actual exposures were minimal; a scientist was shown on television making measurements with a Geiger counter in streets near the plant and interpreting very low environmental readings as the effects of a serious fallout incident. Actually, the one bit of good news was that the radioactivity was contained and the risk of cancer from radiation exposure to any member of the public was minimal. While the press had continuously played upon the chances that the reactor might melt or explode, the fact that no uranium was found in the coolant solution indicated that the fuel never melted or began to melt. But core damage was extensive. The zirconium tubing holding the fuel together was severely damaged as a result of reactions with water at temperatures of 2,700°F or higher. Perhaps 30 percent of the fuel pellets may have fallen out of place as a result.

The operators faced an enormous cleanup job. The damage to the core and possibly to the piping as a result of countermeasures to the accident were great, and it was learned later that half the reactor core had melted.

The cleanup took ten years at a cost of over one billion dollars, the aftermath of an accident in a plant that would never operate again.

6.7   Chernobyl—The Fear of a Nuclear Catastrophe That Became a Reality

A nuclear disaster of sobering proportions occurred at Chernobyl in the USSR on April 26, 1986. It was a runaway of a 3,200-megawatt (thermal) boiling water, graphite-moderated reactor. It began at 1:23 A.M. The fission rate in the reactor core suddenly shot up to hundreds of times the normal operating level. In a little over a second, the fuel temperature went from 330°C to well beyond the uranium dioxide melting point of 2,760°C (DOE, 1987). The accompanying explosion lifted a 1,000-ton cover plate off the reactor (Wilson, 1987), took off the roof, and blew fuel out. The mixture of hot fuel fragments and graphite led to some 30 fires in and around the reactor. Fission products began escaping in large amounts to the atmosphere from the huge inventory that had accumulated as the result of 5.6×10^{27} fissions over 2 years and 5 months of operations; an inventory of petabecquerel (PBq) proportions (1 PBq = 10^{15} Bq) that included 5,180 PBq (140 MCi) of molybdenum-99, 6,216 PBq (168 MCi) of xenon-133, 4,107 PBq (111 MCi) of tellurium-132, 3,034 PBq (82 MCi) of iodine-131, 229 PBq (6.2 MCi) of cesium-137, and 170 PBq (4.6 MCi) of strontium-90. A massive effort was undertaken to control the fires. The firefighters had to work in an incredibly hazardous environment of intense radiation, high temperature, toxic fumes, and escaping steam. Helicopters dropped 5,000 tons of various materials on the inferno as scientists tried out different approaches to extinguish it—40 tons of boron carbide (to prevent the reactor from going critical again), 800 tons of dolomite (to generate carbon dioxide gas), 1,800 tons of a clay-sand mixture (to smother the fire and filter the escaping radioactivity), and 1,400 tons of lead (to absorb heat by melting and provide a liquid layer that would in time solidify to seal and shield the top of the core vault). They worked frantically for 10 days and succeeded in extinguishing the fires (after 250 tons of graphite had burned up) and curtailing the release of radioactivity to the environment. The radioactivity was contained only after liquid nitrogen had been injected into the passages below the reactor core to cool the reactor sufficiently to prevent evaporation of the fission products.

Two hundred thirty-seven workers involved with the initial emergency response suffered acute radiation sickness and 31 of them died. During the first year, 200,000 workers were employed as liquidators. Their duties included cleaning up around the remains of the reactor, destruction and disposal of contaminated structures, and construction of new roads. They were allowed to accumulate 0.25 Sv (25 rem). Some accumulated their limit in just a few minutes.

When it was all over, 3,700 PBq (100 MCi) of fission products had escaped, and 1,110 PBq (30 MCi) of activity were contaminating the environment within 30 km of the reactor (Anspaugh et al., 1988). The released activity included all the noble gases, half the inventory of iodine-131 and cesium-137 and as much as 5 percent of the more refractory material, such as strontium, cerium, and plutonium. The 100 PBq (2.7 MCi) of cesium-137 ejected was about 10 percent of that released to the atmosphere in all the nuclear weapons tests. The radioactive fallout was worldwide. Most of the population dose was and would be from cesium-137, of which 37 PBq (1 MCi) fell on the European portion of the USSR, 444,000 GBq (12,000 Ci) on the United Kingdom, 1,887,000 GBq (51,000 Ci) on Italy, 281,200 GBq (7,600 Ci) on the USA, and 5,291 GBq (143 Ci) on Israel. Estimates of individual external doses that would be imparted over 50 years were 0.13 mGy (13 mrad) in the UK, 0.47 mGy (47 mrad) in Italy, 0.002 mGy (0.2 mrad) in the USA, and 0.20 mGy (20 mrad) in Israel. The dose from ingestion would produce a comparable dose over the next 50 years. One hundred fifteen thousand persons were evacuated from a 30 km zone around the reactor. Of these, 50,000 received 0.5 Gy (50 rad) or more (Anspaugh et al., 1988), including 4,000 persons subjected to an average dose of 2 Gy (200 rad). It was expected that the incidence of spontaneous fatal acute myeloid leukemia (about 1 in 10,000 per year) would increase by about 150 percent in the heavily exposed group. The direct costs of the accident (loss of the reactor, decontamination, relocation, and medical care) amounted to about 7 billion dollars.

By ten years after the accident, 463 children and adolescents had been treated surgically for thyroid cancer, about 50 times the normal rate. No significant increases had been found in leukemia and solid tumors in regions in which the fallout levels exceeded 550 kBq/m^2 compared with the uncontaminated regions, but more time was needed both for additional studies on the exposed population and for the development of adverse effects.

The determination of remedial measures to apply to contaminated land presented by far the most intractable problem resulting from the accident. Most of the population dose was from contamination of large tracts of land by cesium-137. A special commission recommended that remedial measures be required between 550 and 1,480 kBq/m^2 and relocation required for deposition above 1,480 kBq/m^2. It was estimated that the 550 kBq/m^2 limit would result in a whole-body dose of about 5 mSv per year. The Ukraine and Belarus were the provinces most affected. Seven thousand laboratories were mobilized to perform the measurements required. Two hundred thousand children were given thyroid scans and thousands of whole-body measurements for radiocesium were performed. More than 10,000 km² of land were above the 550 kBq/m². Fifty-four collective farms were taken out of production, and restrictions on the use of extensive forest areas were imposed. The government of Belarus estimated that the 30-year program required to rehabilitate the contaminated areas will cost 235 billion (U.S.) dollars (Eisenbud and Gesell, 1997).

While the Chernobyl disaster is considered by some to prove the folly of using nuclear power, the power plant design was not at all like those used elsewhere in the world. The reactor had a positive void coefficient at low power, meaning that with loss of water, the multiplication of neutrons and power level actually increased. This was a foolhardy design and not found anywhere outside the Soviet Union. Prior to the accident, the operators had deliberately disconnected some of the safety systems designed to prevent a reactor runaway. The building housing the Soviet reactor was not designed to contain any releases of radioactivity at the overpressures produced by an accident, in contrast to the protection provided by the concrete or steel structures in the United States (Ahearne, 1987).

6.8 Nuclear Power from the Perspective of the Three Mile Island and the Chernobyl Accidents

The energy locked up in the uranium nucleus is enormous. A kilogram (2.2 lb) of uranium has a heat content of 950 megawatt days (MWD) or 23 million kilowatt hours (assuming all the uranium is eventually fissioned, as in a breeder reactor). This is equivalent to 2.674 million kilograms (2,674 tons) of coal or 13,529 barrels (U.S., 42 gal) of oil. A 1,000 megawatt (electrical) power station consumes 3 kilograms (0.003 metric tons) per day of uranium if powered by fission; if powered by combustion, it consumes 10,000 tons of coal per day, delivered in 140 railcars, or 40,000 barrels of oil per day, delivered by 1 supertanker per week (Wilson and Jones, 1974). The current technology allows for only a portion of the fuel energy content to be utilized before it must be replaced. This portion of energy used, expressed as the burnup of the fuel, amounts to about 33,000 MWD per metric ton (Uranium Information Centre, (*www.uic.com.au,* September 2001). A typical refueling schedule calls for the annual delivery of 35 tons of uranium enriched to 3.3 percent in uranium-235. The daily emissions from a coal plant to the atmosphere include between 33 and 330 tons of sulfur dioxide, 55 tons of nitrous oxide, over 16,000 tons of carbon dioxide, and large quantities of nitrogen oxides, hydrocarbons, heavy metals, and fly ash, all of which produce deleterious health effects. The sulfur dioxide, although a pollutant in its own right, is converted in the atmosphere to sulfate, which is an even greater problem. A large coal plant increases the sulfate level over a wide area, which may encompass several states, by perhaps 1 μg/m^3. When superimposed on existing pollution levels, typically of the order of 12 μg/m^3 in industrialized areas (associated with ambient SO_2 levels of 80 μg/m^3), the additional emission produces significant increases in asthmatic attacks, aggravates heart and lung diseases, lower respiratory disease in children, and chronic respiratory disease symptoms, and causes a significant increase in premature deaths. The sulfates in the atmosphere also produce an acid rain that is very destructive to fish and in some cases has caused their complete disappearance from lakes. The large quantities of CO_2 introduced into the atmosphere from increasing use of fossil fuels could change the global climate and seriously affect living conditions.

Radioactivity is also released to the atmosphere from coal-burning plants in quantities that have a greater radiological significance than the radioactivity released by a normally operating nuclear plant of equal capacity (McBride et al., 1978). The hydrocarbons released to the atmosphere are known to be carcinogenic but the extent of their impact is unclear. Also generated daily at a coal power plant are 100 truckloads of ash (about 200,000 tons of ash per year) containing such toxic substances as selenium, mercury, vanadium, and benzopyrene. The wastes are dumped close to the surface of the ground, a practice that leads to pollution of the groundwater. The extremely large quantities (in terms of radioactivity, but not in volume) of radioactive wastes produced in a nuclear power plant are also a worry, and they must be carefully controlled and disposed of in sites selected for their isolation from the pathways that could lead them to public consumption.

The main concern in the utilization of nuclear power continues to be the probability of a catastrophic accident at a nuclear power plant. Of the two major accidents that have occurred to date, one, at Three Mile Island (TMI), had no significant radiological consequences for either the public or the environment. On the other hand, the Chernobyl accident was an environmental and economic disaster whose health consequences are still to be resolved.

Was the Chernobyl catastrophe as bad as it could get? The accident occurred in a reactor that was basically unsafe and had minimum protective barriers, in stark contrast to the design of reactors in other countries. Such a reactor would never have been licensed in the United States, where all information about reactor design, construction, and operation is public knowledge and the whole process is subject to intervention by members of the public. In the event of an accident, however severe, engineered safeguards are built into the plant to limit the release of fission products and the dose to the public. The reactor is housed in a strong concrete vessel with steel plate linings, the containment building, designed to withstand the high pressures that would be produced by escaping steam and the impacts of missiles that might be generated in a nuclear excursion. In addition, special systems are incorporated to prevent or limit the pressure buildup by condensing steam discharged into the building. These safeguards apparently worked well at Three Mile Island. Studies determined that no member of the public received more than 1 mGy, even though half the core melted down. But while the radiological impact on the public may have been small, it took ten years and over a billion dollars to clean up a plant that would not operate again, at least as a nuclear plant. Many lessons were learned from the TMI experience leading to the incorporation of additional safeguards in nuclear plants to prevent severe accidents in the future, and the nuclear energy industry has been working hard to demonstrate that with proper designs, training, and operational controls, nuclear power is a viable and desirable technology (Nuclear Energy Institute, *www.nei.org,* September 2001). But no new nuclear plant has been ordered in the United States since TMI, and orders for about 100 plants were canceled. Still, in 2001, there were 438 nuclear power reactors in operation in the world with an electrical generating capacity of 353,000 megawatts, contributing 16 percent of the world's capacity (Uranium Information Centre, *www.uic.com.au/reactors.htm,* September 2001). The countries with the largest nuclear electrical generating capacity were the United States (98,100 MW), France (63,200 MW), and Japan (44,300 MW). There were an additional 36 reactors under construction and 44 on order (none in the U.S. or the member states of the European Union). Nuclear reactors were providing a substantial percentage of the electricity generated (76 percent in France, 57 percent in Belgium, 34 percent in Japan, 31 percent in Germany, 22 percent in the UK, and 20 percent in the U.S.).

Perhaps, in the final analysis, the question to be answered is, "How badly do we need nuclear power?" Can the countries of the world meet their energy needs without nuclear power? Does the global warming produced by the buildup of carbon dioxide in the atmosphere from the combustion of fossil fuels warrant increased reliance on nuclear power, which does not produce greenhouse gases? The answers are not clear. What are the alternatives, their harmful effects, and their economics? Most experts in the field do not believe they know enough to say with confidence whether nuclear power should be abandoned on the basis of economics or risk to the worker or public safety. More experience is needed. To date, injury to life and property everywhere except the former Soviet Union has not been out of line with other industrial operations, and the savings in fossil fuels have been very beneficial and have certainly served to reduce the cost of those fuels. Fortunately the great risks of nuclear power are appreciated. Society realizes that the operations of nuclear facilities must be policed with great care, and a continuing surveillance must be maintained of the effects of radiation on working populations, the public, and the environment. Beyond that, only time and experience can provide the answers for both the optimists and the pessimists.

We conclude this section with some data on the contributions that can be made to our energy supply from alternative energy sources (Table 6.22). Any decisions on energy policy and justifiable risks must begin with these data.

Table 6.22 U.S. energy resources (2000).

Fuel	Annual domestic production	Annual imports	Annual electricity consumption	U.S. reserves
Quantities of fuel				
Oil	2.13×10^9 brl	3.31×10^9 brl[a]	0.195×10^9 brl 109×10^9 kwh	21.8×10^9 brl
Natural gas, liquid contents	0.698×10^9 brl	Included in "oil"		7.91×10^9 brl
Natural gas, dry	19.2×10^{12} ft^3	3.73×10^{12} ft^3	6.33×10^{12} ft^3 596×10^9 kwh	167×10^{12} ft^3
Coal	1.08×10^9 tons		0.991×10^9 tons 1965×10^9 kwh	507×10^9 tons
Uranium, U_3O_8	3.12×10^6 lb	44.9×10^6 lb	15,700 tons ^{238}U 754×10^9 kwh	136,000 tons[b] U 926 tons U^{235}
Wood, waste			64.1×10^9 kwh	
Hydroelectric			295×10^9 kwh	
Geothermal			14.2×10^9 kwh	
Wind			4.9×10^9 kwh	
Solar			0.8×10^9 kwh	
Energy content (quads) of fuel				
Oil	12.4	23.8	0.37	128[c]
Natural gas, liquid contents	2.61	Included in "oil"		46.5
Natural gas, dry	19.6	3.57	2.03	167
Coal	22.7		6.71	12,700
Uranium	8.01		2.57	68.6
Hydro	2.84		0.94	
Geothermal	0.319		0.048	

Source: Based on statistics of the U.S. Energy Information Administration, *www.eia.doe.gov* (accessed Oct., 2001).

a. Includes imports into strategic petroleum reserve.

b. At $30/lb; 900×10^6 lb estimated at $50/lb.

c. Reserve energy contents estimated from conversion factors: 1 quad = 10^{15} BTU = 2.93×10^{11} kwh = 170×10^6 barrels (42 gal) oil = 40×10^6 tons coal = 10^{12} ft^3 nat gas = 14 tons ^{235}U fissioned.

Note: Significant figures shown do not indicate accuracy of data, but are given as published.

7 Nuclear Weapons—Ready for Armageddon

July 16, 1945, saw the first test of a nuclear bomb. The successful test was the triumph of the Manhattan Project, established in 1942 by the United States government, which engaged the efforts of American and British physicists in a race to produce a nuclear bomb before it was developed by the Nazis (Rhodes, 1988).

On August 6, 1945, a nuclear bomb was exploded 1,850 feet over Hiroshima, causing 66,000 immediate deaths and tens of thousands of deaths later from the explosion and the fire storm. On August 9 a second nuclear bomb was exploded 1,850 feet over Nagasaki, causing 40,000 immediate deaths and an additional tens of thousands later. One day after the second explosion, the Emperor of Japan announced a desire to surrender, ending World War II. The war had taken the lives of 7,500,000 Russians, 3,500,000 Germans, 2,200,000 Chinese, 1,219,000 Japanese, 410,000 Yugoslavians, 320,000 Poles, 292,000 Americans, 244,000 British, and 210,000 French. The total, including casualties worldwide, added up to 20,000,000 military and civilian deaths. In addition, the war witnessed a campaign of genocide by the Nazis resulting in the murder of 6,000,000 Jews and 6,000,000 human beings of other nationalities.

The bomb exploded over Hiroshima used 64.1 kg of highly enriched uranium, comprising 80 percent ^{235}U, for the explosive. It was 10 feet long, 2 1/3 feet wide, weighed 9,000 pounds, and had an explosive power designed to be equivalent to 20,000 tons of TNT. It was dubbed "Little Boy." The method of detonation utilized a "gun type" design. Two subcritical halves at opposite ends of a long tube were fired at each other at a speed of 30 m/sec (670 mph) to produce a supercritical configuration and the fission explosion.

The bomb exploded over Nagasaki used 6.2 kg of plutonium for the explosive. It was 10 2/3 feet long, 5 feet wide, weighed 10,000 pounds, and had an explosive power equivalent to 21,000 tons of TNT. It was dubbed "Fat Man." The subcritical plutonium sphere was surrounded by charges which, on detonation, compressed it to produce supercriticality and the explosion.

In August 1949, the Soviet Union detonated its first nuclear weapon, and a nuclear arms race began with the United States. The United Kingdom exploded its first fission device, fueled with plutonium-239, in 1952. France detonated its first nuclear weapon in 1960 and China entered the nuclear club with a successful test in 1964.

The United States began work to build a superbomb, a hydrogen bomb, powered by thermonuclear fusion using deuterium and tritium (isotopes of hydrogen) in 1950. One H-bomb, incorporating a fission bomb to achieve the temperature required for fusion, had the explosive potential of a thousand fission bombs. By the end of 1952 the United States performed a successful test of an H-bomb in the Pacific, obliterating a small island and leaving a crater more than a mile in diameter (Rhodes, 1996). The Soviet Union followed up with its first thermonuclear test in August 1953, with a yield of 400 kilotons. In October 1961, it produced the most powerful explosion ever recorded, a test with an explosive power equivalent to 50 million tons of TNT (50 Mt). The British pulled off the first thermonuclear blast that met their expectations, following some disappointing previous shots, at Christmas Island in November 1957, with a yield of 1.8 Mt. The French conducted their first test of a fission bomb in 1960 and exploded a thermonuclear bomb in 1968. China conducted its first fission bomb test in 1964, with a yield of 22 kt, and its first fusion bomb test in 1967, with a yield of 3.3 Mt. Over the years, an enormous number of nuclear weapons were accumulated by the declared and undeclared nuclear states (for more details, see the Federation of American Scientists website, *www.fas.org*).

On June 18, 1979, the President of the United States of America, Jimmy Carter, and the President of the U.S.S.R. Supreme Soviet, Leonid I. Brezhnev, signed a treaty on the limitation of strategic offensive arms (SALT II). The treaty expressed the "deep conviction that special importance should be attached to the problems of the prevention of nuclear war and to curbing the competition in strategic arms." Implementation of the treaty required ratification by the United States Senate. The treaty was never ratified because of widespread concern that it in fact compromised the national security, but both signatories complied with its provisions.

The treaty placed a ceiling of 2,250 on the number of land-based intercontinental ballistic missile (ICBM) launchers, submarine-launched ballistic missile (SLBM) launchers, heavy bombers, and air-to-surface ballistic missiles (ASBM) allowed each side by January 1, 1985. Under the overall ceiling the number of ICBMs with multiple warheads (MIRVs) would be limited to 820; the number of ICBMs and SLBMs with MIRVs would be limited to 1,200; and the number of multiple-warhead ICBMs and SLBMs and bombers with cruise missiles would be limited to 1,320. The treaty did not regulate the maximum number of nuclear warheads but limited the number of multiple warheads on a single missile to the number already tested, a maximum of 10 on land-based weapons and 14 on those based on submarines. Heavy bombers were restricted to a total of 28 cruise missiles. Each party undertook not to flight-test or deploy new types of heavy ICBMs.

While the SALT treaty was designed to reduce and limit the tremendous stockpiles of nuclear weapons of both signatories (ISS, 1978), the de-

struction that could be wrought by even a single nuclear missile was beyond comprehension. The largest of the land-based ICBMs possessed by
the Soviet Union, the SS-9 and SS-18 with a throw-weight of 6–10 tons,
could carry one 18–25 megaton (Mt) or several smaller megaton warheads.
The largest land-based ICBM possessed by the United States, the Titan II,
had a throw-weight of 3.75 tons and could carry one 5–10 Mt warhead.
Bombers could carry warheads rated between 5 and 400 kt. Nuclear submarines fitted with up to 24 ballistic missiles armed with as many as 10
nuclear warheads per missile, and warheads with the explosive power between 100 and 2,000 kilotons of TNT, patrolled the oceans.

Another major milestone in arms control was reached on January 3,
1993, when the START-2 treaty (Strategic Arms Reduction Treaty) was
signed by U.S. President George Bush and President Boris Yeltsin of the
Commonwealth of Independent States. The treaty was ratified by the U.S.
Senate on January 26, 1996, and by the Russian Duma on April 14, 2000.
It provided for the eventual elimination of all multiple-warheaded (MIRV)
ICBMs. Only ICBMs carrying a single warhead would be allowed. It also
reduced the total number of strategic nuclear weapons deployed by both
countries by two-thirds below pre-START levels—to between 3,000 and
2,500 warheads. Multiple warheads could be deployed on SLBMs, but the
total number was limited to between 1,700 and 1,750.

The limitations and reductions must be completed by December 31,
2007. Table 6.23 lists the weapons arsenals of the five major nuclear
powers.

Yields of strategic nuclear weapons fall roughly into classes of 10 Mt, 1
Mt, 200 kt, and 50 kt. The Hiroshima and Nagasaki bombs were in the 20
kt class; the biggest bombs ever made from conventional explosives contained the equivalent of 10 tons of TNT. Most of the U.S. land-based missiles are approximately 300 kt; the yield of the land-based Russian missiles
is about 500 kt. The submarine-launched missiles range between 100 and
500 kt. Both the U.S. and Russia have missiles that carry several independently targeted warheads (MIRV), but these are to be replaced eventually
by single warheads in accordance with the START II treaty.

What would be the effects of detonating a 1 Mt bomb at an altitude of
1,000 m or so above the ground (giving the greatest blast and thermal effect)? The energy released, equivalent to that from one million tons of
TNT, is 4.6×10^{15} joules. But while the power of a nuclear explosion is
expressed in terms of the blast effect of an equivalent tonnage of TNT, the
uniqueness of this weapon is in the incredible heat of the fireball—a man-
made sun that destroys by heat radiation (35 percent of the total energy released) all that has not already succumbed to the blast and nuclear radia-

Table 6.23 Strategic nuclear weapons of the five major nuclear powers (2001).

Type	U.S.	Russia	U.K.	France	China
Weapons in stockpile	10,500	20,000	185	470	400
ICBM					
Launchers	550	760			>40
Warheads	2,000	3,544			128
Warheads/missile	3–10	1–10			1
Yields (kt)	170–335	550–750			200–5,000
SLBM					
Nuclear submarines	18	17	4	4	1
Launchers	432	348	16/sub	16/sub	12/sub
Warheads	3,456	1,576	185	288	12
Warheads/missile	8	3–10	1–3	6	1
Yields (kt)	100–475	100–500	100	100–300	200–300
Bombs/air-launched missiles					
Bombers	72	78		84	130
Warheads	1,750	898		60	130
Yields (kt)	5–150			300	10–3,000

Sources: NRDC Nuclear Notebook, Bulletin of the Atomic Scientists, March, 2001 (U.S.); May, 2001 (Russia); Sept. 2000 (U.K.), July, 2001 (France); Sept., 2001 (China). Also *www.bullatomsci.org; www.nrdc.org; www.fas.org.*

Note: ICBM, intercontinental ballistic missile; SLBM, submarine-launched ballistic missile; NRDC, National Resources Defense Council; FAS, Federation of American Scientists.

tion. Perhaps 50 percent of the deaths at Hiroshima and Nagasaki were caused by burns from the fireball, whereas 10 percent resulted from radiation sickness. Intense heat produced charring and blackening of trees, wood posts, and fabrics up to 10,000 ft from ground zero. Combustible materials were set afire up to 3,500 feet away. At Hiroshima the burning developed into a fire storm. The storm began about 20 minutes after the bomb burst. Winds blew toward the burning area of the city from all directions, gradually increasing in intensity and reaching a maximum velocity of 30–40 mph some two to three hours after the explosion. The winds were accompanied by intermittent rain. The strong inward draft did limit the spread of the fire beyond the initial ignited area, but virtually everything combustible within the boundary of the storm was destroyed.

The fireball is the result of the production of a tremendous amount of energy in a small volume. It is preceded by shock waves that demolish and crush with far greater effectiveness and range than conventional explosives. The enormous concentration of energy then creates very high temperatures, which at the time of the detonation are thousands of times greater than the 5000°C maximum produced by a conventional explosive weapon

(Glasstone, 1962). At these high temperatures, reaching 100 million degrees, about one-third of the energy is converted to low-energy x rays.[15] This energy is absorbed in the air (by the photoelectric effect) within a distance of a few feet, and it heats the air to thousands of degrees, producing the fireball. The fireball immediately grows in size and rises rapidly, initially at 250 to 350 feet per second (over 200 mph). The fireball from a 1-megaton weapon extends about 440 feet across within 0.7 milliseconds and increases to a maximum diameter of about 7,200 feet in 10 seconds. After a minute, it is 4.5 miles high and has cooled to such an extent that it no longer emits visible radiation; from here on, vapor condenses to form a cloud. If sufficient energy remains in the cloud it will penetrate the tropopause into the stratosphere. The cloud attains its maximum height after about 10 minutes and is then said to be stabilized. It continues to grow laterally, however, to produce the mushroom shape that is characteristic of nuclear explosions. The cloud may continue to be visible for about an hour or more before being dispersed by the winds into the surrounding atmosphere, where it merges with natural clouds in the sky. In contrast to a megaton burst, the typical cloud from a 10-kiloton air burst reaches a height of 30,000 feet with the base at about 15,000 feet. The horizontal extent is also roughly 15,000 feet.

The potential effects of the blast and nuclear radiation have been given intensive study (York, 1976). Wooden homes and most of their inhabitants are destroyed by blast and fire as far as 9 km from detonation. Nuclear fallout sufficient to kill most persons in the open extends to 100 km, and to 25–50 km for persons sheltered in basements of buildings. The way the fallout fans out is shown in Figure 6.8 for 1, 5, and 25 Mt bombs (DCPA, 1973). What would this do to an industrial city large enough to warrant the use of very large weapons? The consequences have been worked out for the city of Detroit (OTA, 1979). First, consider a 1-megaton detonation on the surface, which produces the greatest fallout. The detonation point is the City's Civic Center. The explosion leaves a crater about 1,000 ft in diameter and 200 ft deep, surrounded by a rim of highly radioactive soil about twice this diameter thrown out of the crater. There is not a significant structure standing less than 1.7 miles from the Civic Center

---

15. The energy radiated from a black body as a function of temperature is given by Planck's equation: $J_\lambda = (c/4)(8\pi hc\lambda^5)(e^{E/kt} - 1)^{-1}$ (Glasstone, 1962). J_λ is the energy (ergs) radiated per cm^2 per sec per unit wavelength (in angstroms); E is the energy of the photon of wavelength λ; and the other constants are given in Appendix III. The temperature produced when the bomb is triggered is between 10^7 and 10^8 K. At this temperature most of the thermal radiation is in the 120 to 0.12 keV energy region. It is absorbed in the air with a half-value layer of $E^3/7$ cm, E in keV. The resulting fireball has a surface temperature of about 8,000 K. This radiates mainly in the ultraviolet, visible, and infrared regions of the spectrum, very much like sunlight at the surface of the earth.

6.8 Fallout patterns: peak dose rates (R/hr) and time of peak for 15 mph effective wind. The 1 MT, 5 MT and 25 MT cases cover the yield range of cold war–era warheads. The values are calculations of maximum dose rates that would be measured at three feet above a smooth, infinite plane. The actual dose rates would be less than shown here by a factor of 2 or perhaps more because of the roughness of the surfaces on which fallout had deposited, as well as the limited extent of these surfaces. Also shown, by curved vertical lines, is the time after detonation, in hours, at which the dose rate would attain its maximum. Note that the point 30 miles downwind is always in the heaviest fallout area. *Source:* DCPA, 1973.

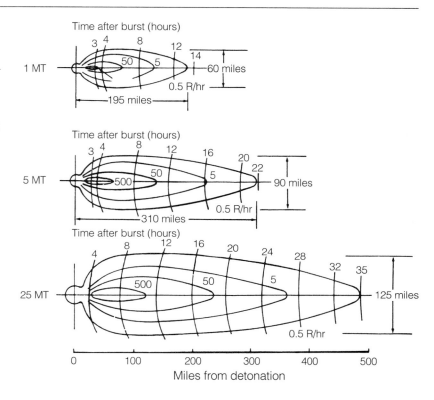

(overpressure 12 psi). During working hours, the entire downtown population of 200,000 is killed. (After working hours, there are 70,000 people around to get killed.) Between 1.7 and 2.7 miles (overpressure 5 psi), the walls are completely blown out of multistory buildings; total destruction is the fate of individual residences; some heavy industrial plants remain functional at the outer distance. Fifty percent of the people are killed, the rest injured. Between 2.7 and 4.7 miles from ground zero there are 50 percent casualties (mostly injuries), and between 4.7 and 7.4 miles, 25 percent casualties (few deaths) with only light damage to commercial structures and moderate damage to residences. The attack causes a quarter-million deaths, plus half a million injuries, destruction of 70 square miles of property (subject to overpressures more than 2 psi), and additional damage from widespread fires. The radioactive fallout pattern depends on the wind velocity and the distance from the explosion. The fallout is most dangerous during the first few days. Possible dose distributions are shown in Figure 6.9. People in the inner contour not in adequate fallout shelters receive

6.9 Main fallout pattern of a 1-megaton surface burst in Detroit. Uniform 15 mph northwest wind. Contours for 7-day accumulated dose (without shielding) of 3,000, 900, 300, and 90 rem. Detonation point assumed at Detroit Civic Center. (*Source:* OTA, 1979.)

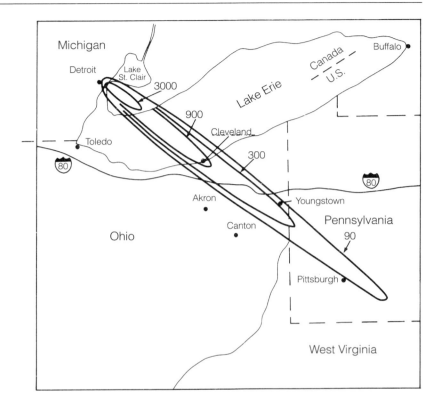

fatal radiation doses in the first week. People in the second contour receive a fatal dose if they spend much time outdoors and incur severe radiation sickness even if they remain indoors but not in the shelter.

If Detroit is subjected to an air burst at 6,000 feet rather than at the surface, the area affected by blast and fire increases. There is no significant fallout and no crater, and the strongest structures may survive in part even directly under the blast. In the event of a 25-megaton air burst at 17,500 feet, the whole metropolitan area is heavily damaged. There are few survivors (1.1 million available to assist 3.2 million casualties). There is virtually no habitable housing in the area and essentially all heavy industry is totally destroyed. As a result, rescue operations would have to be totally supported from outside the area, with evacuation of the survivors the only feasible course.

A full-scale attack could subject the United States to firepower of 6,000 Mt, two-thirds dropped on military targets and one-third on civilian targets, that could result in one hundred million deaths (JCAE, 1959a; Mark,

1976; CFR, 1975). If the attack were limited to strategic targets, perhaps involving 2,000 Mt, the deaths might be reduced to seven million. Some analysts study the consequences of a major nuclear war and prescribe appropriate civil defense measures (Kahn, 1959); others feel that the survivors would envy the dead, that civil defense is meaningless, that every effort should go into the prevention of nuclear war. A warning by the Secretary General of the United Nations presents this precarious state of affairs very well (UNSG, 1968): "There is one inescapable and basic fact. It is that the nuclear armories which are in being already contain large megaton weapons every one of which has a destructive power greater than that of all the conventional explosives that have ever been used in warfare since the day gunpowder was discovered. Were such weapons ever to be used in number, hundreds of millions of people might be killed, and civilization as we know it, as well as organized community life, would inevitably come to an end in the countries involved in the conflict."

PART SEVEN

Exposure to Nonionizing Electromagnetic Radiation

Does the public need to be protected from nonionizing electromagnetic radiation, radiation with frequencies below those able to produce ionization? If so, what amount of exposure is "too much"? What kind of protection is needed? These are questions that trouble many people as the environment becomes ever increasingly flooded with electromagnetic radiation from TV and radio and radar antennas; as millions of people come in close proximity to localized sources of microwave radiation whenever they use their cellular phones; and as alarms are raised about the potential effects of exposure to electromagnetic fields that surround electric power transmission lines.

While a great deal of experience has been gained in providing protection from sources of ionizing radiation, the distribution of the sources and the populations requiring protection are quite different in the case of exposure to sources of nonionizing radiation. Where exposure to ionizing radiation generally occurs at industrial, medical, and research installations and may be controlled and confined to localized regions, nonionizing electromagnetic radiation (hereinafter referred to as electromagnetic radiation) is used to transmit and receive power and information in the home, in the automobile, at work—essentially everywhere in the industrialized world. The electromagnetic energy is imparted not only to electronic receivers designed to detect and process the information but to human beings, who absorb the energy but are not aware of its presence, at least for most parts of the electromagnetic spectrum.

It should be borne in mind that while electromagnetic fields produce physical effects on objects they reach, by far the main effect and the main concern is the information they carry to human beings. This information has an enormous impact on the behavior of multitudes of recipients—behavior that may be of great benefit or great harm to themselves and to society—and the consequences of these behavioral effects can be of much greater significance than any physical effects the fields may produce. The scientific evaluation and control of these effects, however, is beyond the scope of this book, and we shall leave this problem to the domain of sociologists, psychologists, and politicians.

1 ELECTROMAGNETIC FIELDS—QUANTITIES, UNITS, AND MAXWELL'S EQUATIONS

1.1 The Electric Field

The source of electric fields in space is the electron, that tiniest of particles, with a mass $m = 9.109 \times 10^{-31}$ kg and an electric charge $q = 1.602 \times 10^{-19}$ coulomb. It takes 6.24×10^{18} electrons to produce a charge of 1 coulomb. The electron played a central role in our treatment of ionizing radiation.

The magnitude of the electric field at a point is defined as the electrical force in newtons (N) exercised on a positive charge of 1 coulomb (C) at that point, and that force is also exercised in a specific direction. Thus, the electric field at a point has both magnitude and direction, it is a vector.

Mathematically, the production of an electrical field by a distribution of electrical charges is described by one of four differential equations published by James Clerk Maxwell, in his *Treatise on Electricity and Magnetism* in 1873, that describe the nature of electromagnetic fields in terms of space and time.

$$\frac{\partial \varepsilon E_x}{\partial x} + \frac{\partial \varepsilon E_y}{\partial y} + \frac{\partial \varepsilon E_z}{\partial z} = \rho \qquad (7.1)$$

ρ is the density of electric charge at a point in space and can be positive or negative, ε is the permittivity of the medium, and E_x, E_y, and E_z are the components along the x, y, and z axes of the electric field vector E. This equation follows from the observation that electric fields are produced from positive and negative electric charges, and that the generated fields begin and end on charges.

The electric field in air at a distance r from a single point charge of magnitude q is given by Coulomb's law as

$$E = (1/4\pi\varepsilon_0)(q/r^2) \text{ newtons/coulomb}$$

ε_0 is the permittivity of free space (8.85×10^{-12} C^2/N-m^2, so

$$1/4\pi\varepsilon_0 = 8.99 \times 10^9 \text{ N-m}^2/\text{C}^2$$

Note that the intensity falls off inversely as the square of the distance.

Example 7.1   The distance of the electron from the nucleus in the Bohr theory of the hydrogen atom is 0.529×10^{-8} cm. What is the magnitude of the electric field at the orbit of the electron in the ground state?

The nucleus of the hydrogen atom has just one proton with a positive charge of 1.60×10^{-19} C. The electric field at the distance of the electron is (8.99×10^9 N-m^2/C^2)(1.60×10^{-19} C/(0.529×10^{-10} m)^2) = 5.14×10^{11} N/C

Electric field intensity is normally given in units of volts per meter (V/m) rather than newtons per coulomb. Both are SI units and the numerical values are identical, so the electric field has the very large value of 5.14×10^{11} V/m. Table 7.1 gives some examples of electric fields.

There are many ways in which accumulations of charge are built up—rubbing together certain dry materials, as when removing synthetic fabrics from a dryer; atmospheric processes, which may lead to the accumulation of enormous quantities in clouds; and commercial processes, such as xerography, in which electrically charged powders are sprayed on a drum and then transferred to paper. Electrostatic fields are normally not dangerous—they cause shocks that can startle but do not produce injury—but high voltages and large accumulations of charge, as may occur in thunderstorms, may be life-threatening.

Table 7.1   Examples of electric field strengths.

Field	Value (N/C or V/m)
At the electron orbit of hydrogen atom	5×10^{11}
Electric breakdown (lightning) in air	1,000,000
At the charged drum of a photocopier	100,000
High-voltage power line	10,000
Near a charged plastic comb	1,000
Electric blanket	200
In the lower atmosphere	100
Inside the copper wire of household circuits	0.01

Source: Halliday and Resnick, 1988 (in part).

1.2 The Magnetic Field

There are no free magnetic charges corresponding to single electric point charges. Mathematically this is described by one of Maxwell's equations as

$$\frac{\partial B_x}{\partial x} + \frac{\partial B_y}{\partial y} + \frac{\partial B_z}{\partial z} = 0 \qquad (7.2)$$

which is similar to Maxwell's equation for electric fields (7.1) but does not have a term for the charge. The quantity B describes the magnetic field and is called the magnetic flux density. It is defined in section 1.4. The equation indicates that a path along the direction of the magnetic field always follows a closed loop, since it does not start or end on a magnetic charge.

1.3 Maxwell's Equation for Faraday's Law of Induction

Changing magnetic fields produce electric fields, which affect electric charges. Mathematically, this is described by a third Maxwell equation:

$$\frac{\partial E_z}{\partial y} - \frac{\partial E_y}{\partial z} = -\frac{\partial B_x}{\partial t} \qquad (7.3)$$

where the subscripts pertain to the components of the electric and magnetic fields along the x, y, and z axes; there are corresponding equations for B_y and B_z. (Static magnetic fields have no effect on electric charges at rest.)

This equation is the mathematical expression for Faraday's law of induction, describing the production of electrical fields from changing magnetic fields. It provides the theoretical basis for the electrical generator, as well as the production of potentially harmful electrical fields and associated currents in the human body exposed to varying magnetic fields. This equation also states that the E field circles around the B field.

1.4 Maxwell's Equation for Ampere's Law as Modified for the Displacement Current

Magnetic fields arise from electric charges in motion (such as electric currents flowing through a wire) and from changing electric fields. They exercise a magnetic force on other charges in motion or electric currents. This property is described by the fourth of Maxwell's equations, which also serves as the theoretical basis for the action of the electric motor.

$$\frac{\partial B_y}{\partial z} - \frac{\partial B_z}{\partial y} = \mu_0 \left(j_x + \frac{\partial \varepsilon E_x}{\partial t} \right) \qquad (7.4)$$

This equation, without the $\partial \varepsilon E/\partial t$ term, is an expression, in the form of a differential equation, of Ampere's law, which describes the magnetic field around an electrical current. It refers to the projection of the current density J in the x direction (j_x) and states that the resultant magnetic field has components in the y and z directions, that is, perpendicular to the direction of travel of the current. The constant μ_0 is the permeability constant of free space. Similar equations can be written for projections of the current density in the y and z directions.

The $\partial \varepsilon E/\partial t$ term (the x component is shown in Equation 7.4) was added by Maxwell as a generalization of Ampere's law. It relates the production of a magnetic field by a changing electric field to the conduction current. Because it appears along with the conduction current term, $\partial \varepsilon E/\partial t$ is referred to as a *displacement current.*

The magnetic flux density, $B,$ is defined as the force F exerted on a small moving test charge Q moving with a velocity v divided by the product of the charge q and the velocity, $B = F/qv.$ The direction of the force is perpendicular to the direction of the flux density and the velocity (in vector notation, $\boldsymbol{F} = q\boldsymbol{v} \times \boldsymbol{B}$). The direction of $\boldsymbol{B}$ is the same as the direction in which a compass needle would point. The SI unit for B is the tesla (T). From the equation for $F,$ 1 tesla = 1 newton/(coulomb-meter/second) or 1 coulomb/ampere-meter. Another commonly used unit for $\boldsymbol{B}$ is the gauss (G): 1 tesla = 10^4 gauss, $1\,\mu T = 10$ mG.

The magnetic flux density in free space at some distance r from a long straight wire through which a current of electrons i is flowing is given by the equation

$$B = \mu i/2\pi r \tag{7.5}$$

$\mu,$ the permeability constant, has the value $\mu_0 = 4\pi \times 10^{-7}$ T-m /A for free space, which is also very closely equal to the permeability of air and essentially all biological media.

In addition to the magnetic flux density, $B,$ given above, magnetic fields are also described by the magnetic field strength $H.$ It is defined mathematically as a vector point function whose negative line integral over any closed-line path is equal to the current enclosed by the path, irrespective of the permeability of the medium. H is measured in terms of the current per unit distance and has units of amperes/meter (A/m). It is equal to the quotient B/μ_0 in air or biological media. We shall follow a common practice of referring to both the flux density, $B,$ and the magnetic field strength, $H,$ as measures of the magnetic field. The conversion from H to B units is 1 A/m = 12.57 milligauss (mG). Magnetic fields that accompany the transmission of electromagnetic radiation through space are normally expressed in terms of H (A/m), while the nonradiating fields associated with electric currents are normally expressed in terms of B (G or mG).

1.5 The Interactions of Electric and Magnetic Fields in a Medium

Electric *(E)* fields act on the charges in a medium. The nature of the interaction varies, depending on the type of medium, whether it is primarily a dielectric or a conductor, or whether it exhibits the properties of both.

Dielectrics are insulating media in which the charges do not move freely but are bound to each other and either neutralize each other or are slightly separated, forming dipoles. Where the positive and negative bound charges neutralize each other, an E field separates them slightly, producing dipoles that create new fields that did not exist previously. Where electric dipoles exist in dielectrics (as polar molecules), in the absence of an applied E field they are randomly oriented and do not affect their surroundings electrically. When an E field is applied, they tend to align with the field, also producing new fields.

Materials that are primarily conductors contain "free" charges, such as electrons in metallic conductors and positive and negative ions in electrolytes. In the absence of an applied E field, these charges move randomly under thermal excitation, with no preferred direction and thus no current. An applied field causes them to drift in the direction of the field, producing a conduction current. This current produces new fields that did not exist before the E field was applied.

In an ideal dielectric, all the energy required to establish an electric field is stored in the field and recoverable as electromagnetic energy when the field is removed; that is, there is no absorption and production of heat, even in the presence of displacement currents. In conductors, the flow of the current against the resistance of the conductor, often referred to as friction encountered by the moving charges, results in the expenditure of energy and the heating of the conductor. There is no storage of energy in the ideal conductor.

However, the only ideal dielectric is a perfect vacuum, and all conductors store some energy by virtue of their capacitance. There are very good, if not ideal, dielectrics, such as polystyrene, rubber, and (to a lesser extinct) distilled water. There are very good conductors, including solid metals, such as copper, and (to a lesser extent) electrolytes, such as an aqueous solution of potassium chloride. In copper, the charge is carried by electrons; in aqueous potassium chloride, the charge is carried through a water medium by positive potassium ions and negative chlorine ions.

The degree to which electric dipoles are induced or aligned with the applied field in dielectrics is determined by the *permittivity*, ε. The magnitude of the current density, J, as produced by the drift of electric charges in conductors under the action of an electric field, E, is determined by the *conductivity*, σ (substitute σE for j in Maxwell's equation for Ampere's law, Equation 7.4).

Normally, a good conductor is adequately characterized by σ, while a good dielectric is adequately characterized by ε. However, a complex medium like tissue has both conductive and dielectric properties and is described as a "lossy" or leaky dielectric. These properties can be combined in a single quantity called the *complex permittivity*, a complex number

$$\varepsilon = \varepsilon_0(\varepsilon' - j\varepsilon'') \tag{7.6}$$

The complex permittivity replaces the permittivity in the displacement current term in Equation 7.4. The real part, ε', is the dielectric constant, a measure of the polarization. It is not associated with the loss of energy. The imaginary part, ε'', is a measure of the resistive properties, or friction, and accounts for energy losses in the medium. These losses result from the movements of charge in the production of dipoles and in the rotation of existing dipoles, as well as in the flow of conduction currents, and they can produce significant heating in biological tissue. The complex permittivity is based on interaction processes that are frequency dependent and therefore is itself frequency dependent (NCRP, 1981b; HPS, 1997; Brown et al., 1999).

A static magnetic field also exerts forces on moving charges, but since these forces are always perpendicular to the velocity of the charges, no energy is transmitted to them. When B, the magnetic flux density (also called the magnetic field), is changing with time, however, it induces an E field that does transmit energy to charges, as given by Maxwell's equations. The main effect of B on electric charges, therefore, is a secondary one through the induced electric field. The principal direct effect of B fields on materials is to align partially any magnetic dipoles that may be present in the material. Because tissue is essentially nonmagnetic, it contains only a negligible amount of magnetic dipoles. Therefore the direct effect of B on tissue is usually negligible.

2 INTERACTION OF FIELDS FROM ELECTRIC POWER LINES WITH THE BODY

Electrical power is transmitted through transmission lines as alternating currents in which the electrical charges oscillate at 60 cycles per second (60 Hz) in North America and 50 Hz elsewhere. These frequencies are classified as *extremely low frequencies* (ELF), a class which ranges from 30 to 300 cycles per second. The $\partial \varepsilon E / \partial t$ term in equation 7.4 is negligible, which means that B is generated only by currents as given by Ampere's law for the magnetic fields around electrical currents and E is described separately by Coulomb's law for electrical fields generated by electrical charges. The methods used to assess the electric and magnetic fields are those of electric circuit theory. Radiation is insignificant (about 0.000002 percent at a distance of 100 m) and the energy in the electrical and magnetic fields drops off quickly with distance from the current circuits.

If a conductor is placed in a static electric field, its electrons are redistributed on the surface until they produce a field that just cancels the original at all points within the conductor. Transient currents will appear in the conductor while the charge is redistributing, but they will disappear when the field is reduced to zero. If the electric field oscillates at low frequency, as in the vicinity of a power line, it will continue to generate currents and associated electric fields in the conductor, but the fields will be only a very small fraction of the applied field. Since the human body is a good if complex electrical conductor, it also experiences internal electric fields near power lines, but they are very small, less than a millionth of externally applied fields (Repacholi, 1988).

Exposure to a vertical E field from power lines of 10 kV/m would result in a current density in the chest of about 190 nA/cm^2 (Tenforde and Kaune, 1987). This is of the order of the average values of natural current densities due to electrical activity in the brain and heart (IRPA, 1990). Assuming a conductivity σ of 0.5 S/m, this current density would be related to an internal E field of about 4×10^{-3} V/m, a reduction of about 1×10^{-7} times compared with the external field. The coupling of the external and internal fields is proportional to frequency; the current at 60 Hz would be about 20 percent greater than the current at 50 Hz. The effect on the cell membrane is much larger; the electric field strength at the cell membrane will be about 3,000 times larger than the internal electric field (Hitchcock, 1994). Some believe that these induced fields produce biological effects, although they are smaller than the thermal noise in the cell membrane.

The external magnetic lines of force are not perturbed by the body and are much more penetrating than the electric field lines. As the magnetic lines of force cut through the body, they induce currents to flow at right angles to the direction of the field. These currents flow in closed loops, producing circulating eddy currents.

To obtain an idea of the magnitude of the currents that are induced in the body by magnetic induction at power line frequencies and flux densities, assume a copper ring is immersed in a sinusoidal magnetic field and that the magnetic flux ϕ is incident perpendicular to the plane of the ring. The changing magnetic field will induce an electric field, and thus current in the ring. Faraday's law of magnetic induction tells us that the integral of the electric field intensity around a closed curve—that is, the induced emf or voltage, V—is equal to the negative time rate of change of the flux through the loop:

$$V = \int E \cdot \mathrm{d}l = -\mathrm{d}\phi/\mathrm{d}t = -\mathrm{d}/\mathrm{d}t \int\int B \cdot \mathrm{d}S$$

The bold characters represent vectors. Since the induced electric field is constant around the loop and always tangent to it, $\int E \cdot \mathrm{d}l = 2\pi r E$. The flux density B is constant through the loop so $\int\int B \cdot \mathrm{d}S = \phi = \pi r^2 B$. For a sinusoidal magnetic field with frequency ν, $B = B_0 \sin 2\pi\nu t$ and $\mathrm{d}\phi/\mathrm{d}t = \pi r^2 2\pi\nu B_0 \cos 2\pi\nu t$. Thus, $2\pi r E = -\pi r^2 2\pi\nu B$ and (neglecting signs) $E = \pi r \nu B$.

The current density, J, induced in a ring with conductivity σ, is $J = \sigma E$.

For a circular loop of radius 5 cm immersed in a uniform 3 mG magnetic field from a 60 Hz line, and converting to SI units, the induced electric field $E = \pi (0.05 \text{ m})(60/\text{sec})(0.3 \times 10^{-6} \text{ T}) = 2.83 \times 10^{-6}$ V/m.

The conducting ring is not necessary for the induction of the electric field. The field would exist even in the absence of the ring—it would exist in space or in a conducting object like the body. Since the conductivity of tissue averages about 0.2 S/m, circulating current densities (eddy currents) of the order of $(0.2 \text{ S/m})(2.83 \times 10^{-6} \text{ V/m}) = 0.57 \times 10^{-6}$ A/m^2 would be produced in tissue under similar geometrical conditions. Calculations pertaining to exposure of the human body are much more complicated, but give circulating current densities in the range of μA/m^2 from characteristic environmental extremely low frequency (ELF) magnetic fields.

The minimum current density that is required to stimulate excitable cells is about 0.1 mA/cm^2 in the ELF range. Calculations using an ellipsoidal approximation of a man give the estimated magnetic flux density required to induce this current density as 0.07 T at 60 Hz, much larger than magnetic flux densities produced by electric power facilities (Kaune, 1986).

3   THE PHYSICS OF RADIATING ELECTROMAGNETIC FIELDS

When the frequency is high enough, all the terms in Maxwell's equations are significant and the electric and magnetic fields become strongly coupled together. One cannot exist without the other, and Maxwell's equations produced an astonishing revelation: electromagnetic energy can be radiated into space through the utilization of electric circuits.

3.1 The Derivation of Equations for Electromagnetic Waves from Maxwell's Equations

When Maxwell's equations are applied to free space, that is, when ρ and J are both 0, two equations can be derived from them for E and B. These are:

$$\frac{\partial^2 E}{\partial x^2} + \frac{\partial^2 E}{\partial y^2} + \frac{\partial^2 E}{\partial z^2} = \frac{1}{c^2} \frac{\partial^2 E}{\partial t^2} \tag{7.7}$$

$$\frac{\partial^2 B}{\partial x^2} + \frac{\partial^2 B}{\partial y^2} + \frac{\partial^2 B}{\partial z^2} = \frac{1}{c^2} \frac{\partial^2 B}{\partial t^2} \tag{7.8}$$

The solutions of these equations reveal the transmission of electromagnetic radiation through space as waves. The speed of the radiation is given by the constant c and is equal to $1/(\varepsilon_0 \mu_0)^{1/2}$.

Example 7.2 Calculate the value of the speed of electromagnetic radiation through space as obtained from Maxwell's equations. ε_0 is the permittivity of free space (8.85×10^{-12} C^2/N-m^2, μ_0 is the permeability constant ($4\pi \times 10^{-7}$ T-m/A).

$$c = 1/(\varepsilon_0 \mu_0)^{1/2} = 1/(8.85 \times 10^{-12} \times 4\pi \times 10^{-7})^{1/2}$$
$$= 3.0 \times 10^8 \text{ m/sec}$$

This is exactly equal to the speed of light, a coincidence of tremendous significance when it was first recognized. The solutions to Maxwell's equations did not limit the frequency (or the wavelength) of the radiation to any particular region of the spectrum, but all the radiation traveled with the speed of light and represented the propagation in space of fields identical to those associated with electric circuits. The electric and magnetic fields in free space are transverse fields, that is, they are perpendicular to the direction of travel and perpendicular to each other. Thus Maxwell concluded that light also consisted of the "transverse undulations of the same medium which is the cause of electric and magnetic phenomena" (Halliday and Resnick, 1988, p. 847). So did the neighboring regions of the optical spectrum, infrared radiation at the low end and ultraviolet light at the high end. Eight years after Maxwell's death, in 1887, Heinrich Hertz generated radiowaves in the laboratory, and as the years went by radar and microwaves joined the family. With the discovery of radioactivity and the associated radiations, gamma rays and x rays were added to a spectrum that seems to know no bounds with respect to frequency or wavelength (Fig. 7.1)

3.2   Electromagnetic Waves Generated by a Sinusoidal Oscillator

While we have used the magnetic flux density B as a measure of the magnetic field, it can be characterized just as well by the magnetic field strength H. The two quantities differ only by the permeability constant μ, $B = \mu H$, so the solutions of Maxwell's equations can be written in terms of either B or H. In accordance with normal practice, we will express the magnetic field strength of electromagnetic radiation in space in terms of H (see section 1.4).

A case of particular interest is the generation of electric and magnetic

7.1   The electromagnetic spectrum.

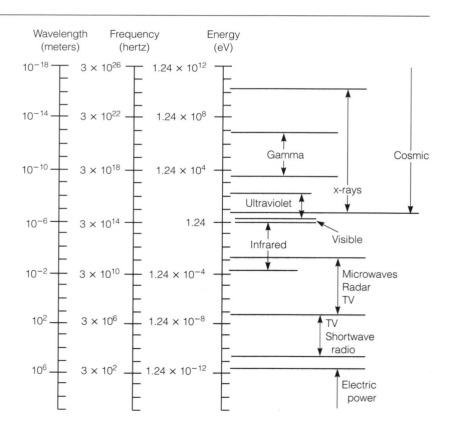

fields by an oscillator with intensities that vary in time as $\sin(\omega t)$, where ω is 2π times the frequency ν. Then the solutions to Maxwell's equations take the following forms in free space:

$$E = E_0\sin(kx - \omega t) \tag{7.9}$$

$$H = H_0\sin(kx - \omega t) \tag{7.10}$$

These equations describe electric and magnetic waves traveling through space with a wavelength $\lambda = 2\pi/k$ and a frequency $\nu = \omega/2\pi$. The relationship between the frequency (ν) and wavelength (λ) of the radiation is $\lambda = c/\nu$; that is, wavelength = speed of light/frequency.

The properties of the electromagnetic waves are shown in Figure 7.2 as plane waves: the E and H fields are perpendicular to each other and perpendicular to the direction of propagation of the wave, and the intensity is a function of only one dimension, shown in the diagram as the x dimension. The waves take this form at large distances from the antenna, that is, at distances much larger than the wavelength. This region is characterized as the *far field* of the antenna.

The wave transmits power, and the magnitude of power density or energy for a wave passing through unit area per unit time is given by the product $EH\sin\theta$, where θ is the angle between the E and the H vectors. Since in free space, the E and H fields are perpendicular to each other, θ is 90° and $\sin\theta = 1$. The average power density is $E_0H_0/2$, where E_0 and H_0 are the maximum values of E and H. The direction of travel of the power density (S) is perpendicular to E and H. For those who know vectors, all this is given very succinctly by the expression $S = E \times H$.

The solution also gives the ratio of E to H in free space, a quantity

7.2 Propagation of plane electromagnetic waves in space. The wave is traveling in the x direction with speed c. The electric and magnetic fields are perpendicular to the direction of travel and perpendicular to each other.

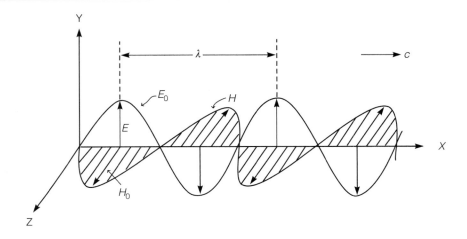

known as the characteristic impedance, Z_0, that is, $Z_0 = |E/H|$. The magnitude is 120π or 377 ohms and the power density is $E^2/377 = 377H^2$. While the power density falls off inversely as the square of the distance from a point isotropic source, the E and H fields fall off inversely as the distance.

The characteristic of the radiation is much different close to the antenna, that is, in the *near field* (see section 5.1 on antennas).

3.3   Relationships of Photons and Waves

The electromagnetic radiation that appears here as wave solutions to Maxwell's equations was treated previously in the sections on x rays and gamma rays as particles with energies high enough to produce ionization. The wave and particle properties of the electromagnetic radiation are related through Planck's constant *(h)*: photon energy equals Planck's constant times frequency of waves, or

$$E = h\nu, \quad \nu = E/h, \quad h = 4.135 \times 10^{-15} \text{ eV-sec}$$

Since, for waves, speed equals the frequency times the wavelength, or

$$c = \lambda\nu, \quad \nu = c/\lambda, \quad E = hc/\lambda$$

E (in eV) = $(4.135 \times 10^{-15}$ eV-sec $\times 2.998 \times 10^8$ m/sec$)/\lambda$ (in meters) = $1.240 \times 10^{-6}/\lambda$ (in meters) = $1,240/\lambda$ (in nanometers). In words, the energy of photons in electron volts equals 1,240 divided by the wavelength (of waves) in nanometers.

It may be easier to remember this relationship, with only a slight decrease in accuracy, as E (eV) = $1,234/\lambda$ (nm).

Example 7.3   An FM station broadcasts at a frequency of 100 MHz. What is the wavelength (when radiation is considered as waves)? And what is the energy (when radiation is considered as photons)?

For ν = 100 MHz (10^8/sec), $\lambda = 3 \times 10^8/10^8 = 3$ m $= 3 \times 10^9$ nm. The energy $E = 1,240/3 \times 10^9 = 4.13 \times 10^{-7}$ eV.

---

4   ABSORPTION OF ELECTROMAGNETIC RADIATION IN THE BODY

All the interactions of electromagnetic radiations with matter, their reflection from surfaces and their penetration and absorption in media, are the result of encounters between the oscillating electric and magnetic components of their fields and the electrons, atoms, and molecules of the media through which the waves pass. Of particular interest in studying the health effects of exposure to electromagnetic fields (EMF) is how far the fields penetrate into the body and how much power is absorbed locally in different parts of the body. The penetration and absorption depend strongly on the frequency of the radiation, its wavelength relative to the dimensions of different regions and organs in the body, its polarization, and the electrical and magnetic properties of the medium.

At frequencies below 100 kHz, energy absorption from ambient electric and magnetic fields is low and heating of the tissues is not a consideration. Between about 300 Hz and 1 MHz, however, currents are induced in the body whose effects, as manifested by nerve and muscle stimulation and involuntary movements, must be considered. At 100 kHz, heating effects begin to be significant, with increasing absorption (and heating) at a select number of frequencies (resonances) between 20 and 300 MHz and significant local, nonuniform absorption between 300 MHz and several GHz. Above 10 GHz, absorption (and heating) occurs mainly in the skin.

4.1 Penetration of EMF into the Body

The degree of penetration of EMF into the body, which depends on the rate at which the energy of the radiation is absorbed by the medium, is described by the *penetration depth,* that depth at which the magnitude of the E and H fields have decayed to $1/e$ (36.8 percent) of their value at the surface. It can be derived from the complex permittivity discussed in section 1.5. It is also the value at which the power has decreased by $(1/e)^2$, or 13.5 percent of its value at the surface. Calculations for various frequencies used in industry, research, or medicine (diathermy) are given in Table 7.2 for body tissues of high and low water content. Brain, lung, and bone marrow have properties that lie between the tabulated values for the two listed groups.

4.2 Induced and Contact Currents

Radiofrequency (RF) fields might induce circulating currents throughout the human body when the wavelengths are greater than 2.5 times the body length. These induced currents, associated with the E field, flow through the body to ground, where they may be measured as the short circuit current through the feet. High currents might produce high local rates of absorption and heating. At low frequencies (less than 100 MHz), the human body becomes increasingly conductive and contact currents might cause shock and burns, while internal RF-induced currents might reach locally high values.

Table 7.2.   Penetration of electromagnetic fields into the body.

Frequency (MHz)	Wavelength in air (m)	Muscle, skin, and tissues with high water content		Fat, bone, and tissues with low water content	
		Wavelength (m)	Penetration depth (m)	Wavelength (m)	Penetration depth (m)
1	300	4.36	0.913		
27.12	11.06	0.681	0.143	2.41	1.59
433	0.693	0.0876	0.0357	0.288	0.262
915	0.328	0.0446	0.0304	0.137	0.177
2,450	0.122	0.0176	0.0170	0.0521	0.112
10,000	0.0300	0.00464	0.00343	0.0141	0.0339

Source: After NCRP, 1981b

5   Specifying Dose to Tissue from Electromagnetic Fields

5.1   The Production of Heat as the Main Biological Effect

The mechanisms of the absorption of electromagnetic radiation lead to the production of heat. Exposure assessment in a given situation requires evaluation of the rate of absorption of the energy in order to ensure that it is not being absorbed at a greater rate than the body can dispose of it, both as a whole and for individual body parts, and the temperature is not raised to the point where injury results. The analysis is much more complicated than the evaluation of absorbed dose from ionizing radiation. There, the focus was on the action between particles (photons or charged particles) and targets (atoms or electrons) and the absorption rate depended only on the flux of particles and the attenuation coefficient.

For interactions of radiofrequency radiation in the body, the wave properties of the radiation govern the interaction. These waves encompass wavelengths ranging from micrometers to meters depending on the frequency. The intensity of the interactions varies throughout the body. It depends on the electrical properties of the medium, as expressed by the complex permittivity and permeability terms, which in turn depend on the chemical composition. It depends on the dimensions of the body parts, cells, organs, structures (eyes, liver, heart, limbs, head, and so on), and is strongly dependent on the relationship of the wavelength of the radiation and the dimensions of the exposed parts. All are governed by a complicated set of equations. The one redeeming feature is that the only significant biological effect of the radiation appears to be to raise the temperature of the affected parts. The effects of the temperature on the body are well understood and appropriate limits should prevent any harmful effects. But even that premise is under dispute.

5.2 Resonance—A Special Concern in Exposure to Radiofrequency Radiation

Much of the technology utilizing alternating electromagnetic fields involves the concept of resonance. The dictionary defines resonance as *the enhancement of the response of an electric or mechanical system to a periodic driving force when the driving frequency is equal to the natural undamped frequency of the system.* A television receiver is tuned to accept only signals from a specific channel by adjusting the capacitance and inductance of the elements in the tuner to have a natural frequency that corresponds to the frequency of the desired channel. The frequency of the television signal is the *resonant frequency* and it produces a very strong signal in the tuner, while the response to other signals is insignificant. Antennas for radar and other high-frequency radiations are designed with physical dimensions that favor certain wavelengths, meaning that their dimensions have a certain ratio to the wavelengths of the incoming radiation. Musical instruments operate on the principle of resonant frequencies. String instruments like the violin and piano produce different notes through vibrations of strings whose lengths (varied in the violin by pressing the finger against a specific point on the string) vibrate at their resonant frequencies. The design of nuclear reactors includes detailed considerations of the resonant energies that characterize the capture of neutrons in uranium and other materials.

When assessing the affects of the exposure of the body to microwave radiation, it is necessary to evaluate the occurrence of resonance absorption. This is primarily a geometrical effect, in which the wavelengths of the radiation are comparable to, and have specific ratios to, the dimensions of structures in the body. Resonance absorption for wavelengths that resonate with the whole body, or other wavelengths that resonate with different structures in the body, result in enhanced absorption and have the potential for serious injury from overheating.

5.3 The Specific Absorption Rate—The Basic Quantity for Assessment of Exposure to Radiofrequency Radiation

The basis of exposure assessment for RF fields is that the health risk is caused only by the heating of the exposed tissue, which depends on the rate of absorption of the RF energy. Thus, the basic dosimetric quantity pertaining to RF fields is the *specific absorption rate (SAR),* defined as the rate energy is absorbed per unit mass at a specific point in a medium. It is a function of the square of the internal electric field, E, the material composition as given by ε_0 and ε'', and the frequency of the radiation.

The equation for the energy absorption rate (power) is

$$P = \omega\varepsilon_0\varepsilon'' \mid E \mid^2 = \sigma \mid E \mid^2 \tag{7.9}$$

where ω equals 2π times the frequency; and ε'' and σ are direct measures of the "lossiness" of the tissue, that is, the energy absorbed by the material for a given E field.

The average specific absorption rate in an exposed individual is highly dependent on the wavelength relative to the size of the individual and hence on the frequency of the radiation. For a far-field illuminated body in free space, human whole-body resonance is established when the body length is about 36–40 percent of a wavelength. It reaches a maximum in an adult at a resonant frequency of approximately 80 MHz, the resonant absorption peak for a full-size human (40 MHz if the subject is grounded). It is 250 MHz for a small child. The maximum energy absorption occurs when the E field is parallel to the body's long axis. Energy absorption is minimized when the H field is parallel to the body.

The concept of the SAR may be used across the entire RF spectrum, but it is most meaningful between approximately 3 MHz and 6 GHz. At higher frequencies, heating of superficial tissues is more important than the whole-body SAR. At the lower frequencies, induced currents are more important.

6   Devices That Produce Electromagnetic Fields

6.1   Antennas

Devices for transmitting electromagnetic radiation into space take many forms, depending on the wavelength of the radiation. Much of the transmission of radiation in commerce is effectuated through the dipole antenna. The elements involved in the transmission of radiation by a dipole antenna and the electrical field generated are illustrated in Figure 7.3.

The electric and magnetic fields are generated by an oscillator and are carried by a transmission line from the oscillator to the antenna, shown in the figure as an electric dipole antenna. Electric charges flow back and forth along the two antenna branches at the frequency set by the oscillator, producing the effect of an electric dipole whose electric dipole moment varies sinusoidally with time. The electric and magnetic field lines are

7.3 Transmission of radiation by a dipole antenna. The frequency is determined by the inductance L and the capacitance C of the oscillator. The closed loops are electric field lines and the black-and-white dots indicate magnetic field lines, which are emerging from and entering into the plane of the figure (*Top:* From D. Halliday and R. Resnick, *Fundamentals of Physics,* 1988; *bottom:* H. H. Skilling, *Fundamentals of Electric Waves,* 1948, John Wiley & Sons, by permission.)

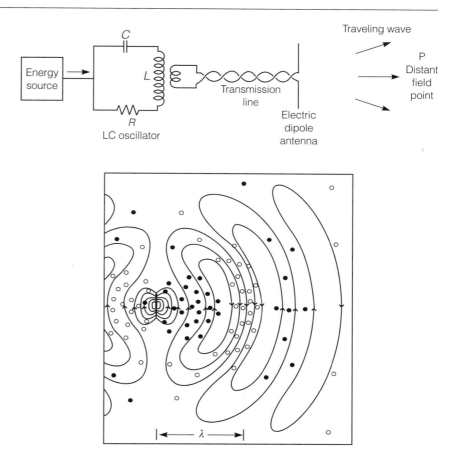

shown in the figure to radiate out into space, traveling away from the antenna with the speed of light. The pattern depicted in the figure is taken to hold beyond a minimum distance from the antenna equal to $2D^2/\lambda$, where D is the largest dimension of the antenna and λ is the wavelength at the transmitted frequency. This region is known as the *far field,* and waves in this region have the properties of plane waves—the E and B fields are equal and perpendicular to each other and to the direction of travel, as discussed in section 3.1 and Figure 7.2. The intensity of the wave traveling in any direction is inversely proportional to the square of the distance from the source and is proportional to the square of the sine of the angle from an axis perpendicular to the axis of the dipole; thus it is a maximum radiating out at 0 degrees and zero at 90 degrees.

The near-field region can be much more complicated. The E and H

fields are not in phase and can differ greatly in intensity at different portions of the field. There may be almost pure E fields in some regions and almost pure H fields in others. In the far field, just one of the quantities, E, H, or S need be measured, to specify the other two. In the near field, it is necessary to measure E and H separately, and the power density is no longer an appropriate quantity to use in expressing exposure restrictions (as it is in the far field).

Antennas may be designed to radiate power in all directions (isotropic radiators), or the power may be directed in a specific direction with a limited beam area by the use of reflectors. Thus, the power density in the far field at a distance, d, from an isotropic source with radiated power P is equal to $P/4\pi d^2$, a simple inverse square relationship. If the radiated power is concentrated in a specific direction, the power density for an isotropic source is increased by the factor G, the gain of the antenna.

6.2 Cellular Phone Networks

Unlike earlier mobile telephone systems, where a large area (large cell) is covered by a single high-power transmitter and a few radio channels, cellular radio supports a large number of simultaneous calls by dividing a service area into small cells, each of which is covered by a low-power transmitter. Each small cell is allocated a portion of the total number of radio channels available to the network operator, and adjacent cells are assigned different sets of channels. Cells and their groups of radio channels are systematically placed throughout the service area so that the same groups of channels can be used over and over again as many times as necessary, provided the spacing between co-channel cells (cells assigned the same groups of channels) is adequate to preclude interference.

Each cell contains a base station. Each base station contains a number of antennas mounted on an elevated structure and an equipment shelter containing receivers, low-power transmitters, and switching equipment. Cell sites are connected by coaxial cable, fiber optic cable, or microwave radio to a central location called a mobile telephone switch office (MTSO). The MTSO interfaces with the public switched telephone network.

Cell site transmitters operate at frequencies between 869 and 894 MHz; mobile transmitters operate at frequencies between 824 and 849 MHz. These frequencies are just above ultra high frequency (UHF) TV frequencies, which range from 470 MHz (channel 14) to 800 MHz (channel 69). PCS (Personal Communications Service) type systems operate at specific bands of frequencies between approximately 1,700 and 2,000 MHz, depending on the type of license held by the network operator.

Transmitters servicing cellular phones are located on towers, water tanks, rooftops, and even penthouses and parapets (see section 10 for measurements of fields). Typical heights for towers and other structures are 15–60 m. Any given site can contain up to 21 transmitting antennas (radio channels) that look like poles 3–5 m long, each one usually accompanied by 2 additional nonradiating antennas to receive signals from mobile units. The FCC permits an effective radiated power (ERP) of up to 500 W per radio channel, depending on the tower height, although the majority of cellular base stations in urban and suburban areas operate at ERPs in the range of 100 to less than 10 W per channel. The signal is essentially directed toward the horizon in a relatively narrow beam in the vertical plane (*www.fcc.gov/oet;* FCC, 1999).

6.3   Magnetic Resonance Imaging (MRI)

The nuclei of hydrogen atoms, the protons, have a small magnetic field which is associated with a quantity called the spin. When placed in a strong magnetic field, they will line up with that field, just as a magnet would, and will assume one of two energy states, with a difference depending on the intensity of the magnetic field. Most of them will be in the lower energy state, where the south pole of the proton is pointed in the direction of the north pole of the magnet and the north pole is in the direction of the south pole of the magnet. Photons of energy *(E)* exactly equal to the difference in energy levels will cause them to flip from the lower state to the upper state. Since the energy of photons is given by the expression *h*ν, where *h* is Planck's constant and ν is the frequency, they can be raised to the higher energy level by absorption of photons generated by a radiofrequency source with frequency exactly equal to *E/h*. In clinical MRI, ν is typically between 15 and 80 MHz for hydrogen imaging. After the radiofrequency signal is terminated, the nuclei will return to the lower state with the emission of radiowaves. This emission is the basis of a number of applications in analytical chemistry and medical practice, in particular in magnetic resonance imaging (MRI)

If a magnetic field whose strength varies across an object is applied, the energy and hence the frequency required to raise the hydrogen nuclei to the higher energy state also varies with position. Hence the frequency of the RF signal emitted by a nucleus as it returns to its lower state also varies with position and thus will mark the position from which it is emitted. This makes it possible to obtain an image of the major hydrogen-containing components of the human body, in particular the location of water and fat.

Thus magnetic resonance imaging is characterized by static magnetic fields, changing (gradient, time-varying) magnetic fields, and pulsed radiofrequency energy fields (Sprawls, 2000). The components of an MR imaging system include a primary magnet, gradient magnets, a radiofrequency transmitter and receiver, a computer control system, a patient handling system, and the image display and analysis system (Fig. 7.4).

The primary magnet produces a very strong, static magnetic field. The magnet may be an electromagnet, permanent magnet, or superconductor, and is quite massive, weighing 3 tons for the electromagnet, 8 tons for the superconductor, and approximately 100 tons for the permanent magnet.

Impressed upon this magnetic field is a gradient field, which is used to localize the signal produced from the patient. The gradient magnet systems consist of room-temperature auxiliary coils that must conduct several hundred amperes of current and be turned on and off in milliseconds. Three sets of gradient coils provide for 3D localization of tissue sources of signal. It is the movement of these coils that is responsible for the pounding noise accompanying the production of the image. The RF transmitter forms the frequency spectrum of each excitation pulse and is used to locate and form the shape of a specific slice or voxel element. The receiver coils and RF receivers must accept the range of frequencies determined by the selected nucleus and the field strength of the magnet itself.

The computer control system is designed to control the exact timing and magnitude of the application of the gradients, RF fields, the waveforms used, and all other details of each imaging protocol. Imaging proto-

7.4 Magnetic resonance imaging system (From P. Sprawls, *Physical Principles of Medical Imaging*, 2000, Aspen Publishers, Inc., by permission.)

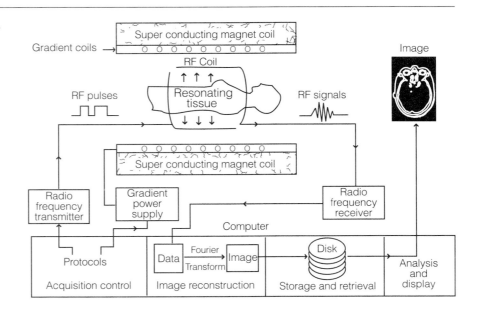

cols vary with the portion of the body studied and the specific character of the desired diagnostic information. The computer performs the calculations needed to produce the image and function information from the RF signals directly measured by the receiver system. Much of the diagnostic decision making is based on the video image, but film hard copy is the basic information source (Fullerton, 1997).

6.4 Video Display Terminals

The video display terminal (VDT) is an indispensable tool in both home and workplace. At its heart is the cathode ray tube, whose operation is based on the tracing of patterns on a fluorescent screen by an electron beam impacting the screen at high energy. The tube and the electrical circuitry designed to fulfill its mission, which are basically the same in VDTs as in TV sets (Kavet and Tell, 1991), are the source of radiation emissions that have been given considerable study.

The basic components of a VDT are shown in Figure 7.5. The cathode emits a narrow beam of electrons, which strike a fluorescent screen at energies up to 25,000 electron volts, depending on the high voltage across the tube. The point of impact emits a spot of light, which is caused by the circuitry to traverse the entire screen through a series of horizontal scans from top to bottom and whose brightness and color are controlled by signals applied to the tube by an operator at the keyboard or mouse of a computer.

The beam scans the entire screen in a series of horizontal lines, starting at the top and moving downward. The vertical and horizontal motions of the electron beam are produced by magnetic fields generated by electrical currents in horizontal and vertical deflection coils. The currents have sawtooth waveforms whose rate of rise and quick fall (Fig. 7.5) determine the speed of the horizontal and vertical deflections. The time it takes the beam to sweep across the screen is between 30 and 60 microseconds. A sharp drop in current in the horizontal deflection coil repositions the beam to the beginning of the next line in only 5 to 10 microseconds. This extremely rapid drop in current is taken advantage of to generate high-voltage pulses in the "flyback" transformer, which are rectified to provide the high voltage between the screen and the source of the electrons. The pulses are generated at a frequency between 15,000 and 31,000 times a second (15–31 kHz, approximately equal to 1/sweep time) and are responsible for the radiofrequency electric and magnetic fields. Because the fields are pulsed, they can be described mathematically in terms of a fundamental frequency equal to the sweep frequency and harmonics, which are multiples of the fundamental frequency. Most of the power is in electrical and magnetic fields at the fundamental frequency of 15–31 kHz. The vertical

7.5 Basic components of a video display terminal (From P. Kavet and R. A. Tell, VDTs: field levels, epidemiology and lab studies, *Health Physics* 6 (1):48–49, 1991, by permission.)

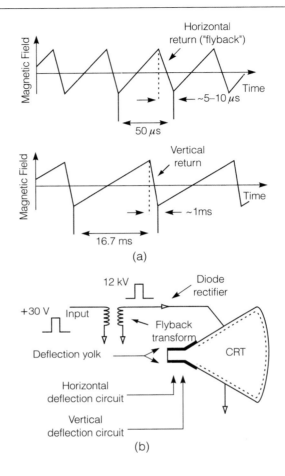

deflection circuitry normally operates at power line frequencies, 60 cycles/second (60 Hz) in the United States. The beam scans the whole screen in about 16 milliseconds, returning from the bottom to the top of the screen in less than 1 millisecond.

Fields with a frequency of 60 Hz measured at the operator position (taken as 30 cm from the screen) arise from the vertical deflection coil and circuitry and range between 1 and 10 V/m root mean square (rms) for the electric field, and between 0.08 and 0.6 A/m rms (0.1 and 0.75 μT) for the magnetic field. The RF magnetic field is generated primarily from the current in the horizontal deflection coil and has a sawtooth shape. The RF electric field arises mainly from the output lead of the flyback transformer; maximum values of measurements reported were less than 10 V/m rms.

The maximum value of the magnetic field was measured as 0.17 A/m rms (about 0.21 μT). These fields in front of a VDT are essentially no different from those in front of a color television set (Kavet and Tell, 1991.

7 Making Measurements of ELF and Radiofrequency Electromagnetic Fields

All it takes to measure a 60 Hz magnetic field is a coil of wire. When the coil is exposed to an alternating magnetic field, an AC voltage is induced in the coil and the amplitude of that voltage is a measure of the strength of the magnetic field. The coil can be designed and calibrated so that a 1 mG magnetic field induces a voltage of 1 mV, which can be read by plugging the coil into a simple millivoltmeter.

Other types of detectors are used as the frequency increases. At frequencies up to 30 MHz, both the magnetic field and the electric field should be measured, and the larger of the calculated equivalent power density values used to compare to the limits. Measurements at radiofrequencies of field strength and power density near antennas (in the near field, such as rooftop installations) are usually made with broadband isotropic field-strength meters (survey instruments). Far-field measurements are made using calibrated antennas, field-intensity meters, and spectrum analyzers (NCRP, 1993c; IEEE, 1999).

Broadband isotropic survey meters obtain a response close to isotropic by summing the outputs from three sets of orthogonal antenna elements, usually dipoles. The sensing elements for E fields of some instruments are diodes operated in their square-law region, giving outputs proportional to the power density. The sensing elements of other instruments are linear arrays of series-connected thermocouple elements with the hot and cold junctions in proximity to minimize the effects of thermal drift.

Narrowband instruments, such as spectrum analyzers with calibrated antennas, provide both amplitude and frequency information. Because of their sensitivity (which arises from their method of detecting fields by tuning in signals), narrowband systems are ideal for far-field measurements of signals well below the maximum permissible exposures (MPEs). They are also used in conjunction with broadband instruments to identify different sources operating in a complex environment where broadband near-field measurements are being made. This is particularly important for determining the contribution of a source, such as a cellular phone station, to preexisting TV, FM, or other sources of electromagnetic radiation.

Static magnetic fields are mapped with a probe whose calibrations are based on the Hall effect (UNEP, 1987). Fluxmeters are also used. These measure the variation of magnetic flux directly as a search coil is moved through the static field.

8   STANDARDS FOR PROTECTION AGAINST ELECTROMAGNETIC FIELDS

Many different government agencies and professional societies have developed recommendations and standards for exposure to electromagnetic radiation. These include the International Radiation Protection Association (IRPA); the International Commission on Non-Ionizing Radiation Protection (ICNIRP, a scientific body having a close association with IRPA but independent of it); the World Health Organization (WHO); the American Conference of Governmental Industrial Hygienists (ACGIH); the Center for Devices and Radiological Health of the U.S. Department of Health and Human Services (CDRH); the Institute of Electrical and Electronic Engineers (IEEE), the American National Standards Institute (ANSI); the National Council on Radiation Protection and Measurements (NCRP); and the Massachusetts Department of Public Health. These organizations all have extensive study and review processes, with the goals of developing standards that enjoy consensus among the professional communities participating in their development and are recognized by the community at large as authoritative and based on a solid creative process. There is widespread agreement among the expert groups that have worked on standards of exposure to EM fields that there is no strong statistical support for a link between exposure to electromagnetic fields and significant health consequences at levels to which the public is exposed. As a result, the differences in the standards recommended by the various groups are based primarily on differences in safety factors and degrees of conservatism applied. Higher safety factors may reassure members of the public concerned about the ubiquitous and ever-increasing levels of electromagnetic radiation in the home and the environment and the perceived potential they have for causing significant health effects. The challenge to standard-setting organizations is to provide a strong defense of standards proposed through clear and comprehensive presentations to the public of relevant information, analyses, and conclusions entering into their recommendations. The standards may be adopted voluntarily by industry in work to which they apply, or they may be incorporated into regulations by government. The journey from consideration of the various recommendations to the promulgation of regulations by government agencies, which are guided by the political process, can be long and arduous.

Because of the fundamentally different nature of the interactions at power line frequencies and radiofrequencies, the standards for exposure are expressed in different forms.

8.1   Power Lines

The fundamental dosimetric quantities at power line frequencies are the current density induced in the body and the internal E field. These effects are induced by exposure to external E or B fields that vary in time. Time-varying E fields induce surface fields and currents, which induce internal E fields and current densities. Time-varying B fields induce internal E fields and current densities. Both ACGIH and ICNIRP recommend limiting the current density to 10 mA/m^2 at 50/60 Hz. It is difficult to measure current density, however, and limits are expressed instead in terms of E and B fields. These are levels derived by mathematical modeling to give the limiting values of the current densities. According to such models, it takes an external magnetic flux density of about 7,000 mG to produce the limiting current density of 10 mA/m^2. This value applies only to the frequency of 50/60 Hz for which it was derived, as the induced current density increases with frequency.

Guidelines for limiting occupational exposures have been published by the American Conference of Governmental Industrial Hygienists (ACGIH), the International Radiation Protection Association (IRPA), the (European) Comité Européen de Normalisation Electrotechnique (CENELEC), the (British) National Radiological Protection Board (NRPB), and the (German) Deutsches Institut für Normung-Verband Deutscher Elektrotechniker (DIN/VDE). The guidelines are presented in Table 7.3 (Bailey et al., 1997). Some standards allow higher exposures for the limbs, such as 250,000 mG (IRPA) and 2.5 times the general limit (DIN/VDE).

8.2   Radiofrequency Standards

Most radiofrequency field standards are based on a limiting specific absorption rate (W/kg) for frequencies above about 1 MHz and extending up to several GHz. For lower frequencies, extending down to 300 Hz, more emphasis is placed on electrostimulation by induced currents (mA/m^2). For frequencies above 10 GHz, the incident energy is absorbed essentially at the surface, and the heating is limited mainly to the skin. Under these conditions, the power density expressed in units of energy per unit area (W/m^2) is the quantity that relates best to biological effects.

The initial safety standard for exposure to radiofrequency radiation was a single value, a power density of 10 milliwatts per square centimeter. It

Table 7.3   Exposure guidelines for power-frequency magnetic B fields (mT) and electric E fields (kV/m).

	Public	Occupational	
		Work-day	Short-term
Organization	24 hr limit	exposure limit	exposure limit
ACGIH, 1995–6		1 mT,[a] 25 kV/m	
CENELEC, 1995		1.3 mT	
DIN/VDE, 1995		1.1 mT	2.1 mT, 2 h/d
			3.5 mT, 1 h/d
IRPA/INIRC, 1990	0.1 mT, 5 kV/m[b]	0.5 mT, 10 kV/m	5mT, 30 kV/m, 2 h/d
NRPB, 1993	1.3 mT, 10 kV/m	1.3 mT, 10 kV/m	

Sources: Bailey et al., 1997; Zeman, 1991.

a. Limits for magnetic fields apply to 60 Hz power transmission. They may be increased by about 20 percent for 50 Hz transmission. Limits are lower (0.1 mT) for workers wearing cardiac pacemakers and are expressed as ceiling values. 1 mT = 10,000 mG.

b. Increased to 1 mT and 10 kV/m for a few hours per day.

was independent of the frequency. The assumption behind this standard was that the heating of body tissues was the only significant consequence of the absorption of RF energy, and if a radiation field of 10 milliwatts per square centimeter impinged on the entire body, which presented a cross-sectional area of about 0.7 m², and if the energy was completely absorbed, it would constitute a heat load of 70 W. This would be the same heat load produced by a 70 watt light bulb inside the body, and it is of the order of the resting metabolic rate. A normal individual could easily sustain it. This limit received consensus status in an ANSI standard in 1966 (ANSI Standard C95.1) and was reaffirmed in 1974. Recognizing that the body could tolerate higher heating rates for a short period of time, the standard allowed short time exposures in excess of the limit, for exposure durations less than six minutes. However, the product of the higher power level and the duration of exposure could not exceed 1 mW-h/cm². For example, an exposure to an RF field at 30 mW/cm² could not last longer than 1/30 of an hour, or two minutes.

As calculational and analytical methods became increasingly sophisticated and detailed, it was possible to take into account the dependence of the distribution of heating over the body and the tissues on frequency. Accordingly, new limits were developed by the ANSI C95 Committee and issued in 1982 which included dependence on the frequency. The occupational limits were still based on limiting the rate of energy absorption in the body, the specific absorption rate, or SAR. The limit chosen was 0.4 W/kg (average) and it applied to whole-body exposure either from continuous wave or pulsed radiation, averaged over any 0.1 hr period.

The safety criteria are based on reported threshold SARs at frequencies between approximately 100 kHz and 6 GHz for the most sensitive reproducible biological endpoints that can be related to human health. For modern standards, this is the disruption of food-motivated learned behavior in laboratory animals. The whole-body-averaged threshold SAR for behavioral disruption reliably occurs between 3 and 8 W/kg across a number of animal species, from small rodents to baboons, and across frequencies from 225 MHz to 5.8 GHz. The corresponding incident power densities range from 80 to 1,400 W/m^2. Contemporary exposure standards and guidelines apply a safety factor of more than 10 to these SARs and limit the whole-body-averaged SAR in humans to 0.4 W/kg in the working environment. These were originally recommended for the general public also. In regulations promulgated in 1983, the Commonwealth of Massachusetts reduced these limits for the general public to 0.08 W/kg for the purpose of regulating environmental fields.

NCRP published recommendations on exposure criteria for radiofrequency electromagnetic fields in 1986, following a comprehensive review of the scientific literature (NCRP, 1986a). The NCRP limits are given in Table 7.4. The report was preceded by an earlier report (NCRP, 1981b), which covered in depth the physical parameters and mechanisms of interaction of radiofrequency fields with matter to provide background essential for the interpretation and understanding of the criteria report. The occupational limits were similar to those published by ANSI (C95) in

Table 7.4 Radio frequency protection guides from the National Council on Radiation Protection and Measurements, 1986.

| Frequency range (MHz) | Occupational[a] | | | Public[b] |
	Equivalent power density[c] (mW/cm^2)	Electric field (V/m)	Magnetic Field (A/m)	Equivalent power density (mW/cm^2)
0.3–3	100	632	1.58	20
3.0–30	$900/f^2$	63.2 $(30/f)$	0.158 $(30/f)$	$180/f^2$
30–300	1.0	63.2	0.158	0.2
300–1500	f/300	3.65 $(f/17.3)$	0.0091f	f/1500
1500–100,000	5.0	141	0.354	1

Source: NCRP, 1986a. These are the same as ANSI C95.1-1982 guides published by the Institute of Electrical and Electronics Engineers, New York, NY.

a. Measured 5 cm or greater from any object in the field and averaged for any 0.1 hr (6 min).

b. The protection guide for the public was set at one-fifth the occupational exposure criterion.

c. (Electric field)^2 /1200π or 12π(magnetic field)^2 , whichever is greater.

Note: f, frequency (MHz); both *f* and *v* are used to denote frequency.

1982, based on the fundamental SAR exposure criterion of 0.4 W/kg and the associated schedule of frequency-dependent power densities. However, NCRP set the averaged exposure criterion for the general public at one-fifth that of the occupationally exposed individuals, at 0.08 W/kg, providing support for the regulatory stand taken by Massachusetts. The 1982 C95 ANSI standard was revised by IEEE in 1991 as IEEE Std C95.1–1991 and adopted by ANSI in 1992. It added a separate, reduced limit for the public at one-fifth the occupational limit, similar to the one recommended by NCRP in their 1986 report, except that this extra safety factor applies only to the resonance frequency range for the human body. IEEE issued another revision of C95.1 in 1999 (IEEE, 1999).

The ICNIRP issued guidelines for limiting EMF exposure that were published in the journal *Health Physics* in 1998. The guidelines were designed to "provide protection against known adverse health effects," defined as effects that caused "detectable impairment of the health of the exposed individual or of his or her offspring." The guidelines superseded guidelines issued in 1988 by INIRC, the IRPA committee that preceded it. The IEEE and ICNIRP standards are presented in Table 7.5.

Dosimetry studies of SAR distributions in various animals and human models have shown that the spatial-peak SAR (the rate at a "hot spot") to whole-body-averaged SAR is about 10 to 1, even under uniform whole-body exposure conditions. By adding peak limits to the SAR, applicable to any 1 gram of tissue in the shape of a cube, as 8.0 W/kg for the working environment and 1.6 W/kg for members of the public, the guidelines are able to take hot spots into account (Petersen et al., 1997).

8.3   Telecommunications Standards

Limits for specific absorption rates (SARs) are used when exposure is to sources held close to the body, such as cellular telephones. FCC regulations (Title 47 in the Code of Federal Regulations) adopted pursuant to the Telecommunications Act of 1996 (Telecom, 1996) require manufacturers to certify that each handset for which equipment authorization is being sought complies with the ANSI/IEEE (1992a) limits on spatial peak SAR for the uncontrolled environment—a maximum of 1.6 W/kg in any 1 g of tissue in the shape of a cube. To test compliance, E is measured in a physical model of a human being and the SAR is calculated from the equation $\sigma E^2/\rho$, where σ and ρ are tissue conductivity and mass density, respectively. Since the extreme near-field conditions apply, it is difficult to generalize on the field intensities that meet the SAR limits. In this case, the standards allow exceeding the guidelines if it can be demonstrated that the appropriate

Table 7.5 Protection guides for exposure to radiofrequency electric and magnetic fields

IEEE, 1999			ICNIRP, 1998		
Frequency (MHz)	Occupational	Public	Frequency (MHz)	Occupational	Public
			25×10^{-6}–820×10^{-6}	0.5/f V/m 0.02/f A/m	0.25/f V/m 0.004/f A/m
			0.00082–0.065	610 V/m 24.4 A/m	
			0.0008–0.003		0.25/f V/m 5 A/m
0.003–0.1	614 V/m 163 A/m 100 mW/cm²	614 V/m 163 A/m 100 mW/cm²	0.003–0.15		87 V/m 5 A/m
0.1–3	614 V/m 16.3/f A/m 100 mW/cm²				
0.1–1.34		614 V/m 16.3/f A/m 100 mW/cm²	0.065–1	610 V/m 1.6/f A/m	
			0.15–1		87 V/m 0.73/f A/m
1.34–3		823.8/f V/m 16.3/f A/m 180/f² mW/cm²	1–10	610/f V/m 1.6/f A/m	87/$f^{1/2}$ V/m 0.73/f A/m
3–30	1842/f V/m 16.3/f A/m 900/f² mW/cm²	823.8/f V/m 16.3/f A/m 180/f² mW/cm²			
30–100	61.4 V/m 16.3/f A/m 1 mW/cm²	27.5 V/m 158.3/$f^{1.668}$ A/m 0.2 mW/cm²	10–400	61 V/m 0.16 A/m 1 mW/cm²	28 V/m 0.073 A/m 0.2 mW/cm²
100–300	61.4 V/m 0.163 A/m 1 mW/cm²	27.5 V/m 0.0729 A/m 0.2 mW/cm²			
300–3,000	f/300 mW/cm²	f/1500 mW/cm²	400–2,000	3$f^{1/2}$ A/m 0.008$f^{1/2}$ V/m f/400 mW/cm²	1.375$f^{1/2}$ V/m 0.0037$f^{1/2}$ A/m f/2000 mW/cm²
3000–15,000	10 mW/cm²	f/1500 mW/cm²			
15000–300,000	10 mW/cm²	10 mW/cm²	2000–300,000	137 V/m 0.36 A/m 5mW/cm²	61 V/m 0.16 A/m 1 mW/cm²

Note: E fields are given in V/m, B fields in A/m; equivalent plane wave power density in mW/cm², and frequency f in MHz.

limits for the whole-body-averaged and spatial-peak SARs are not exceeded (Petersen et al., 1997). The British Department of Health issued a special advisory that, as a precaution, parents should permit children to use cell phones only for calls essential to safety.

Guidelines for the intensities of electric and magnetic fields (V/m, A/m, mW/cm^2) are used to assess environmental exposures from fixed sites, such as around cell-site antennas.

IEEE standard C95.1–1999 lists limits on power densities of 0.53–0.67 mW/cm^2 in uncontrolled areas and 2.67–3.33 mW/cm^2 in occupational and controlled areas for the cellular phone frequency range of 800–1,000 MHz. The FCC has the same values for uncontrolled areas, but does not regulate controlled and occupational areas directly. Exposure values are the mean values obtained by spatially averaging the incident power density over an area equivalent to the vertical cross-section of the human body. FCC guidelines for evaluating the environmental impact of RF radiation in accordance with the National Environmental Policy Act of 1969 (NEPA) are published in Title 47 of the Code of Federal Regulations. Because of their low transmitted power, many wireless transmitter facilities are excluded by the FCC from the requirement to prepare an Environmental Assessment. Mobile devices designed to be used with a separation of at least 20 cm normally maintained between the radiating antennas and the body are subject to routine environmental evaluation for RF exposure prior to equipment authorization if their effective radiated power is 1.5 W or more. The mobile devices include car-mounted antennas and transportable "bag phones" for cellular, personal communications services (PCS), and specialized mobile radio (SMR) applications. PCS facilities subject to routine environmental evaluation are listed in the rules of the FCC (47CFR).

8.4 Microwave Ovens

Special standards for microwave ovens were promulgated by the U.S. Food and Drug Administration between 1971 and 1980. The regulations were issued under authority granted in the Radiation Control for Health and Safety Act of 1968. Limits were set for ovens prior to sale and increased to allow for some leakage as the oven became older. They were:

<1 mW/cm^2 at 5 cm from external surface prior to sale
<5 mW/cm^2 thereafter

Measurements were to be made with the oven loaded with 275 ml H_2O in a glass beaker centered in the oven.

8.5 Video Display Units

Video display units produce local RF fields at frequencies around 15–35 kHz from the vertical and horizontal sweep generators. According to the ICNIRP (formerly the International Non-Ionizing Radiation Committee), the fields are too low to present a health hazard (INIRC/IRPA, 1988). However, public concern has been aroused periodically in the past by reports in the media of health problems associated with exposure to VDT fields, stimulating a considerable amount of research in the field. The results are discussed in section 9.3.1. While the levels from video display terminals are very low, the extent to which they exist has been used by some to promote the use of liquid screen video display terminals, which have insignificant radiation fields.

8.6 Static Magnetic and Electric Fields

Standards for exposure of humans to static magnetic and electric fields are given in Table 7.6. The 200 mT (2 million mG) limit for occupational exposure will induce 10–100 mA/m^2 current density in tissue with no adverse effects. The 20 T/sec (200 million mG/sec) limit for the rate of change of the magnetic field is based on peripheral nerve stimulation. When magnetic fields exceed 3 T (30 million mG), precautions should be taken to prevent hazards from flying metallic objects.

Limits of 0.5 mT (5,000 mG) are specified for protection of cardiac pacemakers and other medical electronic devices. Locations with magnetic flux densities in excess of 0.5 mT should be posted with appropriate warning signs. Implanted ferromagnetic materials can cause injury at a few mT. Analog watches, credit cards, magnetic tapes, and computer disks should not be exposed to fields greater than 1 mT.

8.7 Comparison of Basic Limits for Ionizing and Nonionizing Radiation

The protection standards for both ionizing radiation and nonionizing radiation are derived from a basic standard limiting the energy absorption per unit mass in the exposed individual. The ionizing radiation standard refers to an annual limit; the nonionizing standard to an absorption rate. The two standards are:

$$
\begin{aligned}
\text{Ionizing radiation:} \quad & 0.05 \text{ J/kg per year} \\
\text{Nonionizing radiation:} \quad & 0.4 \text{ W/kg} = 0.4 \text{ J/sec-kg} \\
\text{—Annually:} \quad & = 0.4 \text{ J/sec-kg} \times 3.15 \times 10^7 \text{ sec/yr} \\
& = 12.6 \times 10^6 \text{ J/kg per year}
\end{aligned}
$$

Table 7.6   Limits of exposure to static magnetic and electric fields.

	NRPB	ACGIH	CDRH
Occupational			
Working day, time weighted average	$40/t$ mT, $t > 2$ hr	60 mT	
	$80/t$ kV/m, $t > 2$ hr		
Limbs	$400/t$ mT, $t > 2$ hr	600 mT	
Short term	20 mT, $t < 2$ hr	2,000 mT	
	40 kV/m, $t < 2$ hr		
Limbs	200 mT, $t < 2$ hr	5,000 mT	
Public			
Areas of access	$40/t$ mT, $t > 5$ hr		
	$80/t$ kV/m, $t > 5$ hr		
	8 mT, $t < 5$ hr		
	16 kV/m, $t < 5$ hr		
Residences (continuous exposure)	2 mT		
	3.5 kV/m		
Medical			
Patient (MRI)			4,000 mT
dB/dt			20,000 mT/sec[a]
Occupational (MRI)	20 mT		
	200 mT, $t < 15$ m		
Arms, hands	200 mT		
	2,000 mT, $t < 15$ m		

Sources: Zeman, 2000; Repacholi, 1988b; ACGIH, 1996; www.fda.gov/cdrh/ode/95.html.
Note: E fields given in kV/m, B fields in mT; 1 mT = 10,000 mG = 795.5 A/m.
a. CDRH guidance provides for rates of change of the gradient field greater than 20,000mT/sec if the manufacturer of a magnetic reasonance diagnoster device can demonstrate that they are not sufficient to cause peripheral nerve stimulation by an adequate margin of safety.

The allowable annual limit on energy absorption for nonionizing radiation is 200 million times greater than the limit for ionizing radiation. This demonstrates very clearly that two different mechanisms for producing biological effects are involved.

9   MEDICAL FINDINGS ON HUMANS

9.1   Static Magnetic Fields

High static magnetic fields can exert forces on the moving electrolytes in blood, which make up a current of moving charges, and produce electric currents and flow potentials. A calculation for the aorta (and applicable to other large blood vessels) exposed to a 200 mT static field gives a maximum electric field of 84 mV/m across the lumen of the vessel and a maximum induced current density of 44 mA/m^2. These levels are not expected to produce adverse hemodynamic or cardiovascular effects (ICNIRP, 1991). Magnetic fields up to 2 T (20 million mG) induce flow potentials that produce changes in electrocardiograms, especially of the T-wave segment, but do not significantly influence cardiac performance during brief exposures. At 5 T, magnetoelectrodynamic and magnetohydrodynamic interactions with blood flow may lead to effects on the cardiocirculatory system (UNEP, 1987). Macromolecules and some organized cellular structures possessing a high degree of magnetic anisotropy will rotate in the presence of a static magnetic field. Rotation may cause, for example, alterations in the shape of normal red blood cells and in the alignment of sickled red blood cells, but it does not seem to result in any detectable clinical effects. However, observations like these have led to recommendations for continued studies to assess the effects of prolonged exposure to fields of the order of 1 T on the cardiovascular and central nervous system. Research is also recommended on the potential effects on cellular, tissue, and animal systems resulting from exposure to ultrahigh fields in the range of 2–10 T, because of interest in their use in future NMR devices (Tenforde and Budinger, 1985).

Of interest, though not necessarily of medical significance, is that the movements of personnel in static magnetic fields also induce currents in tissue in accordance with Faraday's law of induction for magnetic fields that vary in time. The movement of a person in a field of 200 mT is calculated to result in an induced current density of between 10 and 100 mA/m^2 in tissue. This current density is not considered to create adverse effects in the central nervous system at frequencies less than 10 Hz.

Visual sensations are believed to result from direct excitation of the optic nerve or retina by currents induced by the repeated and rapid application of magnetic field gradients to the patient during magnetic resonance imaging. Rapid eye motion occurring within high static magnetic fields has also produced visual sensations in volunteers working in and around research systems generating fields of 4 T or more. The current densities necessary to elicit visual stimulation is several orders of magnitude below that estimated to induce ventricular fibrillation or nerve action potentials (Bowser, 1997).

Subjective symptoms reported by workers exposed to high magnetic fields include fatigue, headache, numbness, nausea, and vertigo, but studies have not revealed any long-term effects or increased incidence of disease. The International Commission on Non-Ionizing Radiation Protection concludes that "current scientific knowledge does not suggest any detrimental effect on major developmental, behavioral, and physiological parameters in higher organisms from transient exposures to static magnetic flux densities up to 2 T" (ICNIRP, 1994).

The ICNIRP cautions pregnant women against exposure to magnetic fields, particularly during the first trimester. Its concern is due to the thermal vulnerability of the fetus, and it notes that the safety of MR examinations during pregnancy has not been established. While it concludes from a review of the biological effects of magnetic fields that no adverse health effects are to be expected from short-term (hours) exposure to static fields up to 2 T, it notes that there are many gaps in our knowledge of the biological effects and interaction mechanics of static magnetic fields with tissues. It advises that MR examinations should be performed only when there is a potential clinical advantage to the patient.

Meanwhile, the current state of knowledge brings reassurance that human beings seem to tolerate well high exposures to static electromagnetic fields for short periods of time, and there is no evidence that even fields produced in MR examinations cause permanent biological side effects for short exposures.

9.2   Extremely Low Frequencies, Including Power Lines

For decades, studies of the effects of electromagnetic fields from power lines and other conductors of electric currents have been prompted by a concern for the safety of members of the public as well as of occupationally exposed workers (Hitchcock et al., 1995). We may trace heightened public concern to two widely discussed studies involving childhood leukemia associated with power lines in Denver, Colorado. The first was by Nancy Wertheimer and Ed Leeper in 1979, the second by David Savitz and several colleagues in 1988. Both reported an increase in childhood leukemia in homes close to heavy-duty distribution lines, the big wires found on the tops of many large poles in the street. The Wertheimer and Leeper study reported a relative cancer risk factor of 2.2 for children living in homes near high-current power lines. The Savitz study was marginally significant. Other studies did not show a clear relationship between childhood or adult cancer risk and residential exposure to 50 or 60 Hz fields from power distribution lines (EPA, 1990). Graphic accounts of the alleged effects of power lines on the health of children greatly increased public concern over these findings (Brodeur, 1989a).

The majority of the epidemiological studies published by 1991 indicated an apparent association between chronic exposure to power-frequency electromagnetic fields in lines characterized by certain wiring codes and cancer risk, but methodological deficiencies limited the soundness of their conclusions. In particular, in nearly all of the studies reported, there was no quantitative assessment of power-frequency field exposures. In many of the studies, the sample populations were small, resulting in a large statistical uncertainty in the results. There was often poor matching between the control and the exposed populations. The existence of contributing factors—such as exposure to solvents known to be carcinogenic (in the occupational studies)—was ignored (Tenforde, 1991). Thus, in 1991 there were enough findings and questions to warrant the conduct of additional epidemiological investigations into the risk of cancer from exposure to power-frequency fields. Continued basic research was also needed at the molecular and cellular level to cast more light on the mechanisms through which the radiation affected living systems.

The Oak Ridge Associated Universities published a review report in 1992 (ORAU, 1992) in which they concluded that there was no convincing evidence that ELF fields generated by sources such as household appliances, video display terminals, and local power lines were demonstrable health hazards. Furthermore, they did not feel that additional research into health effects of ELF fields should receive a high priority.

In 1997, the National Research Council, the principal operating agency of the National Academy of Sciences, issued a committee report on the health effects of ELF fields. The report followed an intensive three-year review and evaluation of published studies on the incidence of cancer, primarily childhood leukemia; on reproductive and developmental effects, primarily abnormalities and premature pregnancy termination; and on neurobiologic effects, primarily learning disabilities and behavioral modifications (NAS/NRC, 1997). After reviewing more than 500 studies, the committee concluded that EMFs did have biological effects, such as disruption of chemical signaling between cells in cultures, inhibition of melatonin production, and promotion of bone healing in animals, but no adverse effects on cells or animals were found at the low levels measured in residences. Some epidemiological studies suggested that EMFs from electric blankets and video display terminals could harm the developing fetus, and elevated rates of brain, breast, and other cancers were found in workers in electrical jobs, but "the results were inconsistent and difficult to interpret." The committee concluded that the evidence did not show that exposure to residential and magnetic fields presented a hazard to human health. The committee did find an increase in the incidence of childhood leukemia in homes assigned a high "wire code", which is a measure of the impact of the power lines on the home that takes into account such factors as distance of the home from the power line and the size of the wires close to the home. For example, the highest classification, Class 1, as defined by Wertheimer and Leeper (1979), comprised high-voltage transmission lines, distribution lines with six or more wires (more than one distribution circuit), or a single three-phase distribution circuit with thick wires. However, the wire code ratings exhibited only a "rather weak association with measured residential magnetic fields," while they correlated with many other factors, such as age of home, housing density, and neighborhood traffic density. In addition, no association between the incidence of childhood leukemia and magnetic-field exposure was found in epidemiologic studies that estimated exposure by measuring present-day average magnetic fields. Thus, although the committee did find a correlation between the qualitative designation of "wiring code" and leukemia, it could not implicate magnetic fields as the agent causing the disease.

Examples of the data used to obtain the relative risk values are shown in Table 7.7.

Example 7.4   Verify the relative risk value determined from the Savitz data in Table 7.7.

The ratio of exposed to unexposed controls is 8/251. The expected exposed cases is 8/251 × 90 (unexposed cases) = 2.87. The number of exposed cases is 7. The relative risk is 7/2.87 = 2.44.

The "not guilty" verdict rendered by the NAS/NRC committee to charges that residential ELF fields were hazardous to health was enthusiastically greeted by some scientists but criticized by others. The head of a working group conducting an EMF study for the National Council on Radiation Protection and Measurements said that the summary did not ade-

Table 7.7   Findings of case-control studies of childhood leukemia and power lines.

Location/ Study	Controls		Leukemia cases		Leukemia cases Expected	Relative risk Exposed/Expected
	Unexposed	Exposed	Unexposed	Exposed		
Denver/	131	5[a]	130	6[a]	4.96	1.21
Wertheimer and Leeper, 1979	150	5[b]	143	12[b]	4.77	2.52
Denver/Savitz et al., 1988	251	8	90	7	2.87	2.44

Source: NAS/NRC, 1997.
Notes: Controls refer to selected population that did not have leukemia. Cases are selected because they had or died from leukemia. Exposed are controls or cases selected as having higher than average exposures. In this table, data presented are for subjects at the highest wire code listed, wire codes > high or > ordinary high.
a. Wire codes at birth greater than "ordinary high."
b. Wire codes at death greater than "ordinary high."

quately reflect the body of biological and biomedical knowledge about EMFs. Although all 16 members of the NAS/NRC committee signed the report, three members released a separate statement saying that whether EMFs threatened health was still an open question, and underscored the committee's call for more research. The research was important because, according to the committee, the possibility that some characteristic of the electric or magnetic field was biologically active at environmental strengths could not be totally discounted (Kaiser, 1996).

The National Cancer Institute (NCI) also completed in 1997 a five-year epidemiological study of the link between magnetic-field exposure and childhood leukemia. The study investigated the risk factors for acute lymphoblastic leukemia (ALL), which accounts for 98 percent of all childhood leukemias. The cancer cases were identified through the Childhood Cancer Group, a consortium of over 100 institutions that identify and treat about half the children with cancer in the United States and pool their cases for study. The researchers eventually identified 638 children with childhood leukemia and 610 matched controls in nine Midwestern and mid-Atlantic states and assessed their EMF exposures. They found no association between an increased risk of childhood leukemia and magnetic fields of $0.2 \mu T$ (2 mG).

A year later, an advisory panel of the National Institute of Environmental Health Sciences (NIEHS) voted 19–9 to classify electromagnetic fields as a "possible" human carcinogen. The conclusion was based on the criterion of the International Agency for Research on Cancer that a substance could be labeled a carcinogen after a finding of an association in a population, even in the absence of evidence linking a substance to tumors in laboratory animals. The vote was stated, in the NIEHS report, to be a "conservative, public health decision based on limited evidence." Whatever the cancer risk from EMFs was, it was slight. One recommendation in the report was consistent with a recommendation in every previous report— more research was needed. As one panelist put it, if there was a link between EMFs and cancer, "it's very small, very subtle, and very complex, and something we don't understand at any level" (Kaiser, 1996).

Exposure limits for low-frequency magnetic fields are based primarily on the effects of induced currents of 100–1,000 mA/m^2 on the excitable cell membranes in the nervous system and muscles (ICNIRP, 1991, 1998). The effects can be strongly dependent on frequency and wave shape. A current density of more than 1 A/m^2 in the vicinity of the heart or an electric field strength greater than 5 V/m in tissue may cause ventricular fibrillation (UNEP, 1987).

Various forms of electrical stimulation have been reported to promote bone growth and fracture healing in cases that do not respond to conventional treatment. One approach is to apply a direct current of about 20 μA to the fracture, either by positioning electrodes to the nonunion area through the skin or by implanting by open surgery a helix cathode across the fracture site. A noninvasive procedure using pulsed electromagnetic fields is performed in the home, in daily treatments of 10–12 hours (Bassett et al., 1982). Pulsing electromagnetic fields (PEMFs) and direct current stimulation have been reported to repair fractures and failed arthrodeses with an 80 percent success rate, comparable to the rates for cortical and cancellous bone grafting procedures (Compere, 1982).

The significant effects reported in the literature that are very specific to certain waveforms and frequencies support the need for more research. For example, the frequencies of electromagnetic fields of 1 to 10 Hz are in the frequency range of human brain wave currents, and there is some evidence from tests on monkeys that fields as low as 1 to 1,000 V/m at these frequencies can cause small changes in the speed and precision with which the brain and nerves respond to external stimuli when the field is on. These studies indicate that peculiar electromagnetic fields may have effects other than heating; they may introduce strange electrical signals into the communication networks in the body that can have deleterious effects on physical and mental health. These are just conjectures based on the present state of knowledge, but they provide an incentive for continuing investigation.

To the extent that it is possible to generalize about the vast body of knowledge that is accumulating on the effects of static and low-frequency electric fields, it appears that AC electric fields are harmless below about 1 kV/m and are probably incapable of causing permanent harm even for relatively long-time exposures up to 100 kV/m. In the range between 1 kV/m and 100 kV/m it is evident that temporary effects can occur, including minor blood chemistry changes and mild physiological effects, such as headaches and slight impairment of nerve and muscle function.

9.3   Radiofrequencies

There is no lack of literature on the effects of radiofrequency radiation on human beings. A comprehensive report of the National Council on Radiation Protection and Measurements on exposure standards for workers and the general public lists 64 pages of references in support of its recommendations (NCRP, 1986a). The 1970s were a particularly fruitful time for the production of findings and for defining critical areas for continued research. The work cited in the NCRP report covers such basic studies as the effects on molecules, chromosomes, and cells; continues on to the effects on reproduction, growth, and development and specialized effects on different organ and tissue systems in animals, including mammals; and then presents an extensive review of the effects on humans, including epidemiological studies to determine the risk of cancer. There was no evidence of congenital anomalies in human beings following exposure in utero to RF other than some case reports, although birth defects were produced in laboratory animals. There were no reports of RF-induced cancer in human beings, but there were suggestions that RF radiation may play a role in cancer promotion. Analyses of occupational statistics among workers with potential exposure to electric and magnetic fields in two studies reported increased proportional mortality and morbidity ratios for leukemia, but the data were not based on actual exposure measurements. The report concludes with the recommendations on exposure standards that are cited in section 7.

The Environmental Protection Agency conducted a critical evaluation of the scientific literature on the biological effects of radiofrequency radiation through its Health Effects Research Laboratory (Elder and Cahill, 1984). Among the conclusions was that human data were limited and incomplete but did not indicate any obvious relationship between prolonged low-level RF-radiation exposure and increased mortality or morbidity, including cancer incidence. Research on experimental animals also did not produce convincing evidence that RF radiation was a primary cancer inducer. The International Non-Ionizing Radiation Committee of the International Radiation Protection Association (IRPA) published the proceedings of a workshop on nonionizing radiation that gave a comprehensive overview of the field and included reviews of the effects of radiofrequency and extremely low frequency electromagnetic radiation (Repacholi, 1988a). The workshop was held to address the concerns of IRPA as expressed in the preface to the proceedings: "Non-ionizing radiation protection issues continue to receive wide media coverage, which in many cases includes distortions of fact and misinterpretations, or gives incorrect impressions of our actual knowledge of this field. Since quite widespread concerns are evident in the general public and workers exposed to various forms of non-ionizing radiation, it was felt necessary to provide basic information on our state of knowledge of biological effects and to give an assessment of the potential human health hazards from these radiations." Some experimental results on the biological effects of radiofrequency radiation on animals were presented, but apparently the presenters could not find any significant epidemiological data on production of cancer in humans. Nonthermal interactions in humans were reported for low-intensity RF fields modulated at extremely low frequencies, including changes in the calcium ion efflux from brain tissue in vitro and in vivo and alterations in electroencephalograms and behavior. These responses were similar to those elicited by extremely low frequency fields alone. One review noted that both EPA and NIOSH had proposed governmental limits in the United States, but "because of some controversies, these may not become official limits in the near future, if ever."

The World Health Organization (WHO) published a review by an international group of experts of the data on the effects of electromagnetic field exposure on biological systems pertinent to the evaluation of human health risks (WHO, 1993). In a summary of the data, it stated, "The few epidemiological studies that have been carried out on populations exposed to RF fields have failed to produce significant associations between such exposures and outcomes of shortened life span, or excesses in particular causes of death, except for an increased incidence of death from cancer, where chemical exposure may have been a confounder. In some studies, there was no increase in the incidence of premature deliveries or congenital malformations, while other studies produced indications that there was an association between the level of exposure and adverse pregnancy outcome. Such studies tend to suffer from poor exposure assessment and poor ascertainment and determination of other risk factors."

In summary, while some investigators claim an association between certain levels of exposure to radiofrequencies with cancer, there is no well-documented evidence to back up this conclusion. Most of the reports of effects other than those that might arise from excessive heating of body tissues are speculative. On the other hand, it is impossible to prove that any agent that acts on the body—chemical, physical, or biological—is completely harmless, so long as there is a natural occurrence of the adverse health effect of concern. Therein lies the dilemma of panels and committees that are charged to review studies and report to the public on the health effects of exposure to radiofrequency fields—to radar beams and radiowaves and microwave towers and cellular phones.

Unless and until studies provide sound arguments to support a change in approach, the rationale in setting standards will continue to be that the main effect of radiofrequency radiation incident on the body is to cause tissue heating resulting from absorption of the energy; the heating can be quite nonuniform; and control measures must take these hot spots into account to limit the heating to safe and sustained levels.

Environmental RF fields produce unusual electrical effects, but these do not normally result in a detrimental effect on health. At frequencies below about 100 MHz, RF fields can induce significant charges on ungrounded or poorly grounded metallic objects, such as cars. A person who approaches close to such an object may draw nothing more than a startling spark discharge. Touching the object will result in the flow of current through the body to ground, with a sensation of warmth, or even a burn if the current density is high enough. Modest values of such currents will result in stimulation of electrically excitable tissues, which can be felt as a tingling or pricking sensation and cause pain or burns at higher intensities (Hitchcock, 1994; WHO, 1993).

Clinical complaints from addicted users of cellular phones include headaches, memory problems, and dizziness, which some attribute to the radiofrequency radiation, but the implications of phone use for health effects remain nebulous. In response to public concern over the health effects of radiation from cellular phones, the British Department of Health convened an expert panel to study the problem. The panel concluded that the balance of evidence to date indicated that mobile phones did not harm health. However, there were ample indications that cell-phone emissions could induce biological changes. The health significance of these changes remained open to interpretation, but could include "potential adverse health effects" (Raloff, 2000).

Research on health effects of radiofrequency fields is focusing on non-thermal effects, which could possibly affect health in the long term. This approach requires modeling the human body in ways quite distinct from the traditional approach. It may entail looking at the body as a finely tuned electrical system that is sensitive to the physical characteristics of the electromagnetic radiation to which it is exposed—to the waveform, the frequency, the whole complex character of the electrical field carried by the radiation—but that has developed mechanisms, through evolution, to take corrective action against adverse influences of electrical fields, within limits. This approach is characterized by its own difficulties in interpreting the findings of research and applying them to develop standards for protection of the public health; in other words, it is much less amenable to practical implementation by regulatory agencies.

The fact is that the lack of evidence of significant morbidity attests to the effectiveness of current protection standards. Controls based on limiting the absorption of energy in tissues, whether from ionizing or nonionizing radiation, in accordance with the recommendations of professional standards-setting bodies appear to work quite well. But there is some scientific support to continue studies on the biological effects of RF fields and radiation—both through continuing epidemiological studies of the public health and through basic research on mechanisms of interactions. There is still much to learn about electrical interactions in the body.

9.3.1 The Special Case of Video Display Terminals

A small fraction of epidemiological studies have associated working at VDT work stations with an increased risk of health effects, primarily spontaneous abortion and congenital malformations. In general these studies related health effects to time spent at the terminal rather than specific measurements of the electric and magnetic fields; nor did they rule out other possible causes, such as potential psychological stress associated with repetitive tasks, physical inactivity associated with a sedentary job, and smoking. In any event, most of the studies do not show an association, and the general conclusion is that VDT use per se is not associated with increased risk of adverse reproductive outcomes (Kavet and Tell, 1991; Delpizzo, 1994).

Early concerns were expressed in the 1970s over exposure from x radiation from VDTs in newsrooms, as the press adopted this new technology. Extensive studies by the Occupational Safety and Health Administration (OSHA) dismissed the radiation exposure as minuscule. Concern shifted to the production of cataracts from radio wave emissions from VDT transformers when two young *New York Times* employees discovered that they had developed cataracts at the same time. This initiated a study by the National Institute for Occupational Safety and Health (NIOSH) that found the radiation levels were too low and of too low a frequency to have such an effect. In 1980, four out of seven babies born to Toronto *Star* employees who had worked at VDTs while pregnant were discovered to be deformed. The Ontario Ministry of Labour conducted an investigation and reported that there were no measurable traces of x-ray or microwave radiation and no identifiable chemical hazards. The actual outcome of these early studies was to focus attention on headaches, backaches, and eye problems not linked to radiation (Marshall, 1981).

In 1989, the *New Yorker* devoted three issues to extensive articles on the hazards of electromagnetic fields. The third article in the series, thirty pages in length, was concerned with VDTs, citing reports on cataracts, adverse pregnancy outcomes, neurological problems, defects in the embryonic development of chicks, and genetic effects at power-line frequencies (Brodeur, 1989b). In the conclusion to the article, the author cited the action of a nonprofit organization established by the Ford Foundation, the Fund for the City of New York, which designed its new offices so that all VDT operators sat at least 28 inches from their own terminals and about 40 inches from other terminals.

That same year, NIOSH was concluding a major epidemiological study on the effect of VDT radiation on the incidence of miscarriages. NIOSH conducted the study despite the absence of strong scientific evidence that VDTs caused a hazard. With 10 million women in the United States using VDTs, the number was enough to warrant looking into it further—even though current scientific evidence didn't point to a problem, it didn't mean that a problem might not exist (Kiell, 1989). The results of the study were published in 1991. The conclusion, based on following the outcome of 882 pregnancies of female telephone operators who used VDTs at work, was that "the use of VDTs and exposure to the accompanying electromagnetic fields were not associated with an increased risk of spontaneous abortion" (Schnorr, 1991).

The World Health Organization reviewed the outcomes of studies of women working with VDTs during pregnancy (WHO, 1993). It concluded that epidemiological studies failed to show an effect of magnetic fields from VDTs on the outcome of the pregnancies.

The electrical fields from VDTs, ranging between 1 and 10 V/m rms at the operator position, are similar to those found in the home and are not considered to be of any consequence. The interactions of the high-energy electrons in high-voltage rectifiers and at the screen of the cathode ray tube produce a low level of x radiation, which is well shielded in current sets. The primary interest in the radiation emissions is concentrated on the magnetic fields produced by the deflection coils and their sawtooth and pulsed patterns, which is felt by some to have greater biological effects than the sine-wave pattern characteristic of alternating electrical fields. Thus, many laboratory studies have been performed on various living subjects, including chick embryos and rodents. These studies demonstrated that exposure to low levels of radiation—both the extremely low frequencies characteristic of power lines and the more complex radiofrequency radiation characteristic of VDTs—produces biological effects, but the implications of these effects for human health are only speculative.

Since miscarriages are very common, with one out of every six pregnancies ending in a miscarriage, it is not improbable that findings of elevated cases of miscarriage relative to a national average occur by chance. To the extent that there may be some relationship between sitting long hours at a VDT and an increase in adverse pregnancy outcomes, factors other than radiation could be responsible, such as stress and poor ergonomics.

In summary, reviews by neither professional societies nor governmental agencies have incriminated VDTs as a significant health hazard.

10   Effects on Animals—Basic Research

Studies of the effects of living near power lines are not limited to humans. One study (Reif et al., 1995) looked at the possible linkage between residential ELF fields and cancer in pets. The study population was 230 dogs hospitalized with cancer, including 93 animals with canine lymphoma. Animals living in homes close to power lines with significant magnetic fields faced double to triple the cancer risk of animals in homes with buried power lines, dependent on how much time the animal spent outside. The most powerful statistical association occurred in 10 dogs whose homes were located very near a large, "primary" power distribution line. After adjusting for potentially confounding variables, the researchers found that the dogs had 13.4 times the lymphoma risk of animals from homes with buried power lines.

Static fields at levels up to 2 tesla (20 million milligauss) have not been found to produce adverse behavioral or physiological changes in mammals (Bowser, 1997). ELF fields have demonstrated effects on ion mobility and transcription in in-vitro studies (Hitchcock, 1994), but they do not seem to be genotoxic. Animal studies have shown that exposure to ELF fields might affect behavior and melatonin concentrations, influence ocular phosphenes, and possibly act as a stressor in rodents, as indicated by effects on the Harderian gland. Exposures of mice and of chick embryos to magnetic fields characteristic of those from VDTs have yielded both positive and negative results (Kavet and Tell, 1991). No consistent observations have been reported on reproduction, development, or cancer production. In most areas of study, more research is necessary, both to independently replicate and to extend findings.

11 EXPOSURES FROM ENVIRONMENTAL FIELDS

The earth is surrounded by a static magnetic field that can be represented by the field of a magnetic dipole located near its center. The field's horizontal intensity, its deviation from geographic north (declination), and its angle with a horizontal plane (dip, or inclination) are of strong interest because of practical applications in navigation (including the navigation of birds), communication, and prospecting. Tables of the U.S. Coast and Geodetic Survey (*Handbook of Chemistry and Physics,* Chemical Rubber Publishing Company) provide specifics. For example, measurements of horizontal intensity vary from 14 μT in Maine to 30 μT in Puerto Rico; the dip varies from 39 degrees in Hawaii to 77 degrees in North Dakota; and the declination varies from 1.5 degrees east in Miami, Florida, to 21 degrees west in Eastport, Maine. The total field varies from about 33 μT at the equator to 67 μT at the North Magnetic Pole (Bloxham and Gubbins, 1989). The values also vary over time—between the years 1580 and 1820 the direction of the compass needle in London changed by 35 degrees. While the earth's magnetic field is static, low-frequency, time-varying fields, generally less than 0.3 μT, occur ubiquitously in the home and office environments. Much higher values occur in industry and medical practice. Solar flares produce fields of about 1 to 3 μT (10–30 mG). Cosmic rays and terrestrial radioactivity produce ions in the atmosphere and a resultant normal field intensity of 100–200 V/m. Electric field strengths during local thunderstorms can be as high as 100,000 V/m, but 1,000 V/m is more usual (Grandolfo and Vecchia, 1985; Hansson Mild and Lovstrand, 1990).

11.1 Broadcasting: The Dominant Source of RF Radiation in the Environment

The environment bustles with radiowaves, which are assigned to different frequency bands according to their use. AM radio occupies a band between 535 and 1605 kHz (561 m to 187 m). FM radio broadcast frequencies are between 88 and 108 MHz (3.4 m and 2.8 m). VHF TV occupies the bands 54–72, 76–88, and 174–216 MHz. UHF TV frequencies are between 470 and 890 MHz. Communications frequencies of fixed systems (microwave relay, satellite communications) are between 0.8–15 GHz (38 cm to 2 cm), and radar systems are also included in this band, extending from 1–15 GHz. Frequencies for mobile systems (CB radios, walkie-talkies, cellular phones) are assigned between 27 and 900 MHz (11 m to 33 cm).

The EPA conducted an extensive survey of exposure to VHF and UHF broadcast radiation (30 to 3,000 MHz) and reported that 95 percent of the population was exposed to less than 0.0001 mW/cm^2 and no more than 1 percent of the population was exposed to ambient fields in excess of 0.001 mW/cm^2 (Tell and Mantiply, 1980). FM radio broadcasting was responsible for most of the continuous exposure of the population. Exposure near transmitting facilities is higher. EPA measured levels slightly over 0.1 mW/cm^2 near major TV or FM transmitters. These were the highest levels recorded, approaching limits for exposing the public and in the region where small, temporary psychological and biochemical effects have been observed in animals.

FM and TV broadcast antennas may be stacked on high towers or located on tall buildings in metropolitan areas. Tower maintenance personnel commonly report a sensation of warmth when climbing energized broadcast towers. Workers may be exposed to high-field strengths, and they may also be in areas susceptible to spark discharge and sustained contact currents. When work is performed on hot (energized) AM towers, it is possible that values of body current would exceed the recommended exposure criteria at some locations.

11.2 Radar Installations for Civilian and Military Purposes

Police radar units operate at frequencies of 10,525 MHz, 24,150 MHz, and 35,000 MHz. The radiated power is less than 100 mW, typically less than 35 mW. Typical local levels of exposure are less than 1 mW/cm² at distances greater than 30 cm from the transmitter. However, exposures in front of the antenna and within approximately 30 cm of the radiating surface to small areas of the body for periods of time exceeding 6 minutes may approach applicable protection guides.

Evaluations of commercial radar (airport surveillance, airport approach traffic control, weather tracking radar, etc.) have not revealed potential overexposure during normal operation.

11.3   Transmitters for Cellular Phone Systems

Measurements were made within 140 m of omnidirectional antennas mounted on a 66 m high lattice-type tower at a cellular base-station. The power density per 100 W effective radiated power (ERP), averaged over the various channels operating during each measurement interval, varied between about 5×10^{-8} mW/cm² to 5×10^{-6} mW/cm² (Petersen and Testagrossa, 1992). The highest measured values considered representative of exposures in uncontrolled environments, produced by continuous operation of nineteen 100 W ERP channels, were of the order of 0.0002 mW/cm² (Petersen et al., 1997).

Radiation levels from sector antennas mounted on the tops of buildings are of considerable interest to the occupants of the buildings. Sector antennas mounted on the sides of a penthouse at a height of approximately 2.6 m above the roof of a tall office building produced power densities per 100 watt ERP channel between 0.010 and 0.1 mW/cm² within about 5 meters of the antenna, dropping down to about 0.002 mW/cm² at 10 meters.

Measurements were made in several apartments on the top floor of an apartment building of the radiation levels produced by three sector antennas mounted on the outside of a parapet just above the apartments. For a single antenna at 100 W ERP, power levels ranged between 0.0000003 and 0.00002 mW/cm². If each of the three transmitting antennas was operating at capacity (19 channels, 100 W ERP per channel), the corresponding maximum power density would be of the order of 0.00038 mW/cm², still far below safety criteria (Petersen et al., 1997).

11.4   Power Lines

Power is typically generated at about 10 kV at the power plant, then stepped up to voltages ranging from about 69 to 765 kV for transmission over large distances to substations, where a step-down transformer brings it back to 5–35 kV, and then to a local distribution step–down transformer, where it is reduced to 230 V and delivered to commercial and residential end users. Electrical field intensities can range from 5,000 to 10,000 V/m under a 500 kV line, with magnetic fields of 5–50 μT (50–500 mG).

The current in the lines depends on the load, which varies considerably with varying demands during the day and season. Large industrial loads may generate harmonic currents and harmonic magnetic fields as well.

11.5   Home and Office

One study found that appliances, residential grounding systems, and power lines were major sources of residential fields, while internal wiring was a minor source (Douglas, 1993). Other researchers found that the strongest factor influencing exposure at home was the presence or absence of overhead lines at voltages of 132 kV or above within 100 m of the home (Merchant et al., 1994). Wiring systems are not major contributors to B fields if the hot and neutral conductors are close together and the currents are balanced. Unbalanced currents exist if the current enters and exits the building by different paths. This can occur if the neutral conductor is connected to an earth electrode, such as metallic water pipes, driven ground rods, or structural foundation steel. Sources that projected the highest magnetic flux density over distance were vacuum cleaners, microwave ovens, and small, hand-held tools and appliances. Although the local B field may be high, B fields from appliances contribute little to whole-body exposure, but they are major contributors to exposure of the extremities (Hitchcock et al., 1995). Following are results of surveys of specific items in the home.

Fluorescent fixtures. These lights produce magnetic fields of 2–40 mG at 30 cm, 0.1–3 mG at 1 m.

Electric blankets. The electric field intensity (AC) one foot under an electric blanket is about 200 V/m, but because of its relatively close coupling with the human body, it can induce currents equivalent to those induced in a person standing erect in a field of about 1 kV/m, or one-tenth the field directly under a 765 kV transmission line.

Hand-held electric appliances and tools. Fields near a hair dryer or soldering gun may exceed 1 mT (10 G). Even an electric shaver can have a magnetic field of 0.5 to 1 mT (5 to 10 G) around it.

Heavy duty appliances. Certain electrical appliances, such as electric lawnmowers, hedge trimmers or drills, particularly the older models, may generate fields of 33 kV/m.

Video display terminals. VDTs produce ELF fields (50–60 Hz), 1–10 V/m and 0.1–0.7 μT (1–7 mG) at the operator; and VLF fields, 17–30 kHz, 0.1 μT (1 mG), from the high-voltage (flyback) transformer, which is part of the horizontal deflection system. VDTs themselves are affected by AC fields as low as 1 μT: the screen "flutters" or "wobbles." The interference is typically caused by adjacent AC power equipment, such as transformers, conduits, and electrical switching panels. The easiest solution is relocation of the VDT, since magnetic shielding is very expensive. Computers can also be very sensitive to electronic interference. Failure of magnetic storage media can occur with DC fields greater than 0.5 mT (Smith, 1997).

TV sets. Fields of 0.04–2 μT (0.4–20 mG) at 30 cm and 0.01–0.15 μT (0.1–1.5 mG) at 1 m, are produced by televisions. Stronger fields, 0.1 mT (1 G), may occur near color receivers.

Cellular Phones. Hand-held cellular phones operate at a very low power compared with two-way radios and similar devices. Analog phones operate between 0.006 and 0.6 W, as controlled by the base station. (Mobile cellular phones such as car phones operate at higher levels, up to approximately 3 W). The average power is even lower in digital phones, which employ time-division multiple access (TDMA) techniques by operating one-third of the time during a conversation. Because of the lower power of hand-held cellular phones, environmental levels are very low. The fields were calculated for a 2 W 900 MHz phone of the type primarily in use in Great Britain, at 2.2 cm from the antenna. The maximum value of the electric field was about 400 V/m and the maximum magnetic field was about 1 μT. The maximum intensity was roughly about 20 mW/cm^2 (IEGMP, 2000). Hand-held phones are held against the face, however, and can produce SARs in the brain close to the FCC standard maximum of 1.6 W/kg (Foster and Moulder, 2000).

12   EFFECTS OF ELECTROMAGNETIC INTERFERENCE ON PACEMAKERS

Magnetic fields associated with MRI and NMR units could potentially cause the reed switch to move inside a pacemaker. This effect, combined with other electrical effects (induced currents) and the potential to cause physical movement of the pacemaker, dictate that individuals with pacemakers should stay away from these types of equipment. The American Conference of Governmental Industrial Hygienists has set a threshold limit value of 0.5 mT for those individuals with pacemakers and other ferromagnetic implants (ACGIH, 1996).

The effect of cellular phones on pacemakers is still under study (Smith, 1997; FDA, 1995). However, it appears that only phones using digital technology, and not analog technology, have the potential to cause interference in pacemakers. The FDA has also reported an electromagnetic interference effect on infusion pumps, incubators, apnea monitors, ventilators, and oxygen monitors when placed within 0.5 m of an analog cellular phone (FDA, 1996). Other devices under study for cellular phone interactions include implanted automatic defibrillators and hearing aids.

13 Exposures to Patients and Staff from Medical Devices

As in the case of ionizing radiation, exposures of patients in medical diagnosis and treatment with nonionizing radiation are much greater than allowed by health and safety standards for workers and the public. Magnetic fields used to promote healing of bone fractures average about 3,000 mG, with peaks of 25,000 mG. Diathermy treatments at 2,450 MHz typically dissipate 125 to 250 W in parts of the body for 15–20 minutes.

13.1 Magnetic Resonance Imaging (MRI)

Static magnetic fields vary between 0.15 T (1,500 G) and 1 T (10,000 G). A patient with an aneurysm clip implanted in her head died when the magnetic field of an MRI unit dislocated the clip and caused an acute intracerebral hemorrhage. Previous information on the type of clip implanted in the patient indicated that the clip would not deflect in a magnetic field of up to 1.89 T. However, the clip removed from the patient after autopsy exhibited deflection within 1.8 m of the 1.5 T MRI unit. Medical personnel who perform MRI examinations need to be familiar with safety precautions for the protection of the patient (ICNIRP, 1991).

Transient repeated and rapid application of magnetic field gradients during MR imaging induces voltages and currents in accordance with Faraday's law of induction. The resultant heating is negligible but direct effects of the induced current are possible. In typical multi-slice mode, whole-body MR imaging, magnetic field gradient switching typically yields field variations of 1.5–5 T per second at a distance of 25 cm from the center of the magnet. Newly developed echo planar sequences utilize much greater time rates of change of the gradient magnetic fields. Recent studies have reported what appears to be peripheral muscle stimulation in humans beyond a threshold of 60 T per second.

There is a strong frequency dependence in determining the threshold of current required to produce biological effects. Once the threshold is passed, the possible nonthermal effects include stimulation of nerve or muscle cells, induction of ventricular fibrillation, increase in brain mannitol space, epileptogenic potential, stimulation of visual flash sensation (induction of so-called "magnetophosphenes"), and bone healing. The threshold currents required for nerve stimulation and/or ventricular fibrillation are known to be much higher than the current estimated to be induced during routine clinical MR imaging. Seizures have been reported but may have been due to auditory, visual, or other stimuli associated with the examinations rather than from induced voltages or currents (ACR, 1996).

Most of the transmitted RF power is converted into heat. Currents from oscillating radiofrequency magnetic fields (as from gradient switching) are unable to excite tissue because of their low periodicity (less than 1,000 Hz). High-frequency (>1,000 Hz) currents induced by RF fields are capable of heating tissue via resistive losses. The little quantitative data on temperature increases in patients suggest that MRI ohmic heating is maximal at the surface of the body and approaches zero at the core.

Burns have resulted from secondary local heating due to the inadvertent formation of conductive loops by placement of gating wires, physiological monitoring fingertip attachments, or coil leads over the patient's skin. It is recommended that no unused coils be left in the imaging volume and that for active coils, no loops be formed. No exposed wires or conductors should ever directly touch the patient's tissue (Bowser, 1997).

High noise levels are produced as gradient coils are energized and de-energized dozens of times every second in the presence of the static magnetic field. The amplitude of this noise tends to remain between 65 and 95 dB, but can be quite variable. The patient should use protective devices to avoid hearing loss.

Designers of new installations must take into account steel in the environment that could affect static magnetic fields; radiofrequency interference; induced interference from power lines, transformers, generators, or motors; passing magnetic objects such as cars, forklifts, and materials of construction; and vibrations.

14   OCCUPATIONAL EXPOSURE TO ELECTROMAGNETIC RADIATION

Induction heating. Conductive materials are heated by eddy currents, which are induced when alternating magnetic fields are applied to the material. Induction heating is used mainly for forging, annealing, tempering, brazing, and soldering. The current in the coil producing the magnetic field can be very large; 5,000 A is not unusual. Induction heaters may operate at frequencies as low as 50–60 Hz up to 27 MHz. In reports of safety evaluations, most RF units operated between 250 kHz and 488 kHz (Hitchcock, 1994). Magnetic flux densities were measured between 0.9

and 65 mT at distances of 0.1–1 m from induction heaters at frequencies between 50 Hz and 10 kHz (Lovsund et al., 1982).

Dielectric heating. The frequency ranges from a few to 120 MHz and more. Today the total power used for dielectric heating probably exceeds that installed for broadcasting throughout the world. RF sealers and heaters are used to heat, melt, or cure such materials as plastic, rubber, or glue; in the manufacture of plastic products such as toys, rain apparel, and plastic tarpaulins; in wood lamination (for glue setting); and drying operations in the textile, paper, plastic, and leather industries. Dielectric heaters operate between 10 and 70 MHz but usually operate at frequencies designated for industrial, scientific, and medical (ISM) use of 13.56, 27.12, and 40.68 MHz, with 27.12 MHz encountered most commonly. A number of workplace evaluations have demonstrated the potential for overexposure (Hitchcock, 1994).

Welding. Electric welding, arc welding, resistance welding, and electroslag refining use 60 Hz. Operating currents may be in the tens of thousands of amperes. Magnetic fields between 0.1 to 10 mT (1–100 Gauss) are measured in electric welding processes, with the highest values near spot-welding machines (Hitchcock et al., 1995).

High-voltage lines. Some of the highest exposures to E fields are received by high-voltage line crews (230–500 kV), high-voltage substation maintenance electricians, and 500 kV substation operators (Chartier et al, 1985). Examples of E fields reported for power plants are 15,000 V/m and 29,000 V/m, with corresponding B fields of 2,000 μT and 4,200 μT. E fields at transmission lines are 10,000–12,000 V/m for an 800 kV line and 7,000 V/m for a 500 kV line (Hitchcock et al., 1995). Job classifications in electric utilities receiving the highest exposures are reported to be substation operators and utility electricians (Bracken, 1993).

15 BEYOND MICROWAVES

The electric and magnetic fields and electromagnetic radiation discussed in previous sections were produced by electric charge distributions, either static or in motion. The penetration of the fields into the body and their biological effects depended on the electrical properties of the medium and the dynamic characteristics of the charge, which ranged from static fields to fields vibrating with frequencies up to 300,000 MHz. The radiation at the highest frequencies covered penetrates only a slight distance into the skin, and its main effect is to produce warming of the skin.

To go to higher radiation frequencies, we need to turn to atoms and molecules as sources of infrared radiation, visible light, and ultraviolet light. The radiations are emitted as a result of transitions between electronic, vibrational, and rotational energy levels. The total power radiated and the spectrum of the radiation depend on the temperature. The spectrum for a "black body," that is, a theoretically perfect absorber of all radiation, is given by Planck's radiation law:

$$W_\lambda = \frac{C_1}{\lambda^5 (e^{-C_2/\lambda T} - 1)} \text{W/cm}^2\text{-}\mu\text{m}$$

where W_λ is the radiant energy emitted per second per unit area of surface per unit wavelength; C_1 and C_2 are Planck's first and second radiation constants and equal to 3.74×10^4 and 1.438×10^4, respectively; λ is the wavelength in μm; and T is the temperature in degrees K. If W_λ is plotted against λ for constant T, the curve rises to a maximum and then falls. The equation for the wavelength corresponding to the maximum in the radiant power as a function of the black-body temperature is derived as $\lambda_m = C/T$ nanometers, where T is the temperature in degrees K, and $C = 2.90 \times 10^6$ nm-K. From this equation, $\lambda_m = 9,666$ nm for the radiation from the earth, assuming a black-body temperature of 300 K; $\lambda_m = 3,625$ nm for a red hot object at 800 K; and $\lambda_m = 483$ nm for the surface of the sun at 6,000 K. Thus the frequency of the radiation at earth temperature is 3×10^{17} nm/sec $\div$ 9,666 nm = 3.1×10^{13}/sec or 31,000,000 MHz, considerably above the frequency that can be produced with a microwave generator.

These frequency and wavelength regions are of great interest, both because of their biological effects and their place in technology. Thus, electromagnetic radiation in the wavelength range between 1,000,000 nm (1 mm) and 100 nm is categorized as optical radiation, further divided into infrared (IR) radiation (1,000,000 to 780 nm); visible (light) radiation (780–380 nm); and ultraviolet (UV) radiation (400–100 nm). These bands were defined by the International Commission on Illumination (the visible and ultraviolet bands were deliberately defined to overlap because the spectral limits of the photobiological effects are not sharply defined). Studies of the biological effects in this energy range are directed mostly to the eyes and the skin. The applications in technology are many—in research, in medicine, and in industry. Laser sources, characterized by monochromatic coherent emissions of enormous intensity, are capable of being collimated into beams of light of microscopic dimension and focused on spots about the size of a wavelength. Surgeons found that they could cut tissue with laser energy and produce local coagulation that controlled bleeding at the same time. Laser beams are used to spot-weld detached retinas. Their applications vary from telephone communication over optical fibers to hydrogen fusion research.

As wavelengths decrease, the spectrum enters the ultraviolet region (400 nm to 100 nm). UV photons from 400 to 240 nm penetrate into a cell and are energetic enough to damage proteins and DNA, disrupting normal DNA metabolism, interrupting transcription (copying of DNA into RNA, required for protein synthesis), halting replication (required for cell division), and producing mutations that may lead to cancer (Alberts et al., 1989). A cell in the skin of a lightly pigmented person exposed to one minimal erythemal dose of UV (which causes just perceptible skin reddening with well-defined borders of the exposed area 24 hours after the exposure) may have 600,000 pyrimidine dimers induced in its DNA. The ability of humans to withstand such high damage levels indicates that human cells have highly efficient repair mechanisms, except for extra-sensitive individuals with certain genetic defects (Friedberg et al., 1995). In addition to producing sunburn, a short-term effect, and skin cancer, a long-term effect, UV exposure also alters the skin, especially after chronic exposure: skin thickening, roughening, and premature skin aging are all hazards of the farmer, sailor, and golfer. The major source of concern is the sun. Although it is 93 million miles away, the sun is largely responsible for 40 percent of the new cases of skin cancer that occur in the United States each year. The great majority are nonmelanoma skin cancers, but about 34,000 new cases of malignant melanoma of the skin are diagnosed each year, and the incidence is increasing at about 4 percent each year (HSPH, 1996). Other significant sources of ultraviolet light include fluorescent lighting and halogen lamps.

Accidents and personal injury may result from the use of lasers and other strong sources of ultraviolet, visible light and infrared radiation. The exposure assessment and protection measures for these exposures occupy their own realm in radiation protection and are beyond the scope of this book. Details may be found in a publication of the National Safety Council (Plog et al., 1996).

PART EIGHT

Current Issues in Radiation Protection: Where the Experts Stand

The path to take in dealing with the various complex problems in radiation protection that are brought to the attention of the public is often not clear. Resources are finite and when they are not sufficient to satisfy all the demands for measures to protect public health, priorities must be assigned. When experts are in sharp disagreement over the risks associated with different types and levels of radiation exposure and the control measures that are reasonable, how can lay persons make up their minds? What measures should they take to avoid or mitigate personal risks, or to help shape public policy? Sometimes the only recourse is to identify experts in whom they can place their trust and to accept their evaluations and advice.

This concluding section contains statements on major issues in radiation protection by recognized authorities, taken from the technical literature, periodicals, books, and testimony before congressional committees. In contrast to the presentation up to this point, where the effort was to limit the text to factual material, here we turn our attention to opinions. The criterion for selection of the opinions was that they came from knowledgeable and concerned people. The decision as to the merit of an opinion is left to the reader.

1 ON ELECTROMAGNETIC FIELDS

After a three-year study of the risk of cancer, neurobehavioral problems, or reproductive and developmental disorders from exposure to electromagnetic fields from power lines and electric appliances, a panel of the National Research Council publishes its conclusion.

The current body of evidence does not show that exposure to these fields presents a human-health hazard. (NAS/NRC, 1997)

After a five-year investigation, a report of the National Cancer Institute finds no link between magnetic-field exposure and childhood leukemia.

The results are very clear. They're negative. (NCI, 1997)

A prominent epidemiologist advises prudence in the control of exposure to electromagnetic fields.

It appears to me that the evidence linking EMF and cancer is strong enough that prudent persons will minimize those EMF exposures which cost little (in effort and money) to avoid, but will postpone until we have more data decisions on those EMF exposures which are difficult or expensive to avoid. New electric appliances and new power lines should be designed so as to minimize EMF exposure of people. Inflammatory newspaper reports on EMF and cancer (both exaggeration and denial) should be avoided. (Archer, 1997)

An editorial in the *New England Journal of Medicine* says enough is enough.

It is time to stop wasting our research resources on the EMF–cancer question. (Campion, 1997)

A policy statement of the American Physical Society downplays the risk of exposure to electromagnetic fields.

The scientific literature and the reports of reviews by other panels show no consistent, significant link between cancer and power line fields. This literature includes epidemiological studies, research on biological systems, and analyses of theoretical interaction mechanisms. No plausible biophysical mechanisms for the systematic initiation or promotion of cancer by these power line fields have been identified. Furthermore, the preponderance of the epidemiological and biophysical/biological research findings have failed to substantiate those studies which have reported specific adverse health effects from exposure to such fields. While it is impossible to prove that no deleterious health effects occur from exposure to any environmental factor, it is necessary to demonstrate a consistent, significant, and causal relationship before one can conclude that such effects do occur. From this standpoint, the conjectures relating cancer to power line fields have not been scientifically substantiated. These unsubstantiated claims, however, have generated fears of power lines in some communities, leading to expensive mitigation efforts, and, in some cases, to lengthy and divisive court proceedings. The costs of mitigation and litigation relating to the power line cancer connection have risen into the billions of dollars and threaten to go much higher. The diversion of these resources to eliminate a threat which has no persuasive scientific basis is disturbing to us. More serious environmental problems are neglected for lack of funding and public attention, and the burden of cost placed on the American public is incommensurate with the risk, if any.   (APS, 1995)

An award-winning former *New York Times* science writer warns about the hazards of exposure to electromagnetic fields.

A whole new EMF era is dawning with virtually no safeguards in place and with all these myriad questions unanswered. "Wireless" America is looming on the horizon. It will alter our ecosystem in a way never experienced before. The stakes may be higher than we know at this juncture. Or they may turn out to be lower than a summation of the research contained in this book indicates. But erring on the side of caution has never proven an ill-advised course of action and is certainly a more intelligent approach than the reckless abandon with which we have thus far embraced many modern technologies.   (Levitt, 1995)

2   ON DEFINING AND REGULATING THE HAZARDS OF EXPOSURE TO IONIZING RADIATION

2.1   On the Validity of the Linear No-Threshold (LN-T) Theory, That the Effects of Radiation Are Directly Proportional to the Dose Down to Zero Dose

The director of the National Radiological Protection Board, the official agency in the United Kingdom to advise the government on radiation protection and radiation hazards, states that the protagonists in the LN-T controversy are fighting the wrong battles and overlooking emerging intelligence on the fundamental nature of the cause of cancer.

Some 5,000 to 10,000 DNA damage events per hour per cell occur because of thermodynamic instability and attack by chemical radicals . . . Because a lot of single strand damage to DNA occurs spontaneously in cells, argue the threshold people, a small increment from a low dose of radiation is insignificant for risk.

This argument fails to recognize the very low abundance of spontaneous double strand damage and the critical importance of these lesions and their misrepair. A single radiation track traversing the nucleus of a target cell—the lowest dose and dose rate possible—has a finite probability, although very low, of causing the specific damage to DNA that results in a tumor initiating mutation at the DNA level, therefore, no basis exists for a threshold below which the risk of tumor induction would be zero . . .

The real issue to be decided between scientists, regulators, and the public is not a threshold for risk but the acceptability of risk. They should join forces to determine acceptability in different circumstances—in work and public environments, and under normal and accidental conditions. (Clarke, 1996)

An authority on the theory of the biological effects of radiation expresses his frustration with the influence of political considerations on the acceptance of the linear no-threshold hypothesis.

What may be proved [with respect to LN-T] is proportionality between transformation and dose but not proportionality between cancer and dose . . .

There are serious doubts concerning the accuracy of the principal source of epidemiological data which are the survivors of the Japanese A-bombing. It has become apparent that the DS-86 dosimetry study seriously underestimates the contribution of neutron doses in the dominant Hiroshima data . . . Thus, there remain uncertainties but it is evident that LN-T lacks theoretical or experimental support.

In general, one wishes to set limitations of exposure to environmental hazards on the known (small) frequency of the untoward effects involved. In the case of radiation the claimed frequencies are fictional. It is unknown whether at current limits (and larger earlier ones) there are some effects (negative or positive). Agencies or organizations regulating the undetectable hazards claim to be responsibly conservative but they are in fact quite irresponsible especially when (in ALARA) they insist that minute radiation doses must be avoided. It is not only a waste of billions of dollars in "clean-up" operations. There is irrationality and, in the case of some future minor reactor accident, the prospect of a hysteric reaction.   (Rossi, 2000)

2.2 The Exemption from Regulatory Control of Radiation Levels Below Which Causation of Cancer Is Considered Insignificant

The American Nuclear Society questions the basis of regulatory control of radiation exposure, namely that any radiation exposure, no matter how small, increases detrimental health effects.

It is the position of the American Nuclear Society that there is insufficient scientific evidence to support the use of the Linear No Threshold Hypothesis (LNTH) in the projection of the health effects of low-level radiation.

Given this situation, an independent group of reputable scientists, medical experts and health researchers should be established to conduct an open scientific review of all data and analyses on the subject of LNTH. Based on the conclusions of this review group, a separate group composed of stakeholders should make recommendations on whether adjustments to current radiation protection guidelines should be made immediately to reflect current information.

In addition, it is the ANS position that new research on low-level radiation health effects, spanning several disciplines, should be initiated. Meritorious existing research within the disciplines should continue to receive funding.   (ANS, 1999)

3   ON REDUCING POPULATION RADIATION EXPOSURE FROM MEDICAL AND DENTAL X RAYS

Representatives of the Commission of the European Communities, the World Health Organization, and the International Commission on Radiological Units and Measurements discuss radiation protection with regards to medical radiation at an international conference on dosimetry in diagnostic radiology (excerpts taken from Kramer and Schnuer, 1992).

> We have learnt . . . how the practice of diagnostic imaging differs throughout the world. The figures . . . from Japan impressed me very much . . . I could hardly believe the figure they have of four thousand (diagnostic procedures) per thousand inhabitants . . . You have seen that in the UK health care is provided with four hundred procedures and I know that in Switzerland the average number is one thousand two hundred . . . That shows that something should be done towards the reduction of radiological procedures . . . I know an initiative in the UK where some of these recommendations [of the WHO] were transformed into posters and displayed on the wall particularly in the emergency rooms of hospitals or in other places where doctors can see them. In this way they have some advice as to how to handle a patient.   (N. Racoveanu, World Health Organization)

> There is also our communication problem with doctors and radiographers . . . In Denmark we have taken legislative steps in order to impose constraints on the medical profession. It seems not to be a very pleasant way of proceeding and we are not happy about it. I think we should try to communicate the problems of radiation protection and doses to patients to the medical profession by other means and I hope the Commission will support

this action. (O. Hjardemaal, National Institute of Radiation Hygiene of Denmark)

I have to underline that usually doctors are not thinking in radiation dose. They are thinking in other directions of optimization . . . Optimization for them is to make a good diagnosis. We have a lot of difficulties in demonstrating to them that good diagnosis also can be obtained with lower dose . . . The radiologists like to see in the image amplifier some clinical finding. They have the possibility of seeing the picture on the screen for a longer time without the dose to the patient but in all those cases he must be trained or be educated about this technique. Therefore one of the most efficient measures is good training of the radiologists. We have introduced in Germany a system of quality control of the radiographs. The clinical radiographic pictures are examined by a commission of experts from the point of view of diagnostic information, quality, development and exposure. If radiologists present too many bad radiographs they are not paid for them and they will be advised to participate in training and education. (F. Stieve, Munich)

The director of the Health Physics Division, Oak Ridge National Laboratory, and first president of the International Radiation Protection Association, appeals for greater controls in the use of medical and dental x rays in testimony before the U.S. Congress.

Although I have been very proud of some of the actions taken by ICRP, NCRP and FRC which have led to the reduction of population exposure, I feel very strongly that more specific recommendations should be made to prevent the exposure of a patient by a physician untrained in the use of ionizing radiation, radiation physics and in radiation protection. I believe values of maximum permissible dose should be given for certain types of diagnostic exposure that would be expected to remove some of the disparity that at present results in a difference of more than an order of magnitude in dose delivered to the skin of a patient by various dentists for the same dental information or some x-ray technicians for the same chest x-ray examination. I feel the Federation Radiation Council has been particularly conspicuous by its lack of action in regard to patient exposure.

I have no doubt that measures discussed later in this paper, if properly applied, would reduce the average diagnostic per capita dose to 10% of its present value. (Morgan, 1967)

The president of the National Council on Radiation Protection and Measurements comments in an interview on the significance of exposure of the public to x rays. When asked by the interviewer if x radiation to the public added up to a dangerous level, he replied:

No, but it adds up to some potential harm. I am deeply concerned over the possible deleterious effects of radiation which is received by people without any positive return of benefit—that is, unnecessary radiation. Now, I'm not hysterical about it, but I am concerned, because this is a source of damage that you simply don't have to put up with.   (Taylor, 1961)

The director of the Office of Radiation Control, New York City Department of Health, discusses the problem of the prescription of excess x rays.

The problem of excessive use of unnecessarily repeated examinations are abuses that could not easily be regulated under any circumstances. Popular feeling and professional education has and will probably continue to be the only effective controls.

And I don't think we should overlook popular feeling. Those of you who have been in the field a long time know that it was once the practice of pediatricians to fluoroscope babies and young children every month and when they had the annual checkup. When we questioned this practice, pediatricians would say, "Well, the parents expect it. They think if I don't fluoroscope the patients, they are not getting a complete examination."

Well, times changed. We had fallout, then we had the National Academy of Sciences Report pointing out the dangers of x-rays, and now today pediatricians say that when they want to fluoroscope or x-ray a child, they often encounter parent's resistance. As a result, there is very little routine fluoroscopy being practiced today on young patients. I think this sort of result of the population's education is most important in any improvements we want to make.

I think this feeling, this reaction of the public is going to be helpful to us. But I think we have to do something to stimulate it properly without frightening patients. (Blatz, 1970)

A professor of radiology deplores the unnecessary administration of x rays.

Even more serious . . . is the growing tendency of medicine to use radiological methods unwisely and for purposes other than to answer clinical questions for which these methods are uniquely suited. I speak here of the increasing extent to which our services are used without careful consideration of clinical benefit and cost, without adequate evaluation of the patient, and often as a means merely to provide medicolegal protection. We radiologists have generally assumed that all radiological procedures are of clinical benefit, favorably influencing the clinical course of the individuals on whom they have been performed, and that all examinations are valuable regardless of cost. Recent studies have shown that these assumptions are all too often unfounded and that there is urgent need for research critically evaluating the clinical benefits of radiological procedures and the conditions under which they may be optimally applied.

The system of medical care in the United States has tended to avoid investigations of this sort, investigations which measure quality of performance and which suggest alternative methods which potentially yield greater information at lower cost. However, society demands that such research be undertaken and, if I dare a prediction, I forecast that in the next decade investigations which fall into the general category of "quality control" will receive major emphasis.   (Morgan, 1972)

A radiologist comments on overuse of medical x rays.

My problem is to Stop OVERUSE. X-rays are being used too much for diagnosis in our country, and at a rate of increase that is scandalous. The last time I looked into the matter we were exposing annually 11,000 acres of x-ray film every year and were clamoring for more. Now we are trying to slow down and it isn't easy—old habits, ripe prejudices, and the love of money being what they are. In passing, it has always seemed interesting to me, that after all the years of pleas and lamentations of geneticists and the indispensable alarms of Ralph Nader, our present clear call for restraint comes not from the laboratory, it comes from the Treasury. There has been prodigious waste, and we can no more afford prodigality in radiology than we can in any other branch of science or government.   (McClenahan, 1976)

The United Nations points out the high contribution of medical radiation to population dose.

Medical exposures contribute the highest man-made per capita doses in the population, are given with high instantaneous dose rates and cause the highest individual organ doses short of accidental exposures. From the radiation protection point of view, they also offer the largest scope for implementing methods of dose reduction without loss of the information required. (UNSCEAR, 1977, p. 301)

4   ON THE SAFETY OF NUCLEAR POWER

A specialist in environmental and radiological health writes on pollution from power sources.

In any case, present experience indicates that continuous release of gaseous wastes from either the pressurized water reactors or the boiling water reactors presents a lower order of hazard than that of coal-fired plants.

Nuclear energy has a critically important role in combating the growing assault on our atmosphere. Still, even with nuclear energy completely supplanting fossil fuels for new plants built late in this century, much more must be done. What then can the nuclear energy industry do to aid our fight for clean air? The answer is implicit in the very advantages claimed by nuclear power. Unquestionably, the potential for massive pollution exists in the fission products produced by a nuclear reactor; in the absence of effective control to restrict the emission of radioactivity, the nuclear program could have become a leading contributor to atmospheric pollution. The key word is control. Essentially every phase of design, site selection, construction, and operation of a nuclear power plant is under the strict surveillance and control of responsible and technically competent review boards. The same tight control is overdue for other actual and potential polluters and must surely come into being, hopefully soon.   (Fish, 1969)

Two journalists warn dramatically against nuclear power.

Thus, when atomic power advocates are asked about the dangers of contaminating the environment they imply that the relatively small amounts of radioactive materials released under "planned" conditions are harmless.

This view is a myth . . .

Efforts are of course being made toward effective handling of the waste problems, but many technical barriers must still be overcome. It is unlikely they will all be overcome by the end of the century, when waste tanks will boil with 6 billion curies of strontium-90, 5.3 billion curies of cesium-137, 6.07 billion curies of prometheum-147, 10.1 billion curies of cerium-144, and millions of curies of other isotopes . . .

The burden that radioactive wastes place on future generations is cruel and may prove intolerable . . .

What must be done to avert the period of the peaceful atom? . . . The only course may be to turn boldly away from atomic energy as a major source of electricity production, abandoning it as this nation has abandoned other costly but unsuccessful technological enterprises.   (Curtis and Hogan, 1969)

An eloquent friend of nuclear power belittles the waste disposal problem.

People just don't like the idea of radioactive wastes being put out of the way for thousands of years . . . They fear this danger not because it is great, but because it is new . . . Radioactive poisons underground, threatening somehow to get into your food—no matter how absurdly small the probability, it's new, it's a danger that wasn't there before.

The hell it wasn't. There are some 30 trillion cancer doses under the surface of the United States—the deposits of uranium and its daughters. They are not sealed into glass, they are not in salt formations, they are not deliberately put where it is safest, they occur in random places where Mother Nature decided to put them. And they do occasionally get into water and food, and they do occasionally kill people . . . The mean number of Americans killed by ingesting uranium or its daughters from natural sources is 12 per year.

"There is nothing we can do about those 30 trillion cancer doses," some people say when they first learn about them, "but at least we need not add any more to them."

But we add nothing. We take uranium ore out of the unsafe places where Nature put them, and after we extract some of its energy, we put the wastes back in a safer place than before, though we put them back in fewer places in more concentrated form . . . Plutonium, with its half-life of almost 25,000 years slows the decay process, but it remains there only as an impurity that failed to be recovered for further use as a valuable fuel. And what if the Luddites have their way and dispose of the plutonium unused? Like the proverbial man who killed his parent and then demanded the Court's mercy on the grounds that he was an orphan, they want to waste plutonium and then scare people with the long half-life of nuclear wastes.   (Beckmann, 1976)

A distinguished Russian physicist argues for the development of nuclear power.

The development of nuclear technology has proceeded with much greater attention on the problems of safety techniques and preservation of the environment than the development of such branches of technology as metallurgy, coke chemistry, mining, chemical industry, coal power stations, modern transportation, chemicalization of agriculture, etc. Therefore, the present situation in nuclear power is relatively good from the point of view of safety and possible effects on the environment. The ways to improve it further are also quite clear. The basic peculiarity that distinguishes nuclear technology from that using chemical fuels is the high concentration and small volume of the dangerous by products and the small size of the process as a whole. This makes it easier to solve the safety and environmental problems for a nuclear power station than it is for a power station using coal, oil, etc.

Therefore I assert that the development of nuclear technology is one of the necessary conditions for the preservation of the economic and political independence of every country—of those that have already reached a high development stage as well as of those that are just developing. For the countries of Western Europe and Japan, the importance of nuclear technology is particularly great. If the economy of these countries continues to be in any important way dependent on the supply of chemical fuels from the USSR or from countries which are under her influence, the West will find itself under constant threat of the cutting off of these channels. This will result in a humiliating political dependence. In politics, one concession always leads to another and where it will finally lead is hard to foresee. (Sakharov, 1978)

The need for extraordinary safety and control measures for the nuclear power industry is voiced by a well-known radiation scientist in a "Journal of Politics."

No more important engineering challenge exists today than making sure that the reactors coming into use conform to a rigid set of codes so that the public safety is assured for the coming decades of nuclear power.

I do not make the charge that the AEC is imposing an unsafe system of nuclear power on the nation; I submit that the public record is not visible to substantiate public confidence in the AEC's assurance.

The nation needs power, clean power, and I believe it is not beyond our technological capabilities to design, site and operate nuclear power plants *and* insure the public safety. But as we, meaning all of us, enter into the nuclear decades, it is essential that the record is clear—that we, not just a few experts in a closed community, audit the nuclear books and lay the basis for public confidence in our nuclear future. (Lapp, 1971)

5   On the Hazards of Nuclear Weapons Tests and Underground Explosions

5.1   Hazards to the Public from Fallout from Atmospheric Testing of Nuclear Bombs

A Nobel laureate in chemistry, and a leader of the opposition to bombs testing, writes a letter to the *New York Times* on the genetic damage from carbon-14 in fallout.

A straightforward calculation based on the above assumptions leads directly to the conclusion that one year of testing at the standard rate of 30 megatons a year (two 15-megaton bombs, similar to the one detonated by the United States on March 1, 1954) will ultimately be responsible for the birth of 230,000 seriously defective children and also for 420,000 embryonic and neonatal deaths . . .

As other people have pointed out, these numbers will represent a minute fraction of the total number of seriously defective children and of embryonic and neonatal deaths during coming centuries. But I feel that each human being is important, and that it is well worthwhile to calculate the numbers of individual human beings who will be caused to suffer or to die because of the bomb tests, rather than to talk about "negligible effects," "undetectable increase," "extremely small fraction."   (Pauling, 1958)

A university professor comments on the policy-making process regarding nuclear bomb testing.

In sum, here are the tasks which the fallout problem imposes upon us. Research into the hazards of fallout radiation needs to be more fully and widely published so that the scientific community will be constantly aware of the changes which worldwide radiation is making in the life of the planet and its inhabitants. This knowledge must be at the ready command of every scientist, so that we can all participate in the broad educational campaign that must be put into effect to bring this knowledge to the public. If we succeed in this we will have met our major duty, for a public informed on this issue is the only true source of the moral wisdom that must determine our Nation's policy on the testing and the belligerent use of nuclear weapons.

There is a full circle of relationships which connects science and society. The advance of science has thrust grave social issues upon us. And, in turn, social morality will determine whether the enormous natural forces that we now control will be used for destruction—or reserved for the creative purposes that alone give meaning to the pursuit of knowledge.   (Commoner, 1958)

5.2   Safety of the Use of Nuclear Explosives Underground for Large-Scale Excavation or Development of Natural Resources

The Chairman of the U.S.S.R. State Committee for the Utilization of Atomic Energy affirms the safety of nuclear explosions for peaceful purposes, in an interview for *Pravda.*

Question: When nuclear explosions are being discussed, the question of radiation inevitably arises. Of course, each of us understands that people's security is insured, but is there no danger from the side effects of irradiation?

Answer: In our country there is an effective radiation security service. It has the right to veto any work if the slightest doubt arises. This concerns not only nuclear explosions for peaceful purposes, but also the use of atomic power stations and work with radioactive isotopes. In whatever form you may come into contact with atomic science and technology, you inevitably feel the presence of the radiation security service.

So far as nuclear explosions for peaceful purposes are concerned, since they are carried out deep under the ground there is, naturally, no escape of radioactive products to the surface.   (Petrosyants, 1969)

The Chief Judge of the United States District Court for the District of Colorado in a ruling rejects a suit to block a project for the release of natural gas by an underground nuclear explosion (Project Rulison) for reasons of public safety.

> Now, as to the . . . nuisance claim, I think the law is clear, as the government attorney pointed out, that action which is the direct result or an incident which is the direct result of an authorized activity, that is, authorized by the Congress of the United States, cannot be a nuisance in the legal sense. It certainly can be a nuisance, all right, but I mean, in a strictly legal sense, and that's the one we are concerned with here, of course, is the legal sense.
>
> There is evidence that the government has expended something in the neighborhood of a half a million dollars on this project up to this point, and that if the continuation of the project, that is the detonation schedule, isn't permitted, there is something in the evidence that suggests that the daily expense to the Commission would be something in the neighborhood of $31,000 per day.
>
> Now, I have gone over much of the material that has been submitted, I have gone over all of the affidavits, the letters, copies of letters, the Exhibit F series, and it would take me some days, probably up beyond the target date, to understand what is all involved, but I am impressed with the fact that the government has up to this point exercised extreme caution and care to protect the persons, the animal life, the plant life, the water supply and any other things that may be adversely affected by the detonation of this device.
>
> I think it is fair to say that certainly an experiment such as this necessarily carries some risk with it. Any experiment in a new area where you are dealing with materials such as this is bound to carry some risk. An airplane flying over this building may be a risk; if it should fall it might do considerable damage both to the property and to the people. In congested areas such as New York City, with LaGuardia and Kennedy and New York all close by there, and literally hundreds of airplanes coming and going a day, flight over the densely populated areas of the city, there are risks there, and they are not experiments. This is a day to day happening.   (Arraj, 1969)

6 On the Consequences to Civilization of an All-Out Thermonuclear War

An authority in the field of national defense claims that effective defense measures can be taken against thermonuclear war.

> The general belief persists today that an all-out thermonuclear war would inevitably result in mutual annihilation, and that nothing can be done to make it otherwise. Even those who do not believe in total annihilation often

do believe that the shock effect of the casualties, the immediate destruction of wealth, and the long-term deleterious effects of fallout would inevitably jeopardize the survival of civilization.

A study recently carried out by the author and a number of colleagues at Rand, and privately financed by the Rand Corporation, has reached conclusions that seriously question these beliefs. While a thermonuclear war would be a catastrophe, in some ways an unprecedented catastrophe, it would still be limited catastrophe. Even more important, the limits on the magnitude of the catastrophe might be sharply dependent on what prewar measures had been taken. The study suggests that for the next 10 or 15 years, and perhaps for much longer, feasible combinations of military and nonmilitary defense measures can come pretty close to preserving a reasonable semblance of prewar society.   (Kahn, 1959)

A distinguished scientist and member of the General Advisory Committee of the Atomic Energy Commission speaks in favor of preparedness for nuclear war.

There are relatively simple things we can do in preparation for the time of disaster which will make a tremendous difference in our response as individuals and as a nation.

The most effective way to reduce war casualties is to not have the war; and the national policy is to work continually toward conditions which lead to a lasting, just peace for all men.

We are led, when we review the history of man, ancient and modern, to the conclusion that it is wise to take out some insurance for our protection in the event that something goes wrong and peaceful international relations come to an end.

The nature of the effects of modern nuclear weapons and the range over which these effects can produce casualties may provoke the question: "Is there really anything we can do?" My answer to this question is, "Yes." (Libby, 1959)

The following statement was made by Albert Einstein in February 1950, shortly after the announcement by President Truman that the United States would engage in an all-out effort to develop a hydrogen bomb.

The arms race between the United States and the Soviet Union, initiated originally as a preventive measure, assumes hysterical proportions. On both sides, means of mass destruction are being perfected with feverish haste and behind walls of secrecy. And now the public has been advised that the production of the hydrogen bomb is the new goal which will probably be accomplished. An accelerated development toward this end has been solemnly proclaimed by the President. If these efforts should prove successful, radio-active poisoning of the atmosphere, and hence, annihilation of all life on earth will have been brought within the range of what is technically possi-ble. The weird aspect of this development lies in its apparently inexorable character. Each step appears as the inevitable consequence of the one that went before. And at the end, looming ever clearer, lies general annihilation. (Nathan and Norden, 1960)

A former director of the Livermore Radiation Laboratory, responsible for the development of nuclear weapons, writes about "the ultimate absur-dity" in the arms race between the United States and the Soviet Union.

As we have seen, deployment of MIRV (multiple independently targetable reentry vehicles) by both sides, coupled with advances in accuracy and reli-ability, will put a very high premium on the use of the frightful launch-on-warning tactic and may place an even higher premium on a preemptive strike strategy. Under such circumstances, any fixed land-based-missile sys-tem must be able to launch its missiles so soon after receipt of warning that high-level human authorities cannot be included in a decisionmaking pro-cess without seriously degrading the system, unless perhaps such authorities have been properly preprogrammed to produce the "right" decision in the short time that might be available to them . . . Thus we seem to be heading for a state of affairs in which the determination of whether or not doomsday has arrived will be made either by an automatic device designed for the pur-pose or by a preprogrammed President who, whether he knows it or not, will be carrying out orders written years before by some operations analyst.

Such a situation must be called the ultimate absurdity.   (York, 1970)

7   A PERSONAL STATEMENT

I wrote the following statement for the first edition of *Radiation Protection* in 1971, in the midst of the "cold war." In that era two antagonistic super-powers were constantly at the ready to inflict inconceivable destruction on each other. At this writing the tension between those two superpowers has subsided but the nuclear weapons are still in place, if somewhat reduced in number. The statement is still relevant, though it might be edited to men-tion the potential for the employment of chemical and biological weapons of mass destruction, in addition to nuclear weapons. Now we must also concern ourselves with acts not only by superpowers but by smaller nations and even by well-funded terrorist groups.

We are in danger, programmed to destroy ourselves by the very nuclear arsenals established in pursuit of self-preservation. To prevent nuclear war, we will need all the wisdom with which the human intellect is endowed. But will the dimensions of the problem respond to the intellect alone? I believe we also need those values that emanate from the human soul—values that speak of conduct constrained by a sense of accountability to our Creator, of the sanctity of life, of the search for eternal truths, and of the pursuit of justice; values that call for people to transcend political and religious boundaries and to form bonds of friendship and respect throughout the world. Here is the way to security that is the antithesis to the deployment of weapons of mass destruction. Here lies the hope for effecting their neutralization and ultimate obsolescence.

Problems

A few practice problems are included here to assist workers who need to demonstrate proficiency in calculations pertaining to work with radioactive materials. Formulas used in solving the problems are listed in Part III, section 4. References to the sections of the text that cover the subject matter of the problems or present similar examples are given after each problem. Data needed to solve these problems are given in Appendix II.

1. Shielding beta particles.
Determine the thickness of aluminum (density = 2.7 g/cm^3) needed to stop all the beta particles from a ^{90}Sr source. Why should aluminum be used over a "heavier" metal, like lead? (Ref.: Part Two, Table 2.1 and Example 2.1, section 6.1.3.)

2. Shielding gamma photons.
Determine the thickness of concrete required to reduce the intensity of the 1.33 MeV gamma radiation from ^{60}Co to 1/100 of its value when unshielded. (Ref.: Part Two, Figs. 2.10, 2.13.)

3. Determining equivalent thicknesses of different materials for shielding.
Calculate the thickness of lead that is equivalent in shielding effectiveness to a concrete block 20 cm thick for 0.66 MeV gamma photons from ^{137}Cs. The densities of lead and concrete are 11.3 and 2.3, g/cc respectively. (Ref.: Part Two, Figs. 2.10, 2.13.)

4. Calculating the thickness of shielding required for reducing the dose rate to a given level.
Determine the thickness of lead shielding required to reduce the dose

rate from a 30 mCi ^{137}Cs source to 0.05 mSv/hr (5 mrem/hr) at 30 cm from the source. (Ref.: Part Two, Fig. 2.13.)

5. Determining exposure rate by "rule of thumb" and by use of the specific gamma rate constant.

(a) Calculate the exposure rate in mR/hr at 15 cm from a vial containing 185 MBq (5 mCi) of ^{131}I. (Ref.: Part Two, section 21.2.) Use methods of Examples 2.18 and 2.19. Use data in Table 2.2.

(b) Calculate the exposure rate at 15 cm if the vial in *(a)* is shielded by 2 mm lead (half-value layer of lead for ^{131}I gamma photons is 3 mm). (Ref.: Part Two, section 21.3.)

6. Illustrating that even though a radionuclide may be stored in a lead container, the dose rate at the surface of the container can be quite high, especially if the container is small.

Assume that 370 MBq (10 mCi) of ^{24}Na in 100 μl of solution are contained in a small bottle. The bottle is kept in a lead container with an inner diameter of 1.5 cm and walls 1.2 cm thick.

(a) Calculate the dose rate at the surface of the container and the dose to the hands if the lead container is held in the hands for 30 sec. Assume the photons are incident perpendicular to the surface.

(b) Calculate the dose equivalent rate in mSv/hr (mrem/hr) at 30 cm from the axis of the container. (Ref.: Part Two, section 21.3.)

7. Illustrating how permissible working time may be controlled to comply with standards of exposure.

A technician in a pharmaceutical company handles routinely 18.5 GBq (500 mCi) of ^{131}I, 3.7 GBq (100 mCi) of ^{198}Au, and 0.925 GBq (25 mCi) of ^{42}K, all stored together in a hood. When he works in front of the hood, his mean body position is 60 cm from the active material.

How long can the technician work per week in front of the hood without additional shielding? (Ref.: Part Two, Table 2.10, and accompanying discussion; section 21.2.)

8. Calculating the increased working distance needed to allow a specified working time.

At what distance from the sources described in Problem 7 must the technician stand to allow a working time of 5 hr per week? (Ref.: Part Two, Table 2.10, and accompanying discussion; section 21.2. For additional discussion on inverse square law, see Part Three, section 3.2.2.)

9. Determining the size of a potion of a radioactive pharmaceutical to administer to a patient.

A patient is to be given 0.37 MBq (10 μCi) of ^{131}I on Friday morning at 10 A.M. for a thyroid function test. The assay of the master solution was 0.074 MBq/ml (2 μCi/ml) on the preceding Monday at 2 P.M. How many milliliters from the master solution must be given to the patient? (Ref.: Part Two, section 21.4.)

10. Evaluating the dose from inhalation of a radioactive aerosol in a radiation accident.

The liquid contents of a plastic bottle containing 10 mCi of ^{131}I are vaporized in a fire in a closed room 3 m × 4 m × 3 m high. A person attempting to put out the fire breathes the vapor for 10 min. Calculate the exposure to the thyroid: (a) in megabecquerels deposited, (b) in milligrays. Assume a breathing rate of 1,200 l/hr; an average uptake by the thyroid gland of 27 percent of the inhaled activity; a 20 g mass for the thyroid; effective half-life of 7.6 days; and a gamma dose equal to 10 percent of the beta dose.

How would you evaluate the seriousness of this exposure? What would you recommend regarding subsequent handling of this person? (Ref.: Part Two, section 21.1; Part Three, section 1.5.)

11. Calculating whole-body beta dose.

An investigator proposes the use of ^{14}C-labeled alanine to study its turnover in obese patients. He plans to inject 1.85 MBq (50 μCi) into each subject. The alanine is rapidly eliminated with a biological half-life of 120 min.

(a) What value may we use for the effective half-life?

(b) Calculate the total dose received by the patient for an irradiated mass of 70,000 g.

(c) What restrictions, if any, would you place on the choice of patients? (Ref.: Part Three, section 1.5.)

12. Calculating the beta dose to a tissue.

A patient receives an injection in the body of 37 MBq (1 mCi) of ^{32}P. Thirty percent concentrates rapidly in the bones. The patient weighs 70 kg, and the bones constitute 10 percent of the body weight.

(a) Calculate the initial average dose rate to the bones in mGy/hr.

(b) Calculate the total dose to the bones as a result of this injection. (Ref.: Part Three, section 1.5.)

13. Another exercise in a beta-dose calculation for a tissue.

A certain compound is tagged with ^{35}S, a pure beta emitter. After administration it is found that 1/5 is promptly excreted, 1/5 is concentrated in the skin (mass 2 kg), and the remainder uniformly distributed through the soft tissues (60 kg). The effective half-life in the body is 18 days. A medical institution set maximum committed doses of 1 mSv for the whole body and 50 mSv for the skin for studies on human subjects. What is the maximum permissible tracer dose? (Ref.: Part Three, section 1.5.)

14. Checking for dose to reproductive organs in administration of a beta emitter.

It is proposed to use ^{35}S-labeled sulfate in measurements of extracellular fluid. The protocol calls for the injection of 3.7 MBq (100 μCi).

Calculate the dose to the testes as a result of this test if 0.2 percent of the administered activity localizes in the testes with a biological half-life of 627 days. Assume a mass of 80 g for the testes. (Ref.: Part Three, section 1.5.)

15. Obtaining metabolic data for a human use study.

An investigator must determine the biological half-life of a radioactive compound he is proposing to administer to patients in an experimental study. The radionuclide is ^{35}S. He administers 0.074 MBq (2 μCi) of the compound to a volunteer. He collects all the urine and feces for 30 days for radioassay and finds that the total activity in the excreta at the end of the thirtieth day is 0.0296 MBq (0.8 μCi).

(a) What would the activity in the patient be after 30 days if none were excreted?

(b) What is the actual activity remaining in the patient?

(c) What is the biological half-life, assuming it was constant during the period of collection of the excreta? (Ref.: Part Three, section 1.4.)

16. Dose calculations associated with human use of a gamma emitter.

A patient is given 74 MBq (2 mCi) of microaggregated albumin tagged with ^{99m}Tc for a liver scan.

(a) What is the gamma dose rate at 1 m from the patient at the time the isotope is administered? Neglect attenuation in the patient and assume a point source.

(b) What is the gamma dose to the liver of the patient if 80 percent of the administered activity is taken up by the liver? Neglect the contribution from activity external to the liver.

Biological half-life of microaggregated albumin = 4 hr; weight of liver = 1,700 g; absorbed fraction = 0.15. (Ref.: Part Three, sections 3.3, 3.4.)

17. Assaying a sample with a G-M counter.

A G-M counter was used to measure the calcium-45 content of a labeled sample by comparison with an aliquot from a ^{45}Ca standard solution.

The standard solution was obtained from the National Bureau of Standards and had an activity of 7,380 dis/sec/ml at the time it was shipped, 23 days prior to the day of the measurement. An aliquot of 0.2 ml of this solution was diluted to 10 ml, and 1 ml of the solution was evaporated to dryness and counted on the second shelf. A 1 ml aliquot of the sample was evaporated to dryness and also counted on the same shelf. The following counts were obtained in two minutes: Sample, 5,400; Standard, 8,100; Background, 45. Find the activity of the sample in dis/min per ml. (Ref.: Part Four, section 1.9.)

18. An exercise in the interpretation of counting data.

A patient is given 0.185 MBq (5 μCi) of ^{131}I in a thyroid uptake test. Twenty hours later, the patient's neck is examined with a scintillation detector. A cup containing a potion identical to the one given to the patient is counted in a phantom at the same time and at the same distance.

The data obtained are as follows:

Distance from end of crystal to front of source	35 cm
Counting time of patient	2 min
Counts from patient	2,734
Counting time of administered solution	2 min
Counts from solution	5,734
Counting time for background	10 min
Counts from background	3,670

(a) Calculate the net thyroid count rate and its standard deviation.

(b) Calculate the fractional uptake in the thyroid and its standard deviation.

(c) Express the fractional uptake with limits at the 95 percent confidence interval.

(d) What would be the percent error introduced in the uptake values if the true distance to the patient was 1.5 cm less than the value recorded? (Ref.: Part Four, sections 6.6, 6.7.)

19. Determining the significance of the release of a radioactive gas to the air in terms of radiation standards.

Thirty-seven GBq (1 Ci) of ^{85}Kr is released to a room 12 ft × 15 ft × 8 ft.

(a) Calculate the average concentration of ^{85}Kr in the room in Bq/cc, assuming all the ^{85}Kr remains in the room.

(b) How many times maximum permissible concentration is this, assuming occupational exposure for a 40-hour week?

(c) How long could a worker remain in the room without the maximum weekly exposure being exceeded?

(d) How long could a member of the public remain in the room without the maximum weekly exposure being exceeded? (Ref.: Part Five, section 7, Table 5.4.)

20. Calculating the permissible release rate of a radioactive gas through a hood.

An investigator is interested in releasing 7.4 GBq (200 mCi) of ^{133}Xe through a hood. The air velocity into the hood is 125 ft/min, through an opening 15 in. high and 3 ft wide. Determine the permissible release rate so that the concentration in the effluent from the hood stack does not exceed maximum allowable concentrations averaged over a 24 hr period. (Ref.: Part Five, section 18. 1, Table 5.4.)

21. Monitoring a contaminated area with a G-M counter.

An end-window G-M tube with the cap off gave a counting rate of 10,800 c/min in a survey of a contaminated area. The counting rate with the cap on, at the same location, was 7,000 c/min. The background counting rate was 50 c/min. What was the counting rate due to beta contamination? (Ref.: Part Five, section 21.1.)

Answers to Problems

1. 0.41 cm.

2. 35 cm (based on HVL = 5.4 cm, "good geometry").

3. 3.6 cm (based on HVL = 3.15 cm for concrete and 0.57 cm for lead; "good geometry").

4. 2.5 cm (based on HVL = 0.57 cm, "good geometry"). NCRP, 1976, Report 49, fig. 11, gives a value of about 2.8 cm for an attenuation factor of 0.045 in broad-beam geometry.

5. *(a)* 42.1 mR/hr (rule of thumb); 48.9 mR/hr, Γ. *(b)* AF = 0.63; 0.63 × 48.9 = 30.8 mR/hr.

6. Assume activity originates from a point source at center of container. Distance from source to surface (hand) is 1.95 cm. $\Gamma = 0.5$ R-cm^2/hr-MBq. HVL = 1.5 cm. *(a)* 28.0 R/hr; 232 mR (approx. 232 mrad). *(b)* 280 mSv/hr (28,000 mrem/hr) × $(1.95)^2/(30)^2$ = 11.8 mSv/hr (118 mrem/hr) approx.

A more accurate result is obtained by considering each photon energy separately. From Fig. 2.19, in Part Two, $\Gamma = 11.7$ for 2.754 MeV and 7.2 for 1.37 MeV. The half-value layers are 1.48 cm and 1.13 cm, respectively, and the attenuation factors 0.55 and 0.46, respectively. The surface dose rate is 25.5 R/hr and the dose in 30 sec = 213 mR.

7. Dose rate is 379 mR/hr (3.79 mSv/hr, approx.). Permissible weekly dose is 50 mSv/52 weeks, = 0.962 mSv, giving a permissible working time of 0.962/3.79 = 0.254 hr or 15 min.

8. 268 cm.

9. 6.95 ml.

10. *(a)* 0.555 MBq deposited. *(b)* 790 mGy β dose plus an additional 10 percent γ dose, or 869 mGy total.

11. *(a)* 120 min. *(b)* 0.00215 mGy.

12. *(a)* 0.633 mGy/hr; *(b)* 244 mGy, based on effective average life of 15.9 day.

13. Assume 1 MBq administered. Then committed dose is 1.75 mSv to skin and 0.175 mSv to whole body. The maximum tracer activities allowed are 28.6 MBq based on skin limit and 5.71 MBq based on whole-body limit. 5.71 MBq = maximum tracer dose.

14. 6.88 mGy.

15. *(a)* 0.0581 MBq (1.57 μCi). *(b)* 0.0285 MBq (0.77 μCi). *(c)* number of effective half-lives to give fractional decay of 0.385 = 1.37. T_e^b = 21.9 day; T_b^b = 29.3 day.

16. *(a)* 0.12 mR/hr. *(b)* 1.3 mGy (130 mrad).

17. 5,355/min-ml.

18. *(a)* 1,000 $\pm$ 26.8/min. *(b)* 0.40 $\pm$ 0.012. *(c)* 0.40 $\pm$ 0.024. *(d)* 35 cm recorded, actual distance 33.5 cm. Since patient uptake is measured with reference to 35 cm, patient reading must be reduced by $(33.5/35)^2$, or patient uptake is too high by $(35/33.5)^2$, or 9.2 percent.

19. *(a)* 907 Bq/cc (2.45 $\times$ 10^{-2} μCi/cc). *(b)* 245. *(c)* 9.8 min. *(d)* 0.29 min.

20. 354 MBq/day (9,565 μCi/day).

21. 3,800 c/min.

Data on Selected Radionuclides

Radionuclide	Half-life and type of decay	Major radiations, energies (MeV), percent of disintegrations and equilibrium dose constant, Δ (g-mGy/MBq-hr)				Γ(R-cm^2/hr-MBq); Pb HVL (cm)
		β_{max} (%); β_{av}; e$^-$; x (np)	Δ	γ, x-ray	Δ	
^{3}H	12.3 yr, β^-	0.0186 (100%); 0.0057 av	3.27			
^{11}C	20.3 min, β^+	.980 (100%); .394 av	226	0.511 (200%)	587	
^{14}C	5730 yr, β^-	.156 (100%); .0493 av	28.4			
^{13}N	10.0 min, β^+	1. 19 (100%); .488 av	281	.511 (200%)	588	
^{15}O	124 sec, β^+	1.70 (100%); .721 av	415	.511 (200%)	588	
^{18}F	109 min, EC, β^+	.633 (97%); .250 av	139	.511 (194%)	571	
^{22}Na	2.602 yr, EC, β^+	1.821 (0.06%); .836 av	0.270	1.275 (100%)	734	0.32 R/hr; 1.0 cm
		.546 (91%); .216 av	112.5	.511(181%)	533.4	
		e$^-$ (Auger)	0.0270			
^{24}Na	15.0 hr, β^-	1.392 (100%); .555 av	319	3.860 (0.08%)	1.76	0.50 R/hr; 1.5 cm
				2.754 (100%)	1584	
				1.369 (100%)	788	
^{32}P	14.3 day, β^-	1.71 (100%); .695 av	400			
^{35}S	87.0 day, β^-	.167 (100%) .0488 av	28.1			
^{40}K	1.270×10^6 yr, EC, β^-	1.30 (89.5%); .556 av	286	1.46 (10.3%)	86.9	
		e$^-$ (Auger)	0.0811			
^{42}K	12.4 hr, β^-	3.52 (82%); 1.56 av	731	1.52(18%)	158	0.038/hr; 1.2 cm
		2.00 (18%); .822 av	85.4	.313 (0.17%)	0.297	
		1.68 (0.18%) .699 av	0.703			
^{43}K	2.44 hr, β^-	1.2–1.4 (12.4%); 0.67 av	187	0.167(73%)		0.15/hr
		.825 (82%); .297 av		.373 (90%)		
		e$^-$ 0.37 (0.04%)	0.108	All (191%)	535	
^{45}Ca	163 day, β^-	.257 (100%); .077 av	44.5			
^{47}Ca	4.53 day, β^-	1.985 (18%); .816 av	197	1.30 (75%)		
		.688 (82%); .240 av		All (89%)	613	
^{51}Cr	27.7 day, EC	e$^-$ (Auger)	2.16	.32 (10.2%)	18.8	0.0043 R/hr; 0.2 cm
		x (np)	0.568			

Radionuclide	Half-life and type of decay	Major radiations, energies (MeV), percent of disintegrations and equilibrium dose constant, Δ (g-mGy/MBq-hr)				Γ(R-cm²/hr-MBq); Pb HVL (cm)
		β_{max} (%); β_{av}; e⁻; x (np)	Δ	γ, x-ray	Δ	
⁵⁴Mn	312 day, EC	e⁻ 0.83 (.02%)	0.108	.835 (100%)	480	0.13
		e⁻ (Auger)	2.35			
		x (np)	0.730			
⁵⁵Fe	2.70 yr, EC	e⁻ (Auger)	2.49			
		x (np)	0.838			
⁵⁹Fe	45.0 day, β⁻	.467 (52%); .150 av	67.8	1.292 (44%)	329	0.17 R/hr; 1.1 cm
		.273 (46%); .081 av		1.099 (55%)	351	
		e⁻ (IC)	0.0811	.1922 (2.9%)	4.49	
⁵⁷Co	270 day, EC	e⁻ 0.1 (3.51%)	2.41	.136 (10.4%)	8.81	0.024 R/hr
		e⁻ (IC)	3.59	.122 (86%)	60.4	
		e⁻ (Auger)	4.78			
		x (np)	2.68			
⁵⁸Co	71.3 day, EC, β⁺	.474 (16%); 0.201 av	17.9	.811 (990/0)	472	0.15 R/hr
		e⁻ (IC)	0.135	.511 (31%)	91.2	
		e⁻ (Auger)	2.19			
		x (np)	0.838			
⁶⁰Co	5.26 yr, β⁻	.313 (100%) .0941 av	54.6	1.33 (100%)	767	0.36; 1.1 cm
				1.17 (100%)	674	
⁶⁵Zn	243 day, EC, β⁺	.325 (1.5%); .141 av	1.27	1.11 (51%)	325	0.073 R/hr; 1.0 cm
		e⁻ (Auger)	2.86	.511 (3%)	8.81	
		x (np)	1.65			
⁶⁷Ga	78.1 hr, EC	e⁻ 0.2 (0.4%)	0.351	.394 (4%)	11.1	.030 R/hr
		e⁻ (IC)	16.4	.300 (16%)	27.9	
		e⁻ (Auger)	4.00	.185 (24%)	28.4	
		x (np)	2.57	.092 (41%)	22.1	
⁶⁸Ga	68.3 min, EC, β⁺	1.8980 (88%); 0.835 av	422	1.0774 (3.2%)	22.4	
		.8200 (1.3%); .352 av	2.62	.5110 (178%)	525	
		e⁻ (Auger)	0.270			
		x (np)	0.162			
⁷⁵Se	120 day, EC	e⁻ 0.1-0.4 (3%)	2.59	.400 (11%)	27.0	0.054 R/hr
		e⁻ (IC)	1.97	.265 (57%)	129	
		e⁻ (Auger)	3.41	.136 (54%)	56.6	
		x (np)	3.27			
⁸⁵Kr	10.7 yr, β⁻	.672 (99.6%), .246 av	141	.514 (0.42%)	1.24	0.0011 R/hr
		.150 (0.4%); .041 av	0.0811			
⁸⁶Rb	18.6 day, β⁻	1.7720 (91%); .710 av	373	1.0766 (8.8%)	54.2	0.013
		.6920 (8.8%); .230 av	11.6			
⁸⁵Sr	65.1 day, EC	e⁻ .5 (0.01%)	2.22	.514 (99%)	294	0.081 R/hr
		e⁻ (Auger)	3.05			
		x (np)	4.65			
⁹⁰Sr	28.1 yr, β⁻	.546 (100%); .196 av	113			
⁹⁰Y	64 hr, β⁻	2.273 (100%); .931 av	536			
⁹⁹Mo	66.7 hr, β⁻	1.234 (80%); 0.452 av	225	.778 (4.8%)	22.7	0.024 R/hr; 0.74 cm
		.456 (19%); .140 av		.740 (13.7%)	58.2	

Radionuclide	Half-life and type of decay	Major radiations, energies (MeV), percent of disintegrations and equilibrium dose constant, Δ (g-mGy/MBq-hr)				Γ(R-cm²/hr-MBq); Pb HVL (cm)
		β_{max} (%); β_{av}; e⁻; x (np)	Δ	γ, x-ray	Δ	Γ(R-cm^2/hr-MBq); Pb HVL (cm)
		e⁻ 0. 1-0.2 (2%)	1.46	.366 (1.4%)	3.30	
		e⁻ (IC)	0.595	.181 (6.6%)	6.84	
		e⁻ (Auger)	0.162	.141 (5.6%)	4.84	
				.018 (3.8%)	0.405	
⁹⁹ᵐTc	6.03 hr, ISOM	e⁻ 0.1 (12%)	8.43	.1405 (88%)	71.1	0.016 R/hr; 0.03 cm
		e⁻ (IC)	0.946	.0183 (6.6%)	0.784	
		e⁻ (Auger)	0.595			
¹¹¹In	2.81 day, EC	e⁻ 0.1–0.2 (16%)	16.6	.2470 (94%)	134	
		e⁻ (Auger)	4.08	.1720 (90%)	88.7	
		x (np)	0.189	.0230 (70%)	9.32	
				.0263 (14%)	2.14	
¹²³I	13 hr, EC	e⁻ 0.1 (16%)	11.9	.159 (84%)	76.5	
		e⁻ (Auger)	4.32	.2- .8 (2.3%)	6.62	
		x (np)	0.270	.0273 (71%)	11.2	
				.0313 (15%)	2.73	
¹²⁵I	60.14 day, EC	e⁻ (IC)	4.08	.035 (6.7%)	1.35	0.0073; 0.0037 cm
		e⁻ (Auger)	7.19	.027 (115%)	18.1	
		x (np) 22%	0.459	.031 (25%)	4.41	
¹³¹I	8.06 day, β^-	.806 (0.8%); 0.284 av	1.30	.723 (1.7%)	7.22	0.059 R/hr; .3 cm
		.606 (90%); .192 av	99.1	.637 (6.5%)	24.5	
		.25–.333 (9.3%); .090 av	4.78	.364 (82%)	174	
		.1–.6 (2.1%)	3.89	.284 (5.8%)	10.0	
		e⁻ (IC)	1.14	.080 (2.6%)	1.19	
		e⁻ (Auger)	0.216	.030 (3.8%)	0.757	
¹¹³Sn	115 day, EC	e⁻ .2 (.08%)	0.081	.2550 (2.1%)	3.03	0.046 R/hr; .3 cm
		e⁻ (Auger)	3.54	.0241 (61%)	8.43	
		x (np)	0.162	.027 (1.3%)	1.97	
¹³³Xe	5.31 day, β^-	.1006 (98%)	57.6	.0809 (36%)	17.1	0.0027 R/hr
		e⁻ (IC)	18.8	.0308 (39%)	8.59	
		e⁻ (Auger)	2.51			
		x (np)	0.189			
¹³⁷Cs	30 yr, β^-	1. 1760 (5%); 0.427 av	13.3	[.6616(85%)]^a		0.089 R/hr; 0.5 cm
		.5140 (95%); .175 av	95.1			
¹³⁷ᵐBa	2.55 min, ISOM	e⁻ 0.6 (10.2%)	36.9	.6616 (89.8%)	342	
		e⁻ (Auger)	0.351	.032 (5.9%)	1.30	
		x (np)	0.0270			
¹⁹⁸Au	2.69 day, β^-	.9612 (99%); .316 av	180	.6758 (1.1%)	5.59	0.062 R/hr; .3 cm
		e⁻ .3–.6 (4.2%)	8.43	.4117 (96%)	2.26	
		e⁻ (Auger)	0.162	.0708 (1.4%)	1.14	
		x (np)	0.054			
¹⁹⁷Hg	65 hr, EC	e⁻ .1–.2 (1.5%)	1.43	.1915 (0.29%)	0.432	0.0108 R/hr
		e⁻ (IC)	28.5	.0773 (25%)	11.3	
		e⁻ (Auger)	9.08	.068 (56%)	21.9	
		x (np)	0.351	.079 (16%)	7.14	

Radionuclide	Half-life and type of decay	Major radiations, energies (MeV), percent of disintegrations and equilibrium dose constant, Δ (g-mGy/MBq-hr)				Γ(R-cm^2/hr-MBq); Pb HVL (cm)
		β_{max} (%); β_{av}; e$^-$; x (np)	Δ	γ, x-ray	Δ	
^{203}Hg	46.5 day, β^-	.2120 (100%) .0577 av	6.19	.2792 (82%)	131	0.035 R/hr
		e$^-$ .2–.3 (18%)	22.4	.083 (2.9%)	1.35	
		e$^-$ (Auger)	1.03	.072 (9.9%)	4.11	
		x (np)	0.351			

Sources: Dillman and Van der Lage, 1975; ICRU, 1979, Report *32;* Jaeger et al., 1968 (for specific gamma-ray constants, except ^{99}Mo and ^{99m}Tc, which were calculated from MIRD data); Martin and Blichert-Toft, 1970 (for average beta energies); Quimby, Feitelberg, and Gross, 1970 (for lead half-value layers).

Notes: Δ converted from units of g-rad/μCi-hr in 3d edition to g-mGy/MBq-hr by multiplying by 270.27. Γ converted from R-cm^2/hr-mCi to R-cm^2/hr-MBq by dividing by 37. β_{max} is maximum beta-particle energy; β_{av} is average beta-particle energy. Percentage given in parentheses is for beta particles or photon energy listed, two principal energies given. Equilibrium dose constant (Δ) is for group of energies, including those not listed, where sum of dose constants gives dose for all particles or photons emitted per disintegrations. Photons less than 0.015 MeV classified as nonpenetrating (np). Percentage of disintegrations given only for beta particles and internal conversion (IC) electrons with energies equal to or greater than 0.1 MeV (range 14 mg/cm^2). All Auger electrons and IC listings without stated percentages are less than 0.1 MeV. HLV is approximate half-value layer (cm) for lead. ISOM = isomeric transition. EC = electron capture.

a. ^{137}Cs is normally listed as emitting a 0.66 MeV gamma ray in 85 percent of the disintegrations, but the gamma ray is actually emitted in 89.8 percent of the disintegrations of its decay product, ^{137m}Ba (T^h = 2.55 min), which results from 95 percent of the ^{137}Cs disintegrations.

Some Constants, Conversion Factors, and Anatomical and Physiological Data

Electron mass	$m_e = 9.109 \times 10^{-31}$ kg; 0.511 MeV
Proton mass	$m_p = 1.67252 \times 10^{-27}$ kg; 938.256 MeV
Neutron mass	$m_n = 1.67482 \times 10^{-27}$ kg; 939.50 MeV
Alpha particle mass	$m_a = 6.6443 \times 10^{-27}$ kg; 3,727 MeV
Mass unit, unified mass scale	$u = 1.6605402 \times 10^{-27}$ kg; 931.478 MeV
Electron charge	$e = 1.602 \times 10^{-19}$ coulomb (C)
Energy expended by alpha particles per ion pair in air	$W(\alpha) = 34.98$ eV
Energy expended by beta particles per ion pair in air	$W(\beta) = 33.73$ eV
Avogadro's number	$N_a = 6.023 \times 10^{23}$/mole
Planck's constant	$h = 6.626 \times 10^{-34}$ joule-second
Boltzmann constant	$k = 1.381 \times 10^{-23}$ J K^{-1}
Velocity of light in vacuum	$c = 2.998 \times 10^8$ m/sec
Density of dry air at 0°C, 760 mm Hg	0.001293 g/cc
Density of dry air at 20°C, 760 mm Hg	0.001205 g/cc

Multiply	By	To Obtain
feet	30.48	centimeters
cubic feet	2.832×10^4	cubic centimeters
gallons	3.785	liters
pounds	453.5	grams
British thermal units	2.931×10^{-4}	kilowatt hours
British thermal units	251.8	calories
million electron volts	1.603×10^{-13}	joules

Masses of organs and tissues (grams) for reference adult

	Male	Female		Male	Female
Mass of total body	70,000	58,000	Lungs[a]	1,000	800
Total body water (ml)	42,000	29,000	Pulmonary blood	(530)	(430)
Total blood volume (ml)	5,200	3,900	Lung tissue	(440)	(360)
Total blood mass	3,500	4,100	Bronchial tree	(30)	(25)
Total body fat	13,500	10,000	Lymphatic tissue	700	580
Bladder	45	45	Lymphocytes	1,500	1,200
Normal capacity (ml)	200	200	Muscle, skeletal	28,000	17,000
Brain	1,400	1,200	Ovaries		11
Breasts	26	360	Pancreas	100	85
Esophagus	40	30	Prostate	16	
Stomach	150	140	Skeleton[b]	7,000	4,200
Small intestine	640	600	Red marrow	1,800	1,300
Upper large intestine	210	200	Yellow marrow	1,500	1,300
Lower large intestine	160	160	Skin	2,000	1,790
Heart (without blood)	330	240	Spleen	180	150
Kidneys	310	275	Testes or ovaries	35	11
Liver	1,800	1,400	Thyroid	20	17

Source: ICRP, 1975 (report 23).

a. Numbers in parentheses pertain to the constituents of the lung and add up to the mass of the lung.

b. Mass of skeleton excluding marrow.

Physiological data

	Male	Female
Volume air breathed (m³)		
8-hr working, "light activity"	9.6	9.1
8-hr nonoccupational activity	9.6	9.1
8-hr resting	3.6	2.9
Total	23	21
Water intake (ml/day)		
Total fluid intake	1,950	1,400
In food	700	450
By oxidation of food	350	250
Total	3,000	2,100

Source: ICRP, 1975 (report 23).

Selected Bibliography

BASIC TEXTS AND HANDBOOKS

Attix, F. H., and W. C. Roesch, eds. 1968. *Radiation Dosimetry*, vol. I, *Fundamentals*. New York: Academic Press. *See also Radiation Dosimetry*, vol. II, *Instrumentation* (1966); *Radiation Dosimetry*, vol. III, *Sources, Fields, Measurements, and Applications* (1969); *Topics in Radiation Dosimetry*, *Supplement* (1972).

Bernier, D. R., P. E. Christian, and J. K. Langan. 1997. *Nuclear Medicine: Technology and Techniques*. St. Louis: Mosby.

Cember, H. 1996. *Introduction to Health Physics*. Oxford: Pergamon Press.

Evans, R. D. 1955. *The Atomic Nucleus*. New York: McGraw-Hill.

Hall, E. 2000. *Radiobiology for the Radiologist*. Philadelphia: Lippincott Williams and Wilkins.

Hardy, K., M. Melta, and R. Glickman, eds. 1997. *Non-Ionizing Radiation: An Overview of the Physics and Biology*. Proceedings of the Health Physics Society 1997 Summer School. Madison, WI: Medical Physics Publishing.

Hendee, W. R., and E. R. Ritenour. 1992. *Medical Imaging Physics*. St. Louis: Mosby Year Book.

Hitchcock, R. T., and R. M. Patterson. 1995. *Radio-Frequency and ELF Electromagnetic Energies: A Handbook for Health Professionals*. New York: Van Nostrand Reinhold.

Knoll, G. F. 1999. *Radiation Detection and Measurement*. New York: John Wiley.

Mettler, F. A., and A. C. Upton. 1995. *Medical Effects of Ionizing Radiation*. Philadelphia: W. B. Saunders Company.

Radiological Health Handbook. 1970. Compiled and edited by the Bureau of Radiological Health and the Training Institute, Environmental Control Administration. Washington, DC: Government Printing Office.

Shleien, B., L. A. Slaback, Jr., and B. Birky. 1998. *Handbook of Health Physics and Radiological Health*. Baltimore: Lippincott Williams and Wilkins.

Sprawls, P., Jr. 1995. *Physical Principles of Medical Imaging*. Madison, WI: Medical Physics Publishing.

Tsoulfanidis, N. 1995. *Measurement and Detection of Radiation.* New York: McGraw-Hill.

Webster, E. W. 1988. Radiologic physics. In *Radiology,* ed. J. Taveras and J. Serrucci, vol. I. Philadelphia: J. B. Lippincott.

PUBLICATIONS AND INTERNET SITES OF SOCIETIES AND AGENCIES

Health Physics, the official journal of the Health Physics Society, is an extremely valuable source of information in radiation protection. The Health Physics Society's website is *www.hps.org.*

The Radiation Internal Dose Information Center (RIDIC), a program of the Oak Ridge Institute for Science and Education (ORISE), provides up-to-date information on internal dose estimates and internal dosimetry techniques applied to the practice of nuclear medicine. Their web site is *www.orau.gov/ehsd/ridicint.htm.*

The Medical Internal Radiation Dose Committee (MIRD) of the Society *of* Nuclear Medicine publishes a series of pamphlets giving methods and data for absorbed dose calculations. Pamphlet 10 (1975) gives extensive data on radionuclide decay schemes for use in radiation dose calculations. Pamphlet 11 (1975) gives "S" factors for many radionuclides. The pamphlets may be purchased from MIRD Committee, 404 Church Avenue, Suite 15, Maryville, TN 37801. Information on MIRD may be obtained from the Society of Nuclear Medicine web site, *www.snm.org.*

The National Council on Radiation Protection and Measurements (NCRP) issues reports providing information and recommendations based on leading scientific judgment on matters of radiation protection and measurement. Available reports are given on their web site, *www.ncrp.com,* and are available from NCRP Publications, 7910 Woodmont Avenue, Suite 800, Bethesda, MD 20814.

The International Commission on Radiation Units and Measurements (ICRU) issues reports concerned with the definition, measurement, and application of radiation quantities in clinical radiology and radiobiology. Their web site is *www.icru.org.* The reports are available from ICRU, 7910 Woodmont Avenue, Suite 800, Bethesda, MD 20814.

The International Commission on Radiological Protection (ICRP) issues reports dealing with the basic principles of radiation protection. The reports may be obtained from Pergamon Press, Maxwell House, Fairview Park, Elmsford, NY 10523. The ICRP website is *www.icrp.org.*

The International Atomic Energy Agency issues many publications pertaining to the nuclear science field, including the proceedings of symposia (Proceedings Series, Panel Proceedings Series), a Safety Series covering topics in radiation protection, a Technical Reports Series, a Bibliographical Series, and a Review Series. Publications may be ordered from UNIPUB, Inc., P.O. Box 433, New York, NY 10016. Their web site is *www.iaea.org.*

National consensus standards relating to radiation protection provide valuable information and guidance, and they are often incorporated into the regulations of the Nuclear Regulatory Commission. The major national organization issuing such standards is the American National Standards Institute (ANSI), 1430 Broadway, New York, NY 10018. Their web site is *www.ansi.org.*

The U.S. Nuclear Regulatory Commission issues Regulatory Guides that describe methods acceptable to the NRC staff for implementing specific parts of the Commission's regulations. They are published and revised continuously. Interested persons should request current information from the U.S. Nuclear Regulatory Commission, Washington, DC 20555, Attention: Director, Office of Nuclear Regulatory Research. The complete set of available guides, as well as other useful publications, are given at their web site, *www.nrc.gov.*

The Federation of American Scientists conducts studies on issues pertaining to nuclear technology and public policy, including the development, testing, and arsenals of nuclear weapons. Its web site is *www.fas.org,* which also contains links to other nuclear-related web sites.

The National Institute of Standards and Technology provides a variety of radiological data bases including nuclear physics data, radiation dosimetry data, x-ray and gamma-ray data (including attenuation coefficients and photon cross-sections) and physical constants. Its website is physics.nist.gov/PhysRefData/.

The National Nuclear Data Center at Brookhaven National Laboratory provides compilations of data in the fields of low- and medium-energy nuclear physics and links to other related sites. Some of the fields covered include neutron, charged particle, and photonuclear reactions; nuclear structure; and nuclear decay data. Its web site is *www.nndc.bnl.gov/nndc/.*

The Department of Physics and Health Physics of Idaho State University provides information on radiation-protection matters through its Radiation Information Network, *www.physics.isu.edu/radinf.* It has many links to other sources of radiation information.

The Division of Environmental Health and Safety of the University of Illinois at Urbana–Champaign provides practical material and tutorials on the safe handling of radioactive materials, as well as on laser safety and analytical x-ray machines. Its site is *www.ehs.uiuc.edu.*

In addition to publications of agencies like those cited above, the literature of manufacturers and suppliers of commercial equipment often includes not only information on the availability of equipment but extensive educational material and practical information on methods and applications.

SITES OF REGULATORY AGENCIES

www.doe.gov	Department of Energy
www.epa.gov	Environmental Protection Agency
www.fcc.gov	Federal Communications Commission; nonionizing radiation
www.fda.gov/cdrh	Center for Devices and Radiological Health of the Food and Drug Administration

www.nrc.gov	Nuclear Regulatory Commission
www.crcpd.org	Conference of Radiation Control Program Directors
www.state.nj.us/dep/rpp/index.htm	New Jersey radiation protection program

Links provided from the home page generally provide an efficient way to access desired information. Web browsers like Google *(www.google.com)* also lead to valuable resources.

References

AAPM. 2001. *Reference Values for Diagnostic Radiology.* Report of American Association of Physicists in Medicine Task Group. Submitted for publication, 2002.

Abelson, P. H. 1987. Municipal waste [Editorial]. *Science* 236:1409.

Abrams, H. L. 1978. Factors underlying the overutilization of radiologic examinations. Paper presented at the National Conference on Referral Criteria for X-Ray Examinations. Summarized in the weekly news bulletin issued by the Harvard University News Office for the Medical Area, *Focus,* Nov. 30, 1978.

ACGIH. 1983. *Air Sampling Instruments for Evaluation of Atmospheric Contaminants,* 6th ed. Cincinnati: American Conference of Governmental Industrial Hygienists.

——— 1994. *1994–1995 Threshold Limit Values and Biological Exposure Indices,* 6th ed. [Updated yearly.] Cincinnati: American Conference of Governmental Industrial Hygienists.

——— 1996. *Threshold Limit Values for Chemical Substances and Physical Agents and Biological Exposure Indices.* Cincinnati: American Conference of Governmental Industrial Hygienists.

ACR. 1996. *Standards 1996: Techniques and Indications for MR Safety and Sedation.* Reston, VA: American College of Radiology.

Adams, E. J., et al. 1997. Estimation of fetal and effective dose for CT examinations. *Brit. J. Radiol.* 70:272–278.

Adams, J. A. S., and W. M. Lowder, eds. 1964. *The Natural Radiation Environment.* Chicago: University of Chicago Press.

AEC. 1972. *The Safety of Nuclear Power Reactors (Light-Water Cooled) and Related Facilities.* U.S. Atomic Energy Commission Report WASH-1250. Washington, DC: National Technical Information Service.

Agosteo, S., et al. 1995. Radiation Transport in a Radiotherapy Room. *Health Phys.* 68(1):27–34.

Ahearne, J. F. 1987. Nuclear power after Chernobyl. *Science* 236:673–679.

Ahrens, L. H. 1965. *Distribution of the Elements in Our Planet*. New York: McGraw-Hill.

AIHA. 1991. *Ultraviolet Radiation*. Akron, OH: American Industrial Hygiene Association.

Akselrod, M. S., et al. 1996. A thin layer Al_2O_3:C beta TL detector. *Radiat. Prot. Dosimetry* 66:105–110.

Alberts, B., D. Bray, et al. 1989. *Molecular Biology of the Cell*. New York: Garland.

Alcox, R. W. 1974. Patient exposures from intraoral radiographic examinations. *J. Am. Dent. Assoc.* 88:568–579.

Altshuler, B., N. Nelson, and M. Kuschner. 1964. Estimation of lung tissue dose from the inhalation of radon and daughters. *Health Phys.* 10:1137–1161.

Altshuler, B., and B. Pasternack. 1963. Statistical measures of the lower limit of detection of a radioactivity counter. *Health Phys.* 9:293–298.

Andersen, P. E., Jr., et al. 1982. Dose reduction in radiography of the spine in scoliosis. *Acta Radiol. (Diagn) Wstockhh.* 23(3A):251–253.

Anderson, L. 1997. Biological and health effects of ELF fields: Laboratory studies. In *Non-Ionizing Radiation: An Overview of the Physics and Biology*, ed. K. Hardy, M. Meltz, and R. Glickman. Proceedings of the Health Physics Society 1997 Summer School. Madison, WI: Medical Physics Publishing.

Anderson, L. E., and W. T. Kaune. 1993. Electric and magnetic fields at extremely low frequencies. In *Nonionizing Radiation Protection*, ed. M. J. Suess, L. E. Anderson, and D. A. Benwell-Morison. Geneva: World Health Organization.

Anderson, T. T. 1978. Radiation exposures of Hanford workers; a critique of the Mancuso, Stewart and Kneale report. *Health Phys.* 35:739–750.

Andersson, I. O., and J. Braun. 1963. *A Neutron Rem Counter with Uniform Sensitivity from 0.025 eV to 10 MeV. Neutron Dosimetry*. International Atomic Energy Agency Proceedings Series STI/PUB/69. Vienna: IAEA.

Angelo, D. J., et al. 1994. Exposure rates associated with high level fluoroscopic equipment and data recording modes. *Proceedings of the Twenty-sixth Annual National Conference on Radiation Control*. CRCPD Report 94–9. Frankfort, KY: Conference of Radiation Control Program Directors.

Angevine, D. M., and S. Jablon. 1964. Late radiation effects of neoplasia and other diseases in Japan. *Ann. N.Y. Acad. Sci.* 114:823–831.

ANS. 1999. *Health Effects of Low-Level Radiation*. Position Statement of the American Nuclear Society, April 1999. ANS Document PPS-41. Statement can be found on the ANS website *www.ans.org/PI/lowlevel.html*.

ANSI. 1975. *Radiation Protection Instrumentation Test and Calibration*. American National Standards Institute Standard N323. New York: ANSI.

——— 1978. *Control of Radioactive Surface Contamination on Materials, Equipment, and Facilities to Be Released for Uncontrolled Use*. Draft American National Standard N13.12 (issued on trial basis). New York: ANSI.

——— 1996. *American National Standard for the Safe Use of Lasers*. American National Standards Institute Standard ANSI Z136.1. New York: ANSI.

ANSI/IEEE. 1992a. *IEEE Recommended Practice for the Measurement of Potentially*

Hazardous Electromagnetic Fields—RF and Microwave. IEEE Report C95.3–1991. New York: Institute of Electrical and Electronics Engineers.

——— 1992b. *Standard for Safety Levels with Respect to Human Exposure to Radio Frequency Electromagnetic Fields, 3 kHz to 300 GHz* (IEEE Std. C95.1–1991). New York: Institute of Electrical and Electronics Engineers.

ANSI/HPS. 1999. *Standard and Volume Radioactivity Standards for Clearance.* American National Standards Institute Standard ANSI/HPS N13.12–1999. Prepared by a working group of the Health Physics Society and approved by the American National Standards Institute. McClean, VA: Health Physics Society.

Anspaugh, L. R., R. J. Catlin, and M. Goldman. 1988. The global impact of the Chernobyl reactor accident. *Science* 242:1513–1519.

APS. 1995. *Statement on Power Line Fields and Public Health.* National Policy Statement of the American Physical Policy, adopted by Council April 23, 1995, available on APS website, *www.aps.org/statements/95.2.html.*

Archer, V. E. 1997. EMF and cancer. *HPS Newsletter,* November, pp. 7–8.

Archer, V. E., J. D. Gillam, and J. K. Wagner. 1976. Respiratory disease mortality among uranium miners. *Ann. N.Y. Acad. Sci.* 271:280–293.

Armstrong, B., et al. 1994. Association between exposure to pulsed electromagnetic fields and cancer in electric utility workers in Quebec, Canada, and France. *Am. J. Epidemiol.* 140:805–820.

Arnold, L. 1992. *Windscale 1991.* London: Macmillan.

Arraj, H. A. 1969. Ruling, Richard Crowther et al., plaintiffs, vs. Glenn T. Seaborg et al., defendants. Reprinted in JCAE, 1969b, pp. 734–739.

ASTM. 2001. *Radon Mitigation for Existing Construction.* American Society for Testing and Materials Standard E2121 (in process). West Conshohocken, PA: ASTM.

Attix, F. H., and W. C. Roesch, eds. 1966. *Radiation Dosimetry,* vol. 2, *Instrumentation.* New York: Academic Press.

——— eds. 1968. *Radiation Dosimetry,* vol. 1, *Fundamentals.* New York: Academic Press.

Ayres, J. W. 1970. *Decontamination of Nuclear Reactors and Equipment.* New York: Ronald Press.

Baedecker, P. A. 1971. Digital methods of photopeak integration in activation analysis. *Analyt. Chem.* 43:405–410.

Bailey, W. H., S. H. Su, et al. 1997. Summary and evaluation of guidelines for occupational exposure to power frequency electric and magnetic fields. *Health Phys.* 73:433–453.

Bair, W. J., C. R. Richmond, and B. W. Wachholz. 1974. *A Radiobiological Assessment of the Spatial Distribution of Radiation Dose from Inhaled Plutonium.* U.S. Atomic Energy Commission Report WASH-1320. Washington, DC: U.S. Government Printing Office.

Bair, W. J., and J. M. Thomas. 1975. Prediction of the health effects of inhaled transuranium elements from experimental animal data. In *Transuranium Nuclides in the Environment.* Proceedings of a symposium. Vienna: IAEA.

Bair, W. J., and R. C. Thompson. 1974. Plutonium: biomedical research. *Science* 183:715–722.

Baird, P. A., et al. 1988. Genetic disorders in children and young adults: A population study. *Am. J. Hum. Genet.* 42:677–693.

Bale, W. F. 1951. *Hazards Associated with Radon and Thoron.* Memorandum to files, March 14, Division of Biology and Medicine, Atomic Energy Commission, Washington, DC.

Bale, W. F., and J. Shapiro. 1955. Radiation dosage to lungs from radon and its daughter products. *Proceedings of the U.N. International Conference on Peaceful Uses of Atomic Energy* 13:233.

Baranov, A. E., et al. 1989. Bone marrow transplantation following the Chernobyl nuclear accident. *New Engl. J. Med.* 321:205–212.

Baranov, A. E. et al. 1994. Hematopoietic recovery after 10 Gy acute total body irradiation. *Blood* 83:596–599.

Barkas, W. H. 1963. *Nuclear Research Emulsions.* New York: Academic Press.

Barnes, G. T., and R. E. Hendrick. 1994. Mammography accreditation and equipment performance. *Radiographics* 14:29–138.

Bassett, C. A., et al. 1982. Pulsing electromagnetic field treatment in ununited fractures and failed arthrodeses. *JAMA* 247:623–628.

Bathow, G., E. Freytag, and K. Tesch. 1967. Measurements on 6.3 GeV electromagnetic cascades and cascade-produced neutrons. *Nucl. Phys.* B2:669–689.

Baum, J. W., A. V. Kuehner, and R. L. Chase. 1970. Dose equivalent meter designs based on tissue equivalent proportional counters. *Health Phys.* 19:813–824.

Baxt, J. H., S. C. Bushong, S. Glaza, and S. Kothari. 1976. Exposure and roentgen-area-product in xeromammography and conventional mammography. *Health Phys.* 30:91–94.

Beck, H. L., W. M. Lowder, and B. G. Bennett. 1966. *Further Studies of External Environmental Radiation.* U.S. Atomic Energy Commission Report HASL 170. Washington, DC: AEC.

Beck, T. J., and M. Rosenstein. 1979. *Quantification of Current Practice in Pediatric Roentgenography for Organ Dose Calculations.* U.S. Department of Health, Education, and Welfare Publication (FDA) 79–8078. Washington, DC: Government Printing Office.

Becker, K. 1966. *Photographic Film Dosimetry.* London and New York: Focal Press.
———— 1973. *Solid State Dosimetry.* Cleveland: CRC Press.

Beckett, Brian. 1983. *Weapons of Tomorrow.* New York: Plenum Press

Beckmann, P. 1976. *The Health Hazards of Not Going Nuclear.* Boulder, CO: Golem Press

Beierwaltes, W. H., M. T. J. Hilger, and A. Wegst. 1963. Radioiodine concentration in fetal human thyroid from fallout. *Health Phys.* 9:1263–1266.

Bell, W. J., and P. H. Anthes. 1998. Technical and management challenges in the mitigation of extremely high radon concentrations in a Massachusetts school: A case study. Paper at the American Association of Radon Scientists and Technologists 1998 International Radon Symposium, Cherry Hill, NJ. Northampton, MA: Massachusetts Department of Public Health Radon Program.

Bennett, B. G. 1974. *Fallout ^{239}Pu Dose to Man.* U.S. Department of Energy Environmental Measurements Laboratory Report HASL 278, pp. 41–63. Springfield, VA: National Technical Information Service.

Bennett, B. G., and C. S. Klusek. 1978. Strontium-90 in the diet—results through 1977. U.S. Department of Energy Environmental Measurements Laboratory Report EML-342. *Envir. Quart.,* July 1.

Bernhardt, J. H. 1992. Non-ionizing radiation safety: Radiofrequency radiation, electric and magnetic fields. *Phys. Med. Biol.* 37:807–844.

Blatz, H. W. 1970. Regulatory changes for effective programs. *Second Annual National Conference on Radiation Control.* U.S. Department of Health, Education, and Welfare Report BRH/ORO 70–5. Washington, DC: Government Printing Office.

Bloxham, J., and D. Gubbins. 1998. The evolution of the earth's magnetic field. *Sci. Am.* 251(12):68–75.

Boice, J. D., ed. 1997. Radiation epidemiology: Past and present. In *Implications of New Data on Radiation Cancer Risk.* Proceedings of the Thirty-Second Annual Meeting of the National Council on Radiation Protection and Measurements, Proceedings No. 18. Washington, DC: NCRP.

Boice, J. D., and C. E. Land. 1979. Adult leukemia following diagnostic x-rays? *Am. J. Public Health* 69:137–145.

Boice, J. D., and R. R. Monson. 1977. Breast cancer in women after repeated fluroscopic examinations of the chest. *J. Natl. Cancer Inst.* 59:823–832.

Bond, V. P. 1979. *The Acceptable and Expected Risks from Fast Neutron Exposure.* Position paper, Nov. 27., to National Council on Radiation Protection and Measurements. Washington, DC: NCRP.

Bond, V. P., E. P. Cronkite, S. W. Lippincott, and C. J. Shellabarger. 1960. Studies on radiation-induced mammary gland neoplasia in the rat, III. Relation of the neoplastic response to dose of total-body radiation. *Radiat. Res.* 12:276–285.

Bond, V. P., and L. E. Feinendegen. 1966. Intranuclear ^{3}H thymidine: Dosimetric, radiological and radiation protection aspects. *Health Phys.* 12:1007–1020.

Bourgeois, M., et al. 1992. Reducing transmitted radiation in dental radiography. *Health Phys.* 62:546–552.

Bowser, C. L. 1997. MRI safety and planning issues. In *Non-Ionizing Radiation: An Overview of the Physics and Biology.* ed. K. Hardy et al. Proceedings of Health Physics Society 1997 Summer School. Madison, WI: Medical Physics Publishing.

Bracken, T. D. 1993. Exposure assessment of power frequency electric and magnetic fields. *Am. Ind. Hyg. Assoc. J.* 54:165–177.

Braestrup, C. B. 1969. *Past and Present Status of Radiation Protection: A Comparison.* U.S. Department of Health, Education, and Welfare, Consumer Protection and Environmental Control Administration Report, Seminar Paper 005. Washington, DC: Government Printing Office.

BRH. 1969. *Population Dose from X-Rays, U.S., 1964.* U.S. Department of Health, Education, and Welfare, Bureau of Radiological Health Report, Public Health Service Publication no. 2001. Washington, DC: Government Printing Office.

———— 1970. *Radiological Health Handbook.* Compiled and edited by the Bureau of Radiological Health and the Training Institute, Environmental Control Administration. Washington, DC: Government Printing Office.

———— 1973. *Population Exposure to X-Rays, U.S., 1970.* Public Health Service x-ray exposure study conducted under the direction of the Bureau of Radiological Health. Department of Health, Education, and Welfare Publication (FDA) 73–8041. Washington, DC: Government Printing Office.

———— 1974. *Suggested Optimum Survey Procedures for Diagnostic X-Ray Equipment.* Prepared by the Conference of Radiation Control Program Directors and the Bureau of Radiological Health, U.S. Department of Health, Education, and Welfare. Washington, DC: Bureau of Radiological Health.

———— 1976. *Gonad Doses and Genetically Significant Dose from Diagnostic Radiology, U.S., 1964 and 1970.* Report of the Bureau of Radiological Health. Washington, DC: Government Printing Office.

———— 1978a. Bureau of Radiological Health, unpublished information. Personal communication from A. B. McIntyre, Chief, Nuclear Medicine Studies Staff, Division of Radioactive Materials and Nuclear Medicine.

———— 1978b. *Ninth Annual National Conference on Radiation Control: Meeting Today's Challenges, June 19–23, 1977.* Proceedings issued as HEW Publication (FDA) 78–8054. Washington, D.C.: Government Printing Office.

———— 1978c. X-ray referral conference could establish pattern for future action. News item on National Conference on Referral Criteria for X-Ray Examinations. *BRH Bulletin* 7(20):1–4.

Brill, A. B., M. Tomonaga, and R. M. Heyssel. 1962. Leukemia in man following exposure to ionizing radiation. *Ann. Intern. Med.* 56:590–609.

Brodeur, P. 1989a. Annals of Radiation: The Hazards of Electromagnetic Fields, I. Power Lines. *The New Yorker,* June 12, 1989.

———— 1989b. Annals of Radiation: The Hazards of Electromagnetic Fields, III. Video-Display Terminals. *The New Yorker,* June 26, 1989.

Bross, I. D. J., M. Ball, and S. Falen. 1979. A dosage response curve for the one rad range: Adult risks from diagnostic radiation. *Am. J. Public Health* 69:130–136.

Bross, I. D. J., and N. Natarajan. 1972. Leukemia from low-level radiation. *New Engl. J. Med.* 287:107–110.

———— 1977. Genetic damage from diagnostic radiation. *JAMA* 237:2399–2401.

Brown, R. F., J. W. Shaver, and D. A. Lamel. 1980. *The Selection of Patients for X-ray Examinations.* U.S. Department of Health, Education, and Welfare Publication (FDA) 80–8104. Rockville, MD: Bureau of Radiological Health.

Brown, R. M., et al. 1990. Oxidation and dispersion of HT in the environment: The August 1986 field experiment at Chalk River. *Health Phys.* 58(3):171–181.

Brown, B. H., et al. 1999. *Medical Physics and Biomedical Engineering.* Philadelphia: Institute of Physics Publishing.

Buckton, K. E., P. A. Jacobs, W. M. Court Brown, and R. Doll. 1962. A study of the chromosome damage persisting after x-ray therapy for ankylosing spondylitis. *Lancet* 2:676–682.

Burchsted, C. A., J. E. Hahn, and A. B. Fuller. 1976. *Nuclear Air Cleaning Handbook.* Oak Ridge National Laboratory Report ERDA 76–21. Springfield, VA: National Technical Information Service.

Burgess, W. A., and J. Shapiro. 1968. Protection from the daughter products of radon through the use of a powered air-purifying respirator. *Health Phys.* 15: 115–121.

Bustad, L. K., ed. 1963. The biology of radioiodine. Proceedings of the Hanford Symposium held at Richland, WA, June 17–19. *Health Phys.* 9:1081–1426.

Butts, J. J., and R. Katz. 1967. Theory of RBE for heavy ion bombardment of dry enzymes and viruses. *Radiat. Res.* 30:855–871.

Cagnon, C. H., S. H. Benedict, et al. 1991. Exposure rates in high-level-control fluoroscopy for image enhancement. *Radiology* 178:643–646.

Calzado, A., E. Vano, et al. 1992. Improvements in the estimation of doses to patients from 'complex' conventional x ray examinations. *Radiat. Prot. Dosimetry* 43:201–204.

Cameron, J. R., N. Suntharalingam, and G. N. Kenney. 1968. *Thermoluminescent Dosimetry.* Madison: University of Wisconsin Press.

Campion, E. 1997. Editorial. *New Engl. J. Med.,* July 3, 1997.

Carter, J. 1978. Radiation protection guidance to federal agencies for diagnostic x-rays. Presidential memorandum. *Fed. Regist.* 43(22):4377–4380.

Carter, L. J. 1976. National environmental policy act: Critics say promise unfulfilled. *Science* 193:130–139.

Carter, M. W., and A. A. Moghissi. 1977. Three decades of nuclear testing. *Health Phys.* 33:55–71.

Castronovo, F. P. 1993. An attempt to standardize the radiodiagnostic risk statement in an institutional review board consent form. *Investigative Radiology* 28(6):533–538.

Cember, H. 1996. *Introduction to Health Physics.* New York: McGraw-Hill.

CFR. 1975. *Analysis of Effects of Limited Nuclear Warfare.* Report of Committee on Foreign Relations, U.S. Senate. Washington, DC: Government Printing Office.

———— 1987. Standards for protection against radiation. United States Nuclear Regulatory Commission rules and regulations. *Code of Federal Regulations,* Title 10, Part 20. Washington, DC: Government Printint Office.

Chamberlain, A. C., and E. D. Dyson. 1956. The dose to the trachea and bronchi from the decay products of radon and thoron. *Brit. J. Radiol.* 29:317–325.

Chartier, V. L., T. D. Bracken, and A. S. Capon. 1985. BPA Study of occupational exposure to 60-Hz electric fields. *IEEE Trans. Power Apparat. Systems* PAS-104:733–744.

Chilton, A. B., and C. M. Huddleston. 1963. A semiempirical formula for differential dose albedo for gamma rays. *Nucl. Sci. Eng.* 17:419.

Chilton, A. B., J. K. Shultis, and R. E. Faw. 1984. *Principles of Radiation Shielding*. Englewood Cliffs, NJ: Prentice-Hall, Inc.

Christman, E. A., and E. J. Gandsman. 1994. Radiation safety as part of a comprehensive university occupational health and safety program. *Health Phys.* 66:581–584.

Clark, H. M. 1954. The occurrence of an unusually high-level radioactive rainout in the area of Troy, N.Y. *Science* 119:619.

Clarke, R. H. 1988. *Statement of Evidence to the Hinkley Point C Inquiry*. Statement by the director of the National Radiological Protection Board (United Kingdom), Report NRPB-1.

———— 1995. ICRP recommendations applicable to the mining and minerals processing industries and to natural sources. *Health Phys.* 69:454–460.

———— 1996. The threshold controversy [guest editorial]. *HPS Newsletter,* 24(8), August.

Classic, K. L., and R. J. Vetter. 1999. Reorganization and consolidation of a safety program. *Health Phys.* 76(Suppl.):S27-S31.

Clayton, R. F. 1970. *Monitoring of Radioactive Contamination on Surfaces*. International Atomic Energy Agency Technical Reports Series no. 120. Vienna: IAEA.

Cloutier, J. R., J. L. Coffey, W. S. Synder, and E. E. Watson, eds. 1976. *Radiopharmaceutical Dosimetry Symposium*. Proceedings of a conference held at Oak Ridge, TN, April 26–29. U.S. Department of Health, Education, and Welfare Publication (FDA) 76–8044. Washington, D.C.: Government Printing Office.

CMR. 1983. Fixed facilities which generate electromagnetic fields in the frequency range of 300 kHz to 100 GHz and microwave ovens. *Code of Massachusetts Regulations,* Title 105, Part 122.010.

Cohen, B. L. 1976. Impacts of the nuclear energy industry on human health and safety. *Am. Sci.* 64:550–559.

———— 1977a. Hazards from plutonium toxicity. *Health Phys.* 32:359–379.

———— 1977b. High-level radioactive waste from light-water reactors. *Rev. Mod. Phys.* 49:1–20.

Cohen, S. C., S. Kinsman, and H. D. Maccabee. 1976. *Evaluation of Occupational Hazards from Industrial Radiation: A Survey of Selected States*. U.S. Department of Health, Education and Welfare Publication (NIOSH) 77–42. Washington, DC: Government Printing Office.

Cole, L. A. 1993. *Element of Risk: The Politics of Radon*. Washington, DC: AAAS Press.

Coleman, M., et al. 1983. Leukemia incidence in electrical workers. *Lancet* 1:982–983.

Comar, C. L., ed. 1976. *Plutonium: Facts and Inferences*. Palo Alto, CA: Electric Power Research Institute.

Commoner, B. 1958. The fallout problem. *Science* 127:1023–1026. Reprinted in JCAE, 1959b, pp. 2572–2577.

Compere, L. L. 1982. Electromagnetic fields and bones [Editorial]. *JAMA* 247:623–628.

Conard, R. A. 1976. Personal communication, reported in UNSCEAR, 1977. See also R. A. Conard et al., *A Twenty-Year Review of Medical Findings in a*

Marshallese Population Accidentally Exposed to Radioactive Fallout (1975). Brookhaven National Laboratory Report BNL 50424. Springfield, VA: National Technical Information Service.

Conlon, F. B., and G. L. Pettigrew. 1971. *Summary of Federal Regulations for Packaging and Transportation of Radioactive Materials.* U.S. Department Health, Education, and Welfare Report BRH/DMRE 71–1. Washington, DC: Government Printing Office.

Conti, E. A., G. D. Patton, and J. E. Conti. 1960. Present health of children given x-ray treatment to the anterior mediastinum in infancy. *Radiology* 74:386–391.

Courades, J. M. 1992. The objectives of the directive on radiation protection for patients. *Radiat. Prot. Dosimetry* 43:7–10.

Court Brown, W. M., and R. Doll. 1965. Mortality from cancer and other causes after radiotherapy for ankylosing spondylitis. *Brit. Med. J.* 5474(Dec. 4):1327–1332.

Court Brown, W. M., R. Doll, and A. B. Hill. 1960. Incidence of leukaemia after exposure to diagnostic radiation in utero. *Brit. Med.* J. 5212(Nov. 26):1539–45.

Cowan, G. A. 1976. A natural fission reactor. *Sci. Am.* 235:36–48.

Cowser, K. E., W. J. Boegly, Jr., and D. G. Jacobs. 1966. ^{85}Kr and Tritium in an Expanding World Nuclear Industry. In U.S. Atomic Energy Commission Report ORNL 4007, pp. 35–37.

Crawford, M. 1987. Hazardous waste: Where to put it? *Science* 235:156–157.

CRCPD. 1988. *Average Patient Exposure Guides 1988.* Conference of Radiation Control Program Directors, Inc. Publication 88–5. Available from Office of Executive Secretary, 71 Fountain Place, Frankfort, KY 40601.

———— 1994. *Proceedings of the 26th Annual Meeting of the Conference of Radiation Control Program Directors.* Frankfort, KY: CRCPD.

Crocker, D. G. 1984. Nuclear reactor accidents—the use of KI as a blocking agent against radioiodine uptake in the thyroid—a review. *Health Phys.* 46:1265–79.

Curtis, R. 1997. Second cancers following radiotherapy for cancer. In *Implications of New Data on Radiation Cancer Risk.* Proceedings of the Thirty-Second Annual Meeting of the National Council on Radiation Protection and Measurements, Proceedings No. 18. Washington, DC: NCRP.

Curtis, R., and E. Hogan. 1969. The myth of the peaceful atom. *Nat. Hist.* 78:6.

Dahl, A. H., R. Bostrom, R. G. Patzer, and J. C. Villforth. 1963. Patterns of ^{131}I levels in pasteurized milk network. *Health Phys.* 9:1179–1186.

Darby, S. C., R. Doll, et al. 1987. Long-term mortality after a single treatment course with x rays in patients treated for ankylosing spondylitis. *Brit. J. Cancer* 55:179–90.

Davies, S. 1968. Environmental radiation surveillance at a nuclear fuel reprocessing plant. *Am. J. Public Health* 58:2251.

Davis, A. L. 1970. Clean air misunderstanding. *Phys. Today* 23(May):104.

Davis, F. G., J. D. Boice, et al. 1989. Cancer mortality in a radiation-exposed cohort of Massachusetts tuberculosis patients. *Cancer Res.* 49:6130–6136.

Davis, R. K., G. T. Stevenson, and K. A. Busch. 1956. Tumor incidence in normal Sprague-Dawley female rats. *Cancer Res.* 16:194–197.

DCPA. 1973. What the planner needs to know about fallout. *DCPA Attack Environment Manual.* Defense Civil Preparedness Agency Manual CPG 2–1A6.

DeBeeck, J. 0. 1975. Gamma-ray spectrometry data collection and reduction by simple computing systems. *Atom. Energy Rev.* 13:743–805.

DeLawter, D. S., and T. Winship. 1963. Follow-up study of adults treated with roentgen rays for thyroid disease. *Cancer* 16:1028–1031.

Delgado, J. M. R., et al. 1982. Embryological changes induced by weak, extremely low frequency electromagnetic fields. *J. Anat.* 134:533–551.

Delpizzo, V. 1994. Epidemiological studies of work with video display terminals and adverse pregnancy outcomes (1984–1992). *Am. J. Ind. Med.* 26:465–480.

Dennis, R., ed. 1976. *Handbook on Aerosols.* U.S. Energy Research and Development Administration Report TID-26608. Springfield, Va: National Technical Information Service.

Dertinger, H., and H. Jung. 1970. *Molecular Radiation Biology.* New York: Springer-Verlag.

DeVoe, J. R., ed. 1969. *Modern Trends in Activation Analysis,* vols. 1 and 2. National Bureau of Standards Special Publication 312. Washington, DC: Government Printing Office.

Dillman, L. T. 1969. Radionuclide decay schemes and nuclear parameters for use in radiation dose estimation. Medical Internal Radiation Dose Committee Pamphlets no. 4 and 6. *J. Nucl. Med.,* vol. 10, suppl. no. 2 (March 1969), and vol. 11, suppl. no. 4 (March 1970).

Dillman, L. T., and F. C. Van der Lage. 1975. *Radionuclide Decay Schemes and Nuclear Parameters for Use in Radiation-Dose Estimation.* NM/MIRD Pamphlet no. 10. New York: Society of Nuclear Medicine.

DiNunno, J. J., et al. 1962. *Calculation of Distance Factors for Power and Test Reactor Sites.* U.S. Atomic Energy Commission Report TID-14844. Washington, DC: Office of Technical Services.

DOE. 1987. *Health and Environmental Consequences of the Chernobyl Nuclear Plant Accident.* Department of Energy Report DOE/ER-0332. Springfield, VA: National Technical Information Service.

Doida, Y., T. Sugahara, and M. Horikawa. 1965. Studies on some radiation induced chromosome aberrations in man. *Radiat. Res.* 26:69–83.

Donn, J. J., and R. L. Wolke. 1977. The statistical interpretation of counting data from measurements of low-level radioactivity. *Health Phys.* 32:1–14.

DOT. 1983. *A Review of the Department of Transportation (DOT) Regulations for Transportation of Radioactive Materials.* Compiled and edited by A. W. Grella. Washington, DC: Department of Transportation, Office of Hazardous Materials.

Douglas, J. 1993. EMF in American homes. *EPRI J.* 18(3):18–25.

Dudley, R. A. 1966. Dosimetry with photographic emulsions. In *Radiation Dosimetry,* vol. 2, *Instrumentation.,* ed. F. Attix, W. Roesch, and E. Tochlin. New York: Academic Press.

Edsall, J. T. 1976. Toxicity of plutonium and some other actinides. *Bull. Atom. Sci.* 32:27–37.

Eisenbud, M. 1963. *Environmental Radioactivity.* New York: McGraw-Hill.

———— 1968. Sources of radioactivity in the environment. In Proceedings of a Conference on the Pediatric Significance of Peacetime Radioactive Fallout. *Pediatrics* suppl. 41:174–195.

Eisenbud, M., and T. Gesell. 1997. *Environmental Radioactivity—From National, Industrial, and Military Sources.* New York: Academic Press.

Elder, J. A., and D. F. Cahill, eds. 1984. *Biological Effects of Radiofrequency Radiation.* U.S. Environmental Protection Agency Report EPA-600/8–83–026F. Research Triangle Park, NC: EPA Health Effects Research Laboratory.

Elkind, M. M., and G. F. Whitmore. 1967. *The Radiobiology of Cultured Mammalian Cells.* New York: Gordon and Breach.

Emery, R. J., et al. 1995. Simple physical, chemical and biological safety assessments as part of a routine institutional radiation safety survey program. *Health Phys.* 69(2):278–280.

EPA. 1974. *Plutonium Accidents.* Supplemental information from the Atomic Energy Commission Division of Biomedical and Environmental Research to Environmental Protection Agency Plutonium Standards Hearings, Washington, DC, Dec. 10–11. Washington, DC: GPO.

———— 1976. *Radiation Protection Guidance for Diagnostic X-Rays.* Report of the Interagency Working Group on Medical Radiation, U.S. Environmental Protection Agency. Federal Guidance Report no. 9 EPA 520/4–76–019. Washington, DC: EPA.

———— 1977. *Radiological Quality of the Environment in the United States, 1977,* ed. K. L. Feldman. U.S. EPA Report EPA 520/1–77–009. Washington, DC: EPA.

———— 1978. *Provisions for Hazardous Waste Regulation and Land Disposal Controls under the Resource Conservation Act of 1976.* Summary (SW-644) prepared by the Office of Solid Waste, U.S. Environmental Protection Agency. Washington, DC: EPA.

———— 1988. *Limiting Values of Radionuclide Intake and Air Concentration and Dose Conversion Factors for Inhalation, Submersion and Ingestion.* Federal Guidance Report No. 11. Washington, DC: Office of Radiation Programs, EPA.

———— 1990. *Evaluation of the Potential Carcinogenicity of Electromagnetic Fields.* U.S. Evironmental Protection Agency Document No. EPA/600/6–90/005B. Washington, DC: EPA.

———— 1991. *Radon-resistant Construction Techniques for New Residential Construction.* EPA report EPA/625/2–91/032. Research Triangle Park, NC: EPA Air and Engineering Research Laboratory.

———— 1992a. *A Citizen's Guide to Radon,* 3d ed. Washington, DC: Government Printing Office.

———— 1992b. *Technical Support Document for the 1992 Citizen's Guide to Radon.* Washington, DC: EPA.

———— 2000. *Home Buyer's and Seller's Guide to Radon.* Washington, DC: Government Printing Office.

Epp, E. R., and H. Weiss. 1966. Experimental study of the photon energy spectrum of primary diagnostic x-rays. *Phys. Med. Biol.* 11:225–238.

Etherington, H., ed. 1958. *Nuclear Engineering Handbook.* New York: McGraw-Hill.

Evans, R. D. 1955. *The Atomic Nucleus.* New York: McGraw-Hill.

———— 1963. Statistical fluctuations in nuclear processes. In *Methods of Experimental Physics,* vol. 5B, ed. L. C. Yuan and C-S. Yo. New York: Academic Press.

———— 1967. The radiation standard for bone seekers—evaluation of the data on radium patients and dial painters. *Health Phys.* 13:267–278.

———— 1968. X-ray and γ-ray interactions. In *Radiation Dosimetry,* vol. 1, ed. F. H. Attix and W. C. Roesch. New York: Academic Press.

———— 1974. Radium in man. *Health Phys.* 27:497–510.

Evans, R. D., and C. Goodman. 1940. Determination of the thoron content of air and its bearing on lung cancer hazards in industry. *J. Ind. Hyg. Toxicol.,* 22:89.

Evans, R. D., A. T. Keane, and M. M. Shanahan. 1972. Radiogenic effects in man of long-term skeletal alpha-irradiation. In *Radiobiology of Plutonium,* ed. B. J. Stover and W. S. S. Jee. Salt Lake City: J. W. Press.

Evans, R., and F. Mettler. 1985. National CT use and radiation exposure: United States, 1983. *Am. J. Roentgenol.* 144:1044.

Fabrikant, J. I. 1980. *The BEIR-III Report and Its Implications for Radiation Protection and Public Health Policy.* Lawrence Berkeley Laboratory (University of California) Report LBL-10494. Berkeley, CA: Lawrence Berkeley Laboratory.

Failla, G. 1960. Discussion submitted by Dr. G. Failla. *Selected Materials on Radiation Protection Criteria and Standards: Their Basis and Use.* Printed for the use of the Joint Committee on Atomic Energy, 86th Congress, 2d sess. Washington, DC: Government Printing Office.

FCC. 1999. *Questions and Answers about Biological Effects and Potential Hazards of Radiofrequency Electromagnetic Fields.* Federal Communications Commission, Office of Engineering and Technology, OET Bulletin 56. Washington, DC: FCC.

FDA. 1990. Safe Medical Devices Act of 1990, Public Law 101–629. *Federal Register,* November 26, 1991, 56:60024.

FDA. 1994. Avoidance of serious x-ray induced skin injuries to patients during fluoroscopically-guided procedures. *FDA Public Health Advisory* <www.fda/cdrh/fluor>. Rockville, MD: CDRH/FDA.

———— 1995. *Update on Cellular Phone Interference with Cardiac Pacemakers.* U.S. Food and Drug Administration, Center for Devices and Radiological Health Facts-On-Demand Shelf No. 1083. Washington, DC: FDA.

———— 1996. *Medical device electromagnetic interference issues, problem reports, standards, and recommendations.* U.S. FDA Center for Devices and Radiological Health Facts-On-Demand Shelf No. 1086. Washington, DC: FDA.

———— 1999. Radioactive drugs for certain research uses. *Code of Federal Regulations,* Title 21, Part 361.1. Revised April 1, 1999.

Feely, H. W., and L. E. Toonkel. 1978. Worldwide deposition of ^{90}Sr through

1977. Environmental Measurements Laboratory (Department of Energy) Report EML-344. *Envir. Quart.,* October 1.

Feychting, M., and A. Ahlbom. 1993. Magnetic fields and cancer in children residing near Swedish high-voltage power lines. *Am. J. Epidemiol.* 138:467–481.

Field, R. W., D. J. Steck, et al. 2000. Residential radon gas exposure and lung cancer. The Iowa Radon Lung Cancer Study. *Amer. J. Epidemiol.* 151:1091–1102.

Findeisen, W. 1935. Uber das Absetzen Kleiner, in der Luft suspendierter-Teilchen in der menschlichen Lunge bei der Atmug. *Arch. Ges. Physiol. (Pflugers),* 236:367.

Fish, B. R. 1967. *Surface Contamination.* Proceedings of a symposium held at Gatlinburg, TN, June 1964. New York: Pergamon Press.

——— 1969. The role of nuclear energy in the control of air pollution. *Nucl. Safety* 10:119–130.

Fitzgerald, J. J. 1969. *Applied Radiation Protection and Control.* New York: Gordon and Breach.

Fleischer, R. L., P. B. Price, and R. M. Walker. 1975. *Nuclear Tracks in Solids.* Berkeley: University of California Press.

Floderus, B. T., et al. 1993. Occupational exposure to electromagnetic fields in relation to leukemia and brain tumors: A case-control study in Sweden. *Cancer Causes Control* 4:465–476.

Focht, E. F., E. H. Quimby, and M. Gershowitz. 1965. Revised average geometric factors for cylinders in isotope dosage, I. *Radiology* 85:151.

Foderaro, A. 1971. *The Elements of Neutron Interaction Theory.* Cambridge, MA: MIT Press.

Foster, K. R., and J. E. Moulder. 2000. Are mobile phones safe? *IEEE Spectrum* 37(8):23–28.

Fowler, E. B., R. W. Henderson, and M. F. Milligan. 1971. *Proceedings of Environmental Plutonium Symposium.* Los Alamos Scientific Laboratory Report LA-4756 (UC-41). Los Alamos, NM: Los Alamos Scientific Laboratory.

Fraumeni, J. F., Jr., and R. W. Miller. 1967. Epidemiology of human leukemia: Recent observations. *J. Natl. Cancer Inst.* 38:593–605.

FRC. 1961. *Background Material for the Development of Radiation Protection Standards.* Federal Radiation Council Report no. 2. Washington, DC: Department of Health, Education, and Welfare.

——— 1963. *Estimates and Evaluation of Fallout in the United States from Nuclear Weapons Testing Conducted through 1962.* Report of the Federal Radiation Council, FRC Report 4. Washington, DC: Government Printing Office.

——— 1964. *Background Material for the Development of Radiation Protection Standards.* Federal Radiation Council Report no. 5. Washington, DC: Department of Health, Education, and Welfare.

Frey, A. H., ed. 1994. *On the Nature of Electromagnetic Field Interactions with Biological Systems.* Austin: RG Landes Company, Medical Intelligence Unit.

Friedberg, E. C., G. C. Walker, and W. Siede. 1995. *DNA Repair and Mutagenesis.* Washington, DC: ASM Press.

Frigerio, N. A., and R. S. Stowe. 1976. Carcinogenic and genetic hazard from background radiation. In *Biological and Environmental Effects of Low Level Radiation.* International Atomic Energy Agency Proceedings Series, STI/PUB/409. Vienna: IAEA.

Fullerton, G. D. 1997. Physical Concepts of Magnetic Resonance Imaging. In *Non-Ionizing Radiation: An Overview of the Physics and Biology,* K. Hardy et al., ed. Madison, WI: Medical Physics Publishing.

Fulton, J. P., et al. 1980. Electrical wiring configurations and childhood leukemia in Rhode Island. *Am J. Epidemiol.* 111:292–296.

Fyfe, W. S. 1974. *Geochemistry.* London: Oxford University Press.

Gallini, R. E., S. Belletti, et al. 1992. Adult and child doses in standardized x ray examinations. *Radiat. Prot. Dosimetry* 43:41–47.

Gamow, G. 1961. *The Creation of the Universe.* New York: Bantam Books.

Gandhi, O. P., ed. 1991. *Biological Effects and Medical Applications of Electromagnetic Energy.* Englewood Cliffs, NJ: Prentice Hall.

Geiger, K. W., and C. K. Hargrove. 1964. Neutron spectrum of an ^{241}Am-Be (α-n) source. *Nucl. Phys.* 53:204–208.

Gesell, T. F., G. Burke, and K. Becker. 1976. An international intercomparison of environmental dosimeters. *Health Phys.* 30:125–133.

Gilkey, A. D., and E. F. Manny. 1978. *Annotated Bibliography on the Selection of Patients for X-Ray Examination.* U.S. Department of Health, Education, and Welfare Publication (FDA) 78–8067. Washington, DC: Government Printing Office.

Gitlin, J. N., and P. S. Lawrence. 1964. *Population Exposure to X-rays.* U.S. Public Health Service Publication no. 1519. Washington, DC: Government Printing Office.

Glasstone, S., ed. 1962. *The Effects of Nuclear Weapons.* Washington, DC: Government Printing Office.

Glickman, R. D. 1997. Biological responses to laser radiation. In *Non-Ionizing Radiation: An Overview of the Physics and Biology,* ed. K. Hardy et al. Madison, WI: Medical Physics Publishing.

Gofman, J. W., and E. O'Connor. 1985. *X-Rays: Health Effects of Common Exams.* San Francisco: Sierra Club Books.

Goldhagen, P. 2000. Overview of aircraft radiation exposure and recent ER-2 measurements. *Health Phys.* 79(5):526–544.

Goldman, M. I. 1968. United States experience in management of gaseous wastes from nuclear power stations. *Treatment of Airborne Radioactive Wastes* (Proceedings of a symposium held by the International Atomic Energy Agency). Vienna: IAEA.

Goldsmith, W. A., F. F. Haywood, and D. G. Jacobs. 1976. Guidelines for cleanup of uranium tailings from inactive mills. In *Operational Health Physics: Proceedings of the Ninth Midyear Topical Symposium of the Health Physics Society,* ed. P. L. Carson, W. R. Hendee, and D. C. Hunt. Boulder: Health Physics Society.

Graham, S., et al. 1966. Preconception, intrauterine and postnatal irradiation related to leukemia. In *Epidemiological Approaches to the Study of Cancer and Other*

Chronic Diseases. National Cancer Institute Monograph no. 19. Washington, DC: Government Printing Office.

Grandolfo, M., and P. Vecchia. 1985. Natural and man-made environmental exposures to static and ELF electromagnetic fields. In *Biological Effects and Dosimetry of Static and ELF Electromagnetic Fields,* ed. M. Grandolfo, S. M. Michaelson, and A. Rindi. New York: Plenum Press.

Gray, H. 1977. *Anatomy, Descriptive and Surgical,* ed. T. P. Pick and R. Howden, revised American ed., from the 15th English ed. New York: Bounty Books.

Grodstein, G. W. 1957. *X-ray Attenuation Coefficients from 10 keV to 100 MeV.* U.S. National Bureau of Standards Report, NBS Circular 583. Washington, DC: Government Printing Office.

Grove, R. D., and A. M. Hetzel. 1968. *Vital Statistics Rates in the United States, 1940–1960.* National Center for Health Statistics publication. Washington, DC: Government Printing Office.

Grubner, 0., and W. A. Burgess. 1979. Simplified description of adsorption breakthrough curves in air cleaning and sampling devices. *Am. Ind. Hyg. Assoc. J.* 40:169–179.

Guth, A. H. 1997. *The Inflationary Universe.* Reading, MA: Addison-Wesley.

Halitsky, J. 1968. Gas diffusion near buildings. In *Meteorology and Atomic Energy,* ed. D. H. Slade. U.S. Atomic Energy Commission Document, TID-24190. Springfield, VA: National Bureau of Standards.

Hall, P., A. Mattsson, and J. D. Boice, Jr. 1996. Thyroid cancer following diagnostic iodine-131 administration. *Radiat. Res.* 145:86–92.

Hallen, S., et al. 1992. Dosimetry at x ray examinations of scoliosis. *Rad. Prot. Dosimetry* 43(1–4):49–54.

Halliday, D. 1996. *Fundamentals of Physics,* extended ed. New York: John Wiley & Sons.

Halliday, D., and R. Resnick. 1988. *Fundamentals of Physics.* New York: John Wiley & Sons.

Hammond, S. E. 1971. Industrial-type operations as a source of environmental plutonium. *Proceedings of Environmental Plutonium Symposium.* Los Alamos Scientific Laboratory Report LA-4756. Available from Clearinghouse for Federal Scientific and Technical Information, Springfield, VA 22151.

Hanford, J. M., E. H. Quimby, and V. K. Frantz. 1962. Cancer arising many years after irradiation of benign lesions in the neck. *JAMA* 181:404–410.

Hankins, D. E. 1968. *The Multisphere Neutron-Monitoring Technique.* Los Alamos Scientific Laboratory Report LA-3700. Available from Clearinghouse for Federal Scientific and Technical Information, Springfield, VA 22151.

Hansson, M., and K. G. Lovstrand. 1990. Environmental and professionally encountered electromagnetic fields. In *Biological Effects and Medical Applications of Electromagnetic Energy,* ed. O. P. Gandhi. Englewood Cliffs, NJ: Prentice Hall.

Haque, A. K. M. M., and A. J. L. Collinson. 1967. Radiation dose to the respiratory system due to radon and its daughter products. *Health Phys.* 13:431–443.

Hardy, E. P., Jr., ed. 1970. *Fallout Program Quarterly Summary Report. U.S.* Atomic Energy Commission Report HASL-227.

———— 1978. Sr-90 in diet. Environmental Measurements Laboratory Report EML-344 (UC-11). *Envir. Quart.,* Oct. 1.

Hardy, E. P. Jr., and J. Rivera, eds. 1965. *Fallout Program Quarterly Summary Report.* U.S. Atomic Energy Commission Report HASL-161. Washington, DC: U.S. Atomic Energy Commission.

Harley, J. H. 1952. *A Study of the Airborne Daughter Products of Radon and Thoron.* Ph.D. thesis, Rensselaer Polytechnic Institute, Troy, NY.

———— 1953. Sampling and measurement of airborne daughter products of radon. *Nucleonics* 11(7):12–15.

———— ed. 1972. *EML Procedures Manual.* Contains the procedures used currently by the Environmental Measurements Laboratory (New York) of the Department of Energy. EML Report EML-300. [28th edition published in 1997.] Available free as CD-ROM and on-line *www.eml.doe.gov.*

Harley, J. H., and B. S. Pasternack. 1972. Alpha absorption measurements applied to lung dose from radon daughters. *Health Phys.* 23:771–782.

Harrison, E. R. 1968. The early universe. *Phys. Today* 21:31–39.

HASL. 1970. *Fallout Program Quarterly Summary Report, July 1.* U.S. Atomic Energy Commission Health and Safety Laboratory Report HASL-223. New York, NY.

Hatcher, R. B., D. G. Smith, and L. L. Schulman. 1979. *Building Downwash Modeling—Recent Developments.* Paper presented at Air Pollution Control Association 72nd annual meeting. Lexington, Ma: Environmental Research and Technology.

Hayden, J. A. 1977. Measuring plutonium concentrations in respirable dust. *Science* 196:1126.

Hayes, E. T. 1979. Energy resources available to the United States, 1985–2000. *Science* 203:233–239.

Healy, J. W. 1971. *Surface Contamination: Decision Levels.* Los Alamos Scientific Laboratory Report LA-4558-MS. Springfield, Va: National Technical Information Service.

———— 1975. The origin of current standards (plutonium). *Health Phys.* 29:489–494.

———— 1977. *An Examination of the Pathways from Soil to Man for Plutonium.* Los Alamos Scientific Laboratory Report (informal) LA-6741-MS. Springfield, Va: National Technical Information Service.

Heath, R. L. 1964. *Scintillation Spectra Gamma Spectrum Catalog.* U.S. Atomic Energy Commission Report IDO 16580, vol. 1. Washington, DC: U.S. Atomic Energy Commission.

Hempelmann, L. H. 1968. Risk of thyroid neoplasms after irradiation in childhood. *Science* 160:159–163.

Hempelmann, L. H., W. J. Hall, M. Phillips, et al. 1975. Neoplasms in persons treated in x-rays in infancy: Fourth survey in 20 years. *J. Natl. Cancer Inst.* 55:519–530.

Hemplemann, L. H., J. W. Pifer, G. J. Burke, et al. 1967. Neoplasms in persons treated with x-rays in infancy for thymic enlargment. *J. Natl. Cancer Inst.* 38: 317–341.

Hendee, W. R. 1995. History and status of x-ray mammography. *Health Phys.* 69:636–648.

Hendee, W. R., E. L. Chaney, and R. P. Rossi. 1977. *Radiologic Physics, Equipment and Quality Control.* Chicago: Year Book Medical Publishers.

Hendee, W. R., and E. R. Ritenour. 1992. *Medical Imaging Physics.* St. Louis: Mosby Year Book.

Hendrick, R. E. 1993. Mammography quality assurance. *Cancer* 72(Suppl):1466–1474.

Hendricks, M. 1988. VDTs on trial: Do video display terminals pose a health hazard? *Sci. News* 134:174–175.

Hillenkamp, F. 1989. Laser radiation tissue interaction. *Health Phys.* 56:613–616.

Hine, G. J., and G. L. Brownell, eds. 1956. *Radiation Dosimetry.* New York: Academic Press.

Hine, G. J., and R. E. Johnston. 1970. Absorbed dose from radionuclides. *J. Nucl. Med.* 11:468.

Hitchcock, R. T. 1994. *Radio-Frequency and Microwave Radiation.* Fairfax, VA: American Industrial Hygiene Association.

Hitchcock, R. T., S. McMahan, and G. C. Miller. 1995. *Extremely Low Frequency (ELF) Electric and Magnetic Fields.* Fairfax, VA: American Industrial Hygiene Association.

Hoenes, G. R., and J. K. Soldat. 1977. *Age-Specific Radiation Dose Commitment Factors for a One-Year Chronic Intake.* USNRC Report NUREG-0172. Springfield, VA: National Technical Information Service.

Hoffman, D. A., et al. 1989. Breast cancer in women with scoliosis exposed to multiple diagnostic x rays. *J. Natl. Cancer Inst.* 81:1307–1312.

Holaday, D. A., D. E. Rushing, et al. 1957. Control of radon and daughters in uranium mines and calculations on biologic effects. U.S. Public Health Service Publication No. 494. Washington, DC: USPHS.

Holtzman, R. 1977. Comments on "Estimate of natural internal radiation dose to man." *Health Phys.* 32:324–325.

Howe, D. B. 1993. Staff and patient dosimetry in a hospital health physics program. In *Hospital Health Physics: Proceedings of the 1993 Health Physics Society Summer School,* ed. G. G. Eicholz and J. J. Shonka. Richland, WA: Research Enterprises Publishing Segment.

Howe, G. R. 1995. Lung cancer mortality between 1950 and 1987 afer exposure to fractionated moderate-dose-rate ionizing radiation in the Canadian fluoroscopy cohort study and a comparison with lung cancer mortality in the atomic bomb survivors study. *Radiat. Res.* 142:295–304.

HPA. 1961. Depth dose tables for use in radiotherapy: A survey prepared by the scientific subcommittee of the Hospital Physicists Association. *Brit. J. Radiol.,* suppl. no. 10.

HPS. 1973. *Health Physics in the Healing Arts.* Proceedings of the Seventh Midyear

Topical Symposium of the Health Physics Society. Washington, DC: Government Printing Office.

———— 1988. *Surface Radioactivity Guides for Materials, Equipment and Facilities to Be Released for Uncontrolled Use.* Standard developed by the Health Physics Society, 8000 Westpark Drive, Suite 400, McLean, VA 22102.

———— 1990. *Perspectives and Recommendations on Indoor Radon.* Position Statement of the Health Physics Society. Available at *www.hps.org* and in *HPS Newsletter,* 19(Jan. 1991).

———— 1991. *Assessing Nonionizing Radiation Hazards.* Selected proceedings of the Health Physics Society Summer School. *Health Phys.* 61:1–6.

———— 1997. *Non-Ionizing Radiation: An Overview of the Physics and Biology,* ed. K. Hardy, M. Melta, and R. Glickman. Proceedings of the Health Physics Society 1997 Summer School. Madison, WI: Medical Physics Publishing.

———— 1999. *Surface and Volume Radioactivity Standards for Clearance.* American National Standards Institute Standard ANSI/HPS N13.12–1999. Prepared by a working group of the Health Physics Society and approved by the American National Standards Institute. McLean, VA: Health Physics Society.

HSPH. 1996. *Harvard Report on Cancer Prevention,* vol. 1: *Causes of Human Cancer.* A publication of the Harvard Center for Cancer Prevention of the Harvard School of Public Health. *Cancer Causes Control 7* (suppl. 1, November).

Hubbell, J. H. 1969. *Photon Cross Sections, Attenuation Coefficients, and Energy Absorption Coefficients from 10 keV to 100 GeV.* National Bureau of Standards Report NSRDS-NBS 29. Washington, DC: Government Printing Office.

———— 1974. *Present Status of Photon Cross Section Data 100 eV to 100 GeV.* Invited keynote paper presented at the International Symposium on Radiation Physics, Calcutta, India. Reprint available from author, National Bureau of Standards, Washington, DC 20234.

———— 1977. Photon mass attenuation and mass energy-absorption coefficients for H, C, N, O, Ar, and seven mixtures from 0.1 keV to 20 MeV. *Radiat. Res.* 20:58–81.

———— 1982. Photon mass attenuation and energy-absorption coefficients from 1 keV to 20 MeV. *Int. J. Appl. Radiat. Isotopes* 33:1269.

Hubbell, J. H., and M. J. Berger. 1966. *Photon Attenuation and Energy Absorption Coefficients: Tabulations and Discussion.* National Bureau of Standards Report 8681. Washington, DC: National Bureau of Standards.

Hubbell, J. H., W. H. McMaster, N. Kerr Del Grande, and J. H. Mallett. 1974. X-ray cross sections and attenuation coefficients. In *International Tables for X-Ray Crystallography,* vol. 4, ed. J. A. Ibers and W. C. Hamilton. Birmingham, England: Kynoch Press.

Hubbell, J. H., et al. 1975. Atomic form factors, incoherent scattering functions, and photon scattering cross sections. *J. Phys. Chem. Ref. Data* 4:471–538.

Huda, W., and K. Gordon. 1989. Nuclear medicine staff and patient doses in Manitoba (1981–1985). *Health Phys.* 56:277–285.

Hughes, D. J., and R. B. Schwartz. 1958. *Neutron Cross Sections.* Brookhaven National Laboratory Report BNL 325. (Supplements were issued in 1965 and 1966.) Washington, DC: Government Printing Office.

Hutchison, G. B. 1968. Leukemia in patients with cancer of the cervix uteri treated with radiation. *J. Natl. Cancer Inst.* 40:951–982.

IAEA. 1962. *Safe Handling of Radioisotopes.* Vienna: International Atomic Energy Agency.

———— 1967. *Basic Safety Standards for Radiation Protection.* International Atomic Energy Agency Safety Series no. 9. STI/PUB/147. Vienna: IAEA.

———— 1971a. *Handbook on Calibration of Radiation Protection Monitoring Instruments.* International Atomic Energy Agency Technical Reports Series no. 133. Vienna: IAEA.

———— 1971b. *Rapid Methods for Measuring Radioactivity in the Environment: Proceedings of a Symposium.* Vienna: IAEA.

———— 1973. *Radiation Protection Procedures.* International Atomic Energy Agency Safety Series no. 38. Vienna: IAEA.

———— 1979. *Manual on Decontamination of Surfaces.* International Atomic Energy Agency Safety Series no. 48. Vienna: IAEA.

Ichimaru, M., and T. Ishimaru. 1975. *Leukemia and Related Disorders: A Review of Thirty Years Study of Hiroshima and Nagasaki Atomic Bomb Survivors. Research Supplement, 1975* (Japan). Chiba: Japan Radiation Research Society.

ICNIRP. 1991. Protection of the patient undergoing a magnetic resonance examination: Recommendations of the International Commision on Non-Ionizing Radiation Protection of the International Radiation Protection Association. *Health Phys.* 61:923–928.

———— 1994. Guidelines on limits of exposure to static magnetic fields: Recommendations of the International Commission on Non-Ionizing Radiation Protection. *Health Phys.* 66:100–106.

———— 1997. Guidelines on limits of exposure to broad-band incoherent optical radiation (0.38 to 3 μm). *Health Phys.* 73:539–554.

———— 1998. Guidelines for limiting exposure to time-varying electric, magnetic and electromagnetic fields (up to 300 GHz). *Health Phys.* 74:494–522.

ICRP. 1951. Recommendations of the International Commission on Radiological Protection. *Am. J. Roent.* 65:99.

———— 1955. Recommendations of the International Commission on Radiological Protection (1953). *Brit. J. Radiol.,* suppl. 6.

———— 1960. Report of ICRP committee II on permissible dose for internal radiation (1959), with bibliography for biological, mathematical and physical data. *Health Phys.* 3:1–233.

———— 1963. Report of the RBE committee to the International Commission on Radiological Protection and on Radiological Units and Measurements. *Health Phys.* 9:357–386.

———— 1964. *Recommendations of the International Commission on Radiological Protection* (as amended 1959 and revised 1962). ICRP Publication 6. New York: Pergamon Press.

———— 1965. *Recommendations of the International Commission on Radiological Protection.* ICRP Publication 9. New York: Pergamon Press.

———— 1966a. Deposition and retention models for internal dosimetry of the human respiratory tract: Report of the Task Group on Lung Dynamics of the International Commission on Radiological Protection. *Health Phys.* 12:173–207.

———— 1966b. *The Evaluation of Risks from Radiation.* ICRP Publication 8. Oxford: Pergamon Press.

———— 1966c. *Radiation Protection; Recommendations . . . Adopted September 17, 1965.* ICRP Publication 9. Oxford: Pergamon Press.

———— 1968a. *A Review of the Radiosensitivity of the Tissues in Bone.* ICRP Publication 11. New York: Pergamon Press.

———— 1968b. *Report of Committee IV on Evaluation of Radiation Doses to Body Tissues from Internal Contamination Due to Occupational Exposure.* ICRP Publication 10. Oxford: Pergamon Press.

———— 1969. *Radiosensitivity and Spatial Distribution of Dose.* ICRP Publication 14. New York: Pergamon Press.

———— 1969b. *Protection against Ionizing Radiation from External Sources.* ICRP Publication 15. New York: Pergamon Press.

———— 1970. *Protection of the Patient in X-Ray Diagnosis.* ICRP Publication 16. New York: Pergamon Press.

———— 1971. *Protection of the Patient in Radionuclide Investigations.* ICRP Publication 17. New York: Pergamon Press.

———— 1972. The Metabolism of Compounds of Plutonium and Other Actinides. ICRP Publication 19. New York: Pergamon Press.

———— 1975. *Report of the Task Group on Reference Man.* ICRP Publication 23. New York: Pergamon Press.

———— 1976a. *Limits for Intakes of Radionuclides by Workers.* Dosimetric data for uranium. Report of ICRP Committee 11.

———— 1976b. *Protection against Ionizing Radiation from External Sources.* Republication in one document of ICRP Publications 15 (1969) and 21 (1971). New York: Pergamon Press.

———— 1977a. *The Handling, Storage, Use and Disposal of Unsealed Radionuclides in Hospitals and Medical Research Establishments.* ICRP Publication 25. New York: Pergamon Press.

———— 1977b. *Recommendations of the International Commission on Radiological Protection.* ICRP Publication 26. New York: Pergamon Press.

———— 1979–1989. *Limits for Intakes of Radionuclides by Workers.* ICRP Publication 30. Parts 1–4 and supplements published in *Annals of the ICRP* 2(3/4); 4(3/4); 5; 6(2/3); 7; 8(1–4); 19(4).

———— 1980. *Biological Effects of Inhaled Radionuclides.* ICRP Publication 31. *Annals of the ICRP* 4(1/2).

———— 1981. *Limits for Inhalation of Radon Daughters by Workers.* ICRP Publication 32. New York: Pergamon Press.

———— 1985a. *Protection of the Patient in Radiation Therapy.* ICRP Publication 44. Oxford: Pergamon Press.

———— 1985b. Statement from the 1985 Paris meeting of the International Commission on Radiological Protection. *Health Phys.,* 48:828.

———— 1987a. *Lung Cancer Risk from Environmental Exposures to Radon Daughters.* ICRP Publication 50. Oxford: Pergamon Press.

———— 1987b. *Protection of the Patient in Nuclear Medicine.* ICRP Publication 52. New York: Pergamon Press.

———— 1987c. *Radiation Dose to Patients from Radiopharmaceuticals.* ICRP Publication 53. New York: Pergamon Press.

———— 1988a. *Individual Monitoring for Intakes of Radionuclides by Workers: Design and Interpretation.* ICRP Publication 54. New York: Pergamon Press.

———— 1988b. *Radiation Dose to Patients from Radiopharmaceuticals.* ICRP Publication 53. Oxford: Pergamon Press.

———— 1989. *Age-dependent Doses to Members of the Public from Intake of Radionuclides,* Part 1. ICRP Publication 56. *Annals of the ICRP* 20(2).

———— 1991a. *1990 Recommendations of the International Commission on Radiological Protection.* ICRP Publication 60. *Annals of the ICRP* 21(1–3).

———— 1991b. *Annual Limits on Intake of Radionuclides by Workers Based on the 1990 Recommendations.* ICRP Publication 61. *Annals of the ICRP* 21(4).

———— 1993a. *Age-dependent Doses to Members of the Public from Intake of Radionuclides,* Part 2: *Ingestion Dose Coefficients.* ICRP Publication 67. *Annals of the ICRP* 23(3/4).

———— 1993b. *Protection against Radon-222 at Home and at Work.* ICRP Publication 65. New York: Pergamon Press.

———— 1993c. *Summary of the Current ICRP Principles for Protection of the Patient in Diagnostic Radiology. Annals of the ICRP* 22(3).

———— 1994a. *Dose Coefficients for Intakes of Radionuclides by Workers.* ICRP Publication 68 (supersedes Publication 61). *Annals of the ICRP* 24(4).

———— 1994b. *Human Respiratory Tract Model for Radiological Protection.* ICRP Publication 66. *Annals of the ICRP* 24(1–3).

———— 1995a. *Age-dependent Doses to Members of the Public from Intake of Radionuclides,* Part 3: *Ingestion Dose Coefficients.* ICRP Publication 69. New York: Pergamon Press.

———— 1995b. *Age-dependent Doses to Members of the Public from Intake of Radionuclides.* Part 4, *Inhalation Dose Coefficients.* ICRP Publication 71. *Annals of the ICRP* 25(3/4).

———— 1995c. *Dose Coefficients for Intakes of Radionuclides by Workers.* ICRP Publication 68. *Annals of the ICRP* 24(4).

———— 1996a. *Age-dependent Doses to Members of the Public from Intake of Radionuclides,* Part 5. *Compilation of Ingestion and Inhalation Dose Coefficients.* ICRP Publication 72. *Annals of the ICRP* 26(1).

———— 1996b. *Radiological Protection and Safety in Medicine.* ICRP Publication 73. *Annals of the ICRP* 26(2).

———— 1998. *Individual Monitoring for Internal Exposure of Workers.* ICRP Publication 78 (supersedes Publication 54). *Annals of the ICRP* 27(3/4).

———— 1999. *Protection of the Public in Situations of Prolonged Radiation Exposure.*

[The application of the Commission's system of radiological protection to controllable radiation exposure due to natural sources and long-lived radioactive residues.] ICRP Publication 82. New York: Elsevier Science.

ICRU 1962. *Physical Aspects of Irradiation.* Recommendations of the International Commission on Radiological Units and Measurements, Report 10b. Published by the National Bureau of Standards as Handbook 85. Washington, DC: Government Printing Office.

———1969. *Neutron Fluence, Neutron Spectra and Kerma.* ICRU Report 13. Washington, DC: ICRU.

——— 1970. *Linear Energy Transfer.* ICRU Report 16. Washington, DC: ICRU.

——— 1971a. *Radiation Protection Instrumentation and Its Application.* ICRU Report 20. Washington, DC: ICRU.

——— 1971b. *Radiation Quantities and Units.* ICRU Report 19. [A Supplement to Report 19 was issued in 1973.] Washington, DC: ICRU.

——— 1976. *Conceptual Basis for the Determination of Dose Equivalent.* ICRU Report 25. Washington, DC: ICRU.

——— 1979. *Methods of Assessment of Absorbed Dose in Clinical Use of Radionuclides.* ICRU Report 32. Washington, DC: ICRU.

——— 1980. *Radiation Quantities and Units.* ICRU Report 33. Washington, DC: ICRU.

——— 1986. *The Quality Factor in Radiation Protection.* ICRU Report 40. Bethesda, MD: ICRU.

IEEE. 1992. *IEEE Standard for Safety Levels with Respect to Human Exposure to Radio Frequency Electromagnetic Fields, 3 kHz to 300 GHz.* (IEEE Std. C95.1–1991.) New York: Institute of Electrical and Electronics Engineers.

——— 1999. *IEEE Standard for Safety Levels with Respect to Human Exposure to Radio Frequency Electromagnetic Fields, 3 kHz to 300 GHz.* (IEEE Std. C95.1–1999 Edition.) New York: Institute of Electrical and Electronics Engineers.

IEGMP. 2000. *Mobile Phones and Health.* A report of the Independent Expert Group on Mobile Phones (UK), Sir William Stewart, Chairman. [Available at *www.iegmp.org.uk.]* Chilton, UK: National Radiological Protection Board.

Iinoya, K., and C. Orr, Jr. 1977. Filtration. In *Air Pollution,* vol. 4, *Engineering Control of Air Pollution,* ed. A. C. Stern. New York: Academic Press.

INIRC. 1991. Protection of the patient undergoing a magnetic resonance examination. *Health Phys.* 61:923–928.

INIRC/IRPA. 1988. Alleged radiation risks from visual display units: Statement of the International Non-Ionizing Radiation Committee of IRPA. *Health Phys.* 54(2):231–232.

IRPA. 1988. Guidelines on limits of exposure to radiofrequency electromagnetic fields in the frequency range from 1000 kHz to 300 Ghz: International Radiation Protection Association. *Health Phys.* 54:115–123.

——— 1990. Interim guidelines on limits of exposure to 50/60 Hz electric and magnetic fields: International Radiation Protection Association. *Health Phys.* 58:113–122.

ISS. 1978. *The Military Balance, 1978–1979.* London: International Institute for Strategic Studies.

Jablon, S., and H. Kato. 1970. Childhood cancer in relation to prenatal exposure to atomic bomb radiation. *Lancet,* Nov. 14, pp. 1000–1003.

Jacobi, W. 1964. The dose to the human respiratory tract by inhalation of short-lived ^{222}Rn- and ^{220}Rn-decay products. *Health Phys.* 10:1163–1174.

——— 1972. Relations between the inhaled potential alpha energy of ^{222}Rn and ^{220}Rn-daughters and the absorbed alpha-energy in the bronchial and pulmonary region. *Health Phys.* 23:3–11.

Jacobi, W., and K. Eisfeld. 1981. Internal dosimetry of radon-222, radon-220 and their short-lived daughters. *Proceedings of the Special Symposium on Natural Radiation Environment, 19–23 January 1981.* Bombay, India: Bhabha Atomic Research Centre.

Jacobs, D. G. 1968. *Sources of Tritium and Its Behavior upon Release to the Environment.* U.S. Atomic Energy Commission Report in Critical Review Series. Reprinted in JCAE, 1969a.

Jaeger, R. G., E. P. Blizard, A. B. Chilton, et al., eds. 1968. *Engineering Compendium on Radiation Shielding,* vol. 1, *Shielding Fundamentals and Methods.* Berlin: Springer-Verlag.

James, A. C., J. R. Greenhalgh, and A. A. Birchall. 1980. A dosimetry model for tissues of the respiratory tract at risk from inhaled radon and thoron daughters. In *Radiation Protection: A Systematic Approach to Safety. Proceedings of the 5th Congress of the International Radiation Protection Association. Jerusalem, March 1980,* vol. 2, 1045–1048. Oxford: Pergamon Press.

Jaworowski, Z. 1969. Radioactive lead in the environment and in the human body. *Atom. Energy Rev.* 7:3–45.

JCAE. 1959a. *Biological and Environmental Effects of Nuclear War: Hearings.* U.S. Congress, Joint Committee on Atomic Energy, 86th Cong., 1st sess., June 22–26. Washington, DC: Government Printing Office.

——— 1959b. *Fallout from Nuclear Weapons Tests: Hearings.* U.S. Congress, Joint Committee on Atomic Energy, 86th Cong., 1st sess., May 5–8. Washington, DC: Government Printing Office.

——— 1969a. *Environmental Effects of Producing Electric Power: Hearings.* U.S. Congress, Joint Committee on Atomic Energy, 91st Cong., 1st sess., Oct. 28–31, Nov. 4–7. Washington, DC: Government Printing Office.

——— 1969b. *Nuclear Explosion Services for Industrial Applications: Hearings.* U.S. Congress, Joint Committee on Atomic Energy, 91st Cong., 1st sess., May 8, 9, and July 17; 2d sess., Jan. 27–30, Feb. 24–26. Washington, DC: Government Printing Office.

Johansen, K. S. 1992. The strategy of a European health policy of the WHO member states. *Radiat. Prot. Dosimetry* 43:1–5.

Johns, H. E. 1969. X-rays and teleisotope γ rays. In *Radiation Dosimetry,* vol. 3, ed. F. H. Attix and E. Tochilin. New York: Academic Press.

Johnson, C. J., R. R. Tidball, and R. C. Severson. 1976. Plutonium hazard in respirable dust on the surface of soil. *Science* 193:488–490.

Johnson, D. W., and W. A. Goetz. 1986. Patient exposure trends in medical and dental radiography. *Health Phys.* 50:107–116.

Jones, A. R. 1962. Pulse counters for γ dosimetry. *Health Phys.* 8:1–9.

Jonsson, A., et al. 1995. Computed radiograpy in scoliosis: Diagnostic information and radiation dose. *Acta Radiol.* 36(4):429–33.

Kahn, B., R. L. Blanchard, H. L. Krieger, et al. 1970. *Radiological Surveillance Studies at a Boiling Water Nuclear Power Reactor.* U.S. Department of Health, Education, and Welfare Report BRH/DER 70–1. Washington, DC: Government Printing Office.

Kahn, H. 1959. Major implications of a study of nuclear war. Statement for *Hearings on Biological and Environmental Effects of Nuclear War,* pp. 908–922. See JCAE, 1959a.

Kahn, F. M. 1997. *The Physics of Radiation Therapy.* Baltimore: Lippincott, Williams & Wilkins.

Kaiser, J. 1996. Panel Finds EMFs Pose No Threat. *Science* 274(Nov. 8):910.

Kalifa, G., et al. 1998. Evaluation of a new low-dose digital x-ray device: First dosimetric and clinical results in children. *Pediatr. Radiol.* 28(7):557–561.

Kathren, R. 1987. External beta-photon dosimetry for radiation protection. In *The Dosimetry of Ionizing Radiation,* vol. 2, ed. by K. Kase, B. Bjarngard, and F. Attix. New York: Academic Press.

Katz, R., S. C. Sharma, and M. Homayoonfar. 1972. The structure of particle tracks. *Topics in Radiation Dosimetry: Radiation Dosimetry,* suppl. 1, ed. F. H. Attix. New York: Academic Press.

Kavet, R., and R. A. Tell. 1991. VDTs: Field levels, epidemiology, and laboratory studies. *Health Phys.* 61:47–57.

Kearsley, E. 1999. 1999 NCRP annual meeting. *HPS Newsletter,* June 1999.

Kellerer, A. M. 1996. Radiobiological challenges posed by microdosimetry. *Health Phys.* 70(6):832–836.

Kellerer, A. M., and H. H. Rossi. 1972. The theory of dual radiation action. *Curr. Top. Radiat. Res. Quart.* 8:85–158.

———— 1978. A generalized formulation of dual radiation action. *Radiat. Res.* 75:471–488.

Kelley, J. P., and E. D. Trout. 1971. Physical characteristics of the radiations from 2-pulse, 12-pulse, and 1000-pulse x-ray equipment. *Radiology* 100:653–661.

Kemeny, J. G., chairman. 1979. *The Need for Change: The Legacy of TMI. Report of the President's Commission on the Accident at Three Mile Island.* Washington, DC.

Kenny, P. J., D. D. Watson, and W. R. Janowitz. 1976. Dosimetry of some accelerator produced radioactive gases. In Cloutier et al., 1976.

Kereiakes, J. G., et al. 1976. Pediatric radiopharmaceutical dosimetry. In Cloutier et al., 1976.

Kernan, W. J. 1963. *Accelerators.* Booklet in a series entitled "Understanding the Atom." Available from United States Atomic Energy Commission, P.O. Box 62, Oak Ridge, TN.

Kiell, M. 1989. Do VDTs create a miscarriage risk? *Health Physics Society's Newsletter,* October 1989, p. 1.

Kling, T. F., Jr., et al. 1990. Digital radiography can reduce scoliosis x-ray exposure. *Spine* 15(9):880–885.

Knoll, G. F. 1999. *Radiation Detection and Measurement.* New York: Wiley.

Knox, H. H., and R. M. Gagne. 1996. Alternative methods of obtaining the computed tomography dose index. *Health Phys.* 71:219–224.

Kobayashi, Y., and D. V. Maudsley. 1974. *Biological Applications of Liquid Scintillation Counting.* New York: Academic Press.

Kogutt, M. S., et al. 1989. Low dose imaging of scoliosis: Use of a computed radiographic imaging system. *Pediatr. Radiol.* 20(1–2):85–86.

Kohn, H. I., J. C. Bailar III, and C. Zippin. 1965. Radiation therapy for cancer of the cervix: Its late effect on the lifespan as a function of regional dose. *J. Natl. Cancer Inst.* 34:345–361.

Korff, S. A. 1955. *Electron and Nuclear Counters.* New York: Van Nostrand.

Kramer, H. M., and K. Schnuer, eds. 1992. *Dosimetry in Diagnostic Radiology.* Proceedings of a seminar held in Luxembourg, March 19–21, 1991. Published as a special issue of *Radiat. Prot. Dosimetry* 43(1–4).

Krey, P. W. 1974. Plutonium-239 contamination in the Denver area. *Health Phys.* 26:117–120.

———— 1976. Remote plutonium contamination and total inventories from Rocky Flats. *Health Phys.* 39:209–214.

Krey, P. W., and E. P. Hardy. 1970. *Plutonium in Soil around the Rocky Flats Plant.* Health and Safety Laboratory Report HASL-235.

Krey, P. W., et al. 1976a. Interrelations of surface air concentrations and soil characteristics at Rocky Flats. In *Atmosphere Surface Exchange of Particulate and Gaseous Pollutants, 1974.* Proceedings of a symposium held at Richland, WA, Sep. 4–6, 1974, coordinated by R. J. Engelmann and G. A. Sehmel. Richland, WA: Pacific Northwest National Laboratory.

———— 1976b. *Plutonium and Americium Contamination in Rocky Flats Soil, 1973.* Report HASL-304, issued by Environmental Measurments Laboratory, Department of Energy, New York, NY 10014.

Krey, P. W., E. P. Hardy, and L. E. Toonkel. 1977. *The Distribution of Plutonium and Americium with Depth in Soil at Rocky Flats.* Report HASL-318, issued by Environmental Measurements Laboratory, Department of Energy, New York, NY 10014.

Krzesniak, J. W., O. A. Chomicki, et al. 1979. Airborne radioiodine contamination caused by ^{131}I treatment. *Nuklearmedizin* 18:246–251.

Kuehner, A. V., J. D. Chester, and J. W. Baum. 1973. Portable mixed radiation dose equivalent meter. In *Neutron Monitoring for Radiation Protection Purposes,* vol. 1, pp. 233–246. Vienna: International Atomic Energy Agency.

Landahl, H. D. 1950. Removal of air-borne droplets by the human respiratory tract: The lung. *Bull. Math. Biophys.* 12:43.

Langham, W. H. 1971. Plutonium distribution as a problem in environmental science. *Proceedings of Environmental Plutonium Symposium.* Los Alamos Scientific Laboratory Report LA-4756. Los Alamos, NM: Los Alamos Scientific Laboratory.

Lanza, F., O. Gautsch, and P. Weisgerber. 1979. *Contamination Mechanisms and Decontamination Techniques in Light Water Reactors.* Report EUR 6422 EN. Ispra, Italy: Commission of the European Communities.

Lanzl, L. H. 1976. State and federal regulatory measurement responsibilities around medical facilities. In NBS, 1976.

Lanzl, L. H., J. H. Pingel, and J. H. Rust. 1965. *Radiation Accidents and Emergencies in Medicine, Research, and Industry.* Springfield, IL: Charles C Thomas.

Lapp, R. E. 1971. How safe are nuclear power plants? *New Republic* 164:18–21.

Lassen, N. A. 1964. Assessment of tissue radiation dose in clinical use of radioactive gases, with examples of absorbed doses from ^{3}H, ^{85}Kr, and ^{133}Xe. *Minerva Nucl.* 8:211–217.

Laws, P. W., and M. Rosenstein. 1978. A somatic dose index for diagnostic radiology. *Health Phys.* 35:629–642.

——— 1980. *Quantitative Analysis of the Reduction in Organ Doses in Diagnostic Radiology by Means of Entrance Exposure Guidelines.* U.S. Department of Health, Education, and Welfare Publication (FDA) 80B8107. Washington, DC: Government Printing Office.

Lea, D. E. 1955. *Actions of Radiations on Living Cells.* Cambridge, England: Cambridge University Press.

Levitt, B. B. 1995. *Electromagnetic Fields: A Consumer's Guide to the Issues and How to Protect Ourselves.* New York: Harcourt Brace & Company.

Levy, A. R., et al. 1994. Projecting the lifetime risk of cancer from exposure to diagnostic ionizing radiation for adolescent idiopathic scoliosis. *Health Phys.* 66(6):621–633.

Lewis, E. B. 1970. Ionizing radiation and tumor production. In *Genetic Concepts and Neoplasia.* Baltimore: Williams & Wilkins.

Lewis, F. 1967. *One of Our H-Bombs Is Missing . . .* New York: McGraw-Hill.

Li, L. B., et al. 1995. Occupational exposure in pediatric cardiac catheterization. *Health Phys.* 69(2):261–264.

Libby, W. F. 1955. *Radiocarbon Dating.* Chicago: University of Chicago Press.

——— 1959. Statement in *Hearings on Biological and Environmental Effects of Nuclear War,* pp. 923–932. See JCAE, 1959a.

Lidsey, J. E. 2000. *The Bigger Bang.* Cambridge, UK: Cambridge University Press.

Lilienfeld, A. M., et al. 1978. *Evaluation of Health Status of Foreign Service and Other Employees from Selected European Posts (NTIS PB288163).* Springfield, VA: National Technical Information Service.

Lindell, B. 1968. Occupational hazards in x-ray analytical work. *Health Phys.* 15:48–86.

Lindop, P. J., and J. Rotblat. 1961. Long term effects of a single whole body exposure of mice to ionizing radiations. *Proc. Roy. Soc. (London)* B154:332–360.

Lister, B. A. J. 1964. *Health Physics Aspects of Plutonium Handling.* A series of lectures given during a visit to Japan, March. Atomic Energy Research Establishment Report AERE-L 151. London: H. M. Stationery Office.

Little, J. B. 1993. Biologic Effects of Low-Level Radiation Exposure, in *Radiol-*

ogy: Diagnosis—Imaging—Intervention, ed. J. M. Taveras, Philadelphia: J. B. Lippincott.

————— 2000. Cancer etiology: Ionizing radiation. In *Cancer Medicine,* ed. J. Holland and E. Frei. Hamilton, Ontario: B. C. Decker.

Loevinger, R. 1956. The dosimetry of beta sources in tissue: The point-source function. *Radiology* 66:55–62.

London, S. J., et al. 1991. Exposure to residential electric and magnetic fields and risk of childhood leukemia. *Am. J. Epidemiol.* 134:923–937.

London, S. J., et al. 1994. Exposure to magnetic fields among electrical workers in relation to leukemia risk in Los Angeles County. *Am. J. Ind. Med.* 26:47–60.

Loomis, D. P., D. A. Savitz, and C. V. Anath. 1994. Breast cancer mortality among female electrical workers in the United States. *J. Natl. Cancer Inst.* 86:921–925.

Lorenz, E., J. W. Hollcroft, E. Miller, et al. 1955. Long-term effects of acute and chronic irradiation in mice. 1. Survival and tumor incidence following chronic irradiation of 0–11 R per day. *J. Natl. Cancer Inst.* 15:1049–1058.

Lovsund, P., et al. 1982. ELF magnetic fields in electrosteel and welding industries. *Radio-Science* 17(5s):355–385.

Lowder, W. M., and W. J. Condon. 1965. Measurement of the exposure of human populations to environmental radiation. *Nature* 206:658–662.

Lucas, J. N., et al. 1995. Dose response curve for chromosome translocations measured in human lymphocytes exposed to ^{60}Co gamma rays. *Health Phys.* 68(6):761–765.

Luftig, S., and L. Weinstock. 1997. *Establishment of Cleanup Levels for CERCLA Sites with Radioactive Contamination.* Memorandum OSWER No. 9200.4–18, United States Environmental Protection Agency. Available at *www.epa.gov/radiation/cleanup/rad.arar.pdf.*

Lytle, C. D., W. H. Cyr, et al. 1993. An estimation of squamous cell carcinoma risk from ultraviolet radiation emitted by flourescent lamps. *Photoderm., Photoimmunol. and Photomed.* 9:268–274.

Mabuchi, K., E. Ron, and D. L. Preston. 1997. Cancer incidence among Japanese atomic-bomb survivors. In *Implications of New Data on Radiation Cancer Risk.* Proceedings of the Thirty-Second Annual Meeting of the National Council on Radiation Protection and Measurements, no. 18. Washington, DC: NCRP.

Machta, L. 1963. Meteorological processes in the transport of weapon radioiodine. *Health Phys.* 9:1123–1132.

MacMahon, B. 1962. Prenatal x-ray exposure and childhood cancer. *J. Natl. Cancer Inst.* 28:1173–1191.

Mancuso, T. F. 1978. *Study of Lifetime Health and Mortality Experience of Employees of ERDA Contractors.* Testimony prepared for hearing of the Subcommittee on Health and Environment of the House of Representatives on Feb. 8 (U.S. Congress). Washington, DC: GPO.

Mancuso, T., A. Stewart, and G. Kneale. 1977. Radiation exposures of Hanford workers dying from cancer and other causes. *Health Phys.* 33:369.

————— 1978. Reanalysis of data relating to the Hanford study of the cancer risks of radiation workers. *JAEA International Symposium on the Late Biological Effects of Ionizing Radiation.* Vienna: International Atomic Energy Agency.

Manor, R. 1997. Litigating DC to daylight. In *Non-Ionizing Radiation: An Overview of the Physics and Biology. Health Physics Society 1997 Summer School,* ed. by K. Hardy, M. Meltz and R. Glickman. Madison, WI: Medical Physics Publishing.

Marbach, W. 1984. Are VDT's health hazards? *Newsweek,* Oct. 29, 1984, p. 122.

Mark, J. C. 1976. Global consequences of nuclear weaponry. *Ann. Rev. Nucl. Sci.* 26:51–57.

Marks, H. S., ed. 1959. *Progress in Nuclear Energy, Series X. Law and Administration.* New York: Pergamon Press.

Marshall, E. 1981. FDA sees no radiation risk in VDT screens. *Science* 212:1120–1121.

Martin, M. J., and P. H. Blichert-Toft. 1970. Radioactive atoms, auger electron, α-, β-, γ-, and x-ray data. *Nucl. Data Tables* 8:1–198.

Maruyama, T., Y. Kumamoto, et al. 1992. Determinations of organ or tissue doses and collective effective dose equivalent from diagnostic x ray examinations in Japan. *Radiat. Prot. Dosimetry* 43:213–216.

Masse, F. X., and M. M. Bolton, Jr. 1970. Experience with a low-cost chair-type detector system for the determination of radioactive body burdens of M.I.T. radiation workers. *Health Phys.* 19:27–35.

Matanoski, G. M., et al. 1993. Leukemia in telephone linemen. *Am. J. Epidemiol.* 137:609–619.

Mathieu, I., et al. 1999. Recommended restrictions after ^{131}I therapy: Measured doses in family members. *Health Phys.* 76(2):129–136.

Maxon, H. R., S. R. Thomas, E. L. Saenger, C. R. Buncher, and J. G. Kereiakes. 1977. Ionizing irradiation and the induction of clinically significant disease in the human thyroid gland. *Am. J. Med.* 63:967–978.

Mayneord, W. V., and C. R. Hill. 1969. Natural and manmade background radiation. In *Radiation Dosimetry,* vol. 3, ed. H. H. Attix and E. Tochilin. New York: Academic Press.

Mays, C. W. 1975. Estimated risk from [239]Pu to human bone, liver, and lung. *Biological and Environmental Effects of Low-Level Radiation.* Symposium proceedings. Vienna: International Atomic Energy Agency.

McBride, J. P., R. E. Moore, et al. 1978. Radiological impact of airborne effluents of coal and nuclear plants. *Science* 202:1045–1050.

McClenahan, J. L. 1976. A radiologist's view of the efficient use of diagnostic radiation. In *Assuring Radiation Protection: Proceedings of the 7th Annual National Conference on Radiation Control.* Washington, DC: Government Printing Office.

McCullough, E. C., and J. R. Cameron. 1971. Exposure rates from diagnostic x-ray units. *Health Phys.* 20:443–444.

McCullough, E. C., and J. T. Payne. 1978. Patient dose in computed tomography. *Radiology* 129:457–463.

McDonnel, G. M. 1977. Computerized axial tomographic scanners—use, potential, and control. In *Eighth Annual National Conference on Radiation Control. Radiation Benefits and Risks: Facts, Issues, and Options,* May 2–7, 1976. HEW Publication (FDA) 77–8021. Rockville, MD: Bureau of Radiological Health.

McDowell, W. J., F. G. Seeley, and M. T. Ryan. 1977. Penetration of HEPA filters by alpha recoil aerosols. In *Proceedings of the Fourteenth ERDA Air Cleaning Conference.* Springfield, VA: National Technical Information Service.

McGinley, P. H. 1992. Photoneutron production in the primary barriers of medical accelerator rooms. *Health Phys.* 62(4):359–362. See also, Photoneutron fields in medical accelerator rooms with primary barriers constructed of concrete and metals. *Health Phys.* 63(6):698–701.

McGregor, D. H., et al. 1977. Breast cancer incidence among atomic bomb survivors, Hiroshima and Nagasaki, 1950–1969. *J. Natl. Cancer Inst.* 59:799–811.

McGregor, R. G., et al. 1980. Background concentrations of radon and radon daughters in Canadian homes. *Health Phys.* 39:285–289.

McKeever, S. W. S., B. G. Markey, and L. E. Colyott. 1995. Time-resolved optically stimulated luminescence from α-Al_2O_3:C. *Radiation Measurements* 24:457–463.

McKlveen, J. W. 1980. X-ray exposures to dental patients. *Health Phys.* 39:211–217.

McLaughlin, J. E., Jr., and H. Blatz. 1955. Potential radiation hazards in the use of x-ray diffraction equipment. *Ind. Hyg. Quart.* 16:108–112.

Mercer, T. T. 1973. *Aerosol Technology in Hazard Evaluation.* New York: Academic Press.

———— 1976. The effect of particle size on the escape of recoiling RaB atoms from particulate surfaces. *Health Phys.* 31:173–175.

Merchant, C. J., D. C. Renew. and J. Swanson. 1994. Exposures to power-frequency magnetic fields in the home. *J. Radiol. Prot.* 14:77–87.

Mettler, F. 1987. Diagnostic radiology: Usage and trends in the United States. *Radiology* 162:263–266.

Mettler, F. A., and R. D. Moseley. 1985. *Medical Effects of Ionizing Radiation.* Orlando, FL: Grune and Stratton.

Mettler, F. A., A. G. Williams, et al. 1985. Trends and utilization of nuclear medicine in the United States: 1972–1982. *J. Nucl. Med.* 26:201.

Milham, S. 1982. Mortality from leukemia in workers exposed to electrical and magnetic fields. *New Engl. J. Med.* 307:249.

Miller, S. W., and F. P. Castronovo, Jr. 1985. Radiation exposure and protection in cardiac catheterization laboratories. *Am. J. Cardiol.* 55:171–176.

Miller, W. B., and V. D. Baker. 1973. The low exposure capabilities of electronic radiography. See HPS, 1973, pp. 399–402.

Millman, J., and C. C. Halkias. 1972. *Integrated Electronics: Analog and Digital Circuits and Systems.* New York: McGraw-Hill.

MIRD. 1975. *Radionuclide Decay Schemes and Nuclear Parameters for Use in Radiation Dose Estimation.* Medical Internal Radiation Dose Committee Pamphlet 10. New York: Society of Nuclear Medicine.

Modan, B., D. Baidatz, H. Mart, et al. 1974. Radiation induced head and neck tumors. *Lancet* 1:277–279.

Moeller, D. W., J. M. Selby, D. A. Waite, and J. P. Corley. 1978. Environmental surveillance for nuclear facilities. *Nucl. Safety* 19:66–79.

Moeller, D. W., and D. W. Underhill. 1976. *Final Report on Study of the Effects of Building Materials on Population Dose Equivalents.* Prepared for U.S. Environmental Protection Agency by Department of Environmental Health Sciences, Harvard University School of Public Health, Boston, MA.

Moore, H. E., E. A. Martell, and S. E. Poet. 1976. Sources of polonium-210 in atmosphere. *Envir. Sci. Technol.* 19:586–591.

Moore, H. E., and S. E. Poet. 1976. Background levels of ^{226}Ra in the lower troposphere. *Atmos. Envir.* 10:381–383.

Morgan, K. Z. 1967. Reduction of unnecessary medical exposure. *Radiation Control for Health and Safety Act of 1967: Hearings on S. 2067.* U.S. Congress, Senate, Committee on Commerce, 90th Cong., 1st sess., part 1, Aug. 28–30. Washington, DC: GPO.

———— 1975. Suggested reduction of permissible exposure to plutonium and other transuranium elements. *Am. Ind. Hyg. Assoc. J.* 36:567–575.

Morgan, K. Z., and J. E. Turner, eds. 1967. *Principles of Radiation Protection.* New York: Wiley.

Morgan, R. H. 1972. Radiological research and social responsibility [editorial]. *Radiology* 102:459–462.

Moulder, J. 2000. *Cellular Phone Antennas (Base Stations) and Human Health.* Medical College of Wisconsin website *www.mcw.edu/gcrc/cop.html.*

Murray, R., P. Heckel, and L. H. Hempelmann. 1959. Leukemia in children exposed to ionizing radiation. *New Engl. J. Med.* 261:585–589.

Nachtigall, D. 1967. Average and effective energies, fluence-dose conversion factors and quality factors of the neutron spectra of some (a, n) sources. *Health Phys.* 13:213–219.

Nachtigall, D., and G. Burger. 1972. Dose equivalent determinations in neutron fields by means of moderator techniques. In *Topics in Radiation Dosimetry: Radiation Dosimetry,* suppl. 1, ed. F. H. Attix. New York: Academic Press.

Nair, I., M. Granger-Morgan, and H. K. Florig. 1989. *Biological Effects of Power Frequency Electric and Magnetic Fields.* Publication No. OTA-BP-E-53. Washington, DC: Office of Technology Assessment.

Najarian, T., and T. Colton. 1978. Mortality from leukemia and cancer in shipyard nuclear workers. *Lancet* 1:1018.

NAS/NRC. 1956. *The Biological Effects of Atomic Radiation.* Summary Reports from a Study by the National Academy of Sciences. Washington, DC: NAS.

———— 1961a. *Effects of Inhaled Radioactive Particles.* National Academy of Sciences/National Research Council Publication 848. Washington, DC: NAS.

———— 1961b. *Long-Term Effects of Ionizing Radiations from External Sources.* National Academy of Sciences/National Research Council Publication 849. Washington, DC: NAS.

———— 1980. *The Effects on Populations of Exposure to Low Levels of Ionizing Radi-*

ation. Report of the Advisory Committee on the Biological Effects of Ionizing Radiations (BEIR Committee), NAS/NRC. Washington, DC: NAS.

———— 1988. *Health Risks of Radon and other Internally Deposited Alpha Emitters.* Report of the Committee on the Biological Effects of Ionizing Radiation, BEIR IV. Washington, DC: National Academy Press.

———— 1990. *Health Effects of Exposure to Low Levels of Ionizing Radiation.* Report of the Committee on the Biological Effects of Ionizing Radiation, BEIR V. Washington, DC: National Academy Press.

———— 1997. *Possible Health Effects of Exposure to Residential Electric and Magnetic Fields.* Committee on the Possible Effects of Electromagnetic Fields on Biologic Systems, National Research Council. Washington, DC: National Academy Press.

———— 1999a. *Health Effects of Exposure to Radon.* Report of the Committee on the Biological Effects of Ionizing Radiation, BEIR VI. Washington, DC: National Academy Press.

———— 1999b. *Risk Assessment of Radon in Drinking Water.* Report of the Committee on Risk Assessment of Exposure to Radon in Drinking Water, National Research Council. Washington, DC: National Academy Press.

Nathan, 0., and H. Norden, eds. 1960. *Einstein on Peace.* New York: Simon and Schuster.

NATO. 1989. *Control and Evaluation of Exposure to Personnel to Radiofrquency Radiation.* North Atlantic Treaty Organization Standardization Agreement, STANAG 2345, Edition 2, first draft, in *Radiofrequency Radiation Standards: Biological Effects, Dosimetry, Epidemiology, and Public Health Policy,* ed. B. J. Klauenberg et al. New York: Plenum Press.

NBS. 1953. *Regulation of Radiation Exposure by Legislative Means.* National Bureau of Standards Handbook 61. Washington, DC: Government Printing Office.

———— 1957. *Protection against Neutron Radiation up to 30 Million Electron Volts.* National Bureau of Standards Handbook 63. Washington, DC: Government Printing Office.

———— 1964. *Safe Handling of Radioactive Materials.* National Bureau of Standards Handbook 92. Washington, DC: Government Printing Office.

———— 1968. *Modern Trends in Activation Analysis.* Proceedings of an international conference. National Bureau of Standards Special Publication 312, vols. 1 and 2. Washington, DC: Government Printing Office.

———— 1976. *Symposium on Measurements for the Safe Use of Radiation,* ed. S. P. Fivozinsky. National Bureau of Standards Report SP 456. Washington, DC: NBS.

NCI. 1997. *Study Finds Magnetic Fields Do Not Raise Children's Leukemia Risk.* National Cancer Institute press release, available at *www.nci.nih.gov.* Also reported in *New England Journal of Medicine,* July 3, 1997, and *Science,* July 4, 1997.

NCRP. 1954. Permissible Dose from External Sources of Ionizing Radiation. Recommendations of the National Committee on Radiation Protection, National Bureau of Standards Handbook 59. Washington, DC: NBS.

———— 1960. *Protection against Radiation from Sealed Gamma Sources.* National

Council on Radiation Protection and Measurements Report 24. Bethesda, MD: NCRP Publications; also *www.ncrp.com*.

———— 1964. *Safe Handling of Radioactive Materials.* NCRP Report 30. Issued by National Bureau of Standards as Handbook 92. Washington, DC: Government Printinng Office.

———— 1968. *Medical X-Ray and Gamma-Ray Protection Up to 10 MeV: Equipment Design and Use.* NCRP Report 33. Bethesda, MD: NCRP Publications; also *www.ncrp.com*.

———— 1970a. *Dental X-Ray Protection.* NCRP Report No. 35. Bethesda, MD: NCRP Publications; also *www.ncrp.com*.

———— 1970b. *Medical X-Ray and Gamma-Ray Protection for Energies Up to 10 MeV: Structural Shielding Design and Evaluation Handbook.* NCRP Report 34. Bethesda, MD: NCRP Publications; also *www.ncrp.com*.

———— 1971a. *Basic Radiation Protection Criteria.* NCRP Report 39. Bethesda, MD: NCRP Publications; also *www.ncrp.com*.

———— 1971b. *Protection against Neutron Radiation.* NCRP Report 38. Bethesda, MD: NCRP Publications; also *www.ncrp.com*.

———— 1975. *Natural Background Radiation in the United States: Recommendations of the National Council on Radiation Protection and Measurements.* NCRP Report 45. Bethesda, MD: NCRP Publications; also *www.ncrp.com*.

———— 1976. *Structural Shielding Design and Evaluation for Medical Use of X-Rays and Gamma Rays of Energies Up to 10 MeV.* NCRP Report 49. Bethesda, MD: NCRP Publications; also *www.ncrp.com*.

———— 1977. *Medical Radiation Exposure of Pregnant and Potentially Pregnant Women.* NCRP Report 54. Bethesda, MD: NCRP Publications; also *www.ncrp.com*.

———— 1978a. *A Handbook of Radioactivity Measurements Procedures.* NCRP Report 58. Bethesda, MD: NCRP Publications; also *www.ncrp.com*.

———— 1978b. *Instrumentation and Monitoring Methods for Radiation Protection.* NCRP Report 57. Bethesda, MD: NCRP Publications; also *www.ncrp.com*.

———— 1979a. *Tritium and Other Radionuclide Labeled Organic Compounds Incorporated in Genetic Material.* NCRP Report 63. Bethesda, MD: NCRP Publications; also *www.ncrp.com*.

———— 1979b. *Tritium in the Environment.* NCRP Report 62. Bethesda, MD: NCRP Publications; also *www.ncrp.com*.

———— 1980a. *Management of Persons Accidentally Contaminated with Radionuclides.* NCRP Report No. 65. Bethesda, MD: NCRP Publications; also *www.ncrp.com*.

———— 1980b. *Radiation Protection in Nuclear Medicine.* A report of NCRP SC #32. Issued in 1982 under the title *Nuclear Medicine—Factors Influencing the Choice and Use of Radionuclides in Diagnosis and Therapy.* NCRP Report 70. Bethesda, MD: NCRP Publications; also *www.ncrp.com*.

———— 1981a. *Radiation Protection in Pediatric Radiology.* NCRP Report No. 68. Bethesda, MD: NCRP Publications; also *www.ncrp.com*.

———— 1981b. *Radiofrequency Electromagnetic Fields—Properties, Quantities and*

Units, Biophysical Interaction, and Measurements. NCRP Report No. 67. Bethesda, MD: NCRP Publications; also *www.ncrp.com.*

———— 1982. *Radiation Protection and New Medical Diagnostic Approaches.* Proceedings of the Eighteenth Annual Meeting of the National Council on Radiation Protection and Measurements. NCRP Proceedings No. 4. Bethesda, MD: NCRP Publications; also *www.ncrp.com.*

———— 1984a. *Evaluation of Occupational and Environmental Exposures to Radon and Radon Daughters in the United States.* NCRP Report 78. Bethesda, Md: NCRP Bethesda, MD: NCRP Publications; also *www.ncrp.com.*

———— 1984b. *Neutron Contamination from Medical Accelerators.* NCRP Report 79. Bethesda, MD: NCRP Publications; also *www.ncrp.com.*

———— 1985b. *SI Units in Radiation Protection and Measurements.* NCRP Report 82. Bethesda, MD: NCRP Publications; also *www.ncrp.com.*

———— 1985a. *General Concepts for the Dosimetry of Internally Deposited Radionuclides.* NCRP Report 84. Bethesda, MD: NCRP Publications; also *www.ncrp.com.*

———— 1986a. *Biological Effects and Exposure Criteria for Radiofrequency Electromagnetic Fields: Recommendations of the National Council on Radiation Protection and Measurements.* NCRP Report 86. Bethesda, MD: NCRP Publications; also *www.ncrp.com.*

———— 1986b. *Mammography—A User's Guide.* NCRP Report 85. Bethesda, MD: NCRP Publications; also *www.ncrp.com.*

———— 1987a. *Exposure of the Population in the United States and Canada from Natural Background Radiation.* NCRP Report 94. Bethesda, MD: NCRP Publications; also *www.ncrp.com.*

———— 1987b. *Recommendations on Limits for Exposure to Ionizing Radiation.* NCRP Report 91. Bethesda, MD: NCRP Publications; also *www.ncrp.com.*

———— 1989a. *Exposure of the U.S. Population from Diagnostic Medical Radiation.* NCRP Report 100. Bethesda, MD: NCRP Publications; also *www.ncrp.com.*

———— 1989b. *Limits for Exposure to "Hot Particles" on the Skin.* NCRP Report 106. Bethesda, MD: NCRP Publications; also *www.ncrp.com.*

———— 1989c. *Medical X-Ray, Electron Beam and Gamma-Ray Protection for Energies Up to 50 MeV (Equipment Design, Performance and Use).* NCRP Report 102. Bethesda, MD: NCRP Publications; also *www.ncrp.com.*

———— 1989d. *Radiation Protection for Medical and Allied Health Personnel.* NCRP Report 105. Bethesda, MD: NCRP Publications; also *www.ncrp.com.*

———— 1991a. *Calibration of Survey Instruments Used in Radiation Protection for the Assessment of Ionizing Radiation Fields and Radioactive Surface Contamination.* NCRP Report 112. Bethesda, MD: NCRP Publications; also *www.ncrp.com.*

———— 1991b. *Misadministration of Radioactive Material in Medicine—Scientific Background.* NCRP Commentary No. 7. Bethesda, MD: NCRP Publications; also *www.ncrp.com.*

———— 1991c. *Some Aspects of Strontium Radiobiology.* National Council on Radiation Protection and Measurements Report 110. Bethesda, MD: NCRP.

———— 1993a. *Evaluation of Risk Estimates for Radiation Protection Purposes.* NCRP Report 115. Bethesda, MD: NCRP Publications; also *www.ncrp.com.*

———— 1993b. *Limitation of Exposure to Ionizing Radiation.* NCRP Report 116. Bethesda, MD: NCRP Publications; also *www.ncrp.com.*

———— 1993c. *Practical Guide to the Determination of Human Exposure to Radio-frequency Fields.* NCRP Report 119. Bethesda, MD: NCRP Publications; also *www.ncrp.com.*

———— 1994. *Considerations Regarding the Unintended Radiation Exposure of the Embryo, Fetus, or Nursing Child.* NCRP Commentary No. 9. Bethesda, MD: NCRP Publications; also *www.ncrp.com.*

———— 1996. *Screening Models for Releases of Radionuclides to Atmosphere, Surface Water, and Ground.* 2 vols. NCRP Report 123. Bethesda, MD: NCRP Publications; also *www.ncrp.com.*

———— 1998. *Operational Radiation Safety Program.* NCRP Report 127. Bethesda, MD: NCRP Publications; also *www.ncrp.com.*

———— 1999a. *Biological Effects and Exposure Limits for "Hot Particles."* NCRP Report 130. Bethesda, MD: NCRP Publications; also *www.ncrp.com.*

———— 1999b. *Radiation Protection in Medicine: Contemporary Issues.* Proceedings of the Thirty-Fifth Annual Meeting. Proceedings No. 20. Bethesda, MD: NCRP Publications; also *www.ncrp.com.*

Nelson, J. L., and J. R. Divine. 1981. *Decontamination Processes for Restorative Operations and as a Precursor to Decommissioning: A Literature Review.* NRC report NUREG/CR-1915; PNL 3706. Washington, DC: Nuclear Regulatory Commission.

Nelson, N. 1988. Letter from Norton Nelson, Chairman, Executive Committee, Science Advisory Board, to Lee M. Thomas, Administrator, U.S. Environmental Protection Agency, File SAB-RAC-88–041.

NEXT. 1976. *Suggested Optimum Survey Procedures for Diagnostic X-Ray Equipment.* Report of the Nationwide Evaluation of X-ray Trends task force, cosponsored by Conference of Radiation Control Program Directors and Bureau of Radiological Health, U.S. Department of Health, Education, and Welfare. Washington, DC: Bureau of Radiological Health.

NIEHS. 1998. *Assessment of Health Effects from Exposure to Power-Line Frequency Electric and Magnetic Fields: Working Group Report,* ed. C. J. Portier and M. S. Wolfe. Research Triangle Park, NC: National Institute of Environmental Health Sciences; also *www.niehs.nih.gov/emfrapid.*

NIRP. 1969. *Straldoser fran radioaktiva amnen i medicinskt bruk-information till sjukhusens isotopkommitteer.* Report of the National Institute of Radiation Protection, Stockholm.

Nishizawa, K., T. Maruyama, et al. 1991. Determinations of organ doses and effective dose equivalents from computed tomographic examinations. *Brit. J. Radiol.* 64:20–28.

NRC 1974. *Termination of Operating Licenses for Nuclear Reactors.* NRC Regulatory Guide 1.86. Washington, DC: NRC.

———— 1975. *Reactor Safety Study: An Assessment of Accident Risks in U.S. Com-*

mercial Nuclear Power Plants. Study sponsored by the U.S. Atomic Energy Commission and performed under the independent direction of Professor Norman C. Rasmussen of the Massachusetts Institute of Technology. U.S. Nuclear Regulatory Commission Report WASH-1400 (NUREG 75/014). Washington, DC: NRC.

———— 1981a. *Radiation Protection Training for Personnel at Light-Water-Cooled Nuclear Power Plants.* USNRC Regulatory Guide 8.27. Washington, DC: NRC.

———— 1981b. *Radiation Safety Surveys at Medical Institutions.* NRC Regulatory Guide 8.23. Washington, DC: NRC.

———— 1984. *Radiation Protection Training for Personnel Employed in Medical Facilities.* USNRC Draft Regulatory Guide. Washington, DC: NRC.

———— 1987a. *Guide for the Preparation of Applications for Medical Programs.* NRC Regulatory Guide 10.8, Rev. 2. Washington, DC: NRC.

———— 1987b. *Interpretation of Bioassay Measurements.* NRC Report NUREG/CR-4884. Washington, DC: NRC.

———— 1988. *Standards for Protection against Radiation.* Final rule (proposed). Nuclear Regulatory Commission paper SECY-88–315, November 10, 1988. Washington, DC: NRC.

———— 1990. *Enforcement Policy for Hot Particle Exposures.* NRC Information Notice No. 90–48. Washington, DC: NRC.

———— 1991. *Standards for Protection against Radiation.* U.S. Nuclear Regulatory Commission. Code of Federal Regulations, Title 10, Part 20 (10CFR20), *Fed. Regist.* 56:23360–23474, No. 98, May 21, 1991; also, *www.nrc.gov* and link in "Reference Library."

———— 1993. *Acceptable Concepts, Models, Equations, and Assumptions for a Bioassay Program.* NRC Regulatory Guide 8.9. Washington, DC: NRC.

———— 1997. *Release of Patients Administered Radioactive Materials.* NRC Regulatory Guide 8.39. Washington, DC: NRC; also, *www.nrc.gov* and link in "Reference Library."

NRPB, 1992, 1994. *Electromagnetic Fields and the Risk of Cancer.* Vol. 3, pp. 1–138; Vol. 5, pp. 77–81. Chilton, Didcot, UK: National Radiological Protection Board.

O'Brien, K., H. Sandmeier, G. Hansen, and J. Campbell. 1980. Cosmic ray induced neutron background sources and fluxes for geometries of air over water, ground, iron, and aluminum. *J. Geophys. Res.* 83:114–120.

Okada, S., et al. 1975. *A Review of Thirty Years of Study of Hiroshima and Nagasaki Atomic Bomb Survivors. J. Radiat. Res.,* suppl. (Japan), 164 pp. Chiba: Japan Radiation Research Society.

O'Kelley, G. D., ed. 1963. *Application of Computers to Nuclear and Radiochemistry.* National Academy of Sciences Report NAS-NS-3107. Washington, DC: NAS.

ONRR. 1976. *Safety Evaluation Report Related to Operation of Three Mile Island Nuclear Station, Unit 2.* Office of Nuclear Reactor Regulation (USNRC) report NUREG-0107. Springfield, VA: National Technical Information Service.

ORAU. 1992. *Health Effects of Low Frequency Electric and Magnetic Fields.* Publication No. ORAU 92/F8. Committee in Interagency Radiation Research and Policy Coordination. Oak Ridge: Oak Ridge Associated Universities.

Orton, C. G. 1995. Uses of therapeutic x rays in medicine. *Health Phys.* 69:662–676.

Orvis, A. L. 1970. *Whole-Body Counting. Medical Radionuclides: Radiation Dose and Effects,* ed. R. J. Cloutier, C. L. Edwards, and W. S. Snyder, 115–132. USAEC Symposium Series 20, Report CONF-691212. Springfield, VA: National Technical Information Service.

Osepchuk, J. M., and R. C. Petersen. 2001. Safety and environmental issues. In *Modern Microwave Handbook,* ed. M. Golio. Boca Raton, FL: CRC Press.

Osterhout, M., ed. 1980. *Decontamination and Decommissioning of Nuclear Facilities.* New York: Plenum Press.

OTA. 1979. *The Effects of Nuclear War.* Office of Technology Assessment, U.S. Congress. Washington, DC: Government Printing Office.

Pagels, H. R. 1982. *The Cosmic Code.* New York: Bantam Books.

Palmer, H. E., and W. C. Roesch. 1965. A shadow shield whole-body counter. *Health Phys.* 11:1913–1919.

Park, R. 2000. *Voodoo Science.* New York: Oxford University Press.

Parker, G. W., and C. J. Barton. 1973. Fission-product release. In *The Technology of Nuclear Reactor Safety,* vol. 2, *Reactor Materials and Engineering,* ed. T. T. Thompson and J. G. Beckerley. Cambridge, MA: MIT Press.

Pass, B., et al. 1997. Collective biodosimetry as a dosimetric "Gold Standard": A study of three radiation accidents. *Health Phys.* 72(3):390–396.

Pauling, L. W. 1958. Letter to the *New York Times,* May 16, 1958. Reprinted in JCAE, 1959b, p. 2462.

Peach, H. G., et al. 1992. *Report of the Panel on Electromagnetic Fields and Health.* Melbourne, Australia: The Victorian Government.

Pearson, E. S., and H. 0. Hartley, eds. 1966. *Biometrika Tables for Statisticians,* vol. 1, p. 227. Cambridge, Eng.: Cambridge University Press.

Pendleton, R. C., C. W. Mays, R. D. Lloyd, and A. L. Brooks. 1963. Differential accumulation of ^{131}I from local fallout in people and milk. *Health Phys.* 9:1253–1262.

Perez, C. A., and L. W. Brady, eds. 1998. *Principles and Practice of Radiation Oncology.* Philadelphia: Lippincott-Raven.

Persson, Lars. 1994. The Auger electron effect in radiation dosimetry. *Health Phys.* 67(5):471–476.

Petersen, R. C., and P. A. Testagrossa. 1992. Radio-frequency electromagnetic fields associated with cellular-radio cell-site antennas. *Bioelectromagnetics* 13:527–542.

Petersen, R. C., A. K. Fahy-Elwood, et al. 1997. Wireless telecommunications: Technology and RF safety issues. In *Non-Ionizing Radiation: An Overview of the Physics and Biology,* ed. K. Hardy. Proceedings of the Health Physics Society 1997 Summer School. Madison, WI: Medical Physics Publishing.

Peterson, H. T., Jr., J. E. Martin, C. L. Weaver, and E. D. Harward. 1969. Environmental tritium contamination from increasing utilization of nuclear energy sources. In JCAE, 1969a, appendix 13, p. 765.

Petrosyants, A. M. 1969. The peaceful profession of the nuclear explosion. Reprinted in JCAE, 1969b, p. 696.

Pfeiffer, S. W. 1965. Mandan milk mystery. *Scientist & Citizen*, 7 (September):1–5.

Phillips, L. A. 1978. *A Study of the Effect of High Yield Criteria for Emergency Room Skull Radiography.* U.S. Department of Health, Education, and Welfare Publication (FDA) 78–8069. Washington, DC: Government Printing Office.

Plog, B. A., et al, eds. 1996. *Fundamentals of Industrial Hygiene.* Itasca, IL: National Safety Council.

Podgorsky, E. B., P. R. Moran, and J. R. Cameron. 1971. Thermoluminescent behavior of LiF (TLD-100) from 77° to 500 °K. *J. Appl. Phys.* 42:2761.

Poet, S. E., and E. A. Martell. 1972. Plutonium-239 and americium-241 contamination in the Denver area. *Health Phys.* 23:537–548.

——— 1974. Reply to "Plutonium-239 contamination in the Denver area," by P. W. Krey. *Health Phys.* 26:120–122.

Preston, D., K. Mabuchi, et al. 1997. Mortality among atomic-bomb survivors, 1950–1990. In *Implications of New Data on Radiation Cancer Risk.* Proceedings of the Thirty-Second Annual Meeting of the National Council on Radiation Protection and Measurements. Proceedings No. 18. Washington, DC: NCRP Publications.

Price, B. T., C. C. Horton, and K. T. Spinney. 1957. *Radiation Shielding.* London: Pergamon Press.

Price, W. J. 1964. *Nuclear Radiation Detection.* New York: McGraw-Hill.

Quimby, E. H., S. Feitelberg, and W. Gross. 1970. *Radioactive Nuclides in Medicine and Biology,* vol. 1, *Basic Physics and Instrumentation.* Philadelphia: Lea and Febiger.

Quittner, P. 1972. *Gamma-Ray Spectroscopy.* London: Adam Hilger.

Racoveanu, N. T., and V. Volodin. 1992. Rational use of diagnostic radiology. *Radiat. Prot. Dosimetry* 43:15–18.

Radge, H. 1997. Brachytherapy for clinically localized prostate cancer. *J. Surgical Oncol.* 64:79–81.

Rallison, M. L., B. M. Dobyns, et al. 1974. Thyroid disease in children: A survey of subjects potentially exposed to fallout radiation. *Am. J. Med.* 56:457–463.

Raloff, J. 1979. Abandoned dumps: A chemical legacy. *Sci. News* 115:348–351.

——— 2000. Two studies offer some cell-phone cautions. *Sci, News* 157:326.

Reddy, A. R., A. Nagaratnam, A. Kaul, and V. Haase. 1976. Microdosimetry of internal emitters: A necessity? In *Radiopharmaceutical Dosimetry Symposium.* See Cloutier et al., 1976.

Reginatto, M., E. Party, et al. 1991. Case studies involving ^{125}I thyroid uptake following "unblocking." *Health Phys.* 60(6):837–840.

Reif, J. S., et al. 1995. Residential exposure to magnetic fields and risk of canine lymphoma. *Am. J. Epidemiol.* 141(4):352–359.

Reissland, J. A. 1978. *An Assessment of the Mancuso Study.* National Radiological Protection Board (England) Report NRPB-R 79. Harwell: NRPB.

Renaud, L. A. 1992. A 5-y follow-up of the radiation exposure to in-room personnel during cardiac catheterization. *Health Phys.* 62:10–15.

Repacholi, M. H., ed. 1988a. *Non-Ionizing Radiations: Physical Characteristics, Biological Effects and Health Hazard Assessment.* Proceedings of the International Non-Ionizing Radiation Workshop. London: IRPA Publications.

———— 1988b. Static electric and magnetic fields. In *Non-Ionizing Radiations: Physical Characteristics, Biological Effects and Health Hazard Assessment,* M. H. Rapacholi, ed. London: IRPA Publications.

RERF. 1987. *US-JAPAN Reassessment of Atomic Bomb Radiation Dosimetry in Hiroshima and Nagasaki, Final Report.* Radiation Effects Research Foundation, Hiroshima, Japan. Available from W. H. Ellett, RERF Office, National Research Council, 2101 Constitution Ave. NW, Washington, DC 20418.

Reuter, F. 1978. Physician and patient exposure during cardiac catheterization. *Circulation* 58(1):134–139.

Rhodes, Richard. 1988. *The Making of the Atomic Bomb.* New York: Simon and Schuster.

———— 1996. *Dark Sun: The Making of the Hydrogen Bomb.* New York: Simon and Schuster.

Ricci, J. L., and D. Sashin. 1976. Radiation exposure in clinical mammography. In *Operational Health Physics.* Proceedings of the Ninth Midyear Topical Symposium of the Health Physics Society. Boulder, CO: Health Physics Society.

Richards, A. G. 1969. Trends in radiation protection. *J. Mich. Dent. Assoc.* 51:18–21.

Ring, J., F. Osborne, et al. 1993. Radioactive waste management at a large university and medical research complex. *Health Phys.* 65(2):193–199.

Roberts, L. 1987. Atomic bomb doses reassessed. *Science* 238:1649–1650.

Robinette, C. D., et al. 1980. Effects upon health of occupational exposure to microwave radiation (radar). *Am. J. Epidemiol.* 112:39–53.

Robinson, A. L. 1978. The new physics: Quarks, leptons, and quantum field theories. *Science* 202:734–737.

Rockwell, T., III, ed. 1956. *Reactor Shielding Design Manual.* New York: McGraw-Hill.

Rodger, W. A., et al. 1978. *"De Minimus" Concentrations of Radionuclides in Solid Wastes.* Atomic Industrial Forum Report AIF/NESP-016. Washington, DC: Atomic Industrial Forum.

Roesch, W. C., and H. E. Palmer. 1962. Detection of plutonium in vivo by wholebody counting. *Health Phys.* 8:773–776.

Rogovin, M. 1980. *Three Mile Island. A Report to the Commissioners and to the Public.* Nuclear Regulatory Commission Special Inquiry Group. M. Rogovin, director. Springfield, VA: National Technical Information Service.

Ron, E. 1997. Cancer risk following radioactive iodine-131 exposures in medicine, In *Implications of New Data on Radiation Cancer Risk.* Proceedings of the Thirty-Second Annual Meeting of the National Council on Radiation Protection and Measurements, Proceedings No. 18. Washington, DC: NCRP Publications.

Ron, E., J. H. Lubin, et al. 1995. Thyroid cancer following exposure to external radiation: A pooled analysis of seven studies. *Radiat. Res.* 141:255–273.

Ronan, C. A. 1991. *The Natural History of the Universe from the Big Bang to the End of Time.* New York: Macmillan.

Rosenstein, M. 1976a. *Handbook of Selected Organ Doses for Projections Common in Diagnostic Radiology.* U.S. Department of Health, Education, and Welfare Publication (FDA) 76–8031. Washington, DC: Government Printing Office.

——— 1976b. *Organ Doses in Diagnostic Radiology.* U.S. Department of Health, Education, and Welfare Publication (FDA) 76–8030. Washington, DC: Government Printing Office.

——— 1988. *Handbook of Selected Tissue Doses for Prjections Common in Diagnostic Radiology.* HHS Publication HHS(FDA) 89-8031.

Rosenstein, M., L. W. Andersen, et al. 1985. *Handbook of Glandular Tissue Doses in mammography.* HHS publication FFDA 85–8239. Rockville, MD: Department of Health and Human Services.

Rosenstein, M., T. J. Beck, and G. G. Warner. 1979. *Handbook of Selected Organ Doses for Projections Common in Pediatric Radiology.* U.S. Department of Health, Education, and Welfare Publication (FDA) 79–8078. Washington, DC: Government Printing Office.

Rosenstein, M. L., and E. W. Webster. 1994. Effective dose to personnel wearing protective aprons during fluoroscopy and interventional radiology. *Health Phys.* 67:88–89.

Rossi, H. H. 1968. Microscopic energy distribution in irradiated matter. In *Radiation Dosimetry*, vol. 1, ed. F. H. Attix and W. C. Roesch. New York: Academic Press.

——— 1976. Interrelations between physical and biological effects of small radiation doses. *Biological and Environmental Effects of Low Level Radiation.* Proceedings of a symposium organized by the International Atomic Energy Agency and World Health Organization. Vienna: IAEA.

——— 2000. LN-T and Politics. *HPS Newsletter,* January, 2000, p. 12.

Rossi, H. H., and C. W. Mays. 1978. Leukemia risk from neutrons. *Health Phys.* 34:353–360.

Rossi, R. P., et al. 1991. Broad beam transmission properties of some common shielding materials for use in diagnostic radiology. *Health Phys.* 61(5):601–608.

Rothenberg, L. N. 1993. Mammography instrumentation: Recent developments. *Med. Progress Technol.* 19:1–6.

Rothenbert, L. N., and K. S. Pentlow. 1994. CT dose assessment. In *Specification, Acceptance Testing and Quality Control of Diagnostic X-Ray Imaging Equipment,* ed. J. A. Seibert, G. T. Barnes, and R. G. Gould. Medical Physics Monograph No. 20. Woodbury, NY: American Institute of Physics.

Rowland, R. E., A. F. Stehney, and H. F. Lucas, Jr. 1978. Dose-response relationships for female radium dial workers. *Radiat. Res.* 76:368–383.

Roy, C. R. 1988. Ultraviolet radiation: Sources, biological interaction and personal protection. In Repacholi, 1998a.

Rundo, J., F. Markun, and N. J. Plondke. 1979. Observations of high concentrations of radon in certain houses. *Health Phys.* 36:729–730.

Russell, W. L. 1967. Factors that affect the radiation induction of mutations in the mouse. *An. Acad. Brasileira de Ciencias* 39:65–75. Reprinted in JCAE, 1969a, pp. 624–634.

Ryan, M. T., K. W. Skrable, and G. Chabot. 1975. Retention and penetration characteristics of a glass fiber filter for ^{212}Pb aggregate recoil particles. *Health Phys.* 29:796–798.

Saenger, E. L., ed. 1963. *Medical Aspects of Radiation Accidents.* Washington, DC: Government Printing Office.

Saenger, E. L., F. N. Silverman, et al. 1960. Neoplasia following therapeutic irradiation for benign conditions in childhood. *Radiology* 74:889–904.

Saenger, E. L., G. E. Thoma, and E. A. Thompkins. 1968. Incidence of leukemia following treatment of hyperthyroidism. *JAMA* 205:855–862.

Sahl, J. D., et al. 1993. Cohort and nested case-control studies of hematopoietic cancers and brain cancer among electrical utility workers. *Epidemiology* 4(2):104–114.

Sakharov, A. D. 1978. Nuclear energy and the freedom of the west. *Bull. Atom. Scientists,* June, pp. 12–14.

Sanders, B. S. 1978. Low-level radiation and cancer deaths. *Health Phys.* 34:521–538.

Sanders, S. M., Jr., and W. C. Reinig. 1968. Assessment of tritium in man. In *Diagnosis and Treatment of Deposited Radionuclides,* ed. H. A. Kornberg and W. D. Norwood. Amsterdam: Excerpta Medical Foundation.

Sanford, J. R. 1976. The Fermi National Accelerator Laboratory. *Ann. Rev. Nucl. Sci.* 26:151–198.

Sankey, R. R. 1993. Brachytherapy: An overview. In *Hospital Health Physics: Proceedings of the 1993 Health Physics Society Summer School,* ed. G. G. Eicholz and J. J. Shonka. Richland, WA: Research Enterprises.

Sashin, D. E., J. Sternglass, A. Huen, and E. R. Heinz. 1973. Dose reduction in diagnostic radiology by electronic imaging techniques. In HPS, 1973, pp. 385–393.

Santiago, P. A. 1994. Enforcement related to implementation of medical quality management programs. In *Proceedings of the 26th Annual National Conference on Radiation Control,* May 22–25, 1994. Frankfort, KY: Conference of Radiation Control Program Directors.

Savitz, D. A., et al. 1988. Case-control study of childhood cancer and exposure to 60-Hz magnetic fields. *Am. J. Epidemiol.* 128:21–38.

Savitz, D. A., and D. P. Loomis. 1995. Magnetic field exposure in relation to leukemia and brain cancer mortality among electric utility workers. *Am. J. Epidemiol.* 141:123–134.

Schiager, K. J., et al. 1996. Consensus radiation protection practices for academic research institutions. *Health Pys.* 71(6):960–965.

Schmidt, A., J. S. Puskin, et al. 1992. EPA's approach to assessment of radon risk. In *Indoor Radon and Lung Cancer: Reality or Myth?*, ed. F. T. Cross. Twenty-Ninth Hanford Symposium on Health and the Environment. Richland, WA: Battelle Press.

Schneider, K., H. Fendel, et al. 1992. Results of a dosimetry study in the European Community on frequent x-ray examinations in infants. *Radiat. Prot. Dosimetry* 43:31–36.

Schnorr, T. M., et al. 1991. Video display terminals and the risk of spontaneous abortion. *New Engl. J. Med.* 324:727–733.

Schuler, G., and E. Stoll. 1991. Too little concern for breast cancer risk in radiation protection estimates? *Health Phys.* 61(3):405–408.

Schwarz, G. S. 1968. Radiation hazards to the human fetus in present-day society. *Bull. N.Y. Acad. Med.* 44:388–399.

Seibert, J. A. 1995. One hundred years of medical diagnostic imaging technology. *Health Phys.* 69:695–720.

Seibert, J. A., G. T. Barnes, and R. G. Gould. 1994. *Specification, Acceptance Testing and Quality Control of Diagnostic X-Ray Imaging Equipment.* AAPM Medical Physics Monograph No. 20. Woodbury, NY: American Institute of Physics.

Seltser, R., and P. E. Sartwell. 1965. The influence of occupational exposure to radiation on the mortality of American radiologists and other medical specialists. *Am. J. Epidemiol.* 81:2–22.

Shambon, A. 1974. *CaSO₄: Dy TLD for Low Level Personnel Monitoring.* U.S. Atomic Energy Commission Health and Safety Laboratory Report HASL 285.

Shapiro, J. 1954. *An Evaluation of the Pulmonary Radiation Dosage from Radon and Its Daughter Products.* Ph.D. thesis, University of Rochester. Issued as University of Rochester Atomic Energy Project Report UR-298. Rochester, NY.

——— 1956a. Radiation dosage from breathing radon and its daughter products. *A.M.A. Arch. Ind. Health* 14:169–177.

——— 1956b. *Studies on the Radioactive Aerosol Produced by Radon in Air.* University of Rochester Atomic Energy Project Report UR-461. Rochester, NY.

——— 1968. *Criteria and Tests for the Evaluation of Hazards from Radioactive Surface Contamination.* Final Report, Research Contract with U.S. Public Health Service. Harvard School of Public Health, Boston, MA.

——— 1970. Tests for the evaluation of airborne hazards from radioactive surface contamination. *Health Phys.* 19:501–510.

——— 1980. The development of the American National Standard, "Control of radioactive surface contamination on materials, equipment and facilities to be released for uncontrolled use." *Proceedings of the 5th International Congress of the International Radiation Protection Association.* Oxford: Pergamon Press.

Shapiro, J., and J. P. Ring. 1999. Benchmarking radiation protection programs. *Health Phys.* 76(Supp.):S23-S26.

Shapiro, J., D. G. Smith, and J. Iannini. 1970. Experience with boron-loaded nuclear track plates as detector elements for spherical moderator type neutron monitors. *Health Phys.* 18:418–424.

Sheline, G. E., S. Lindsay, et al. 1962. Thyroid nodules occurring late after treatment of thyrotoxicosis with radioiodine. *J. Clin. Endocr. Metab.* 22:8–18.

Shleien, B. 1973. *A Review of Determinations of Radiation Dose to the Active Bone Marrow from Diagnostic X-Ray Examinations.* U.S. Department of Health, Education, and Welfare, Bureau of Radiological Health, Report (FDA) 74–8007. Washington, DC: Government Printing Office.

Shope, T. B. 1995. Radiation-induced skin injuries from fluoroscopy. *Radiology* 197(P) Suppl. P449; also *www.fda.gov/cdrh/rsnaii.html.*

Shope, T. B., R. M. Gagne, and G. C. Johnson. 1981. A method for describing the doses delivered by transmission x-ray coomputed tomography. *Med. Phys.* 8:488–495.

Shore, R. E., L. H. Hempelmann, et al. 1977. Breast neoplasms in women treated with x-rays for acute postpartum mastitis. *J. Natl. Cancer Inst.* 59:813–822.

Shultis, J. K., and R. E. Law. 2000. *Radiation Shielding.* Lagrange Park, IL: American Nuclear Society.

Sikov, M. R., R. J. Traub, et al. 1992. *Contribution of Maternal Radionuclide Burdens to Prenatal Radiation Doses.* U.S. Nucear Regulatory Commission Report NUREG/CR-5631. Washington, DC: NRC.

Silk, J. 1994. *A Short History of the Universe.* New York: Scientific American Library.

Silverman, L., C. E. Billings, and M. W. First. 1971. *Particle Size Analysis in Industrial Hygiene.* New York: Academic Press.

Sisefsky, J. 1960. *Autoradiographic and Microscopic Examination of Nuclear Weapon Debris Particles.* Forsvarets Forskningsanstalt Avdelning 4 (Stockholm) Report A 4130–456 (FOAY Rapport).

Shleien, B., L. A. Slaback, Jr., and B. Birky. 1998. *Handbook of Health Physics and Radiological Health.* Baltimore: Lippincott Williams & Wilkins.

Slaback et al. 1997. See Shleien, 1998.

Slack, L., and K. Way. 1959. *Radiations from Radioactive Atoms in Frequent Use.* U.S. Atomic Energy Commission Report. Washington, DC: NRC.

Slade, D. H., ed. 1968. *Meterology and Atomic Energy, 1968.* Available as TID-24190 from Clearinghouse for Federal Scientific and Technical Information, National Bureau of Standards, U.S. Department of Commerce, Springfield, VA 22151.

Sliney, D. H., and S. L. Trokel. *Medical Lasers and Their Safe Use.* New York: Springer-Verlag.

Smisek, M., and S. Cerny. 1970. *Action Carbon.* Amsterdam: Elsevier.

Smith, C. L., et al. 1998. An examination of radiation exposure to clinical staff from patients implanted with ^{137}Cs and ^{192}Ir for the treatment of gynecologic malignancies. *Health Phys.* 74(3):301–308.

Smith, D. G. 1975. *Influence of Meteorological Factors upon Effluent Concentrations on and near Buildings with Short Stacks.* Paper presented at Air Pollution Control Association 69th annual meeting. Boston: Harvard School of Public Health.

——— 1978. *Determination of the Influence of Meteorological Factors upon Building Wake Tracer Concentrations by Multilinear Regression Analysis.* Paper presented at Air Pollution Control Association 71st annual meeting. Boston: Harvard School of Public Health.

Smith, E. M., G. L. Brownell, and W. H. Ellett. 1968. Radiation dosimetry. In *Principles of Nuclear Medicine,* ed. H. N. Wagner, Jr. Philadelphia: Saunders.

Smith, M. H. 1997. Electromagnetic and compatibility issues in public and occupational settings. In HPS, 1997.

Snyder, W. S., B. R. Fish, et al. 1968. Urinary excretion of tritium following exposure of man to HTO—A two exponential model. *Phys. Med. Biol.* 13:547–559.

Snyder, W. S., H. L. Fisher, Jr., M. R. Ford, and G. G. Warner. 1969. Estimates of absorbed fractions for monoenergetic photon sources uniformly distributed in various organs of a heterogeneous phantom. Medical Internal Radiation Dose Committee Pamphlet 5. *J. Nucl. Med.,* suppl. 3, August.

Snyder, W. S., M. R. Ford, G. G. Warner, and S. B. Watson. 1975. *"S," Absorbed Dose per Unit Cumulated Activity for Selected Radionuclides and Organs.* Medical Internal Radiation Dose Committee Pamphlet 11. New York: Society of Nuclear Medicine.

Song, Y. T., and C. M. Huddleston. 1964. A semiempirical formula for differential dose albedo for neutrons on concrete. *Trans. Am. Nucl. Soc.* 7:364.

Spiers, F. W. 1968. *Radioisotopes in the Human Body: Physical and Biological Aspects.* New York: Academic Press.

Spiers, F. W., and J. Vaughn. 1976. Hazards of plutonium with special reference to skeleton. *Nature* 259:531–534.

Sprawls, P., Jr. 2000. *Physical Principles of Medical Imaging.* Gaithersburg, MD: Aspen Publishers, Inc.

Stabin, M. G., J. B. Stubbs, and R. E. Toohey. 1996. *Radiation Dose Estimates for Radiopharmaceuticals.* Nuclear Regulatory Commission Report NUREG/CR-6345. Oak Ridge: RIDIC.

Stannard, 1988. *Radioactivity and Health: A History.* Springfield, VA: NTIS.

Sterlinski, S. 1969. The lower limit of detection for very short-lived radioisotopes used in activation analysis. *Nucl. Inst. Meth.* 68:341–343.

Stewart, A., and G. W. Kneale. 1970. Radiation dose effects in relation to obstetric x-rays and childhood cancers. *Lancet,* June 6, pp. 1185–1187.

Storm, E., and H. I. Israel. 1970. Photon cross sections from 1 KeV to 100 MeV for elements Z = 1 to Z = 100. *Nucl. Data Tables, Sect. A,* 7:565–681.

Straume, T., ed. 1993. *Tritium Dosimetry, Health Risks, and Environmental Fate.* Review Issue. *Health Phys.* 65(6):593–726.

Suess, M. J., and D. A. Benwell-Morison. 1989. *Nonionizing Radiation Protection.* Geneva: World Health Organization.

Suleiman, O. H., S. H. Stern, and D. C. Spelic. 1999. Patient dosimetry activities in the United States: The Nationwide Evaluation of X-ray Trends (NEXT) and tissue dose handbooks. *Applied Radiation and Isotopes* 50:247–259.

Swindell, W., and H. H. Barrett. 1977. Computerized tomography: Taking sectional x-rays. *Phys. Today,* December, pp. 32–41.

Tamplin, A. R., and H. L. Fisher. 1966. *Estimation of Dosage to the Thyroids of*

Children in the United States from Nuclear Tests Conducted in Nevada during 1952–1955. U.S. Atomic Energy Commission Report UCRL-14707.

Tamplin, A. R., and J. W. Gofman. 1970. ICRP publication 14 vs the Gofman-Tamplin report. In JCAE, 1969a, p. 2104.

Taylor, L. S. 1961. Report of interview with Dr. L. S. Taylor entitled, "Is fallout a false scare? Interview with the leading authority on radiation." *U.S. News & World Report,* Nov. 27.

Telecom. 1996. *Telecommunications Act of 1996.* Pub. L. No. 104–104, 110 Stat. 56:1996.

Telegadas, K. 1959. Announced nuclear detonations. Reprinted in JCAE, 1959b, pp. 2517–2533.

———— 1977. An estimate of maximum credible atmospheric radioactivity concentrations from nuclear tests. Health and Safety Laboratory Report HASL-328. *Envir. Quart.,* Oct. 1.

Tell, R. A., and E. D. Mantiply. 1980. Population exposure to VHF and UHF broadcast radiation in the United States. *Proc. IEEE* 68:6–12.

Tenforde, T. S. 1991. Biological interaction of extremely-low-frequency electric and magnetic fields. *Biochem. Bioenergetics*

Tenforde, T. S., and T. F. Budinger. 1985. Biological effects and physical safety aspects of NMR imaging and in vivo spectroscopy. In *NMR in Medicine,* ed. S. R. Thomas and R. L. Dickson. New York: American Association of Physicists in Medicine.

Tenforde, T. S., and W. T. Kaune. 1987. Interaction of extremely low frequency electric and magnetic fields with humans. *Health Phys.* 53:585–606.

Ter-Pogossian, M. 1967. *Physical Aspects of Diagnostic Radiology.* New York: Hoeber.

Theriault, G. P., et al. 1994. Cancer risks associated with occupational exposure to magnetic fields among electric utility workers in Ontario and Quebec, Canada and France: 1970–1989. *Am. J. Epidemiol.* 139:550–572.

Thilander, A., S. Eklund, et al. 1992. Special problems of patient dosimetry in mammography. *Radiat. Prot. Dosimetry* 43:217–220.

Thompson, T. T. 1978. *A Practical Approach to Modern X-Ray Equipment.* Boston: Little, Brown.

Till, J. E., et al. 1980. *Tritium—An Analysis of Key Environmental and Dosimetric Questions.* Oak Ridge National Laboratory Report ORNL/TM-6990. Oak Ridge, TN: Technical Information Center.

Tornqvist, S., et al. 1991. Incidence of leukemia and brain tumors in some "Electrical Occupations." *Brit. J. Ind. Med.* 48:597–603.

Trout, E. D., and J. P. Kelley. 1965. Scattered radiation in a phantom. *Radiology* 85:546–554.

Trout, E. D., J. P. Kelley, and A. C. Lucas. 1960. Determination of half-value layer. *Am. J. Roentgenol. Radium Ther. Nucl. Med.* 84:729–740.

———— 1962. The effect of kilovoltage and filtration on depth dose. In *Technological Needs for Reduction of Patient Dosage from Diagnostic Radiology,* ed. M. L. Janower. Springfield, IL: Charles C Thomas.

Tsoulfanidis, N. 1995. *Measurement and Detection of Radiation.* New York: Taylor and Francis.

UNEP. 1987. *Environmental Health Criteria 69: Magnetic Fields.* United Nations Environment Programme, World Health Organization, International Radiation Protection Association. Geneva: WHO.

UNSCEAR. 2000, 1993, 1988, 1977. *Sources, Effects and Risks of Ionizing Radiation.* 1993 Report of the United Nations Scientific Committee on the Effects of Atomic Radiation to the General Assembly, with annexes. New York: United Nations.

——— 1969, 1966, 1964, 1962, 1958. *Report of the United Nations Scientific Committee on the Effects of Atomic Radiation.* General Assembly, Official Records, 1969, 24th sess., supp. no. 13 (A/7613); 1966, 21st sess., supp. no.14 (A/6314); 1964, 19th sess., supp. no. 14, (A/5814); 1962, 17th sess., supp. no. 16 (A/5216); 1958, 13th sess., supp. no. 17 (A/3838). New York: United Nations.

UNSG. 1968. *Effects of the Possible Use of Nuclear Weapons and the Security and Economic Implications for States of the Acquisition and Further Development of These Weapons.* Report of the Secretary General transmitting the study of his consultation group. United Nations Document A/6858. New York: United Nations.

Upton, A. C., J. Furth, and K. W. Christenberry. 1954. Late effects of thermal neutron irradiation in mice. *Cancer Res.* 14:682–690.

USAEC. 1959. Quarterly statement on fallout by the U.S. Atomic Energy Commission, September. Reprinted in JCAE, 1959b, pp. 2188–2191.

——— 1965. *AEC Licensing Guide, Medical Programs.* Division of Materials Licensing, United States Atomic Energy Commission. Washington, DC: Government Printing Office.

USPS. 1983. *Radioactive Materials.* United States Postal Service Publication 6. Washington, DC: Government Printing Office.

Veinot, K. G., et al. 1998. Multisphere neutron spectra measurements near a high energy medical accelerator. *Health Phys.* 75(3):285–290.

Vennart, J. 1969. Radiotoxicology of tritium and ^{14}C compounds. *Health Phys.* 16:429–440.

Vetter, R. J. 1993. Radiation protection in nuclear medicine. In *Hospital Health Physics. Proceedings of the 1993 Health Physics Society Summer School,* ed. G. G. Eicholz and J. S. Shonka. Richland, WA: Research Enterprises.

Villforth, J. C. 1978. Report on federal activities in meeting today's challenges. *Ninth Annual National Conference on Radiation Control.* June 19–23, 1977. Proceedings, HEW Publication (FDA) 78–8054. Washington, DC: Government Printing Office.

Vives, M., and J. Shapiro. 1966. Utilization of boron-loaded nuclear track plates in a spherical moderator for measuring environmental neutron dose. *Health Phys.* 12:965–967.

Volchok, H. C. 1967. Strontium-90 deposition in New York City. *Science* 156:1487–1489.

Volchok, H. L., M. Schonberg, and L. Toonkel. 1977. Plutonium concentrations in air near Rocky Flats. *Health Phys.* 33:484–485.

Vollmer, R. H. 1994. Report on medical radiation protection. In *Proceedings of the 26th Annual National Conference on Radiation Control,* May 22–25, 1994. Frankfort, KY: Conference of Radiation Control Program Directors.

Wagner, E. B., and G. S. Hurst. 1961. A Geiger-Mueller ?-ray dosimeter with low neutron sensitivity. *Health Phys.* 5:20–26.

Wagner, L. K., and B. R. Archer. 2000. *Minimizing Risks from Fluoroscopic X Rays: Bioeffects, Instrumentation, and Examination.* The Woodlands, TX: R. M. Partnership; also, *www.rmpartnership.com.*

Wagner, L. K., P. J. Eifel, and R. A. Geise. 1994. Potential biological effects following high x-ray dose interventional procedures. *J. Vasc. Inter. Radiology* 5:71–84.

Wagoner, J. K., V. E. Archer, F. E. Lundin, et al. 1965. Radiation as the cause of lung cancer among uranium miners. *New Engl. J. Med.* 273:181–188.

Wall, B. F., and D. Hart. 1992. The potential for dose reduction in diagnostic radiology. *Health Phys.* 43:265–268.

Wanebo, C. K., K. G. Johnson, K. Sato, and T. W. Thorslund. 1968. Breast cancer after exposure to the atomic bombings of Hiroshima and Nagasaki. *New Engl. J. Med.* 279:667–671.

Warren, S., and O. M. Lombard. 1966. New data on the effects of ionizing radiation on radiologists. *Arch. Envir. Health* 13:415–421.

Watt, D. E., and D. Ramsden. 1964. *High Sensitivity Counting Techniques.* Oxford: Pergamon Press.

Weber, J., et al. 1995. Biological dosimetry after extensive diagnostic x-ray exposure. *Health Phys.* 68(2):266–169.

Webster, E. W. 1995. X rays in diagnostic radiology. *Health Phys.,* 69:610–635.

Webster, E. W., N. M. Alpert, and C. L. Brownell. 1974. Radiation doses in pediatric nuclear medicine and diagnostic x-ray procedures. In *Pediatric Nuclear Medicine,* ed. A. E. James, Jr., H. N. Wagner, and R. E. Cooke. Philadelphia: W. B. Saunders.

Weeks, J. L., and S. Kobayashi. 1978. Late biological effects of ionizing radiation: Report on the international symposium held in Vienna. *Atom. Ener. Rev.* 16:327–337.

Weibel, E. R. 1963. *Morphometry of the Human Lung.* Berlin: Springer-Verlag, p. 139.

Weigensberg, I. J., C. W. Asbury, and A. Feldman. 1980. Injury due to accidental exposure to x-rays from an x-ray fluorescence spectrometer. *Health Phys.* 39:237–241.

Weinberg, S. 1977. *The First Three Minutes.* New York: Basic Books.

Weiss, H. A., S. C. Darby, and R. Doll. 1994. Cancer mortality following x-ray treatment for ankylosing spondylitis. *Int. J. Cancer* 59:327–338.

Weisskopf, V. F. 1978. Debate on the arms race. Letter to the editor by V. F. Weisskopf, Professor (emeritus) of Physics, Massachusetts Institute of Technology. *Phys. Today,* Dec. 1, p. 13.

Wertheimer, N. W., and E. Leeper. 1979. Electrical wiring configurations and childhood cancer. *Am. J. Epidemiol.* 11:273–284.

WHO. 1993. *Electromagnetic Fields (300 Hz to 300 GHz).* Environmental Health Criteria 137. Geneva: World Health Organization.

———— 1994. *Ultraviolet Radiation.* Environmental Health Criteria 160. Geneva: World Health Organization.

Wilkening, M. H., W. E. Clements, and D. Stanley. 1972. Radon-222 flux measurements in widely separated regions. In *The Natural Radiation Environment, II,* ed. J. A. S. Adams, W. M. Lowder, and T. F. Gesell. U.S. Energy Research and Development Administration Report CONF-720805-P2. Springfield, VA: National Technical Information Service.

Williams, J. R., and D. I. Thwaites, eds. 1993. *Radiotherapy Physics—In Practice.* Oxford: Oxford University Press.

Wilson, B. W., R. E. Stevens, and L. E. Anderson. 1990. *Extremely Low Frequency Electromagnetic Fields: The Question of Cancer.* Columbus: Battelle Press.

Wilson, R. R. 1977. The tevatron. *Phys. Today,* Oct., pp. 23–30.

———— 1987. A visit to Chernobyl. *Science,* 236:673–679.

Wilson, R., and W. J. Jones. 1974. *Energy, Ecology, and the Environment.* New York: Academic Press.

Winkler, K. G. 1968. Influence on rectangular collimation and intraoral shielding on radiation dose in dental radiography. *J. Am. Dent. Assoc.* 77:95–101.

Winston, J. P., and D. L. Angelo. 2000. The Pennsylvania Computerized Tomography Study. *Health Phys.* 78(Suppl. 2):S67-S71 (Operational Radiation Safety Supplement).

Wood, V. A. 1968. *A collection of Radium Leak Test Articles.* U.S. Department of Health, Education, and Welfare Report MORP 68–1. Springfield, VA: National Technical Information Service.

Wooton, P. 1976. Interim report: Mammographic exposures at the breast cancer detection demonstration project screening centers. *Am. J. Roentgenol.* 127:531.

Wrenn, M. E. 1974. *Environmental Levels of Plutonium and the Transplutonium Elements.* Paper presented as part of the AEC presentation at the EPA Plutonium Standards Hearings, Washington, D.C., Dec. 10–11.

Yalow, R. 1991. Presentation at Science Writers Workshop on "Radon Today: The Science and the Politics," sponsored by the U.S. Department of Energy in Bethesda, MD, in 1991. Quoted in L. A. Cole, *Element of Risk: The Politics of Radon* (Washington, DC: American Association for the Advancement of Science, 1993).

York, H. F. 1970. *Race to Oblivion: A Participant's View of the Arms Race.* New York: Simon and Schuster.

———— 1976. The nuclear "balance of terror" in Europe. *Bull. Atom. Sci.,* May, pp. 9–16.

Zaider, M. 1996. Microdosimetric-based risk factors for radiation received in space activities during a trip to Mars. *Health Phys.* 70(6):845–851.

Zanzonico, P. B., and D. V. Becker. 2000. Effects of time of administration and dietary iodine levels on potassium iodide blockade of thyroid irradiation by ^{131}I from radioactive fallout. *Health Phys.* 78(6):660–667.

Zeman, G. 1991. Guidance on ELF electric and magnetic fields. *HPS Newsletter,* October, 1991.

———— 2000. Magnetic resonance imaging. *HPS Newsletter,* February, 2000.

Zumwalt, L. R. 1950. *Absolute Beta Counting Using End-Window Geiger-Mueller Counters and Experimental Data on Beta-particle Scattering Effects.* U.S. Atomic Energy Commission Report AECU-567. Springfield, VA: National Technical Information Service.

Index